Particle Physics

Cargèse 1985

NATO ASI Series

Advanced Science Institutes Series

A series presenting the results of activities sponsored by the NATO Science Committee, which aims at the dissemination of advanced scientific and technological knowledge, with a view to strengthening links between scientific communities.

The series is published by an international board of publishers in conjunction with the NATO Scientific Affairs Division

A	**Life Sciences**	Plenum Publishing Corporation
B	**Physics**	New York and London
C	**Mathematical and Physical Sciences**	D. Reidel Publishing Company Dordrecht, Boston, and Lancaster
D	**Behavioral and Social Sciences**	Martinus Nijhoff Publishers
E	**Engineering and Materials Sciences**	The Hague, Boston, Dordrecht, and Lancaster
F	**Computer and Systems Sciences**	Springer-Verlag
G	**Ecological Sciences**	Berlin, Heidelberg, New York, London,
H	**Cell Biology**	Paris, and Tokyo

Recent Volumes in this Series

Volume 142—New Vistas in Electro-Nuclear Physics
edited by E. L. Tomusiak, H. S. Caplan, and E. T. Dressler

Volume 143—Atoms in Unusual Situations
edited by Jean Pierre Briand

Volume 144—Fundamental Aspects of Quantum Theory
edited by Vittorio Gorini and Alberto Frigerio

Volume 145—Atomic Processes in Electron–Ion and Ion–Ion Collisions
edited by F. Brouillard

Volume 146—Geophysics of Sea Ice
edited by Norbert Untersteiner

Volume 147—Defects in Solids: Modern Techniques
edited by A. V. Chadwick and M. Terenzi

Volume 148—Intercalation in Layered Materials
edited by M. S. Dresselhaus

Volume 149—Radiative Processes in Discharge Plasmas
edited by Joseph M. Proud and Lawrence H. Luessen

Volume 150—Particle Physics: *Cargèse 1985*
edited by Maurice Lévy, Jean-Louis Basdevant, Maurice Jacob, David Speiser, Jacques Weyers, and Raymond Gastmans

Series B: Physics

Particle Physics

Cargèse 1985

Edited by

Maurice Lévy and Jean-Louis Basdevant
Laboratory of Theoretical Physics and High Energies
Université Pierre et Marie Curie
Paris, France

Maurice Jacob
Theory Division
C.E.R.N.
Geneva, Switzerland

David Speiser and Jacques Weyers
Institute of Theoretical Physics
Université Catholique de Louvain
Louvain-la-Neuve, Belgium

and

Raymond Gastmans
Institute of Theoretical Physics
Katholieke Universiteit Leuven
Leuven, Belgium

Plenum Press
New York and London
Published in cooperation with NATO Scientific Affairs Division

Proceedings of a NATO Advanced Study Institute on
Particle Physics,
held July 15–31, 1985,
in Cargèse, France

Library of Congress Cataloging in Publication Data

NATO Advanced Study Institute of Particle Physics (1985: Cargèse, Corsica)
Particle physics.

(NATO ASI series. Series B, Physics; v. 150)
"Published in cooperation with NATO Scientific Affairs Division."
Bibliography: p.
Includes index.
1. Particles (Nuclear physics)—Congresses. I. Lévy, Maurice, 1922- . II. North Atlantic Treaty Organization. Scientific Affairs Division. III. Title. IV. Series.
QC793.N375 1985 539.7′2 87-2310
ISBN-13: 978-1-4612-9046-9 e-ISBN-13: 978-1-4613-1877-4
DOI: 10.1007/ 978-1-4613-1877-4

Softcover reprint of the hardcover 1st edition 1987
A Division of Plenum Publishing Corporation
233 Spring Street, New York, N.Y. 10013

PREFACE

The 1981 Cargèse Summer Institute on Fundamental Interactions was organized by the Université Pierre et Marie Curie, Paris (M. LEVY and J-L. BASDEVANT), CERN (M. JACOB), the Université Catholique de Louvain (D. SPEISER and J. WEYERS), and the Kotholieke Universiteit te Leuven (R. GASTMANS), which, since 1975 have joined their efforts and worked in common. It was the 24th Summer Institute held at Cargèse and the 8th one organized by the two institutes of theoretical physics at Leuven and Louvain-la-Neuve.

The 1985 school was centered around two main themes : the standard model of the fundamental interactions (and beyond) and astrophysics.

The remarkable advances in the theoretical understanding and experimental confirmation of the standard model were reviewed in several lectures where the reader will find a thorough analysis of recent experiments as well as a detailed comparaison of the standard model with experiment. On a more theoretical side, supersymmetry, supergravity and strings were discussed as well.

The second theme concerns astrophysics where the school was quite successful in bridging the gap between this fascinating subject and more conventional particle physics.

We owe many thanks to all those who have made this Summer Institute possible !

Thanks are due to the Scientific Committee of NATO and its President and to the "Région Corse" for a generous grant.

We wish to thank Miss M-F. HANSELER, Mrs ALRIFRAÏ, Mr and Mrs ARIANO, and Mr BERNIA and all others from Paris, Leuven, Louvain-la-Neuve and especially Cargèse for their collaboration.

We thank Miss B. CHAMPAGNE, and Mrs D'ADDATO for typing the manuscript, and Mr. H. GILSON, M. LAMBIN and P. RUELLE for their help in correcting the proofs.

Mostly however, we would like to thank all lecturers and participants the willingness of the former to answer all questions and the keen interest of the latter provided the stimulus which made (we hope) this institute a success.

D. SPEISER

R. GASTMANS

J. WEYERS

M. LEVY

J-L. BASDEVANT

M. JACOB

CONTENTS

GAUGE BOSON/HIGGS BOSON UNIFICATION, N = 2 SUPERSYMMETRY, GRAND UNIFICATION, AND NEW SPACETIME DIMENSIONS

Pierre Fayet

Laboratoire de Physique Théorique de l'Ecole Normale Supérieure*
24, rue Lhomond, 75231 Paris Cedex 05 - France

ABSTRACT

We discuss the physical consequences of simple (N = 1) and extended (N = 2) supersymmetries, in particular the new relations they provide between massive spin-1 gauge bosons and spin-0 Higgs bosons.

Extended supersymmetric theories predict the existence of spin-0 photons and gluons coupling leptons and quarks to their mirror partners. The anticommutator of the two supersymmetry generators includes a term proportional to the weak hypercharge, and determines the value of the grand-unification mass m_X. In minimal N = 2 supersymmetric GUTs, the proton is expected to be stable.

Such theories may be formulated in a 6-dimensional spacetime. The GUT mass m_X is generated by the additional components of the momentum along the extra compact dimensions. This leads to the possibility of computing m_X proportionately to $\frac{\hbar}{Lc}$, L being the size of an extra dimension.

MOTIVATIONS FOR SUPERSYMMETRY AND EXTENDED SUPERSYMMETRY

Supersymmetric theories associate bosons with fermions in multiplets [1]. Supersymmetry was originally considered, most of the time, as an interesting structure for quantum field theory, without, unfortunately, any application to the description of the physical world. Nevertheless it is possible to make supersymmetry a physically meaningful invariance. This requires the introduction of a whole set of new particles : photinos and gluinos, Winos and Zinos, spin-0 leptons and quarks, etc., which have not been observed yet.

There is, presently, no convincing experimental evidence in favor of supersymmetry. The reasons for considering this new invariance are, still, purely theoretical :

i) better understanding of the Higgs sector of gauge theories, owing to the unification provided by supersymmetry between massive gauge bosons

---

* Laboratoire Propre du Centre National de la Recherche Scientifique, associé à l'Ecole Normale Supérieure et à l'Université de Paris Sud.

and Higgs bosons : spin-0 Higgs bosons now appear as the superpartners of massive spin-1 gauge bosons [2];

ii) connection with gravitation, which is automatically present as soon as the supersymmetry algebra is realized locally (supergravity) [3];

iii) and, as an additional motivation in the framework of grand-unification, hope of a solution to the hierarchy problem.

Gauge boson / Higgs boson unification in N = 2 extended supersymmetry

Grand-unified theories, however, require the existence of a rather large number of Higgs bosons. In a grand-unified theory with only a simple (N = 1) supersymmetry most spin-0 Higgs bosons would remain unmatched with massive spin-1 gauge bosons. In order to get a better understanding of the Higgs sector of GUTs and supersymmetric GUTs we are led to use a more powerful symmetry, N = 2 extended supersymmetry, also called hypersymmetry. In such theories, which are invariant under a set of N = 2 supersymmetry generators Q^1 and Q^2, every massive spin-1 gauge boson may be associated with either *one* or *five* spin-0 Higgs bosons [4] :

1 massive spin-1 gauge boson ↔ 1 spin-0 Higgs boson (e.g. $X^{\pm 4/3}$) (1)

or

1 massive spin-1 gauge boson ↔ 5 spin-0 Higgs bosons (e.g. $W^{\pm}$, Z, or $Y^{\pm 1/3}$). (2)

This leads to an interesting structure for the gauge and Higgs sectors of N = 2 supersymmetric GUTs (cf. Table 2 in Section IV). In an SU(5) theory, each of the $W^{\pm}$, Z and $Y^{\pm 1/3}$'s is associated with 4 spin-1/2 inos and 5 spin-0 Higgs bosons, while the $X^{\pm 4/3}$ is associated with only 2 spin-1/2 Xinos and 1 spin-0 Higgs boson [5].

The GUT mass originates from the extended SUSY algebra

In N = 2 supersymmetry the grand-unification mass appears in an algebraic way, in the anticommutation relations of the two supersymmetry generators. Next to the usual anticommutation relations of one supersymmetry generator with itself :

$$\{Q^1, \overline{Q^1}\} = \{Q^2, \overline{Q^2}\} = -2\not{P} \tag{3}$$

there is another anticommutation relation, which reads :

$$\{Q^1, \overline{Q^2}\} = - \{Q^2, \overline{Q^1}\} = 2Z. \tag{4}$$

Z is a spin-0 symmetry generator called a central charge. It has the dimension of a mass and reads [5] :

Z = Global symmetry generator + Spontaneously-generated term, proportional to the weak hypercharge $Y = 2(Q-T_3)$ (with a coefficient $\sim m_X$). (5)

The eigenvalues of Z for the GUT bosons $X^{\pm 4/3}$ and $Y^{\pm 1/3}$ are precisely equal to $\mp$ the value of the grand-unification mass m_X !

Mirror particles, and spin-0 photons and gluons

N = 2 supersymmetry predicts the existence of mirror leptons and quarks (a source of serious difficulties!) and of spin-0 photons and gluons [4]. The latter are related to the spin-1 photon and gluons, and to *pairs* of spin-1/2 photinos and gluinos. They are described by the same adjoint Higgs field (e.g. a $\underline{24}$ of SU(5)) which is also responsible for the spontaneous breaking of the grand-unification symmetry.

Proton stability

In an N = 2 supersymmetric GUT the mass splittings in the grand-unification multiplets turn out to be of order m_X for the *fermions* (including leptons and quarks) as well as for the gauge bosons and Higgs bosons. This is a consequence of the N = 2 supersymmetry algebra itself (eqs. (4,5)). As a result grand-unification provides us with the following associations :

$$\left\{\begin{array}{l}\text{Light leptons}\\ \text{Heavy quarks of mass} \simeq m_X\end{array}\right. \qquad \left\{\begin{array}{l}\text{Light quarks}\\ \text{Heavy leptons of mass} \simeq m_X\end{array}\right. \tag{6}$$

It follows that, in minimal N = 2 SUSY GUTs, the proton tends to be *absolutely stable*! [6]

6-dimensional theories

Moreover, N = 2 supersymmetric theories in 4 dimensions may be obtained from a N = 1 theory in a 6-dimensional spacetime, the 5th and 6th dimensions being compact. One can then rewrite formulas (3,4) as a single anticommutation relation in the 6nd spacetime, and establish a relation between the grand unification mass and the extra components of the (covariant) momentum along the compact dimensions. The spin-0 symmetry operator Z appears as the 5th component of the covariant 6-momentum $P^{\hat{\mu}} = -i\, \mathcal{D}^{\hat{\mu}}$:

$$Z \rightsquigarrow P^5 = -i\, \mathcal{D}^5. \tag{7}$$

While the $W^{\pm}$ and Z masses are already present in the 6-dimensional theory, *the grand-unification mass* m_X is associated with the large values of the *covariant momenta* carried by the $X^{\pm 4/3}$ or $Y^{\pm 1/3}$ *along the extra compact dimensions* [5]. This leads to the possibility of computing m_X in terms of the lengths of these dimensions :

$$m_X \sim \frac{\hbar}{Lc}. \tag{8}$$

N = 1 SUPERSYMMETRY AND GAUGE BOSON / HIGGS BOSON UNIFICATION

a) The supersymmetry algebra

Supersymmetric theories are invariant under a self-conjugate spin-1/2 supersymmetry generator Q which changes the spin of particles by 1/2 unit, transforming bosons into fermions and conversely [1] :

$$\begin{aligned} Q\,|\,\text{boson}\rangle &= |\,\text{fermion}\rangle \\ Q\,|\,\text{fermion}\rangle &= |\,\text{boson}\rangle \end{aligned} \tag{9}$$

Q satisfies commutation relations with boson fields and anticommutation relations with fermion fields :

$$[Q, \text{boson field}] = \text{fermion field}$$
$$\{Q, \text{fermion field}\} = \text{boson field}. \tag{10}$$

By iterating (10), one finds the equations :

$$[\{Q, \overline{Q}\}, \text{field}] = 2i\,\gamma^{\mu}\,\partial_{\mu}\ (\text{field}) \tag{11}$$

where ∂_{μ} (field) is the 4-derivative of the original (bosonic or fermionic) field considered. This can be rewritten as follows :

$$\{Q, \overline{Q}\} = -2\not{P}. \tag{12}$$

Equation (12), together with

$$[Q, P^{\mu}] = 0 \tag{13}$$

(which expresses that the result of a supersymmetry transformation does not depend on the space-time point where it is performed), defines the supersymmetry algebra. The appearance of the generator of translations P^{μ} in the right-hand side of equation (12) indicates a connection between supersymmetry and space-time. This is at the origin of the relation between supersymmetry and gravitation [3].

b) Minimal content of a N = 1 supersymmetric gauge theory

Any (linear) representation of supersymmetry describes equal numbers of bosonic and fermionic states. They would all have the same masses if supersymmetry remained unbroken. Supersymmetry can only be, at best, a spontaneously broken symmetry.

Spontaneous breaking of global supersymmetry is made difficult by the presence of the Hamiltonian in the algebra (cf. formula (12)). It has the consequence that, whenever a supersymmetric vacuum state exists, it must be stable. This is in sharp contrast with what happens for an ordinary gauge symmetry : the gauge symmetric state can easily be made unstable, gauge invariance being then spontaneously broken.

Remarkably enough,a spontaneous breaking of global supersymmetry, although hard to obtain, is still possible. It generates a massless neutral spin-1/2 Goldstone fermion called a goldstino. This one, however, cannot be identified with a neutrino.

Even if supersymmetry is spontaneously broken, we still need to introduce a rather large number of new particles into the theory. The superpartners of the photon and the W^- should *not* be interpreted as the neutrino and the electron, but as new particles called *photino* and *Winos*. In a similar way the octet of gluons is associated with an octet of spin-1/2 self-conjugate fermions called *gluinos*. Leptons and quarks are associated with spin-0 leptons and quarks (two of them for every Dirac fermion) [1].

Let us consider the massive gauge bosons $W^{\pm}$ and Z. They are, in fact, associated with *two* (Dirac) Winos, and *two* (self-conjugate) Zinos, together with spin-0 particles. The last ones are precisely charged and neutral spin-0 Higgs bosons denoted by $w^{\pm}$ and z, respectively.

Remarkably enough supersymmetry allows one to obtain relations between particles having, not only different spins (1, 1/2 or 0) but also [2] :

- *different electroweak properties* : the $W^{\pm}$ and Z belong to an electroweak triplet and a singlet; the Higgs bosons $w^{\pm}$ and z to electroweak doublets; and the Winos and Zinos are mixings of triplet, singlet and doublet components;

- *couplings of very different strengths* : the gauge bosons couple proportionately to g or g', the Higgs bosons proportionately to (g or g') $\frac{m(\text{fermions})}{(m_W \text{ or } m_Z)}$.

The spontaneous breaking of the SU(2) × U(1) electroweak gauge symmetry makes use of two doublet Higgs superfields instead of only one, and this, for two reasons : i) to have the required degrees of freedom for constructing 2 massive charged Dirac fermions, which will be the two Winos associated with the $W^{\pm}$ under supersymmetry; and ii) to generate masses for both charge + 2/3 quarks on one hand, charge - 1/3 quarks and charged leptons, on the other hand. Then one gets the following minimal scheme, common to all supersymmetric theories irrespectively of the supersymmetry breaking mechanism one considers [1,2] :

Table 1.
Minimal content of a supersymmetric gauge theory.

Spin 1	Spin 1/2	Spin 0	
Gluons Photon	Gluinos Photino		
$W^{\pm}$ Z	2 (Dirac) Winos 2 (Majorana) Zinos	$w^{\pm}$ z	Higgs bosons
	1 (Majorana) higgsino	standard h^o pseudoscalar h'^o	Higgs bosons
	Leptons Quarks	Spin-0 leptons Spin-0 quarks	
+ Gravitation multiplet : Spin -2 graviton, and spin -3/2 gravitino			

The spin-2 graviton and its superpartner, the spin-3/2 gravitino, are present as soon as the supersymmetry algebra is realized locally. (Moreover, if an extra U(1) group is gauged, there is also a second neutral massive gauge boson, U, which acquires a mass while the pseudoscalar h'^o is eliminated; then one has a perfect association between the four massive spin-1 gauge bosons $W^{\pm}$, Z and U, and the four spin-0 Higgs bosons $w^{\pm}$, z and h^o).

c) Supersymmetry breaking and mass spectrum

Supersymmetry is not an apparent symmetry of the physical world, otherwise bosons and fermions would be degenerate in mass, which obviously is not the case.

As we have already said, spontaneous breaking of global supersymmetry generates a massless neutral spin-1/2 particle named a goldstino; it couples leptons and quarks to spin-0 leptons and spin-0 quarks, etc.

In global supersymmetry, one can generate large masses for spin-0 leptons and quarks, at the tree approximation, if and only if the gauge group is extended to include an extra U(1) factor : i.e. it can be SU(3) × SU(2) × U(1) × U(1), or SU(5) × U(1), etc... (The additional neutral gauge boson, U, might possibly be responsible for the somewhat too large value of the asymmetry observed in $e^+e^- \rightarrow \mu^+\mu^-$ scatterings at PETRA).

In locally supersymmetric theories however, there is no goldstino : it is eliminated by the super-Higgs mechanism while the superpartner of the graviton, the spin-3/2 gravitino, acquires a mass. If the gravitino mass $m_{3/2}$ is large, this parameter may be responsible for large mass splittings between ordinary particles and their superpartners : cf. the spectrum represented in Fig. 1.

d) Winos and Zinos : light or heavy ?

Winos and Zinos are a *model-independent prediction* of the *supersymmetric electroweak theory* [1] (cf. Table 1), irrespectively of either grand-unification or supergravity.

The two charged Dirac Winos are always built from

- a *gaugino* field λ^- (a member of an electroweak *triplet*)
- a *higgsino* field Ψ^- (a member of an electroweak *doublet*). (14)

The way the two Winos mix and the masses they acquire are, however, model-dependent : they depend, in particular, on the supersymmetry breaking mechanism considered.

i) Extra U(1) breaking

In theories in which the supersymmetry breaking is triggered by an extra U(1) factor in the gauge group, the two Wino mass eigenstates usually are the *R-eigenstates* [8]

$$\begin{cases} \text{Wino}_1 = \text{gaugino}_L + \text{higgsino}_R & (R = +1) \\ \text{Wino}_2 = \text{higgsino}_L + \text{gaugino}_R . & (R = -1) \end{cases} \tag{15}$$

Their masses verify [8]

$$m^2(\text{Wino}_1) + m^2(\text{Wino}_2) = 2m_W^2 \tag{16}$$

which implies :

$$m(\text{Wino}_{1(2)}) \leq m_W \leq m(\text{Wino}_{2(1)}) . \tag{17}$$

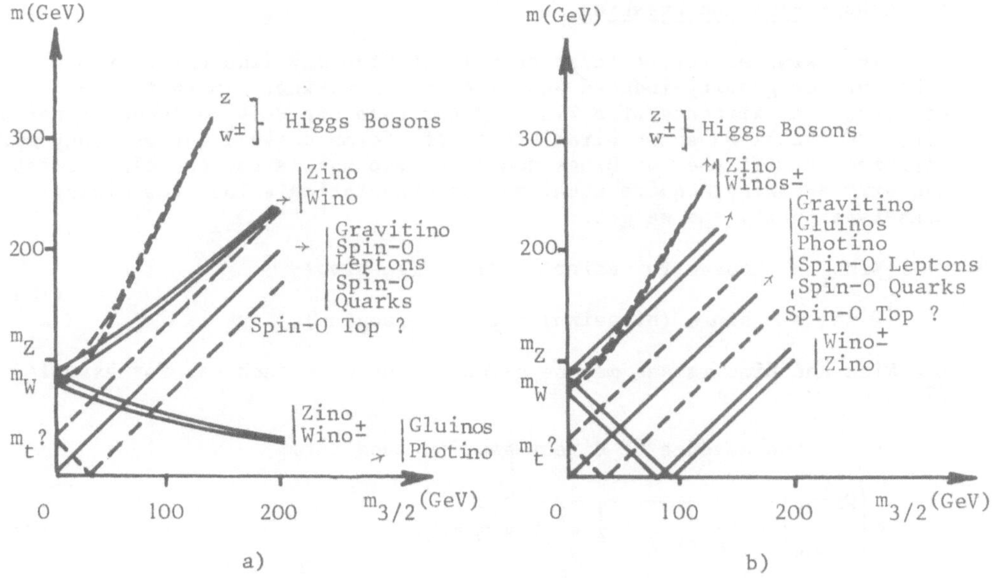

Fig. 1 : Examples of mass spectra for the superpartners of leptons and quarks, photon, gluons, $W^{\pm}$ and Z, in a class of models with gravity-induced supersymmetry breaking [7], as a function of the gravitino mass $m_{3/2}$. Spin-0 leptons and quarks have masses : $|m_{3/2} \pm m_{l(q)}|$; and the spin-0 Higgs bosons $w^{\pm}$ and z associated with the $W^{\pm}$ and Z :

$$\sqrt{m_{W,Z}^2 + 4m_{3/2}^2}.$$

<u>In a)</u> there is no gravitation-induced direct mass term for gauginos, so that the photino and gluinos may be relatively light. The Wino and Zino masses are then given by

$$\sqrt{m_{W,Z}^2 + \frac{1}{4}\, m_{3/2}^2} \pm \frac{1}{2}\, m_{3/2}.$$

<u>In b)</u> the photino and gluinos have the same masses as the gravitino, and the Wino and Zino masses are $|m_{W,Z} \pm m_{3/2}|$. The lightest superpartner could be either the photino (case a), or a Wino or a Zino, a spin-0 t or b quark, or a higgsino, ... (case b). If $m_{3/2}$ becomes relatively large, we lose the hope of observing spin-0 leptons and quarks very soon, but one of the Winos and one of the Zinos may then be accessible.

At the same time the two Majorana Zinos combine (in a R-invariant theory) to form a single Dirac Zino of mass m_Z, carrying R = +1.

Formulas (16,17) allow for the existence of a relatively *light Wino*. Indeed we suggested in 1976 that the newly-discovered τ *lepton* might be identified with a charged Wino [8]. Although this is no longer an acceptable possibility, a relatively light Wino might still exist, provided it is heavier than about 20 GeV/c^2.

ii) Gravity-induced breaking

In a similar way, a relatively light Wino may also exist if one makes use of gravity-induced supersymmetry breaking. In that case, however, the existence of a Wino lighter than the W is no longer a necessity. We still have two Dirac Winos (cf. formula (14)) but now they mix differently. If the two Higgs doublets responsible for the electroweak symmetry breaking acquire equal vacuum expectation values the mixing preserves parity and we get :

$$\begin{cases} \text{Wino}_1 = \cos\omega\,(\text{higgsino}) + \sin\omega\,(\text{gaugino}) \\ \text{Wino}_2 = -\sin\omega\,(\text{higgsino}) + \cos\omega\,(\text{gaugino}). \end{cases} \tag{18}$$

The Wino and Zino masses may be given by formulae such as, for example [7]

i) in the absence of direct gaugino mass terms

$$m\begin{pmatrix}\text{Winos}\\ \text{Zinos}\end{pmatrix} = \sqrt{m^2_{W(Z)} + \frac{1}{4}m^2_{3/2}} \pm \frac{1}{2}m_{3/2}. \tag{19}$$

ii) in the presence of direct gaugino mass terms equal to $\pm\, m_{3/2}$, (which make the photino and gluinos heavy at the tree approximation)

$$m\begin{pmatrix}\text{Winos}\\ \text{Zinos}\end{pmatrix} = |m_{W(Z)} \pm m_{3/2}| \tag{20}$$

or, alternately :

$$m\begin{pmatrix}\text{Winos}\\ \text{Zinos}\end{pmatrix} = \sqrt{m^2_{W(Z)} + m^2_{3/2}}. \tag{21}$$

Formulas (19,20) allow for the existence of a relatively light Wino (or Zino), the other one being heavy. Formulas (21), however, (which also give the mass spectrum of Winos and Zinos in a N = 2 supersymmetric GUT [5]) require *all* Winos and Zinos to be heavier than m_W and m_Z, respectively.

CONSEQUENCES OF SUPERSYMMETRY

a) R-parity and the interactions of the new particles

Ordinary particles can be distinguished from their superpartners by means of a new quantum number, R, associated with a continuous or discrete invariance. In the latter case one talks about R-parity, $(-1)^R$. Ordinary particles all have R = 0, while their superpartners, gravitinos, gluinos and photinos, Winos and Zinos, spin-0 leptons and quarks ... all have R = ± 1.

R-parity [9] may be defined as

$$(-1)^R = (-1)^{2S}\,(-1)^{3(B-L)} \tag{22}$$

and usually does remain unbroken. This is essential in the study of the production and decay of the new particles predicted by supersymmetry. The new particles can only be pair-produced. The decay of an R-odd particle will, ultimately, lead to the lightest R-odd particle. This could be the photino, but it is not necessary (cf. Fig. 1).

As an example one may have, in agreement with R-parity conservation, the following decays, which are represented in Fig. 2 :

spin-0 electron → electron + photino (or goldstino)
spin-0 quark → quark + gluino (photino, or goldstino)
gluino → quark + antiquark + photino (or goldstino)
$W^{\pm}$, Z → pair of spin-0 leptons, or quarks
$\text{Wino}^{\pm}$ → photino (...) + $q\bar{q}'$ (or $l\pm \overset{(-)}{\nu}$). (23)

b) Searching for the new particles

Much work has been done in order to search for the new particles predicted by supersymmetry [1,10].

The first particles one became interested in were those which might have been relatively light, and manifest themselves at not-too-high energies : namely gluinos (and the R-hadrons they could form by combining with quarks, antiquarks and gluons), photinos and goldstinos (or gravitinos), and spin-0 leptons.

The standard signal for gluino production is that gluinos should decay emitting a (light) neutral spin-1/2 particle (photino or goldstino) which would carry away part of the energy-momentum [9]. The present experimental limits indicate that gluinos should be heavier than a few GeV/c^2 [11]. The exact value of the limit depends on the spin-0 quark masses : in beam dump experiments one searches for gluinos by looking at the reinteraction with matter of the photinos emitted in gluino decays (cf. Fig. 2c); but photino interactions depend on the spin-0 quark masses $m_{\tilde{q}}$ as follows :

$$\sigma(\text{photino + nucleon} \rightarrow \text{gluino + hadrons}) \simeq \frac{32\pi\alpha\alpha_s}{9m_{\tilde{q}}^4} < \sum_i Q_i^2 x_i > s. \quad (24)$$

By looking at the production of unstable spin-0 leptons in e^+e^- annihilation one gets the following limits [12,13] :

$$\begin{cases} m(\text{spin-0 electrons}) > 22\ GeV/c^2 \\ m(\text{spin-0 muons}) > 20\ GeV/c^2 \\ m(\text{spin-0 taus}) > 18\ GeV/c^2. \end{cases} \quad (25)$$

For spin-0 electrons the limit can be increased to 25 GeV/c^2 by searching for the production of a single spin-0 electron in association with a photino. A better limit can be obtained by searching for the process [14,15]

$$e^+e^- \rightarrow \gamma + 2\ \text{photinos} \quad (26)$$

induced by the exchanges of spin-0 electrons (see Fig. 3).

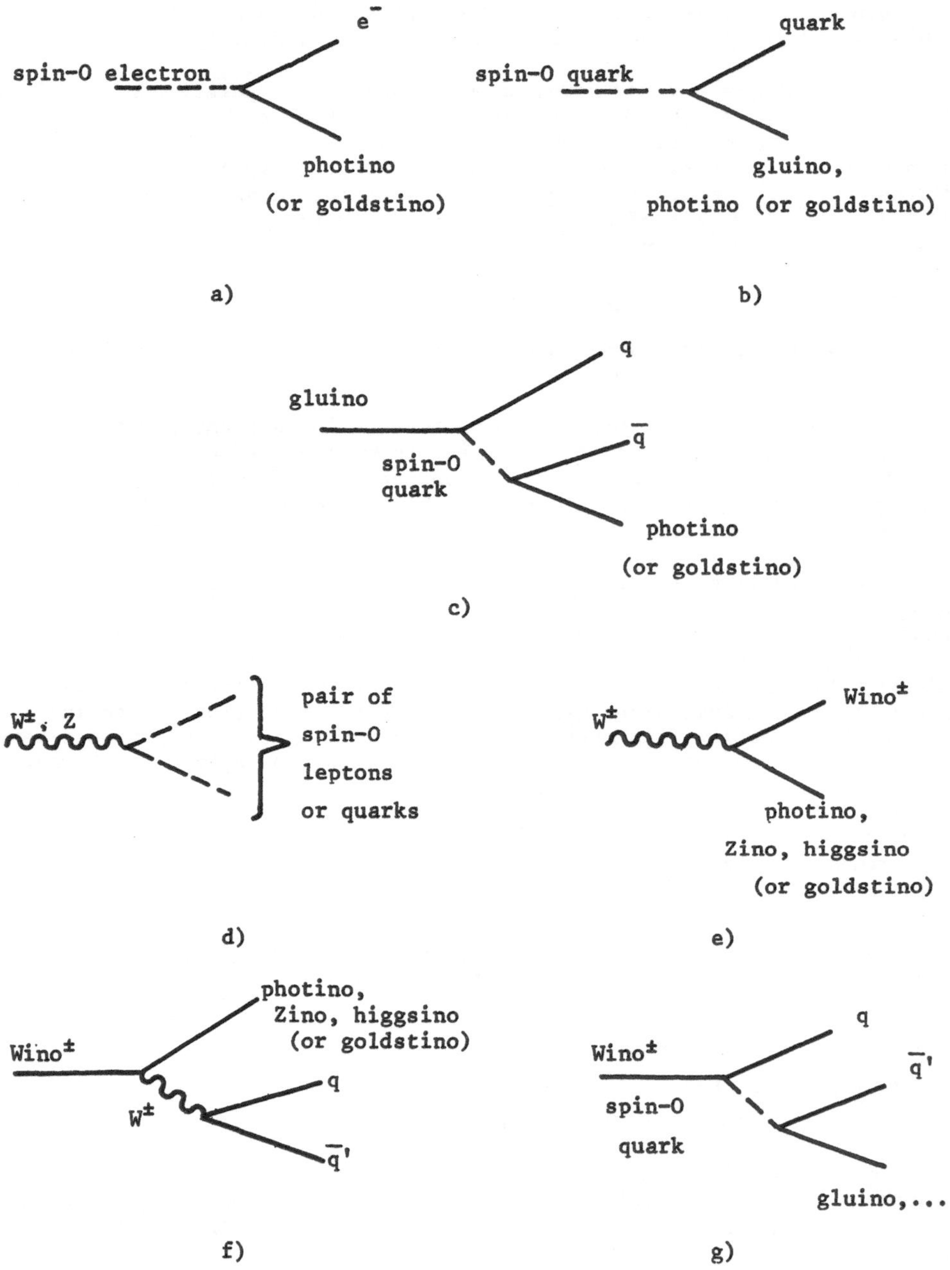

Fig. 2 : Possible decay modes involving the new R-odd particles : spin-0 leptons and quarks, Winos and Zinos, photinos, higgsinos, goldstinos.

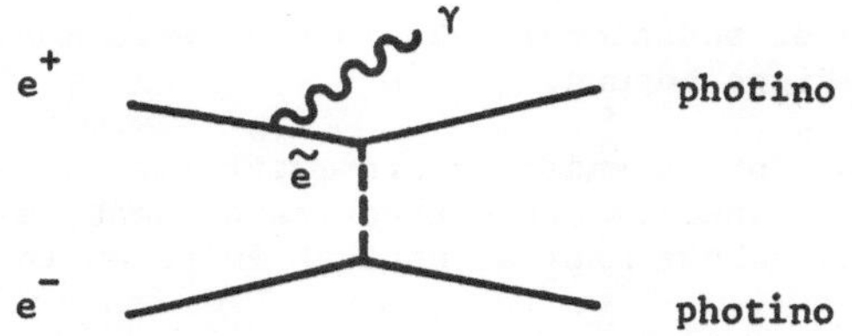

Fig. 3 : Radiative production of a photino pair in e^+e^- annihilation. $\tilde{e}$ denotes a spin-0 electron.

Present experimental results indicate that

$$m(\text{spin-0 electrons}) > 37\ \text{GeV}/c^2 \tag{27}$$

if the two spin-0 electrons are mass-degenerate; if one of them were very heavy, the second could be as light as 30 GeV/c^2 [16]. (To obtain these limits one assumes that the photino is lighter than a few GeV/c^2). This reaction could allow one to detect the effects of spin-0 electrons lighter than about 50 GeV/c^2. Beyond the signal gets rather small and there is a competing background (!) from $e^+e^- \to \gamma\nu\bar{\nu}$.

Moreover, the search for single γ's in e^+e^- annihilations, produced by the reaction

$$e^+e^- \to \gamma + \text{photino} + \text{gravitino} \tag{28}$$

also leads to a *lower* limit on the mass of a spin-3/2 gravitino [14]. From the results of Ref.[16], we infer

$$m_{3/2} > 10^{-6}\ \text{eV}/c^2. \tag{29}$$

After the discovery of the $W^\pm$ and Z bosons one can also search for the production of spin-0 leptons and quarks, Winos and Zinos, photinos and higgsinos in $W^\pm$ and Z decays, provided of course these particles are not too heavy. More generally a systematic search for the effects of the new particles, specially spin-0 quarks, gluinos and photinos, is presently in progress at the CERN $p\bar{p}$ collider [17].

EXTENDED SUPERSYMMETRY AND EXTRA SPACETIME DIMENSIONS

a) General features of N = 2 supersymmetry

As discussed in section 1, one is led to contemplate the possibility for Nature to be invariant, not only under a simple supersymmetry algebra involving a single (N = 1) spinorial generator Q, but under an *extended* supersymmetry algebra involving *N = 2* spinorial generators Q^i : (i = 1,2).

From the theoretical point of view such theories are much more constrained, and do not present yet enough flexibility to be completely realistic. Nevertheless, we can already discuss their main physical properties [4,6] :

i) existence of larger multiplets and therefore of a new set of gravitinos, photinos, gluinos, Winos and Zinos, etc.;

ii) existence of spin-0 photons and spin-0 gluons;

iii) existence of mirror leptons and quarks having V+A charged current weak interactions;

iv) existence of additional relations between massive spin-1 gauge bosons and spin-0 Higgs bosons.

The motivation for extended supersymmetry is not apparent at this stage; indeed $N = 1$ supersymmetric theories of weak, electromagnetic and strong interactions, whose minimal content is given in Table 1, are quite appealing.

The situation, however, changes when one considers grand-unification also : supersymmetric grand-unified theories require a rather large number of spin-0 Higgs bosons, and we no longer have a simple classification, as in Table 1. The Higgs sector gets quite complicated, and we do not know how many Higgses should be there, and to which representations they belong. This is precisely where extended supersymmetry will help.

The reader is probably getting tired of seeing again new unobserved particles show up in the theory, and will wonder why we need to introduce a second octet of gluinos, a second photino, etc. These particles, actually, are already present, although in a hidden way, in a grand-unified theory with only a simple ($N = 1$) supersymmetry. The complex spin-0 Higgs field (such as the $\underline{24}$ of SU(5)) which breaks spontaneously the GUT symmetry describes an octet and a singlet of spin-0 particles which will be interpreted later as spin-0 gluons and spin-0 photons, respectively. Their fermionic partners are a second octet of Majorana fermions ("paragluinos"), similar to the gluinos, and a singlet one ("paraphotino") similar to the photino.

Requiring $N = 2$ extended supersymmetry means, in particular, that there is no essential difference between the by-now familiar octet of gluinos, and the second octet of colored fermions which appears as a consequence of grand-unification. Altogether we get two octets of gluinos, two photinos, as well as a complex octet of spin-0 gluons, and a complex spin-0 photon. This is summarized in Fig. 4.

b) Massless and massive gauge hypermultiplets of N = 2 supersymmetry

The extended supersymmetry algebra reads :

$$\{Q^1, \overline{Q^1}\} = \{\overline{Q^2}, Q^2\} = -2\not{P} \qquad (30)$$

as in formula (12). In addition, the two supersymmetry generators Q^1 and Q^2 also satisfy, together, an anticommutation relation

$$\{Q^1, \overline{Q^2}\} = -\{Q^2, \overline{Q^1}\} = 2Z. \qquad (31)$$

Z is a spin-0 symmetry generator called a central charge, which has the dimension of a mass. Its explicit expression [5] involves, in particular, neutral uncolored grand-unification symmetry generators, such as the weak hypercharge $Y = 2(Q-T_3)$.

With SU(5) as the grand-unification gauge group we find :

$$\{Q^1, \overline{Q^2}\} = -\{Q^2, \overline{Q^1}\} = 2\left[\begin{matrix}\text{Global symmetry}\\ \text{generator}\end{matrix} - \frac{3}{5}\, m_X\, Y\right]. \qquad (32)$$

The central charge Z vanishes for the $W^\pm$ and Z, γ and gluons, but equals

$$Z = \mp\, m_X \qquad (33)$$

for the GUT bosons $X^{\pm 4/3}$ and $Y^{\pm 1/3}$ (which have weak hypercharge $Y = \pm\frac{5}{3}$).

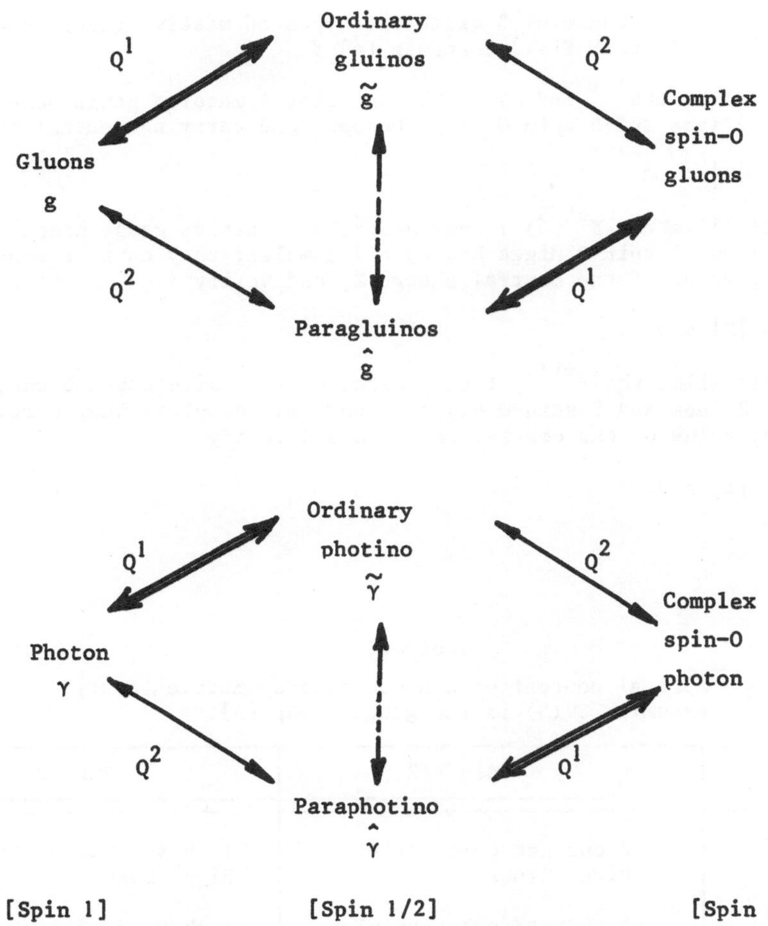

Fig. 4 : Relations between the gluons, the photon, and their spin-1/2 and spin-0 partners, in an N = 2 extended supersymmetric theory. Q^1 denotes the action of the first supersymmetry generator, Q^2 the action of the second one. Spin-0 gluons and spin-0 photons are described by the adjoint Higgs field (e.g. a $\underline{24}$ of SU(5)) which breaks spontaneously the GUT symmetry. (They will appear, subsequently, as the 5th and 6th components of the gluon and photon 6-vector fields, in a 6-dimensional spacetime).

We also find the mass relation

$$m_Y^2 = m_X^2 + m_W^2. \tag{34}$$

In N = 2 extended supersymmetry the minimal particle content gets increased compared with that given in Table 1. It is illustrated in Table 2. As a consequence of extended supersymmetry, almost every Higgs boson now appears as the superpartner of a gauge boson. This association becomes complete if an extra U(1) is gauged.

Note the existence of 3 different types of massive gauge hypermultiplets, with different field contents [4] :

- type I (like the $W^{\pm}$ and Z) : they describe 1 massive gauge boson, 4 spin-1/2 inos and 5 spin-0 Higgs bosons; and carry no central charge :

$$M > |Z| = 0 \tag{35}$$

- type II (like the $X^{\pm 4/3}$) : they describe 1 massive gauge boson, 2 spin -1/2 inos and 1 spin-0 Higgs boson, all complex; they carry a non-vanishing value of the central charge Z, and verify :

$$M = |Z| > 0 \tag{36}$$

- type III (like the $Y^{\pm 1/3}$) : they describe 1 massive gauge boson, 4 spin-1/2 inos and 5 spin-0 Higgs bosons, all complex; they carry a non-vanishing value of the central charge Z and verify :

$$M > |Z| > 0. \tag{37}$$

Table 2.

Minimal content of a N = 2 supersymmetric theory, assuming SU(5) as the gauge group [5].

Spin 1	Spin 1/2	Spin 0
$X^{\pm 4/3}$	2 charged color-triplet Dirac Xinos	1 charged color-triplet Higgs boson
$Y^{\pm 1/3}$	4 charged color-triplet Dirac Yinos	5 charged color-triplet Higgs bosons
$W^{\pm}$	4 charged Dirac Winos	5 charged Higgs bosons
Z	4 Majorana Zinos	5 neutral Higgs bosons
γ	2 Majorana photinos	2 spin-0 photons
gluons	2 color-octet Majorana gluinos	2 color-octet spin-0 gluons
	2 neutral Majorana higgsinos	4 neutral Higgs bosons
	Leptons and quarks + mirror partners	Spin-0 leptons and quarks + Mirrors
+ Gravitation multiplet : Spin-2 graviton, 2 spin-3/2 Majorana gravitinos, 1 spin-1 "graviphoton"		

If an extra U(1) is gauged, the higgsino multiplet is replaced by a neutral massive gauge multiplet describing the new neutral gauge boson U, 4 Majorana Uinos and 5 neutral Higgs bosons. Every spin-0 Higgs boson appears then as the superpartner of a spin-1 gauge boson.

Up to this point, the analysis is essentially model-independent, as was Table 1 for simple (N = 1) supersymmetric theories. The actual mass spectrum of the new particles, however, will depend on the way in which the supersymmetry breaking is performed. If it is induced by gravitation or by dimensional reduction, one tends to get, at the tree approximation

$$m_{\text{gluinos}} = m_{\text{photino}} = m_{3/2} \tag{38}$$

and

$$\begin{cases} m(\text{Winos}) = (m_W^2 + m_{3/2}^2)^{1/2} \\ m(\text{Zinos}) = (m_Z^2 + m_{3/2}^2)^{1/2} \end{cases} \tag{39}$$

up to radiative correction effects.

c) Replication of lepton and quark fields and proton stability in N = 2 SUSY GUTs

A consequence of the appearance of the weak hypercharge operator Y in the anticommutation relation of two different supersymmetry generators (eq. (32)) is the existence of mass splittings $\sim m_X$ in *all* multiplets of the grand-unification group, i.e. in lepton and quark multiplets as well as in the gauge boson multiplet.

As a result the grand-unification symmetry associates light leptons with heavy quarks of mass $\simeq m_X$, and conversely, as indicated in section 1 (eq. (6)). It is therefore necessary to perform a *replication* of representations, in order to describe every single family of quarks and leptons [6].

It follows that the $X^{\pm 4/3}$ and $Y^{\pm 1/3}$ gauge bosons (as well as their Higgs superpartners) do not couple directly light quarks to light leptons. The usual diagram responsible for the standard proton decay mode $p \to \pi^0 e^+$ *does not exist* ! (cf. Fig. 5). A more refined analysis indicates that, in minimal N = 2 SUSY GUTs, the proton is expected to be totally stable [6].

Fig. 5 : *Forbidden* couplings in N = 2 SUSY GUTs : the $X^{\pm 4/3}$ and $Y^{\pm 1/3}$ gauge bosons do *not* couple directly to light leptons and light quarks.

d) New photinos and mirror particles

Spin-0 photons and spin-0 gluons couple ordinary to mirror particles, as indicated in Fig. 6.

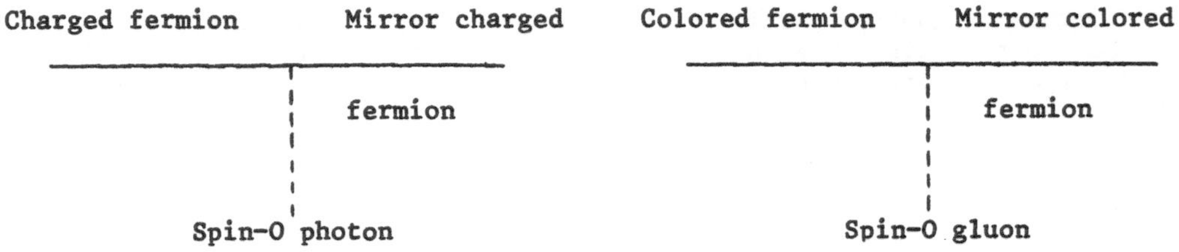

Fig. 6 : Spin-0 photons and gluons have non-diagonal couplings relating leptons and quarks to their mirror partners.

Moreover the second photino (paraphotino $\hat{\gamma}$) couples ordinary leptons and quarks to mirror spin-0 leptons and quarks (and conversely). Like ordinary photinos, paraphotinos $\hat{\gamma}$ could be produced in e^+e^- annihilations, according to the reaction

$$e^+e^- \rightarrow \gamma\hat{\gamma}\hat{\gamma} \tag{40}$$

induced by the exchange of mirror spin-0 electrons ($\tilde{e}_M = \hat{e}$) (cf. Fig. 7) [18]. The non-observation of this process implies for mirror spin-0 electrons the same limit as for spin-0 electrons (presently 37 GeV/c^2 [16]) provided of course we make the rather restrictive assumption that the paraphotino $\hat{\gamma}$ is light.

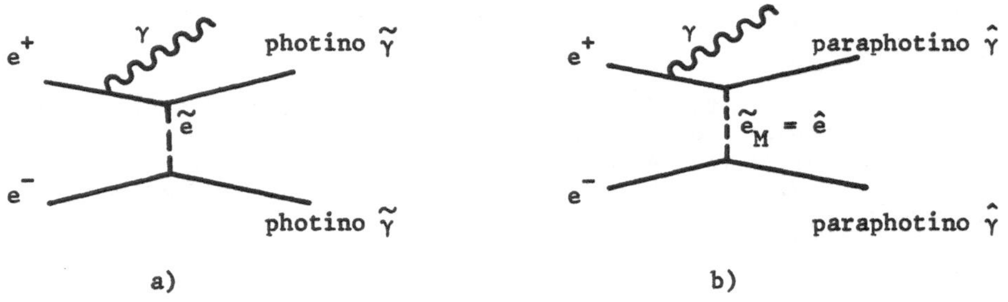

Fig. 7 : Production of photinos ($\tilde{\gamma}$) and paraphotinos ($\hat{\gamma}$) in e^+e^- annihilations. $\tilde{e}$ denotes the two ordinary spin-0 electrons, $\tilde{e}_M = \hat{e}$ the two spin-0 partners of mirror electrons. The paraphotino couples electrons to mirror spin-0 electrons, and conversely.

e) Supersymmetric GUTS in a 5-or-6 dimensional spacetime

An additional interest of extended supersymmetric theories is that they can be formulated in a 6-dimensional spacetime [5,19,20]. The two photinos appearing in Fig. 4, or Table 2, originate from a single Weyl (chiral) spinor in 6 dimensions :

$$\begin{array}{ccc} \text{1 Weyl photino} & \rightarrow & \text{2 Majorana photinos} \\ \text{in 6 dim} & & \text{in 4 dim} \end{array} \tag{41}$$

$$\begin{array}{ccc} \text{1 Octet Weyl gluinos} & \rightarrow & \text{2 Octet Majorana gluinos} \\ \text{in 6 dim} & & \text{in 4 dim.} \end{array} \tag{42}$$

Weyl spinors in 6 dimensions also describe, at the same time, ordinary leptons and quarks as well as their mirror partners.

The $W^{\pm}$ and Z masses are already present in the 6-dimensional spacetime (i.e., they can be generated in a 6 d Poincaré invariant way) [20]. The grand-unification mass m_X, on the other hand, only appears in the 4-dimensional theory : it is associated with the large values of the covariant momenta carried by the grand-unification gauge bosons (e.g. $X^{\pm 4/3}$, $Y^{\pm 1/3}$, in a SU(5) theory) along the extra fifth or sixth dimension [5].

These two extra space dimensions are taken to form a compact space such as a torus or a 2-sphere. This leads to the possibility of computing the grand-unification mass in terms of the lengths of the extra dimensions. In the simplest case, we find :

$$m_X = \frac{10\pi}{3} \frac{\hbar}{L_{5(6)} c}. \tag{43}$$

If we are very naive and assume $m_X \gtrsim 10^{15}$ GeV/c^2, we get from formula (43)

$$L \lesssim 2\ 10^{-28} \text{ cm.} \tag{44}$$

This is to be compared with the Planck length :

$$L_P = \frac{\hbar}{m_P c} = \sqrt{\frac{G_{Newton}\, \hbar}{c^3}} \simeq 1.6\ 10^{-33} \text{ cm} \tag{45}$$

at which quantum gravity effects become essential.

In any case, the extra dimensions, whatever their shapes and sizes, are likely to play an essential role in the determination of the grand-unification mass.

REFERENCES

[1] P. Fayet and S. Ferrara, Phys. Reports 32 (1977) 249;
P. Fayet, Phys. Letters 69B (1977) 489; Proc. Europhysics Study Conf. on the Unification of the Fundamental Particle Interactions (Erice, Italy), ed. Plenum (1980), p. 587; Proc. XXIst Int. Conf. on High Energy Physics (Paris), Journal de Physique C3 (1982) p. 673;
S. Ferrara, Proc. Int. Conf. on High Energy Physics (Brighton, U.K., 1983), p. 522;
J. Ellis, Proc. Int. Symposium on Lepton and Photon Interactions at High Energies (Cornell, U.S.A., 1983) p. 439;
H.P. Nilles, Phys. Reports 110 (1984) 1;

H.E. Haber and G. Kane, Phys. Reports 117 (1985) 75;
S. Dawson, E. Eichten and C. Quigg, Phys. Rev. D31 (1985) 1581; and references therein.
[2] P. Fayet, Nucl. Phys. B237 (1984) 367.
[3] S. Ferrara, D.Z. Freedman and P. van Nieuwenhuizen, Phys. Rev. D13 (1976) 3214;
S. Deser and B. Zumino, Phys. Letters 62B (1976) 335;
P. van Nieuwenhuizen, Phys. Reports 68 (1981) 189.
[4] P. Fayet, Nucl. Phys. B149 (1979) 137; Proc. of the XVIIth Winter School of Theoretical Physics at Karpacz (Poland, 1980), Studies in High Energy Physics, Vol. 3, eds. L. Turko and A. Pękalski (Harwood Acad. Pub., 1981) p. 115.
[5] P. Fayet, Nucl. Phys. B246 (1984) 89; Phys. Letters 146B (1984) 41.
[6] P. Fayet, Phys. Letters 153B (1985) 397.
[7] E. Cremmer, P. Fayet and L. Girardello, Phys. Letters 122B (1983) 41;
P. Fayet, Phys. Letters 125B (1983) 178; 133B (1983) 363.
[8] P. Fayet, Phys. Letters 64B (1976) 159.
[9] G.R. Farrar and P. Fayet, Phys. Letters 76B (1978) 575.
[10] Contributions to these Proceedings.
[11] G.R. Farrar and P. Fayet, Phys. Letters 79B (1978) 442;
J.P. Dishaw et al., Phys. Letters 85B (1979) 142;
G. Kane and J. Léveillé, Phys. Letters 112B (1982) 227;
R.C. Ball et al., Contribution to the XXIst Int. Conf. on High Energy Physics (Paris, 1982); and Phys. Rev. Lett. 53 (1984) 1314;
CHARM Collaboration, Phys. Letters 121B (1983) 429.
W A 66 collaboration, Phys. Letters 160B (1985) 212.
[12] G.R. Farrar and P. Fayet, Phys. Letters 89B (1980) 191.
[13] See, for example, R. Prepost and A. Böhm, contributions to the XXth Rencontre de Moriond at Les Arcs, France (1985) p. 131 and 141, and references therein.
[14] P. Fayet, Phys. Letters 117B (1982) 460.
[15] J. Ellis and J.S. Hagelin, Phys. Letters 122B (1983) 303;
K. Grassie and P.N. Pandita, Phys. Rev. D30 (1984) 22;
T. Kobayashi and M. Kuroda, Phys. Letters 139B (1984) 208;
J. Ware and M.E. Machacek, Phys. Letters 142B (1984) 300.
[16] E. Fernandez et al., Phys. Rev. Lett. 54 (1985) 1118.
[17] See, for example, the proceedings of the EPS International Conference on High Energy Physics, Bari, Italy (1985).
[18] P. Fayet, Phys. Letters 142B (1984) 263.
[19] F. Gliozzi, J. Scherk and D. Olive, Nucl. Phys. B122 (1977) 253;
L. Brink, J. Scherk and J.H. Schwarz, Nucl. Phys. B121 (1977) 77.
[20] P. Fayet, Phys. Letters 159B (1985) 121.

NON-PERTURBATIVE EFFECTS IN SUPERSYMMETRY

G. Veneziano

C.E.R.N.
Division TH
CH-1211 Geneva 23, Switzerland

1. INTRODUCTION AND OUTLINE

In this course I shall discuss some non perturbative aspects of globally supersymmetric (SUSY) gauge theories. As we shall see, these share with their non-supersymmetric analogues interesting non perturbative features, such as the spontaneous breaking of chiral symmetries via condensates. What is peculiar about supersymmetric theories, however, is that one is able to say a lot about non-perturbative effects even without resorting to elaborate numerical calculations : general arguments, supersymmetric and chiral Ward identities and *analytic*, dynamical calculations will turn out to effectively determine most of the supersymmetric vacuum properties.

Such properties turn out to be quite peculiar, unexpected, and at variance with the ones we believe to hold true to ordinary (i.e. non supersymmetric) gauge theories. Which are the reasons for dealing with non-perturbative effects is supersymmetric theories ? The answer is almost obvious : unless a non-perturbative understanding of these theories is available, the possible uses of supersymmetry in particle physics are severely limited. The QCD example is very instructive in this respect: it is because of its (presumed ?) non-perturbative properties of confinement and chiral symmetry breaking that QCD can be seriously taken as a candidate theory of hadrons [1].

This in spite of the fact that perturbative properties, such as asymptotic freedom, make for most of our ability to test QCD right now.

Our perturbative understanding of supersymmetric gauge theories is quite good indeed and, as I shall argue, our non-perturbative understanding has improved considerably in the last few years. The time should be ripe by now to see if and where such theories can play a rôle. An incomplete list of possiblilities could be the following :

a. Composite models of quarks of leptons (supercompositeness)
b. Dynamical SUSY breaking and Witten's hierarchy
c. Local supersymmetry breaking via condensates
d. Some as c) in superstring theory and cancellation of Λ_{cos}.

The outline of the course is as follows :

1. Supersymmetric gauge theories

1.1. Definition of SUSY gauge theories : a resumé
1.2. Perturbative considerations : symmetries and vacua
1.3. Ward identity constraints

2. Dynamical calculations

2.1. Outline of the method
2.2. Supersymmetric Yang Mills (SYM)
2.3. Massless supersymmetric QCD (SQCD)
2.4. Massive SQCD and consistency checks
2.5. Chiral theories and spontaneous SUSY breaking

3. Possible applications

3.1. Supercompositeness
3.2. Supergravity induced breaking
3.3. Superstrings and Λ_{cos}.

1.1. Definition of SUSY gauge theories : a short resumé

P. Fayet [2] has already reviewed at this school supersymmetry and SUSY gauge theories. I shall thus be short in this resumé, and aim mainly at establishing my own notations and conventions.
I shall use two component notations for the spinors :
left handed spinors : $\psi_\alpha \in (1/2, 0)$ of SL(2,C) $(\alpha,\dot\alpha = 1,2)$
right handed spinors : $\overline{\psi}_{\dot\alpha} \in (0,1/2)$ of SL(2,C)

Multiplying two spinors we can get either :

$(1/2,0) \times (1/2,0) = (0,0) + (1,0)$, resp. $\psi_\alpha\psi'_\beta\varepsilon_{\alpha\beta}$; $\psi_\alpha\psi'_\beta + (\alpha \leftrightarrow \beta)$
or :
$(1/2,0) \times (0,1/2) = (1/2,1/2) = \psi_\alpha\overline{\psi}'_{\dot\alpha} = V_{\alpha\dot\alpha} = V_\mu(\sigma^\mu)_{\alpha\dot\alpha}$

I shall also use the language of N = 1 superfields (3).
A SUSY gauge theory is one based on two types of superfields :

i) Chiral (+ antichiral) matter superfields (spin 0 + 1/2)

$\Phi(y,\theta)$; $y = x + i\,\theta\,\sigma_\mu\,\overline{\theta}$;

Defining $\overline{D}_{\dot\alpha} \equiv \partial/\partial\overline{\theta}_{\dot\alpha} + i\ \theta^\alpha(\sigma_\mu)_{\alpha\dot\alpha}\,\partial^\mu$, Φ satisfies $\overline{D}_{\dot\alpha}\,\Phi = 0$

Expanding Φ in powers of θ we find :

$\Phi = \phi(y) + \theta_\alpha\,\psi^\alpha(y) + \theta^2F(y)$ corresponding to :

$$(A + iB) + \theta\ .\ 1/2\ (1 + \gamma_5)\ \psi + \theta^2(F + iG) \qquad (1)$$

in Fayet's notations (2). The h.c. of Φ

$\Phi^* = \phi^*(y^*) + \overline{\theta}\,\overline{\psi} + \overline{\theta}^2\ F^*$

is antichiral :

$D_\alpha\ \Phi^* = 0$

One also has :

$$\phi(y) = \phi(x) + \theta \sigma^{\mu} \bar{\theta} \partial_{\mu}\phi + \theta^2 \bar{\theta}^2 \Box\phi$$

$$\theta\psi(y) = \theta\psi(x) + \bar{\theta} \sigma^{\mu} \partial_{\mu} \psi \; ; \; \theta^2 F(y) = \theta^2 F(x)$$

Φ belongs to the (possibly reducible) representation $\underline{R}$ of $G_{gauge} \equiv G$.
Φ^* belongs to the representation $\bar{R}$ of G.
If R itself is a real representation of G (e.g. $R = r + \bar{r}$; R = adjoint; G = SU(2)), the theory is called vector like. If R is complex the theory is called chiral.

ii) Vector (gauge) superfields (spin 1/2,1)

Then belong to the adjoint repr. of G. In the Wess-Zumino gauge [3]* the relevant superfields take the form :

$$V = \bar{\theta} \sigma_{\mu} \theta A_{\mu} + \bar{\theta}^{2} \theta \lambda + \theta^2 \bar{\theta} \bar{\lambda} + \theta^2\bar{\theta}^2 D$$

$$W_{\alpha} = -1/4 \bar{D}^2 e^{-V} D_{\alpha} e^{V} = i\lambda_{\alpha} + \theta^{\beta}(F_{\mu\nu}(\sigma^{\mu\nu})_{\alpha\beta} + D \varepsilon_{\alpha\beta}) + + \bar{\theta}^2 D_{\mu}(\sigma_{\mu})_{\alpha\dot{\alpha}} \bar{\lambda}^{\dot{\alpha}} \qquad (2)$$

Recall that W_{α} is gauge *covariant* and chiral ($\bar{D}_{\dot{\alpha}} W_{\beta} = D_{\alpha} \bar{W}_{\dot{\beta}} = 0$) while V is neither of these. A_{μ}, λ describe the gluon and the gluino while D is an auxiliary field. The Lagrangian is the sum of three terms :

$$\mathcal{L} = \int d^4 \theta \, \Phi^* e^{V^a T^a} \Phi + (\kappa \int d^2 \theta \, W^a_{\alpha} W^{a,\alpha} + h.c.) + [\int d^2 \theta \, U(\Phi) + h.c.] \qquad (3)$$

We have excluded U(1) factors in the gauge group G implying no Fayet-Illiopoulos term in $\mathcal{L}$. We also recall that, because of integration rules over Grassmann variables $\int d^2\theta() \Rightarrow (\quad)_F$ and $\int d^4\theta() \Rightarrow (\quad)_D$ in Fayet's notations.

The first term in (3) is the gauge covariantization of the matter kinetic term and provides the coupling of matter to gauge fields (T^a are the generators of G in the representation to which Φ belongs). The second term in (3) gives the gauge field kinetic terms. The coefficient κ depends on the normalization of V. If we do not include factors of g in the $e^{V.T}$ (as we did) then

$$\kappa = 1/4g^2 \left(1 + i \frac{\theta_v g^2}{8\pi^2}\right) \qquad (4)$$

where θ_v is the "vacuum angle".
Finally, the 3rd term in $\mathcal{L}$ is the superpotential term (or matter self interaction term). U is a gauge invariant function of Φ (and not of Φ^*). For renormalizability it has to be a polynomial of degree three at most. We are now able to define three particular examples of SUSY gauge theories which will be discussed extensively in the following lectures.

---

* Use of the WZ gauge will not be crucial for us since we shall be dealing mainly with gauge invariant quantities.

a. SUSY Yang-Mills theory (SYM)

This is simply the case in which there are no matter fields Φ. The gauge group G is arbitrary but we shall be mainly concerned with the case G = SU(N).

b. Supersymmetric QCD (SQCD)

Here G = SU(N) and R = Mx(N + $\bar{N}$) i.e. the matter fields are M pairs of fundamental + antifundemental representations (hence M plays the same rôle as the number of flavours is QCD). We shall denoted by Φ^i_α (i = 1 ... M, α = 1 ... N) the superfields belonging to N and by $\tilde{\Phi}^\alpha_j$ those belonging to $\bar{N}$(*). Recall that both Φ and $\tilde{\Phi}$ are chiral, Φ^* and $\tilde{\Phi}^*$ being antichiral. This theory is vectorline (as QCD). This means that we can give mass to all our matter superfields; for this purpose it is enough to introduce the superpotential:

$$U_{mass} = - \sum_{i=1}^{M} m_i \, \Phi^i_\alpha \, \tilde{\Phi}^\alpha_i \tag{5}$$

A more general U_{mass} would seem to have a mass matrix m_{ij} but, without loss of generality, such matrix can be made diagonal by a (non-anomalous) redefinition of the superfields. We shall instead keep the m_i complex for later convenience.

c. Supersymmetric Georgi Glashow (SGG) models

Here we shall stick to G = SU(5) with one or two "families" in the $\bar{5}$ + 10 representation. Hence

$$\Phi : (\bar{5}^\alpha_a \, , \, 10^i_{[\alpha\beta]}) \qquad \begin{array}{l} \alpha,\beta = 1...5 \\ a,i = \text{family labels.} \end{array}$$

Matter fields are chiral and no mass term is possible. For the one family case no superpotential at all can be written down, while for the two family case we can add

$$\mathcal{L}_{Yukawa} = \int d^2\theta \, \eta_i \, \epsilon^{ab} \, \bar{5}^\alpha_a \, \bar{5}^\beta_b \, 10^i_{\alpha\beta} + \text{h.c.} \tag{6}$$

We shall see that these theories have the interesting property of exhibiting dynamical (and non-perturbative) spontaneous supersymmetry breaking.

1.2. Perturbative considerations : symmetries and vacua

We shall now discuss the *perturbative* symmetries and vacua of SUSY gauge theories in order to prepare the ground for the perturbative considerations and to constrast the results. Classically, SUSY gauge theories possess global vector and axial symmetries very much like their non-SUSY analogues. The presence of a massless gluino typically implies an extra symmetry a so-called R-symmetry, which transforms differently fields in the same supermultiplet (this is because $F_{\mu\nu}$ has to be invariant). It can thus be seen, without

(*) In components we shall use the notation : $\Phi = \phi + \theta\psi + \theta^2 F$; $\tilde{\Phi} = \tilde{\phi} + \theta\tilde{\psi} + \theta^2 \tilde{F}$. Of course each superfield, hence each component of it, carries gauge and flavour indices.

too much effort, that, in the 3 cases defined above, to following global symmetries emerge :

a. SYM $\quad G_F^{cl} = U(1)_\lambda \; : \; \lambda \to e^{i\beta}\lambda ; F \to F$ (7)

b. SQCD If $m_i = 0$

$$G_F^{cl} = U(M) \times \tilde{U}(M) \times U(1)_\lambda$$

Here the various factors in G_F^{cl} act as follows :

$$U(M) : \phi_\alpha^i \to U_{ij}\,\phi_\alpha^j \; ; \; UU^+ = 1\!\!1$$

$$\tilde{U}(M) : \tilde{\phi}_i^\alpha \to \tilde{U}_{ij}\,\tilde{\phi}_j^\alpha \; ; \; \tilde{U}\tilde{U}^+ = 1\!\!1 \tag{8}$$

$$U(1)_\lambda : \lambda \to e^{i\beta}\lambda \; ; \; \phi_\alpha^i \to e^{i\beta}\phi_\alpha^i \; ; \; \tilde{\phi}_i^\alpha \to e^{i\beta}\tilde{\phi}_i^\alpha$$

Masses break explicitly G_F^{cl}, but typically preserve a subgroup of it. $U(1)_\lambda$ is not broken, for instance, since

$$\mathcal{L}_{mass} = - \sum_i |m_i|^2 (\phi_i^* \phi_i + \tilde{\phi}_i^* \tilde{\phi}_i) - (\sum_i m_i \psi_i \tilde{\psi}_i + h.c.) \xrightarrow[U(1)_\lambda]{} \mathcal{L}_{mass} \tag{9}$$

$U(M) \times \tilde{U}(M)$ is broken to some $(U(1)_V)^M$ if all m_i are different. However, if $m_i = m$, the unbroken subgroup is $U(M)_V$

$$U(M)_V : \phi_\alpha^i \to U_{ij}\,\phi_\alpha^i \; ; \; \tilde{\phi}_\alpha^i \to U^+\tilde{\phi} \; , \; U^+U = 1\!\!1 \tag{10}$$

Intermediate solutions are also possible, of course.

c. SGG

In the one family case

$$G_F^{cl} = U(1)_\lambda \times U(1)_{\bar{5}} \times U(1)_{10} \tag*{(11)$_1$}$$

While in the two family case, in the absence of a superpotential ($\eta_i = 0$)

$$G_F^{cl} = U(1)_\lambda \times U(2)_{\bar{5}} \times U(2)_{10} \tag*{(11)$_2$}$$

If $\eta_i \neq 0$ it is easy to see that

$$G_F^{cl} \to U(1)_\lambda \times SU(2)_{\bar{5}} \times U(1)' \tag*{(11)$'_2$}$$

where $U(1)'$ is a suitable combination of $U(1)_{\bar{5}}$ and $U(1)_{10}$.

Some of the symmetries in G_F^{cl} suffer from the usual Adler Bell Jackiw anomaly. It is easy to see that :

a. For SYM : $U(1)_\lambda$ is anomalous. The discrete subgroup

$$Z_{2N} : \lambda \to \exp\left(\frac{2i\pi\kappa}{2N}\right)\lambda \tag{12}$$

is not anomalous though and will play a crucial role.

b. <u>For SQCD</u>

The $SU(M) \times S\tilde{U}(M) \times U(1)_V$ subgroup of G_F^{cl} is not anomalous while both $U(1)_A$ and $U(1)_\lambda$ are anomalous. The anomaly free combination :

$$U(1)_X : q_X(\lambda) = -\frac{M}{N} ; q_X(\psi_i, \tilde{\psi}_i) = 1; q_X(\phi_i, \tilde{\phi}_i) = 1 - M/N \qquad (13)$$

will play again a crucial role (together with the Z_{2N} subgroup of $U(1)_\lambda$). Note however that $U(1)_X$ is broken by masses.

c. <u>For SGG</u>

i) In the one family case two independent combinations of the three $U(1)$'s in $(11)_1$ are anomaly free (construct them as an exercise).

ii) In the two family case the two $SU(2)$ factors in $(11)_2$ are anomaly free together with two $U(1)$'s given by the generators

$$\begin{aligned} Q_1 &: Q_1(\bar{5}) = -3\, Q_1(10) \; ; \; Q_1(\lambda) = 0 \\ Q_2 &: Q_2(\lambda) = -10 \; ; \; Q_2(\bar{5}) = -14 \; ; \; Q_2(10) = 8 \end{aligned} \qquad (14)$$

where Q_1 commutes with SUSY while Q_2 is an R-symmetry (the charges indicated are those of the lowest components).

In order to discuss the perturbative vacua of SUSY gauge theories we have to write down their scalar potential. At the classical level this is given by two non negative terms [2]

$$V = V_F + V_D = \sum_i |F_i|^2 + \sum_a D^a D^a \qquad (15)$$

where F and D are expressed in terms of the fundamental scalar fields ϕ_i as

$$F_i = (\partial U/\partial\phi_i)^* \; ; \; D^a = g\, \phi^* \; T^a \, \phi \qquad (16)$$

Supersymmetric vacua satisfy, of course, $F_i = D^a = 0$.
Let us consider again our three examples :

a. <u>SYM</u> - There are no elementary scalar fields here. A *composite* scalar order parameter is $\lambda\lambda \equiv \lambda^a_\alpha \lambda^a_\beta \, \varepsilon^{\alpha\beta}$ has vanishing vacuum expectation value (v.e.v.) to all and has orders in perturbation theory because of chiral symmetry (as it is the case for $\bar{\psi}\psi$ in ordinary massless QCD).

b. <u>SQCD</u> - Here we have to distinguish two cases

i) $m_i \neq 0$. Since $F_{i,\alpha} = (m_i \tilde{\phi}^\alpha_i)^*$, $\tilde{F}^\alpha_i = (m_i \phi^i_\alpha)^*$ setting $F, \tilde{F} = 0$

gives $< \phi^i_\alpha > = < \tilde{\phi}^\alpha_i > = 0$

ii) $m_i = 0$ ($i = 1 \ldots L$); $m_i \neq 0$ ($i = L+1 \ldots M$).

By the previous argument we get $< \phi_i > = < \tilde{\phi}_i > = 0$ ($i = L+1 \ldots M$).

For the remaining (massless) flavours we distinguish the subcases :

ii)$_a$ $L < N : < \phi^i_\alpha > = < \tilde{\phi}^\alpha_i > = v_i \; \delta_{i\alpha} \quad (1 \leq \alpha \leq L)$

$$= 0 \qquad (\alpha > L) \tag{17a}$$

ii)$_b$ $L \geq N : < \phi^i_\alpha > = v_i \; \delta_{\alpha i}; < \tilde{\phi}^\alpha_i > = \sqrt{v_i^2 - c^2} \; \delta_{i\alpha} \quad (1 \leq i \leq N)$

$$< \phi_i > = < \tilde{\phi}_i > = 0 \qquad (i > N) \tag{17b}$$

(c = arbitrary constant)

To these vacua one has to add those obtained from them via transformations of the (explicitely) unbroken subgroup of G_F (e.g. SU(L) x SU(L)). An important feature of the (partially) massless case is the existence of "flat directions" in the vacuum manifold i.e. of directions going off to ∞ (in $\phi,\tilde{\phi}$) along which $V \equiv 0$. This feature, typical of supersymmetry, comes from the fact that SUSY vacua are not just representing G_F but its complexification [4]. Consider for instance the (anomalous) factor $U(1)_A$ in G_F under which $(\phi,\tilde{\phi}) \to e^{i\beta} (\phi,\tilde{\phi})$. This is a classical (and perturbative) symmetry of massless SQCD. The complexification of this $U(1)_A$ simply rescales both ϕ and $\tilde{\phi}$. Although this is not a true symmetry of the theory the vacuum manifold represents it in the sense that rescaling a set of v.e.v.'s gives another (physically inequivalent) vacuum. Since the complexification of G_E is non-compact the vacuum manifold itself is non-compact i.e. it extends to infinity.

c. SGG - We just state without proof that the one family SGG theory has only a trivial vacuum at the origin. The same is true of the two family case if there is a superpotential ($\eta \neq 0$). At $\eta = 0$, this theory has instead non-trivial Higgs vacua and the corresponding flat directions. The (non) existence of flat directions will be argued to have direct consequences on the dynamical breaking of supersymmetry itself.

Up to here our consideration have been essentially classical. In ordinary gauge theories quantum corrections are able, even perturbatively, to change completely the picture of the vacuum (see e.g. Coleman-Weinberg [5]). The opposite is true in supersymmetric theories. A famous non renormalization theorem insures the perturbative persistence of supersymmetric vacua to all orders [6] : in particular the flat directions discussed above will not be lifted to any order of perturbation theory (both in correspondence with good symmetries in G_F and with the anomalous ones). It is precisely in the light of the non-renormalization theorem that our non perturbative considerations acquire all their importance. We shall indeed argue that, non perturbatively, there is a breaking of the just recalled non renormalization theorem and that, as a result, the flat directions are in general lifted up. The fact that such directions did exist classically (and perturbatively) will be partly responsible for the unexpected behaviour of the SQCD vacua.

1.3. Ward identity constraints

We shall now discuss the constraints due to SUSY, as well as to chiral symmetries on v.e.v.'s and, more generally on correlation functions independently of perturbation theory (*). Of course, these constraints will not be sufficient by themselves to determine completely the dynamical properties of the theory; however, they will turn out to be of much value in restricting the possible patterns that can occur.

(*) Being interested is non perturbative effects we shall deal, as in QCD, with gauge invariant (hence composite) operators and correlation functions.

The general strategy will be to assume that at least one SUSY vacuum exists. In some cases one knows by Witten index arguments (discussed later) that such is the case. In other cases we may be led into a contradiction betweem our assumption of a SUSY vacuum and other constaints or dynamical calculations and we shall conclude that the theory under consideration does *not* possess (at least well defined) SUSY vacua.

In a supersymmetric vacuum auxiliary fields cannot take non zero v.e.v.'s since

$$0 = < \delta_\alpha X_{\theta_\beta} > = < X_{\theta^2} > \epsilon_{\alpha\beta} \tag{18}$$

where $\delta_\alpha (\bar{\delta}_{\dot\alpha})$ is the coefficient of the left handed Grassmann parameter $\zeta^\alpha (\bar{\zeta}^{\dot\alpha})$ in the supersymmetric variation (*).

We shall instead consider v.e.v'.s and correlation functions of lowest components of gauge invariant (composite) chiral operators e.g.

$$(W^2)_{\theta=0} \equiv \lambda^a_\alpha \lambda^a_\beta \varepsilon_{\alpha\beta} \; ; \; (T_{ij})_{\theta=0} \equiv \phi^i_\alpha \tilde{\phi}^\alpha_j \quad \text{etc} \ldots \tag{19}$$

Such operators (which we shall generically denote by χ_o) satisfy the property :

$$\bar{\delta}_{\dot\alpha} \; \chi_o = 0 \tag{20}$$

with amounts essentially to the definition of a chiral superfield. Consequently :

$$< \chi_o \bar{\delta}_{\dot\alpha}(\ldots) > = < \bar{\delta}_{\dot\alpha} (\chi_o \ldots) > = 0 \tag{21}$$

A little reminder : Ward identities can be proven in many ways for instance by functional methods. Since

$$< (\ldots) > = \int d\phi \; (\ldots) \; \exp\; (\Gamma(\phi))/\int d\phi \; \exp(\Gamma(\phi))$$

a change of variables $\phi \rightarrow \phi + \delta\phi$ which leaves $d\phi$ invariant (this defines a *non anomalous* transformation) gives :

$$< \delta \; (\ldots) > + < (\ldots) \; \delta\Gamma >_c = 0$$

In massive SQCD a particularly interesting (...) in (21) is given by

$$(\ldots\ldots) = (\bar{T}_{ii})_\theta \tag{22}$$

which gives

$$< \chi_o \; \bar{\delta}_{\dot\alpha} \; (\bar{T}_{ii})_{\bar{\theta}_{\dot\beta}} > = \epsilon_{\dot\alpha\dot\beta} < \chi_o \; (\bar{T}_{ii})_{\bar{\theta}^2} > = 0 \tag{23}$$

(*) We shall assume throughout these lectures that SUSY is anomaly free. This, together with the invariance of the vacuum gives the general constraint $< \delta_\alpha(\ldots) > = < \bar{\delta}_{\dot\alpha} (\ldots) > = 0$ for any gauge invariant (...).

The r.h.s. of (23) is related (*) to $\partial/\partial m_i^* < \chi_o >$ giving :

$$\partial/\partial m_i^* < \chi_o > = \int d^4x < \chi_o \, (\bar{T}_{ii}(x))_{\bar{\theta}^2} > = 0 \qquad (24)$$

We have thus discovered [7] that chiral Green's functions in massive SQCD can only depend on m_i but not on m_i^*; in other words they are analytic functions of the m_i seen as complex parameters.

At this point we combine the above result with the knowledge of the dependence of $< \chi_o >$ upon the *phase* of m_i. This is given by a (non anomalous) chiral Ward identity reading :

$$2i \frac{\partial}{\partial \arg m_i} < \chi_o > = (m_i \frac{\partial}{\partial m_i} - m_i^* \frac{\partial}{\partial m_i^*}) < \chi_o > =$$

$$= < \chi_o \, \delta_A^{(i)} \Gamma > \; = \; - < \delta_A^{(i)} \chi_o > \qquad \text{(see before)} \qquad (25)$$

In order to perform the 2nd and 3rd step in (25) we need a *non anomalous* axial transformation $\delta_A^{(i)}$, broken by m_i alone. This uniquely fixes $\delta_A^{(i)}$ to be :

$$\delta_A^{(i)} \; : \lambda \to e^{-\frac{i\alpha}{2N}} \lambda; \; (\psi_j, \tilde{\psi}_j) \to e^{\frac{i\alpha\delta_{ij}}{2}} (\psi_j, \tilde{\psi}_j);$$

$$(\phi_j, \tilde{\phi}_j) \to e^{\frac{i\alpha}{2}(\delta_{ij} - 1/N)} (\phi_j, \tilde{\phi}_j) \qquad (26)$$

If we now consider operators χ_o which are eigenstates of $\delta_A^{(i)}$ i.e. such that :

$$\delta_A^{(i)} \chi_o = q_A^{(i)}(\chi_o) \cdot \chi_o \qquad (27)$$

then the combination of eqs.(24),(25) and (26) yields :

$$m_i \, \partial/\partial m_i < \chi_o > = - q_A^{(i)}(\chi_o) < \chi_o > \Rightarrow < \chi_o > \sim (m_i)^{-q_A^{(i)}(\chi_o)} \qquad (28)$$

Notice that a result analogous to (25) holds in ordinary QCD (*). It is the combination of it with analyticity (which is *not* valid in QCD) that gives the much stronger result (27).

Let us see an immediate application of eq.(28). Consider a SQCD correlation function $G_{N,M}^{p,q}$ defined by

$$G_{N,M}^{p,q} \equiv T < \lambda\lambda(x_1) \ldots \lambda\lambda(x_p) \, \phi_{i_1} \tilde{\phi}_{j_1}(y_1) \ldots \phi_{i_q} \tilde{\phi}_{i_q}(y_q) > \qquad (29)$$

The operator in the brackets is a particular example of operator χ_o discussed above. Using (26) we find :

(*) This can be proven directly in the bare, regularized, theory and can be converted into an equation in the renormalized theory if a mass independent renormalization scheme (m-independent Z factors) is used.

$$- q_{A_i}(\chi) = p/N + q/N - \sum_k (\delta_{i_k i} + \delta_{i_k i})/2 \qquad (30)$$

Thus

$$G^{p,q}_{N,M} \sim \prod_j m_j)^{(q+p)/N} \prod_k (m_{i_k} m_{j_k})^{-1/2} \qquad (31)$$

Particular cases of (31) are the v.e.v.'s of local operators (order parameters, order condensates)

$$< \lambda\lambda > \sim \prod_j (m_j)^{1/N} \qquad (32)_a$$

$$< \phi_i \tilde{\phi}_i > \sim m_i^{-1} \prod_j (m_j)^{1/N} \qquad (32)_b$$

These equations imply that < λλ > vanishes in the chiral limit of SQCD, while $< \phi \tilde{\phi} >$ is either identically zero or blows up in the same limit.

We shall now argue that even a stronger result can be proven relating the two v.e.v.'s in eq.(32) (and more generally relating correlation functions $G^{p,q}$ with different p's and q's). This is the so-called Kanishi anomaly constraint. It is based on the anomalous commutator [8].

$$\epsilon^{\dot{\alpha}\dot{\beta}} \{ \bar{Q}_{\dot{\alpha}}, \bar{\psi}^i_{\dot{\beta}} \phi^i \} = \bar{F}_i \phi_j + g^2/32\pi^2 \lambda\lambda = - m_i \tilde{\phi}_i \phi_i + $$
$$+ g^2/32\pi^2 \lambda\lambda ; \; i = 1 \ldots M \qquad (33)$$

Such as anomaly can be derived in several ways e.g. following a SUSY generalization of Fujikawa's method [9]. A more direct proof is the one of Konishi himself and is as follows. Let us define the local product $\bar{\psi}\phi$ by a point splitting limit :

$$\{\bar{Q}, \bar{\psi}\phi\} = \lim_{\varepsilon \to 0} \{\bar{Q}, \bar{\psi}(x+\varepsilon) \, e^{2i\varepsilon \; gA} \phi(x-\varepsilon)\}$$

$$= \{\bar{Q}, \bar{\psi}\phi\}_{naive} + 2i\varepsilon_\mu \; g \; \bar{\psi}\phi \; \lambda\sigma^\mu \qquad (34)$$

The extra piece is the insertion indicated in Fig.1a. Naively it is of $O(\varepsilon^\mu)$ but, because of the linear ultraviolet divergence of the diagrams of type (1b), such insertion does not vanish in the limit $\varepsilon^\mu \to 0$ but rather gives a finite constant $(g^2/32\pi^2)$ times the insertion of the local operator λλ (Fig.1c,d).

(*) It holds actually only for the dependence of $< \bar{\psi}_k \psi_k >$ upon $\arg(m_i) - \arg(m_j)$. The dependence upon $\Sigma \arg m_i$ is more difficult and is related to θ_i-dependence and CP violation in QCD.

The superfield generalization of (33) is easy to find :

$$\bar{D}^2 \; (\Phi^* \; e^V \; \Phi) = - m \; \tilde{\Phi} \; \Phi + \frac{g^2}{32\pi^2} \; W^2 \tag{35}$$

and contains eq.(33) as its $\theta = 0$ component. Taking the vacuum expectation value of the anomalous commutator (33) fixes

$$0 = -m_i . < \tilde{\phi}_i \phi_i > + < g^2/32\pi^2 \; \lambda\lambda > \tag{36}$$

This result is not only compatible with the mass dependence (32); it actually says that the proportionality constants in $(32)_{a,b}$ just differ by a factor $g^2/32\pi^2$.

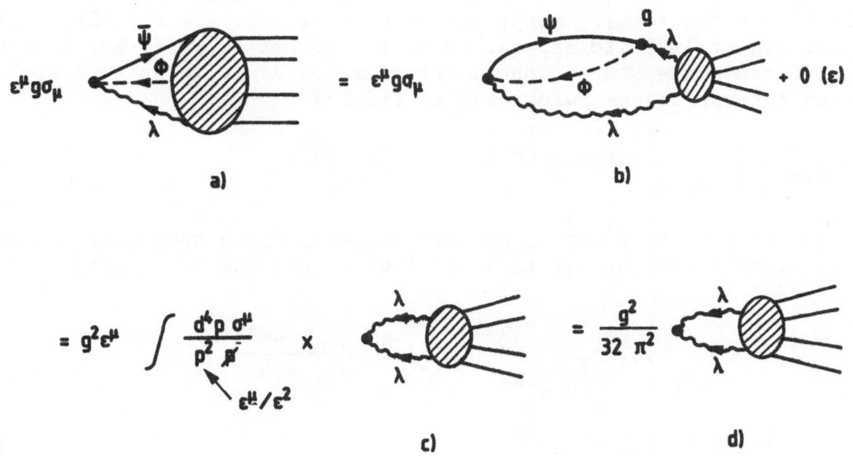

Fig. 1

It follows that either $< \lambda\lambda >$ and $< \phi\tilde{\phi} >$ are identically zero in *massive* SQCD or $\phi\tilde{\phi}$ has to become arbitrarily large as $m \to 0$. Since for m large $< \lambda\lambda >_{SQCD}$ should approach $< \lambda\lambda >_{SYM}$ from general decoupling considerations [10] we conclude that the only way to avoid the blow up of $< \phi_i\tilde{\phi}_i >$ as $m_i \to 0$ is to make $< \lambda\lambda >_{SYM} = 0$. As we shall discuss below such a possibility is very unlikely due to Witten index considerations(and to explicit dynamical calculations). Notice the crucial role of Konishi's anomaly constraints. If there was no anomaly eq.(36) would have been,

$$m_i < \tilde{\phi}_i \phi_i > = 0 \tag{37}$$

implying $<\tilde{\phi}_i\phi_i> = 0$ at $m_i \neq 0$ and no constraint on it at $m_i = 0$. Hence, in the absence of the anomaly, the situation would resemble closely that of perturbation theory ($\phi = \tilde{\phi} = 0$ at $m \neq 0$, undetermined at $m = 0$). It is the anomaly plus the (so far presumed) non-perturbative generation of a non-trivial $<\lambda\lambda>$ that makes the non-perturbative SQCD vacua look so different from the perturbative ones.

There is actually a third crucial property of correlation functions such as $G^{p,q}_{N,M}$: amazingly such correlations have to be x_i, y_j independent! [11]. In order to prove it notice that

$$\partial^{x_i}_{\mu} \; G = (\sigma_\mu)^{\alpha\dot{\alpha}} < 0|T(\chi_o(x_1)\chi_o(x_2) \; \{\overline{Q}^{\dot{\alpha}}, \chi_{\theta_\alpha} \; (x_i)\} \ldots \chi_o \ldots)|0)$$

$$= \sigma_\mu \prod_{k \neq i} < 0| \; T[\overline{Q}, \chi_o(x_k)] \; \chi_\theta(x_i)|0) = 0 \tag{38}$$

At this point one can invoke clustering at large distances or rather the fact that the lowest energy intermediate states are the only ones that survive at infinite separation of the points x_i, y_j (other non-zero energy intermediate states cancel out pairwise as it is easy to see). One thus interprets the value of G as [12,13]

$$G^{p,q}_{N,M} = \sum_{\Omega=\text{vacuum}} < \Omega|\lambda\lambda|\Omega >^p \prod_{k=1}^{q} < \Omega|\phi_{i_k}\tilde{\phi}_{j_k}|\Omega > \tag{39}$$

where in order to be general, we have allowed, for a degenerate vacuum. One can finally combine eqs.(39) and (36) to get the fundamental constraint :

$$G^{p,q}_{N,M} = \delta_{i_1 j_1} \ldots \delta_{i_q j_q} \prod_{k=1}^{q} (m_{i_k})^{-1} \; \frac{< \lambda\lambda >^{p+q}}{(16\pi^2)^q}$$

$$< \lambda\lambda > = c \prod_j (m_j)^{1/N} \tag{40}$$

All these results suggest a simple strategy [13,14] for computing the vacuum properties of SUSY gauge theories i.e.

1- Compute $G^{p,q}_{N,M}$ in regions of x_i, y_i and for values of the masses m_i where the computation is feasible (e.g. short distances or large mass).

2- Use x-independence and/or the known m-dependence in order to extrapolate the result to the region of interest (e.g. large distances to extract vacuum properties or small m to study chiral properties).

3- Use clustering and the constraints (36) to fully determine the condensates. It should be emphasized right now that the above m-dependence arguments are not able to say anything on the strictly massless theory. A priori the $m \to 0$ limit can differ from the $m = 0$ case, which will have to be investigated separately.

A final point I wish to discuss before turning to dynamical calculations is the consistency of the mass dependence (40) with general decoupling considerations [10]. Later on we shall be doing large m calculations and, as explained above, we shall extrapolate the results to arbitrary m. It is essential to check therefore that the m-dependence (40) makes sense even at $m \gg \Lambda$. A priori this looks far from obvious. Take for instance eq.$(32)_a$ for one flavour of mass m and arbitrary N. The dimensional proportionality constant in $(32)_a$ must be a numerical constant times the appropriate power of the only other dimensionful parameter i.e. of the renormalization group invariant scale $\Lambda_{N,1} \equiv \Lambda_{SQCD}\big|^{N}_{M=1}$. Thus :

$$< \lambda\lambda >_{N,M=1} = c(\Lambda_{N,1})^{3-1/N} (m)^{1/N} \tag{41}$$

Eq.(41) appears to imply that $< \lambda\lambda >$ blows up at large m in contrast with decoupling constraints which imply :

$$< \lambda\lambda >_{N,M=1} \xrightarrow[m \gg \Lambda]{} < \lambda\lambda >_{N,M=0} = \Lambda^3_{SYM} \tag{42}$$

where, obviously, Λ_{SYM} does not know about m.

Actually, eqs.(41) and (42) are compatible provided we remember that $\Lambda_{N,1}$ was the invariant scale of SQCD in a mass-independent renormalization scheme. It follows that $\Lambda_{N,1}$ controls the *large* momentum $(q > m)$ behaviour of the SQCD effective (running) coupling :

$$\bar{g}^{-2}_{N,1} = \beta^{(N,1)} \log(q/\Lambda_{N,1}) \; ; \quad q > m \tag{43}$$

The question now is : which is the scale Λ_{SYM} of the SYM theory which is equivalent to our large $\underline{m}$ SQCD in the sense of decoupling ? This is the scale at which $\bar{g}_{N,1}$ actually becomes strong. Now, below $q = m$, $\bar{g}_{N,1}$ does *not* evolve according to (43) but as

$$\bar{g}^{-2}_{N,1} = \beta^{N,0} \log(q/\Lambda_{N,0}) = \beta^{SYM} \log(q/\Lambda_{SYM}) \; (q < m) \tag{44}$$

and all we have to do in order to find the relation between $\Lambda_{N,1}$ and Λ_{SYM} is to impose *continuity* at $q = m$:

$$\beta^{N,1} \log(m/\Lambda_{N,1}) = \beta^{N,0} \log(m/\Lambda_{SYM}) \tag{45}$$

Since $\beta^{N,1}/\beta^{N,0} = (3N-1)/3N$ we get :

$$\Lambda^{3-1/N}_{N,1} (m)^{1/N} = \Lambda^3_{SYM} \tag{46}$$

which make (41) and (42) coincide ! This argument can be easily checked to work in the case of an arbitrary number of light and heavy flavours. We thus conclude that, even if we had no chiral identity arguments for the mass dependence (40), precisely the same result could have been derived from decoupling considerations !

We conclude that we can thus use with confidence eq.(40) in the whole range of values for the masses.

2. Instanton calculations [11-15]

We shall now proceed with some explicit calculations of Green's functions of the type (29). Typically , these correlation functions involve operators of the same chirality (if we mix operators of opposite chirality we loose all the nice properties described above) and hence will be zero to all orders of perturbation theory. The best known examples of non perturbative effects are those due to instantons : hence we shall try to consider correlation function which are instanton dominated . Actually all we can impose on general grounds is for the correlation function to be saturated by gauge field configurations of (Pontriagin) topological number ν :

$$\nu = \int d^4x \frac{g^2}{32\pi^2} F \tilde{F} \tag{47}$$

equal one. Many years ago 't Hooft has shown that the value of ν contributing to the functional integral is related to the total chirality flip in the correlation function. Imposing these constraints we arrive at the conclusion that, in the massless case, the $\nu = 1$ dominated correlation functions (involving lowest components of chiral superfields are restricted to be of the type :

$$G_{N,M} \equiv < \lambda\lambda(x_1) \ldots \lambda\lambda(X_{N-M}) \, \phi_{i_1}\tilde{\phi}_{j_1}(y_1) \ldots \phi_{i_M}\tilde{\phi}_{j_M}(y_M) > \tag{48}$$

where, from now on, the simplified notation

$$\lambda\lambda \equiv g^2/32\pi^2 \, \lambda^a_\alpha \, \lambda^a_\beta \, \varepsilon_{\alpha\beta} \tag{49}$$

will be used. There is a simple way to double check the selection rule.

For a give gauge field configuration $A^{(\nu)}_\mu$ of topological number ν rule. The functional integration over the massless fermions $\lambda, \bar{\lambda}, \psi, \bar{\psi}, \tilde{\psi}, \bar{\tilde{\psi}}$ gives trivially zero by Berezin's integration rule unless the correlation function itself contains as many fermionic fields as there are zero modes of the Dirac operator in the background A_μ

$$\not{D}(A) \, \lambda^{(r)} = 0 \; ; \; \not{D} \, \psi^{(i)} = 0 \; ; \; \not{D} \, \tilde{\psi}^{(j)} = 0 \tag{50}$$

If $\nu = 1$ it is known that there are precisely 2N gluino zero modes as well as one zero mode for each flavour for ψ and the same for $\tilde{\psi}$. For G to receive a $\nu = 1$ contribution one must then have (*)

$$G \sim < (\lambda\lambda)^N \, (\psi\tilde{\psi})^M > \tag{51}$$

Such G however does not satisfy the criteria of the previous section since $\psi\tilde{\psi}$ is *not* the lowest component of a chiral superfield. Nonetheless since there are $\lambda\psi\phi^*$, $\lambda\tilde{\psi}\tilde{\phi}^*$ Yukawa couplings in SQCD one can go over from G of eq.(51) to $G_{N,M}$ of eq.(48) using some perturbative vertices and $G_{N,M}$ does satisfy our criteria.

(*) In the case of ordinary QCD this gives the known 't-Hooft interaction $\sim \det_{ij} \, \psi_{i_1 R} \psi_{j_1 L}$.

The last, most difficult point, consists in justifying the dominance, of small site, single instanton configurations at short distances. We cannot offer a rigorous proof of that, but we shall offer strong plausibility arguments together with checks of selfconsistency of such approximation with all the Ward identity constraints.

Let us now illustrate the technique in a few cases :

a. SYM with N = 2 [11] (For the SU(N) case see Ref.14)
The relevant correlation function is

$$G_2 = < \lambda\lambda(x_1)\ \lambda\lambda(x_2) > \tag{52}$$

In a single instanton background there are 4 gluino zero modes. Integration over the gauge field fluctuations leads as usual to an integration over the instanton collective coordinates (position x_o and size ρ) with a volume

$$d\,V_{inst} = c\ \exp(-8\pi^2/g^2)\ \mu^6\ g^{-8}\ \frac{d^4x_o d\rho}{\rho^5}\cdot \rho^8 \tag{53}$$

where μ is a renormalization scale, g is the corresponding coupling constant and c is a numerical constant. Integration over the fermionic degrees of freedom leads to a fermionic determinant (from the non-zero modes) which cancels the corresponding bosonic $(\text{determinant})^{-1}$ leaving the integration over the (Grassmann) coefficients of the fermionic zero modes. Using

$$\int da\ a = 1 \quad ; \quad a = \text{Grassmann variable} \tag{54}$$

we get

$$G = \int d\,V_{inst} \sum_{\{i_1 \sim . i_4\}} (-1)^p\ \lambda^{i_1}\lambda^{i_2}(x_1)\ \lambda^{i_3}\lambda^{i_4}(x_2) \tag{55}$$

where the index i_n labels the zero modes and the sum runs over permutations. The zero modes can be easily obtained in this case, from supersymmetric and superconformal transformations of the instanton itself and take the form

$$\lambda^{(i)b}_{\alpha,a}(x) = \frac{1}{\pi}\ \rho^2\ f^2(x)\ (\delta_{\alpha b}\ \delta_{ia} - \varepsilon_{ib}\ \varepsilon_{\alpha a})\ ;\ (i = 1,2) \tag{56}$$

$$\hat{\lambda}^{(k)b}_{\alpha,a} = -\frac{1}{\sqrt{2}\pi}\ (\sigma_\mu)^{ki}\ (x - x_o)^\mu\ \lambda^{(i)b}_{\alpha,a} \quad ;\ (k = 1,2) \tag{57}$$

where $f(x) = [(x - x_o)^2 + \rho^2]^{-1}$, i,k are mode indices (giving a total of 4 zero modes) a,b are SU(2) indices and α is the spin index. It is easy to perform the $\sum_p(-1)^p$ in (55) and obtain

$$G_2 = \int d\,V_{inst}\ f^4(x_1)\ f^4(x_2)\ \rho^6 (x_1 - x_2)^2 \tag{58}$$

The origin of the (crucial) factor $(x_1 - x_2)^2$ (which comes out of the integral) rests in the fact that, at $x_1 = x_2$, the four zero modes are all at the same point. Because of the relation (57) only two of them are independent and Fermi statistics forces the antisymmetrized product to vanish as $(x_1 - x_2) \to 0$. Inserting now (53) into eq.(58), the resulting integral can be done using standard Feynman parameter techniques. Before doing that, we remark that the factors ouside the integral reproduce, a part from the $(x_1 - x_2)^2$ factor and powers of g (*),

$$\Lambda^6 = [\mu \exp(-\frac{8\pi^2}{\beta_o g}2)]^{\beta_o} \; ; \; \beta_o = 3N = 6 \quad N = 2 \tag{59}$$

accounting fully for the dimensionality of G_2. The remaining ρ, x_o integrals can only depend on $(x_1 - x_2)$. Because of the explicit $(x_1 - x_2)^2$ factor one finds that there are neither ultraviolet $(\rho \to 0)$ nor inffared $(\rho \to \infty)$ divergences and that the outside factor $(x_1 - x_2)^2$ is just cancelled by a short distance singularity $(x_1 - x_2)^{-2}$ coming from instantons of size $\rho \sim |x_1 - x_2|$. The next result is thus (11) :

$$G_2 = c\,\Lambda^6 \tag{60}$$

Let us now interpret the result (60) in the light of the $Z_{2N} = Z_4$ symmetry discussed earlier (eq. 12). Since such a symmetry is preserved by instantons and $\lambda\lambda$ is invariant under a subgroup Z_2 of Z_4 (given by k = 0,2), while it changes sign under the remaining elements of Z_4 (k = 1,3), it is clear that

$$\sum_{\Omega = \text{SUSY vacum}} < \Omega|\lambda\lambda|\Omega > = 0 \tag{61}$$

The fact that the correlation function G_2 is non zero at large distances implies a long range correlation between the value of $\lambda\lambda$ at x and at y. This means that eq.(61) emerges from a <u>cancellation</u> between different (Z_4 related) vacua and *not* because $\lambda\lambda$ is zero in each vacuum. But this is precisely what we mean by spontaneous symmetry breaking (cf. for instance the case of a ferromagnet). We thus conclude that the result (60) implies :

$$< \Omega_k|\lambda\lambda|\Omega_k > = \sqrt{c}\,\Lambda^3 e^{i\pi k}; \; k = 0,1 \tag{62}$$

on each of the two vacua due to $Z_4 \supset Z_i$ breaking (**). The two supersymmetric vacua are precisely the ones expected on the basis of

---

(*) These are actually right to be absorbed into the 2-loop expressions for Λ

(**) The corresponding SU(N) calculation (14) yields analogously : $(<\lambda\lambda>)^N = c\Lambda^{3N}$ and consequently N-vacua corresponding to the N roots of this equation.

Witten index considerations [16]. Witten has indeed shown that for SU(N) SYM or massive SQCD

$$\Delta_W \equiv T_r\,(-1)^F \equiv (\text{number of bosonic vacua}) - (\#\text{ of fermionic vacua}) = N \qquad (63)$$

Witten obtained this result by putting SYM theory in a suitable box and by going over to a perturbative regime using the fact that Δ_W should be invariant under such modifications of the actual theory. He also a suggested that the interpretation of his result $\Delta_W = N$ could reside in the breaking of Z_{2N} to Z_2 due to gluino condensation. This point of view was further supported by the construction of (super-current anomaly incorporating) effective Lagrangians [17]. We thus see that our dynamical calculation results fully confirm these previous indications.

We now make a small digression. Consider one flavour massive SQCD. Because of Konishis relation we have

$$< \phi_i \tilde{\phi}_j > \; = \delta_{ij}\, m_i^{-1} < \lambda\lambda >_{SQCD} \qquad (64)$$

Since $< \lambda\lambda >_{SQCD}$ should coincide with $< \lambda\lambda >_{SYM}$ at large m, we expect it to be non-zero in massive SQCD as well and, because of the general Ward identity arguments, to have the m dependence (32) when expressed in terms of Λ_{SQCD}. Again, because of Z_{2N} symmetry $< \lambda\lambda >$ will take N possible phases but, once it is fixed, all scalar condensates are fixed as well. Consequently Δ_W will still be N in massive SQCD. Reliable Witten index calculations do not exist for massless SQCD or for chiral (hence massless) theories. We shall now see how our explicit calculations can be carried out in massless SQCD and then we shall come back to massive SQCD, where a few surprises will be in store before achieving full consistency. We shall finally turn to chiral theories.

b. Massless SQCD with N = 2, M = 1

According to our general discussion the interesting correlation function is now

$$G_{2,1} = < \lambda\lambda(x_1)\; \phi\tilde{\phi}(x_2) > \qquad (65)$$

The calculation of this object can be represented graphically as follows :

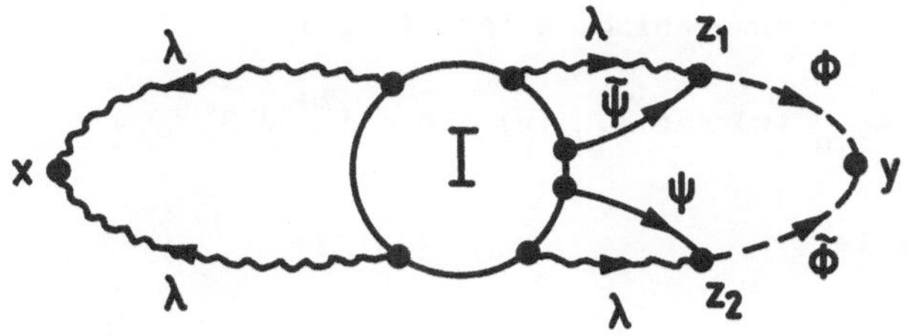

The instanton blob eats up precisely four λ, one ψ and one $\tilde{\psi}$ zero modes. These are provided by the original $\lambda\lambda(x_1)$ and by Yukawa vertices creating $\lambda\psi$ from ϕ and $\lambda\tilde{\psi}$ from $\tilde{\phi}$.

Quite amazingly the calculation can be done explicitly up to the end thanks to some powerful techniques [18] and yields again a constant (13) :

$$G_{2,1} = c' \Lambda^5 \text{ (again } \Lambda^5 \sim \exp(-8\pi^2/g^2)) \tag{66}$$

With (considerable) more work the calculation can be done [4] for generic M and $N \geq M$ with the result :

$$\begin{aligned} G_{N,M} &= \langle \lambda\lambda(x_1) \,..\, \lambda\lambda(x_{N-M})\, \phi_{i_1}\tilde{\phi}_{j_1}(y_1) \,..\, \phi_{i_M}\tilde{\phi}_{j_M}(y_M) \rangle = \\ &= C_{N,M}\, \Lambda^{3N-M}\, \varepsilon_{i_1 i_2 \ldots i_M}\, \varepsilon_{j_1 j_2 \ldots j_M} \end{aligned} \tag{67}$$

What are the conclusions that we should draw from such a result ? The first one is that it is consistent with the Ward identity constraints (x,y independence). The second is that the result is $SU(M) \times \widetilde{SU}(M)$ symmetric but *not* factorized (because of $\varepsilon.\varepsilon$ structure). But this is precisely what we expect if the vacuum is not unique and different vacua are related by symmetry transformations i.e. if there is spontaneous breaking of $SU(M) \times \widetilde{SU}(M)$. There is however a third consequence of (67). Except for the case N = M such equation is not compatible with the combination of vacuum dominance

$$G_{N,M} \sim \sum_{\Omega} (\langle \lambda\lambda \rangle_\Omega)^{N-M} \cdot \langle \phi\tilde{\phi} \rangle_\Omega^M \tag{68}$$

and of Konishi's anomaly according to which

$$\langle \lambda\lambda \rangle \sim m \langle \phi\tilde{\phi} \rangle \xrightarrow[m\to 0]{} 0 \tag{69}$$

Taken together (67) and (69) give (N > M)

$$0 \times (\langle \phi\tilde{\phi} \rangle)^M = \text{constant} \tag{70}$$

implying lack of a SUSY vacuum (except possibly at ∞).
The N = M case is free of these problems. It has the further property that it has another non-vanishing correlation :

$$G = \langle \det \phi_{i,\alpha}(x) \det \tilde{\phi}_{j,\beta}(y) \rangle = c\, \Lambda^{2N} \neq 0 \tag{71}$$

The order parameters

$$X = \det \phi_{i,\alpha} = \varepsilon_{i_1..i_N}\, \varepsilon_{\alpha_1..\alpha_N}\, \phi_{i_1,\alpha_1} \,..\, \phi_{i_N,\alpha_N} \;;\; \tilde{X} = \det\tilde{\phi} \tag{72}$$

are $SU(M) \times \widetilde{SU}(M)$ invariant but have $U(1)_V$ charge. Thus eq.(71) implies the spontaneous breaking of $U(1)_V$ in the massless theory (the opposite is true in the massive case).

c. Massive SQCD

Turning now to massive SQCD we seem to face a puzzle. We have seen that, as a consequence of Ward identities, correlation functions such as $G_{N,M}$ should be given by

$$G_{N,M} = \delta_{i_1 j_1} \ldots \delta_{i_M j_M} \prod_{k=1}^{M} (m_{ik})^{-1} \cdot \prod_j (m_j) \tag{73}$$

Notice that this is very different from the massless result (67) (they only coincide if we take $i_1 = j_1 \neq i_2 = j_2 \neq i_3 = j_3 \ldots$). In particular (73) is factorized in flavour while (67) is not (this is related to the fact that the vacuum degeneracy of the massless situation is removed by the masses and a single vacuum gives factorization). This is all fine. The only puzzle is to understand how the short distance instanton calculation is able to know all these large distance effects. If we take the zero mode $m_i = 0$ calculation and, after adding a mass, we keep considering only the modes with eigenvalue O(m), we find that the final answer is smooth in m and can never reproduce the sudden charge in $G_{N,M}$ which is physically predicted. It was proposed [14] that the solution of this problem lies in the non-zero modes. Superficially, such modes (absent at zero mass) are suppressed at small m by powers of m. It was argued however that, if one kept doing the calculation to the end including non-zero modes, the collective coordinate integrations would give mass singularities of the 1/m type, as to precisely compensate the power of m suppression.

I shall not go into details, but I want to mention that this conjecture - that non-zero modes contribute finite corrections in the massive theory and give back full consistency with Ward identity and Konishi anomaly constraints - has been recently verified [9]. Amazingly, this could be done again only thanks to SUSY.

We consider a large mass limit (m >> Λ or decoupling region) and correlation functions such as

$$G_{2,1} = < \phi\tilde{\phi}(y)\ \lambda\lambda(x) >$$

with $m^{-1} << (x - y) << \Lambda$. The known mass, x,y dependence is the used to obtain $G_{2,1}$ everywhere. We now get two diagrams

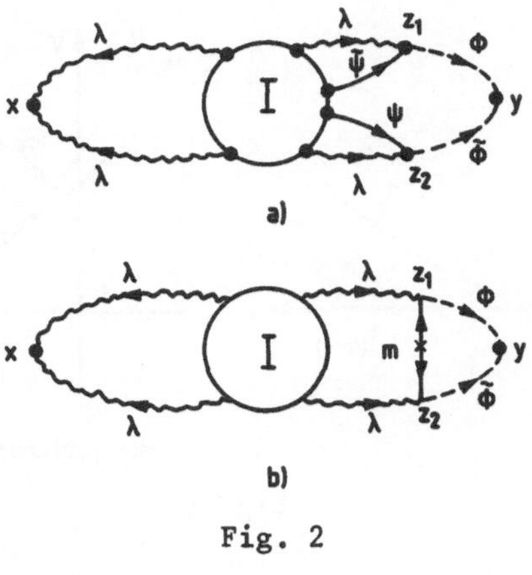

Fig. 2

Because of the large mass z_1 and z_2 have to be (within $\sim m^{-1}$) close to y, hence "far" from x. The instantons that contribute have $\rho \sim x-z \sim x-y >> m^{-1}$. It thus happens that the bosonic and fermionic propagators $(D^2 - m^2)^{-1}_{inst}$, $(\not{D} - m)^{-1}_{inst}$ can be replaced be free propagators $(\Box - m^2)^{-1}$, $(\not{\partial} - m)^{-1}$ and the full calculation can be carried out. The outcome is as follows :

i) In the $m >> \Lambda$ limit, the zero modes give a negligible contribution

ii) The non zero modes give a finite contribution which is flavour diagonal (as in eq.(73)) and has always the correct mass dependence. Because of the similar diagram structure the result automatically satisfies Konishi's constraint as sketched in the figure :

Fig. 3

In conclusion, massive SQCD instanton calculations appear to yield consistent results which imply the following set of non-vanishing expectation values :

$$< \lambda\lambda > = \Lambda^3 \prod_i (m_i/\Lambda)^{1/N} \times \exp(\frac{2i\pi}{N} k)$$

$$k = 1 \ldots N$$

$$< \phi_i \tilde{\phi}_j > = \delta_{ij} \; m_i^{-1} < \lambda\lambda > \qquad (74)$$

providing $\Delta_W = N$.

However the vacuum runs away for $m_i \to 0$ i.e. some $\phi\tilde{\phi}$ order parameters run to ∞ as first observed in an effective Lagrangian approach. We like to think of this phenomenon as due to the original flat directions and to a non-perturbative breakdown of the non-renormalization theorem (see figures).

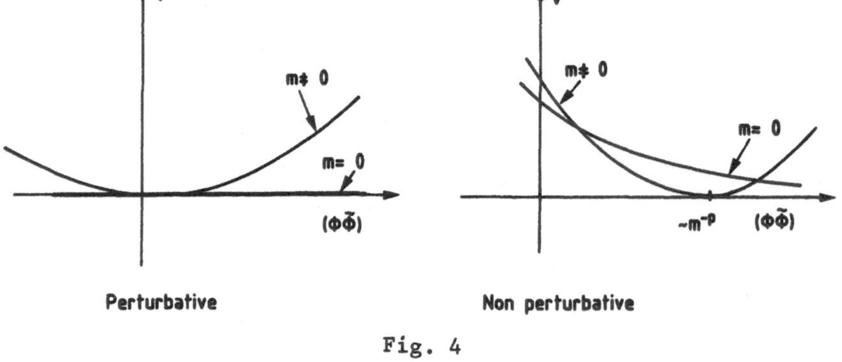

Fig. 4

d. Chiral theories (SU(5) with N_f $(\overline{5} + 10)$)

Calculations are even more elaborate but feasible [21]. One instanton dominated Green's functions are now

$$G_{N_f=1} = < \lambda\lambda(x_1)\ \lambda\lambda(x_2)\ (\lambda\lambda)_{24}(10\ 10\ 10\ \overline{5})_{24}\ (x_3) > \qquad (75)_1$$

$$G_{N_f=2} = < \lambda\lambda(x_1)\ 10^i\ 10^j\ 10^k\ \overline{5}_a(x_2)\quad 10^{i'}\ 10^{j'}\ 10^{k'}\ \overline{5}_{a'}(x_3) > \qquad (75)_2$$

Explicit calculations give in each case a non vanishing constant times the appropriate power of Λ. For the two family case the calculation has been done in the absence of the superpotential (6) with the idea that (unlike the case of mass in SQCD) adding a small Yukawa coupling η will not change the result discontinuously. (one should check that there are no dangerous η^{-1} type singularities at small η).

The non vanishing of the two correlation functions $(75)_{1,2}$ can be shown to be in contradiction with SUSY constraints which force (8,21)

$$< \lambda\lambda > = 0 \qquad N_f = 1,2 \qquad (76)$$

$$< 10\ 10\ 10\ \overline{5} > = 0 \qquad N_f = 2,\ \eta \neq 0 \qquad (77)$$

The possibility of a vacuum at ∞ seems to be excluded now since in both cases (if $\eta \neq 0$) these theories have no vacuum at ∞ to start with (the classical potential blows up at ∞ !) We therefore conclude that the non pertubative effect of instantons has lifted up the perturbative vacuum at the origin *breaking dynamically SUSY* (fig. 6)

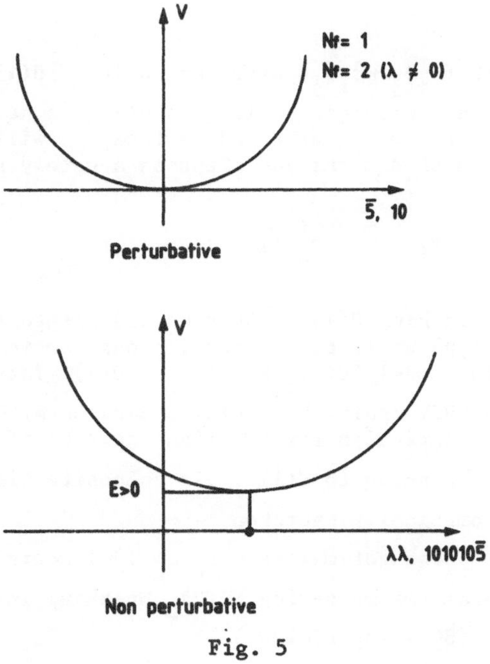

Fig. 5

3. Possible applications

For reasons of time limitations we shall consider only two lines which have been explored.

3.1. Supercompositeness

In this approach one takes up the suggestion of Buchmuller, Peccei and Yanagida [22] that quark and leptons could be the fermionic partners of the Nambu Goldstone bosons of the spontaneously broken global symmetries of a SUSY gauge theory. For this idea to work in practice one needs (at least !) a few things to happen :

i) A suitable SUSY breaking mechanism to lift up the masses of squarks and sleptons
ii) Some chiral symmetry left unbroken in the process that still protects fermion masses in the presence of SUSY breaking.

Recently one attempt was made [23] at constructing a realistic model for one family of composite quarks and leptons and two composite Higgs doublets starting from the understanding of the vacuum structure of SQCD. The attempt failed, but it could be of value in suggesting better alternatives. The origin of the failure is in the too democratic fermion mass protection occuring in the model. Thus some fermionic leptoquarks ($Q = 2/3$, $3B = -L = \pm 1$ objects) are predicted to be in the ($\langle$1GeV energy range, which is experimentally excluded (e.g. by $e^+e^- \to$ hadrons). Roughly the model is as follows. Take as preonic theory a massless SQCD with $N = M = 6$. This theory has an $SU(6) \times SU(6) \times U(1)_v \times U(1)_x$ flavour group. The $U(1)_x$ axial symmetry is broken spontaneously if either $< \lambda\lambda >$ or $< \psi\tilde{\psi} >$ are different from zero. It is not broken by $< \phi\tilde{\phi} > \neq 0$ (see eq.(13) for $M = N$). Actually, SUSY precisely prevents $\lambda\lambda$ and $\psi\tilde{\psi}$ from acquiring v.e.v's and $U(1)_x$ is thus protected by SUSY. The Goldstone superfields (due to $< \phi\tilde{\phi} > \neq 0$) are essentially

$$T_{ij} = \Phi_i \tilde{\Phi}_j \tag{78}$$

and their fermions $(\psi_i \tilde{\phi}_j + \phi_i \tilde{\psi}_j)$ carry one unit of $U(1)_x$ charge. Consequently they are massless as long as $U(1)_x$ is neither explicitly non-spontaneously broken. Quarks and leptons will be such fermions. In order to get rid of squarks and sleptons a purely bosonic SUSY breaking mass term

$$\delta \mathcal{L}_{SB} \approx -\mu_i^2 \phi_i^* \phi_i - \tilde{\mu}_i^2 \tilde{\phi}_i^* \tilde{\phi}_i \tag{79}$$

is added without breaking $U(1)_x$. Squarks and sleptons are seen to acquire masses $O(\mu,\tilde{\mu})$ while quarks and leptons remain massless. One finally adds to the model $SU(3)_c \times SU(2)_L \times U(1)_Y$ interactions and breaks $U(1)_x$ (and SUSY again) by a gluino mass term. One finds, writing down an effective Lagrangian for the light degrees of freedom,

i) $SU(2)_L \times U(1)_Y$ breaking to $U(1)_{el}$ via composite Higgses

ii) $SU(3)_c$ is automatically unbroken

iii) Quarks and leptons get masses $O(\alpha_c\, m_{\tilde{g}} \cdot \varepsilon)$ where $\varepsilon \cong m_W/\Lambda_{SC}$ is a (small ?) parameter measuring $SU(2)_L$ breaking in units of the $SU(6)_{SC}$ scale Λ_{SC} (SC = supercolour).

This sounds all very nice, but, if we look at the fermions $(T_{ij})_\theta$ as a matrix

Pati Salam	colour	1					u	d
		2					u	d
		3					u	d
		4	leptoquarks				ν	e
$SU(2)_L$		5	$\bar{u}$	$\bar{u}$	$\bar{u}$	$\bar{\nu}$	Shiggses	
		6	$\bar{d}$	$\bar{d}$	$\bar{d}$	$\bar{e}$		

one finds that it has more entries than one would like : T_{i4}, T_{4i} (i = 1,2,3) are leptoquarks (lq) and their mass is still protected by $U(1)_x$ so that

$$m_{lq} \sim \alpha_c \, m_g \tag{80}$$

The only way to make m_l, $m_q << m_{lq}$ is to play on ε but one finds that it is not possible to make ε arbitrarily small (ε is controlled by the masses appearing in (79) and we do not want them to exceed Λ_{SC} itself). In spite of the above problems the model is quite interesting in that it seems to represent a first example of containing dynamics in which the lightest composite states are spin 1/2 fermions, fulfilling 't Hooft's original idea [24]. Maybe with more work (and ingenuity) a better model can be constructed, possibly incorporating families which are, after, all, the main motivation for compositeness. Within our scheme families are *not* predicted. They can be added by hand by enlarging the SC and flavour group which is certainly not the most elegant possibility (although it could be the one nature chooses!)

3.2. Supergravity

Affleck, Dine and Seiberg [25] have explored systematically the possibility of inducing breaking of local SUSY (supergravity) by coupling a gauge theory of the SGG-type (representing the Hidden sector) to supergravity. They have found that the resulting scheme has some nice points compared to the usual one (which employs the so-called Polonyi (elementary) field) :

a. No term allowed by symmetries is excluded by hand

b. No strong CP problem (D_u small enough)

c. Hierarchy solve a-la-Witten

Some problems remain :

a. Need a heavy top (as forth generators quark) to induce $SU(2)_L \times U(1)_Y$ breaking by radiative corrections

b. The cosmological constant is put to zero by fine tuning

3.3. Superstring and a possible mechanism to get rid of Λ_{cos}

Recently Dine, Rohm, Seiberg and Witten [26] have pointed out an amusing possibility of cancelling to leading order Λ_{cos} in certain superstring theories while breaking local SUSY through a gluino condensate $< \lambda\lambda > \neq 0$ (it was already known for sometime [27] that $< \lambda\lambda >$ is not an order parameter for global SUSY breaking but it is so for *local* SUSY).

The problem of killing Λ_{cos} is that of achieving $V_{SUGRA} = 0$ at the minimum. It turns out that in the $E_8 \times E_8$ heterotic string theory in D = 10, when reduced to an effective local four dimensional theory through compactification of 6 dimensions, the final form of V_{eff}, including supergravity effects, is semipositive definite and the relevant term takes the form

$$V \cong |c - < \lambda\lambda >|^2 \tag{81}$$

where c is a constant coming from some elementary fiedl expectation value in going from D = 10 to D = 4. Again we expect

$$< \lambda\lambda > \sim \Lambda^3 \sim \mu^3 \exp(-\frac{1}{\beta_o g^2}) \tag{82}$$

but now the rôle of $1/g^2$ is played by a *scalar field*. It turns out then that the scalar field can adjust its v.e.v. so that the gluino condensate exactly cancels c at the minimum yielding $V = \Lambda_{cos} = 0$. There appear to be still problems however to justify a non vanishing value for c [28].

Conclusions

- The non perturbative properties of SUSY gauge theories have recently come under considerable control. A consistent picture of the SUSY vacua has slowly emerged.

- Their structure reveals interesting new phenomena, like the breakdown of the non renormalization theorem, the phenomenon of vacuum runaway in the massless limit of some vector like theories and the spontaneous breaking of SUSY itself in some chiral theories.

- Application of these results to high energy physics has just started in a variety of directions.

- Results are encouraging enough to stimulate further efforts (e.g. in supercompositeness, SUGRA, superstrings).

- Even in their non perturbative regime, supersymmetric theories stand out for their elegance, beauty and predictive power

- It would be just too bad if nature did not take advantage of them !

References

[1] J. Weyers, these proceedings
[2] P. Fayet, these proceedings
[3] We follow essentially J. Bagger and J. Wess, Supersymmetry and Supergravity, Princeton University Press 1983. We shall not be pedantic about factors $\pm$ 1, i or 2 unless they are crucial

[4] Ovrut and J. Wess, Phys. Rev. D25 (1982) 409
[5] S. Coleman and E. Weinberg, Phys. Rev. D7 (1973) 1888
[6] J. Wess and B. Zumino, Phys. Lett. 49B (1974) 52
J. Illiopoulos and B. Zumino, Nucl. Phys. B76 (1974) 1310
S. Ferrara, J. Illiopoulos and B. Zumino, Nucl. Phys. B77 (1977) 413
M.T. Grisaru, M. Rocek and W. Siegel, Nucl. Phys. B159 (1979) 429
[7] G. Veneziano, Phys. Lett. 124B (1983) 357
[8] K. Konishi, Phys. Lett. 135B (1984) 439, and references therein
[9] K. Shizuya and K. Konishi, Univ. of Pisa preprint
[10] T. Appelquist and J. Carazzone, Phys. Rev. D11 (1975) 2856
[11] V.A. Novikov, M.A. Shifman, A.I. Vainshtein, M.B. Voloshin and V.I. Zakharov, Nucl. Phys. B229 (1983) 394; V.A. Novikov, M.A. Shifman, A.I. Vaintein and V.I. Zakharov, Nucl. Phys. B229 (1983) 381, 407
[12] E. Cohen and C. Gomez, Phys. Rev. Letters 52 (1984) 237
[13] G.C. Rossi and G. Veneziano, Phys. Lett. 138B (1984) 195
[14] D. Amati, G.C. Rossi and G. Veneziano, Nucl. Phys. B249 (1985) 1
[15] I. Affleck, M. Dine and N. Seiberg, Phys. Rev. Lett. 51 (1983) 1026, Nucl. Phys. B241 (1984) 493
[16] E. Witten, Nucl. Phys. B188 (1981) 513; B.202 (1982) 253; S. Ceccotti and F. Girardello, Phys. Lett. 110 B (1982) 39
[17] G. Veneziano and S. Yankielowicz, Phys. Lett. 113B (1982) 32
[18] E. Corrigan et al., Nucl. Phys. B140 (1978) 31; B151 (1979) 9
[19] D. Amati, Y. Meurice, G.C. Rossi and G. Veneziano, Nucl. Phys. B263 (1986) 591
[20] T.R. Taylor, G. Veneziano and S. Yankielowicz, Nucl. Phys. B218 (1983) 493
[21] Y. Meurice and G. Veneziano, Phys. Lett. 141B (1984) 69
Y. Meurice, These de doctorat, Louvain-la-Neuve (1985)
For another approach leading to the same conclusions see I. Affleck, M. Dine and N. Seiberg, Phys. Rev. Lett. 52 (1984) 1677; Phys. Lett. 140B (1984) 59
[22] W. Buchmuller, R.D. Peccei and Yanagida, Phys. Lett. 124B (1983) 67
[23] A. Masiero, R. Pettorino, M. Roncadelli and G. Veneziano, Nucl. Phys. B261 (1985) 633
[24] G 't Hooft, Cargèse Proceedings (1979)
[25] I. Affleck, M. Dine and N. Seiberg, Nucl. Phys. B256 (1985) 557
[26] M. Dine, R. Rohm, N. Seiberg and E. Witten, Phys. Lett. 156B (1985) 55
[27] H.P. Nilles, Phys. Lett. 115B (1982) 193
[28] R. Rohm and E. Witten, Princeton Preprint (1985).

SUPERGRAVITY

Julius Wess

Institut für Theoretische Physik
Universität Karlsruhe
D-7500 Karlsruhe, FRG

INTRODUCTION

Supersymmetric particle phenomenology has been thoroughly discussed within the last few years [1]. A surprising development was that supergravity could be a source of supersymmetry breaking. If we start from a supersymmetric model and couple it to supergravity then we construct a theory with a ground state that is usually not supersymmetric. Various models which are phenomenologically acceptable have been based on this idea. Some of these models even generate a hierarchy.

In these four lectures I would like to give an introduction to N = 1 supergravity and particularly I would like to discuss those concepts and ideas which are relevant for the models mentioned above.

Some computations I will discuss in great detail. This should make the reader familiar with the techniques to such a point that he can do his own calculations where I am only indicating the results. Hopefully, these lectures will also help to read the original literature.

In the first lecture I am going to formulate usual gravity in four space time dimensions in such a way that the geometrical concepts can be directly used in superspace. This should make it easier to understand these concepts without having to think about supersymmetry at the same time.

In the second lecture I will study the geometry in superspace. This is the basis for the formulation of N = 1 supergravity. Because we aim at restrictions on supersymmetric theories which follow from supergravity we have to understand supergravity in its most general form. Therefore, I am going to discuss in full detail all the possible constraints on the superspace geometry. From these constraints follow the various supergravity multiplets which will be discussed in the third lecture.

The fourth lecture is devoted to the coupling of chiral matter fields to supergravity. It will be discussed for the old minimal supergravity multiplet only. The same techniques, however, can be used for the other multiplets as well and the details can be found in the literature. For a detailed description of models I also have to refer to the current literature.

METRIC TENSOR AND CHRISTOFFEL-SYMBOL VERSUS VIERBEIN AND CONNECTION

We are used to formulate general relativity in terms of the metric tensor $g_{mn}(x)$ and the Christoffel-symbol $\Gamma_{mn}{}^{\ell}(x)$ [2]. An equivalent formulation can be given in terms of the vielbein and the connection. It is this later formulation which can be generalized to superspace and which then leads to a systematic formulation of supergravity [3].

For the convenience of the reader I am going to exhibit the relation between the two formulations first in ordinary four-dimensional space.

Under general coordinate transformations,

$$x'^m = x^m + \xi^m(x), \tag{1.1}$$

a covariant vector field is supposed to transform like :

$$\delta A_m(x) = -\xi^{\ell}\partial_{\ell}A_m(x) - (\partial_m\xi^{\ell})A_{\ell}(x). \tag{1.2}$$

A covariant derivative is defined with the help of the Christoffel-symbol :

$$\mathcal{D}_mA_n(x) = \partial_mA_n(x) + \Gamma_{nm}{}^{\ell}A_{\ell}(x). \tag{1.3}$$

The Christoffel-symbol can be expressed in terms of derivatives of g_{mn} and in terms of its anti-symmetric part.

$$\begin{aligned}\Gamma_{n\ell,m} = &-\frac{1}{2}(\partial_n g_{m\ell} - \partial_m g_{n\ell} + \partial_{\ell} g_{nm})\\ &+ \Gamma^{(a)}_{n\ell,m} - \Gamma^{(a)}_{nm,\ell} - \Gamma^{(a)}_{\ell m,n}.\end{aligned} \tag{1.4}$$

The antisymmetric part of the Christoffel-symbol transforms like a covariant tensor of rank three, it has no inhomogeneous contribution to its transformation law. It enters as an independent tensorfield and it is consistent to put it equal to zero. This leads to the usual formulation of general relativity in terms of the metric tensor as the only independent dynamical variable.

The other formulation of general relativity starts from a frame as the fundamental geometric object. To each space time point there is associated a set of four four-vectors : the vierbein $e_m{}^a(x)$. These four-vectors define the direction of coordinate lines relative to the differentials dx^m. As a consequence, the components of the four-vectors have to transform like a covariant vector under general coordinate transformations :

$$\delta e_m{}^a = -\xi^{\ell}\partial_{\ell}e_m{}^a - (\partial_m\xi^{\ell})e_{\ell}{}^a. \tag{1.5}$$

The choice of these four four-vectors might be arbitrary to some extent. A different set might describe exactly the same physical situation. All these different sets of frames should be connected by an element of a transformation group, the structure group. Physics suggest the Lorentz group as a structure group :

$$e'_m{}^a(x) = e_m{}^b(x)L_b^{-1a}(x). \tag{1.6}$$

To distinguish the indices which transform under general coordinate transformations or the Lorentz group we take them from the middle or the beginning of the alphabet respectively. The first set we call Einstein index, the second Lorentz index.

The vierbein should not give rise to a singular parametrization of the coordinate lines - therefore, the vierbein is supposed to have an inverse.

$$e_m{}^a e_a{}^n = \delta_m^n$$
$$e_a{}^n e_n{}^b = \delta_a^b. \tag{1.7}$$

The respective transformation law of the inverse vierbein is :

$$\delta e_a{}^m = -\xi^{\ell}\partial_{\ell} e_a{}^m + e_a{}^{\ell}\partial_{\ell}\xi^m + L_a{}^b e_b{}^m. \tag{1.8}$$

The Lorentz index can be lowered and raised by the Lorentz metric η_{ab}. We can, however, not raise or lower the Einstein index - we do not have a metric tensor.

For any other tensor except the vierbein an Einstein index can be transformed into a Lorentz index and vice versa :

$$A_a = e_a{}^m A_m$$
$$A_m = e_m{}^a A_a. \tag{1.9}$$

Equ. (1.9) expresses the fact that the components of a vector can either be given with respect to the coordinate lines defined through the frame or with respect to the differentials.

All tensorial quantities can be defined as Lorentz tensors. To formulate a covariant theory means to make it covariant under local Lorentz transformations. We have to construct the gauge theory of the Lorentz group. As usual, we start with the definition of a covariant derivative (under local Lorentz transformations) :

$$\mathcal{D}_m A_a = \partial_m A_a - \phi_{ma}{}^b A_b. \tag{1.10}$$

Here, the field $\phi_{ma}{}^b$ is called connection instead of Yang Mills potential.

The requirement of covariance of equ. (1.10) leads to the transformation law of the connection :

$$\phi'_{ma}{}^b = L_a{}^c \phi_{mc}{}^d L^{-1}{}_d{}^b + (\partial_m L_a{}^c) L^{-1}{}_c{}^b. \tag{1.11}$$

As usual in gauge theories, the connection is Lie algebra valued :

$$\phi_{mab} = -\phi_{mba} \tag{1.12}$$

for a vector representation of the Lorentz group. Lie algebra valued is, however, meaningful for any representation of the Lorentz group. The concept of covariant derivatives is readily generalized to spinorial quantities as well. This shows the advantage of this formulation when we have to deal with spinors as well - as we have to in supergravity.

A fully Lorentz-covariant derivative can be obtained from (1.10), it is

$$\mathcal{D}_b A_a = e_b{}^m \, \mathcal{D}_m A_a . \tag{1.13}$$

To establish a relation between the two formulations we define metric tensor in terms of the vierbein :

$$g_{mn}(x) = e_m{}^a(x) e_n{}^b(x) \eta_{ab} . \tag{1.14}$$

Starting from $g_{mn}(x)$, the vierbein would be related to this coordinate transformation that transforms g_{mn} into η_{mn}.

The definition of the covariant derivative (1.10) and the relation between an Einstein vector and a Lorentz vector (1.9) allows us to define a covariant derivative of an Einstein vector as well. We shall compare this derivative with the definition (1.3) and we shall obtain a relation between the Christoffel-symbol and the connection.

$$\mathcal{D}_n V_m = e_m{}^a \, \mathcal{D}_n V_a \tag{1.15}$$

is covariant with respect to both indices n and m. We evaluate the definition (1.15) :

$$\begin{aligned} \mathcal{D}_n V_m &= e_m{}^a (\partial_n V_a - \phi_{na}{}^b \, V_b) \\ &= e_m{}^a (\partial_n (e_a{}^\ell V_\ell) - \phi_{na}{}^b \, e_b{}^\ell \, V_\ell) \\ &= \partial_n V_m + e_m{}^a (\partial_n e_a{}^\ell - \phi_{na}{}^b \, e_b{}^\ell) V_\ell . \end{aligned} \tag{1.16}$$

A comparison with (1.3) shows that we should make the following identification :

$$\begin{aligned} \Gamma_{nm}{}^\ell &= e_m{}^a \partial_n e_a{}^\ell - \phi_{nm}{}^\ell \\ \phi_{nm}{}^\ell &= e_m{}^a \phi_{na}{}^b e_b{}^\ell . \end{aligned} \tag{1.17}$$

The antisymmetric part of the Christoffel symbol which transforms like an Einstein tensor is related to the connection as follows :

$$\begin{aligned} \Gamma_{nm}^{(a)\ell} &= \frac{1}{2}(\partial_m e_n{}^a - \partial_n e_m{}^a) e_a{}^\ell \\ &+ \frac{1}{2}(\phi_{mn}{}^\ell - \phi_{nm}{}^\ell) . \end{aligned} \tag{1.18}$$

The corresponding Lorentz tensor is called torsion in the vielbein formalism. It is just the antisymmetrized covariant derivative of the vielbein and, therefore, it is Lorentz and Einstein-covariant :

$$T_{nm}{}^a = \partial_n e_m{}^a - \partial_m e_n{}^a$$

$$+ e_m{}^b \phi_{nb}{}^a - e_n{}^b \phi_{mb}{}^a. \tag{1.19}$$

The torsion is related to the antisymmetric part of the Christoffel symbol. If we put it equal to zero we can solve equ. (1.19) for ϕ_{abc} :

$$\begin{aligned} \phi_{cba} = -\frac{1}{2}\{ & e_c{}^n e_b{}^m (\partial_n e_{ma} - \partial_m e_{na}) \\ & - e_b{}^n e_a{}^m (\partial_n e_{mc} - \partial_m e_{nc}) \\ & + e_a{}^n e_c{}^m (\partial_n e_{mb} - \partial_m e_{nb})\}. \end{aligned} \tag{1.19'}$$

The constraint equation

$$T_{ab}{}^c = 0 \tag{1.20}$$

can be seen as the covariant equation which eliminates the connection as independent dynamical variable in favour of the vierbein.

The dynamics of general relativity is formulated through the curvature. To define the curvature tensor we compute the commutator of two Lorentz-covariant derivatives.

This has to be a Lorentz-covariant tensor. The commutator of two ordinary derivatives is zero. The commutator, therefore, has a part with one derivative and a part without a derivative on the tensor :

$$[\mathcal{D}_a, \mathcal{D}_b] V^c = V^d R_{abd}{}^c - T_{ab}{}^{\ell} \mathcal{D}_{\ell} V^c. \tag{1.21}$$

The first term - without derivative - defines the curvature tensor $R_{abd}{}^c$. From the very definition follows that it is antisymmetric in the first two indices and that it is Lie-algebra-valued in the second set of indices.

The second term - with one derivative - is identical with the torsion as it was defined in equ. (1.19).

Curvature and torsion have been defined through a commutator. The Jacoby identity, which holds for all commutators, implies an identity for curvature and torsion. This is the Bianchi identity :

$$\oint_{abc} [\mathcal{D}_d [\mathcal{D}_b, \mathcal{D}_c]] V^a = 0 \tag{1.22}$$

implies

$$(dbc^a) = \oint_{abc} (\mathcal{D}_d T_{bc}{}^a - R_{dbc}{}^a + T_{db}{}^{\ell} T_{\ell c}{}^a) = 0. \tag{1.23}$$

The symbol $\oint_{abc}$ means cyclic permutations. For a vanishing torsion the BI reduce to the well-known cyclic property of the curvature tensor in its first three indices.

The Lagrangian of the gravitational field is

$$\mathcal{L} = -\frac{1}{2k^2}\det(e_m{}^a)R_{ab}{}^{ab} \tag{1.24}$$

$$k^2 = 8\pi\, G_N.$$

Even without eliminating the connection through (1.19') this Lagrangian would describe the correct dynamics. A variation of (1.24) with respect to the connection would yield (1.20) as an equation of motion and we would be led to equ. (1.19') as solution of this equation.

SUPERSPACE GEOMETRY

Geometry in superspace allows a very systematic formulation of N = 1 supergravity [3].

To introduce superspace we start from the algebra of supersymmetry. The only non-vanishing commutator relation(*) is :

$$\{Q_\alpha,\overline{Q}_{\dot\alpha}\} = 2\sigma^m_{\alpha\dot\alpha}P_m. \tag{2.1}$$

We define a "global" transformation with a parametrization through anti-commuting variables :

$$G(x,\theta,\overline{\theta}) = e^{i\{x^m P_m - \theta^\alpha Q_\alpha - \overline{\theta}_{\dot\alpha}\overline{Q}^{\dot\alpha}\}}. \tag{2.2}$$

The parameter space is called superspace. It is graded

$$\begin{aligned} &Z^M \sim (x^m,\theta^\mu,\overline{\theta}_{\dot\mu}) \\ &Z^M Z^N = (-)^{nm}\, Z^N Z^M. \end{aligned} \tag{2.3}$$

Here n(m) is a function of N(M). This function takes the value zero or one depending on whether N(M) is a vector or a spinor index.

Multiplication of two global transformations induces a motion in the parameter space (superspace) :

$$\begin{aligned} &x^m \to x^m - i\theta\sigma^m\overline{\xi} + i\xi\sigma^m\overline{\theta} \\ &\theta \to \theta + \xi,\ \overline{\theta} \to \overline{\theta} + \overline{\xi}. \end{aligned} \tag{2.4}$$

Here ξ is the parameter of the transformation. The motion (2.4) can be generated by differential operators :

$$\begin{aligned} &D_\alpha = \frac{\partial}{\partial\theta^\alpha} + i\sigma^m_{\alpha\dot\alpha}\overline{\theta}^{\dot\alpha}\partial_m \\ &\overline{D}_{\dot\alpha} = -\frac{\partial}{\partial\overline{\theta}^{\dot\alpha}} + i\theta^\alpha\sigma^m_{\alpha\dot\alpha}\partial_m. \end{aligned} \tag{2.5}$$

As a consequence of their definition the D-operators form a representation of the supersymmetry algebra :

$$\{D_\alpha,\overline{D}_{\dot\alpha}\} = -2i\sigma^m_{\alpha\dot\alpha}\partial_m. \tag{2.6}$$

Supergravity is the gauge theory of supersymmetry. The covariance of the theory should allow an x-dependent parameter ξ. The algebra (2.1) immediately tells us that general coordinate transformations in ordinary

---

(*) For notation and more details see ref. [3].

four-dimensional space time are part of the invariance group.

A natural approach to supergravity is to try to find the gauged supersymmetry transformations (2.1) among the general coordinate transformations in superspace :

$$Z^{M'} = Z^M + \xi^M(Z). \tag{2.7}$$

The generalization of the "flat" space D-algebra (2.6) to curved space will then be the superspace analogue of (1.21) :

$$\begin{aligned}&\{\mathcal{D}_C\mathcal{D}_B - (-)^{bc}\mathcal{D}_B\mathcal{D}_C\}V^A = \\ &(-)^{d(c+b)}V^D R_{CBD}{}^A - T_{CB}{}^D\mathcal{D}_D V^A.\end{aligned} \tag{2.8}$$

The vierbein - now vielbein - formalism in superspace is a systematic procedure to construct representations of the algebra (2.8). The vielbein is a frame in superspace. It has well-defined transformation properties under general coordinate transformation. The structure group has to be chosen on physical grounds we again choose the Lorentz group.

$$\begin{aligned}\delta E_M{}^A(Z) = &- \xi^L \frac{\partial}{\partial Z^L} E_M{}^A \\ &- (\frac{\partial}{\partial Z^M}\xi^L)E_L{}^A + E_M{}^B L_B{}^A.\end{aligned} \tag{2.9}$$

The Lorentz transformation is reducible

$$L_A{}^B = \begin{pmatrix} L_a{}^b & 0 \\ 0 & L_{\underline{\alpha}}{}^{\underline{\beta}} \end{pmatrix}. \tag{2.10}$$

The underlined Greek indices stand for dotted and undotted spinor indices.

The inverse vielbein can be defined as well :

$$E_M{}^A\, E_A{}^N = \delta_M{}^N. \tag{2.11}$$

For the definition of a covariant derivative we have to introduce a connection. It will be Lie-algebra valued :

$$\begin{aligned}&\phi_{MA}{}^B = \begin{pmatrix} \phi_{Ma}{}^b & 0 \\ 0 & \phi_{M\underline{\alpha}}{}^{\underline{\beta}} \end{pmatrix} \\ &\phi_{M\alpha\dot{\alpha},\beta\dot{\beta}} = \sigma^a_{\alpha\dot{\alpha}}\sigma^b_{\beta\dot{\beta}}\phi_{Mab} \\ &\qquad = -2\varepsilon_{\alpha\beta}\phi_{M\dot{\alpha}\dot{\beta}} + 2\varepsilon_{\dot{\alpha}\dot{\beta}}\phi_{M\alpha\beta} \\ &\phi_{M\underline{\alpha}\,\underline{\beta}} = \phi_{M\underline{\beta}\,\underline{\alpha}}.\end{aligned} \tag{2.12}$$

The covariant derivatives are defined as usual :

$$\mathcal{D}_M V_A = \partial_M V_A - \phi_{MA}{}^B V_B$$

$$\mathcal{D}_M V^A = \partial_M V^A + (-)^{mb} V^B \phi_{MB}{}^A \tag{2.13}$$

$$\mathcal{D}_B V_A = E_B{}^M \mathcal{D}_M V_A .$$

The graded commutator of two covariant derivatives represents (2.8) and can be used to define curvature and torsion. The Jacoby identity in its graded version yields the Bianchi identities :

$$(_{DCB}{}^A) = \oint_{DCB} (R_{DCB}{}^A - \mathcal{D}_D T_{CB}{}^A - T_{DC}{}^F T_{FB}{}^A) = 0. \tag{2.14}$$

The structure group is reducible. As a consequence, curvature tensor and torsion can be decomposed in several irreducible parts each of which transforms into itself under general coordinate transformations and Lorentz group. The Bianchi identities (2.14) as well decompose into several independent identities.

To this point we have been following the formalism as it was introduced in the first lecture. Due to the reducibility of the structure group we shall find new features however. One of these features is that the curvature tensor is not an independent quantity. It is related to the torsion and its derivative. This was shown by N. Dragon [4] by solving the Bianchi identities.

For us it has a consequence that we cannot put the torsion to zero entirely, this would lead to a vanishing curvature. The only solution of the constraint zero torsion would be flat space, no dynamical degree of freedom left.

The art of the game is to find those torsion constraints that reduce the dynamical degrees of freedom to a minimal set containing the graviton and the gravitino. A natural way to proceed is to ask for those constraints that allow us to eliminate the connection in favour of the vielbein. Because the structure group is reducible $E_a{}^M$ and $E_{\underline{\alpha}}{}^M$ transform entirely by themselves. One of them can be eliminated as an independent dynamical degree as well. The constraints by which this can be achieved are called natural constraints.

The question for the natural constraints is well posed, however, a very detailed analysis [5] is necessary to find the answer.

This analysis is facilitated by paying some attention to the dimension of the torsion components. It follows from (2.4) that the dimension of θ is the square root of the dimension of x^m. The torsion connects second order derivatives with first order derivatives (2.8). Thus the individual torsion components have a different dimension. In mass dimension :

dim : 0   : $T_{\underline{\alpha}\underline{\beta}}{}^a$

1/2   : $T_{\underline{\alpha}\underline{\beta}}{}^{\underline{\gamma}}$, $T_{\underline{\alpha}b}{}^c$

1   : $T_{\underline{\alpha}a}{}^{\underline{\gamma}}$, $T_{ab}{}^c$

3/2   : $T_{ab}{}^{\underline{\gamma}}$.

We impose constraints on the torsion components with dimension zero. To be in accord with flat space (2.6) we demand :

$$T_{\gamma\beta}{}^{a} = T_{\dot{\gamma}\dot{\beta}}{}^{a} = 0,\ T_{\gamma\dot{\beta}}{}^{a} = 2i\sigma^{a}_{\gamma\dot{\beta}}. \tag{2.15}$$

We shall refer to this set as "trivial" constraints.

To analyse the implications of those constraints we solve the Bianchi identities subject to (2.15). As an exercise and in order to make the reader more familiar with the techniques involved I would like to demonstrate this in some detail.

We start from the B.I. $({}_{\delta\gamma\beta}{}^{a})$:

$$T_{\delta\gamma}{}^{F}T_{F\beta}{}^{a} + T_{\gamma\beta}{}^{F}T_{F\delta}{}^{a} + T_{\delta\gamma}{}^{F}T_{F\beta}{}^{a} = 0. \tag{2.16}$$

Due to (2.15) we find that it is only $T_{\delta\gamma}{}^{\dot{\varphi}}$ which is restricted by (2.16) :

$$T_{\delta\gamma}{}^{\dot{\varphi}}\sigma_{\beta\dot{\varphi}}{}^{a} + T_{\gamma\beta}{}^{\dot{\varphi}}\sigma_{\delta\dot{\varphi}}{}^{a} + T_{\beta\delta}{}^{\dot{\varphi}}\sigma_{\gamma\dot{\varphi}}{}^{a} = 0 \tag{2.17}$$

we multiply (2.17) by $\sigma_{\rho\dot{\rho}}{}^{a}$ and we use

$$\sigma_{\alpha\dot{\alpha}}{}^{a}\sigma_{\beta\dot{\beta}}{}^{a} = -2\varepsilon_{\alpha\beta}\varepsilon_{\dot{\alpha}\dot{\beta}}$$

to find

$$\varepsilon_{\dot{\rho}\dot{\varphi}}(T_{\delta\gamma}{}^{\dot{\varphi}}\varepsilon_{\rho\beta} + T_{\gamma\beta}{}^{\dot{\varphi}}\varepsilon_{\rho\delta} + T_{\beta\delta}{}^{\dot{\varphi}}\varepsilon_{\rho\gamma}) = 0. \tag{2.18}$$

Multiplication of (2.18) with $\varepsilon^{\beta\rho}$ leads to the result :

$$T_{\delta\gamma}{}^{\dot{\rho}} = 0. \tag{2.19}$$

This is a consequence of the natural constraints.
The B.I. $({}_{\dot{\delta}\dot{\gamma}\dot{\beta}}{}^{a})$ will lead to :

$$T_{\dot{\delta}\dot{\gamma}}{}^{\rho} = 0 \tag{2.20}$$

in a completely analogous way.

At a next step we analyse $({}_{\delta\gamma}{}^{\dot{\beta}a})$:

$$T_{\delta\gamma}{}^{F}T_{F\dot{\beta}}{}^{a} + T_{\gamma\dot{\beta}}{}^{F}T_{F\delta}{}^{a} + T_{\dot{\beta}\delta}{}^{F}T_{F\gamma}{}^{a} = 0. \tag{2.21}$$

The first term reduces to :

$$T_{\delta\gamma}{}^{F}T_{F\dot{\beta}}{}^{a} = 2i\sigma_{\varphi\dot{\beta}}{}^{a}T_{\delta\gamma}{}^{\varphi}. \tag{2.22}$$

The second term can be written as follows :

$$\begin{aligned} T_{\gamma\dot{\beta}}{}^{F}T_{F\delta}{}^{a} &= T_{\gamma\dot{\beta}}{}^{\ell}T_{\ell\delta}{}^{a} + T_{\gamma\dot{\beta}\dot{\varphi}}T^{\dot{\varphi}}{}_{\delta}{}^{a} = \\ &= 2i\{\sigma_{\gamma\dot{\beta}}{}^{\ell}T_{\ell\delta}{}^{a} - \sigma_{\delta\dot{\varphi}}{}^{a}T_{\gamma\dot{\beta}}{}^{\dot{\varphi}}\}. \end{aligned} \tag{2.23}$$

The last term in (2.21) demands symmetrization in δ and γ - note that due to its definition (2.8),

$T_{\dot{\beta}\delta}{}^{F} = T_{\delta\dot{\beta}}{}^{F}$. Equ. (2.21) becomes :

$$T_{\delta\gamma}{}^{\varphi}\sigma_{\varphi\dot{\beta}}{}^{a} + \sigma_{\gamma\dot{\beta}}{}^{\ell}T_{\ell\delta}{}^{a} + \sigma_{\delta\dot{\beta}}{}^{\ell}T_{\ell\gamma}{}^{a}$$

$$- T_{\gamma\dot{\beta}}{}^{\dot{\Phi}}\sigma_{\delta\dot{\varphi}}{}^{a} - T_{\delta\dot{\beta}}{}^{\dot{\Phi}}\sigma_{\gamma\dot{\varphi}}{}^{a} = 0. \tag{2.24}$$

We multiply (2.24) by $\sigma_{\alpha\dot{\alpha}}{}^{a}$, use (2.18a) and the definition

$$\sigma_{\gamma\dot{\beta}}{}^{\ell}T_{\ell\delta}{}^{a}\sigma_{\alpha\dot{\alpha}}{}^{a} = - T_{\delta,\gamma\dot{\beta},\alpha\dot{\alpha}} \tag{2.25}$$

to rewrite (2.24) entirely in spinor notation.

$$-2\varepsilon_{\dot{\alpha}\dot{\beta}}T_{\delta\gamma\alpha} - T_{\delta,\gamma\dot{\beta},\alpha\dot{\alpha}} - T_{\gamma,\delta\dot{\beta},\alpha\dot{\alpha}}$$

$$+ 2\varepsilon_{\alpha\delta}T_{\gamma\dot{\beta}\dot{\alpha}} + 2\varepsilon_{\alpha\gamma}T_{\delta\dot{\beta}\dot{\alpha}} = 0. \tag{2.26}$$

For a further evaluation of (2.26) we decompose the torsion components into Lorentz irreducible parts. Underlining of indices indicates symmetrization :

$$T_{\gamma\dot{\beta}\dot{\alpha}} = T_{\gamma\,\underline{\dot{\beta}\dot{\alpha}}} + \varepsilon_{\dot{\beta}\dot{\alpha}}T_{\gamma}$$

$$T_{\gamma\beta\alpha} = T_{\underline{\gamma\beta\alpha}} + \varepsilon_{\alpha\gamma}\Sigma_{\beta} + \varepsilon_{\alpha\beta}\Sigma_{\gamma} \tag{2.27}$$

$$T_{\gamma,\beta\dot{\beta},\alpha\dot{\alpha}} = -2\varepsilon_{\beta\alpha}S_{\gamma\,\underline{\dot{\beta}\dot{\alpha}}} + 2\varepsilon_{\dot{\beta}\dot{\alpha}}S_{\gamma\,\underline{\beta\alpha}}$$

$$+ 2S_{\gamma\,\underline{\beta\alpha}\,\underline{\dot{\beta}\dot{\alpha}}} + 2\varepsilon_{\beta\alpha}\varepsilon_{\dot{\beta}\dot{\alpha}}S_{\gamma}$$

with a further decomposition of S :

$$S_{\gamma\,\underline{\beta\alpha}} = S_{\underline{\gamma\beta\alpha}} + \varepsilon_{\gamma\beta}W_{\alpha} + \varepsilon_{\gamma\alpha}W_{\beta}$$

$$S_{\gamma\,\underline{\beta\alpha}\,\underline{\dot{\beta}\dot{\alpha}}} = S_{\underline{\gamma\beta\alpha}\,\underline{\dot{\beta}\dot{\alpha}}} + \varepsilon_{\gamma\beta}W_{\alpha\,\underline{\dot{\beta}\dot{\alpha}}} + \varepsilon_{\gamma\alpha}W_{\beta\,\underline{\dot{\beta}\dot{\alpha}}}. \tag{2.28}$$

An immediate consequence of (2.26) (symmetrization in all the indices) is :

$$S_{\underline{\gamma\beta\alpha},\underline{\dot{\beta}\dot{\alpha}}} = 0. \tag{2.29}$$

If we symmetrize (2.26) in the indices $(\dot{\alpha}\dot{\beta})$ we learn :

$$(T + W - S)_{\gamma\,\underline{\dot{\beta}\alpha}} = 0. \tag{2.30}$$

Finally, we antisymmetrize (2.26) in $(\dot{\alpha}\dot{\beta})$ and, after some manipulation we obtain the result :

$$(T - 2S)_{\underline{\gamma\beta\alpha}} = 0$$

$$(\Sigma + W + T + S)_{\gamma} = 0. \tag{2.31}$$

These are all the restrictions which follow from the trivial constraints. After this exercise it should be clear how we proceed to obtain a solution of the Bianchi identities with torsion components subject to constraints.

To find the natural constraints which allow us to express the connection as well as the part $E_a{}^M$ of the vielbein in terms of the other parts of the vielbein we first try to solve this problem in the linearized version of the torsion.

We have encountered an expression of the torsion in terms of the vierbein and the connection in equ. (1.18). An analogous expression holds in superspace :

$$T_{NM}{}^A = \partial_N E_M{}^A - (-)^{nm}\, \partial_M E_N{}^A + (-)^{n(b+m)}\, E_M{}^B \phi_{NB}{}^A - (-)^{mb}\, E_N{}^B \phi_{MB}{}^A . \tag{2.32}$$

This expression for the torsion can be linearized by expanding the vielbein around flat space. The equations that characterize the natural constraints can be solved in their linearized version. For a detailed description of this procedure see ref. [5]. The natural constraints that we find this way are :

$$\begin{aligned}
&T_{\alpha\beta}{}^a = 0, \quad T_{\dot\alpha\dot\beta}{}^a = 0, \quad T_{\alpha\dot\beta}{}^a = 2i\sigma_{\alpha\dot\beta}{}^a \\
&T_\gamma{}^{\dot\beta}{}_{\dot\alpha} = (n-1)\delta^{\dot\beta}{}_{\dot\alpha} T_\gamma, \quad T^{\dot\gamma}{}_\beta{}^\alpha = (n-1)\delta_\beta{}^\alpha \overline{T}^{\dot\gamma} \\
&T_{\gamma\beta}{}^\alpha = (n+1)(\delta_\gamma{}^\alpha T_\beta + \delta_\beta{}^\alpha T_\gamma), \quad T^{\dot\gamma\dot\beta}{}_{\dot\alpha} = (n+1)(\delta_{\dot\alpha}{}^{\dot\gamma}\overline{T}^{\dot\beta} + \delta_{\dot\alpha}{}^{\dot\beta}\overline{T}^{\dot\gamma}) \\
&T_{\gamma b}{}^a = 2n\delta_b{}^a T_\gamma, \quad T^{\dot\gamma}{}_b{}^a = 2n\delta_b{}^a \overline{T}^{\dot\gamma} \\
&T_{ab}{}^c = 0.
\end{aligned} \tag{2.33}$$

The torsion component T_γ is the same that we encountered before in the decomposition (2.27) of $T_{\gamma\dot\beta\dot\alpha}$. It remains unrestricted by the "naturalness" of the constraints. The constraints (2.33) depend on one numerical parameter n which is supposed to be real. For each value of n (2.33) solves the requirement we have imposed on the natural constraints.

The Bianchi identities subject to the natural constraints can be solved with exactly the same methods that we have used before. Needless to say, the complete analysis is tedious. As a result we find that all components of the torsion and, therefore, of the curvature as well can be expressed in terms of a few superfields. They are :

$$R,\ R^+,\ G_{\alpha\dot\alpha},\ W_{\underline{\alpha\beta\gamma}},\ \overline{W}_{\underline{\dot\alpha\dot\beta\dot\gamma}},\ T_\alpha \text{ and } \overline{T}_{\dot\alpha}.$$

There are a few relations among these superfields. The most relevant are :

$$\begin{aligned}
&\mathcal{D}_\alpha T_\beta + \mathcal{D}_\beta T_\alpha = 0 \\
&\overline{\mathcal{D}}_{\dot\alpha} R = -2(n+1)\overline{T}_{\dot\alpha} R \\
&(\overline{\mathcal{D}}_{\dot\alpha} + (3n+1)\overline{T}_{\dot\alpha})W_{\underline{\alpha\beta\gamma}} = 0.
\end{aligned} \tag{2.34}$$

As an example of how torsion and curvature are expressed in terms of the above superfields we list a few :

$$T_{\delta,\gamma\dot{\gamma},\dot{\alpha}} = -2i\varepsilon_{\delta\gamma}\varepsilon_{\dot{\gamma}\dot{\alpha}}R^{+}$$

$$R_{\beta\delta\gamma\alpha} = 4(\varepsilon_{\beta\alpha}\varepsilon_{\delta\gamma} + \varepsilon_{\delta\alpha}\varepsilon_{\beta\gamma})R^{+}.$$

A complete list can be found in ref. [5].

SUPERGRAVITY MULTIPLETS

The vielbein is a function of the θ variables. Such a function should be understood as a power series in θ. The coefficients in the expansion are x-dependent fields - the so-called component fields. The transformation law of these component fields could be computed by expanding equ. (2.9) in the θ-variables. Following this procedure, however, would obscure the transformation law of the component fields with respect to the Lorentz group - the reason being that the θ variables carry an Einstein index rather than a Lorentz index. Particular combinations of the component fields would have well defined Lorentz properties.

It is much simpler to modify the formalism from the very beginning in such a way that the Lorentz properties become manifest at each step of the expansion.

A vector superfield has the following transformation properties under general coordinate transformations and Lorentz transformations :

$$\begin{aligned} \delta V^A &= -\xi^M \partial_M V^A + V^B L_B{}^A \\ &= -\xi^M E_M{}^B E_B{}^N \partial_N V^A + V^B L_B{}^A. \end{aligned} \tag{3.1}$$

We now choose $\xi^B = \xi^M E_M{}^B$ as a field independent transformation parameter, we replace the derivative by a covariant derivative (2.13) and subtract the connection part again.

$$\delta V^A = -\xi^B \mathcal{D}_B V^A + V^B\{\xi^C \phi_{CB}{}^A + L_B{}^A\}. \tag{3.2}$$

This is exactly the same transformation law as (3.1). At a next step we make a field dependent redefinition of the Lorentz transformation :

$$\xi^C \phi_{CB}{}^A + L_B{}^A \rightarrow L'_B{}^A. \tag{3.3}$$

This is possible because the connection is Lie algebra valued. Finally, we define the transformation law of a vector superfield as follows :

$$\delta V^A = -\xi^B \mathcal{D}_B V^A + V^B L_B'^A. \tag{3.4}$$

All we have done is a field dependent redefinition of the transformation parameters to obtain a manifest Lorentz covariant expression.

To identify the graviton and the gravitino I am going to follow a suggestion by N. Dragon(*). The covariant derivative $\mathcal{D}_a$ was defined in (2.13).

(*) N. Dragon private communication.

$$\mathcal{D}_a = E_a{}^m\mathcal{D}_m + E_a{}^{\underline{\mu}}\mathcal{D}_{\underline{\mu}}. \tag{3.5}$$

This derivative can be rewritten in a form where the spinorial derivative carries a Lorentz index :

$$\begin{aligned} \mathcal{D}_a &= e_a{}^m\mathcal{D}_m - \frac{1}{2}\psi_a{}^{\underline{\alpha}}\mathcal{D}_{\underline{\alpha}} \\ &= e_a{}^m\mathcal{D}_m - \frac{1}{2}\psi_\alpha{}^{\underline{\alpha}}E_{\underline{\alpha}}{}^M\mathcal{D}_M \\ &= (e_a{}^m - \frac{1}{2}\psi_a{}^{\underline{\alpha}}E_{\underline{\alpha}}{}^m)\mathcal{D}_m - \frac{1}{2}\psi_a{}^{\underline{\alpha}}E_{\underline{\alpha}}{}^{\underline{\mu}}\mathcal{D}_{\underline{\mu}}. \end{aligned} \tag{3.6}$$

Comparing (3.5) with (3.6) leads to the following identification :

$$\begin{aligned} E_a{}^m &= e_a{}^m - \frac{1}{2}\psi_a{}^{\underline{\alpha}}E_{\underline{\alpha}}{}^m \\ E_a{}^{\underline{\mu}} &= -\frac{1}{2}\psi_a{}^{\underline{\alpha}}E_{\underline{\alpha}}{}^{\underline{\mu}}. \end{aligned} \tag{3.7}$$

This equation can be solved for $e_a{}^m$ and $\psi_a{}^{\underline{\alpha}}$. We multiply the last equation by $E_{\underline{\mu}}{}^{\underline{\beta}}$:

$$\begin{aligned} E_a{}^{\underline{\mu}}E_{\underline{\mu}}{}^{\underline{\beta}} &= -\frac{1}{2}\psi_a{}^{\underline{\alpha}}E_{\underline{\alpha}}{}^{\underline{\mu}}E_{\underline{\mu}}{}^{\underline{\beta}} \\ &= -\frac{1}{2}\psi_a{}^{\underline{\alpha}}\{E_{\underline{\alpha}}{}^M E_M{}^{\underline{\beta}} - E_{\underline{\alpha}}{}^m E_m{}^{\underline{\beta}}\}. \end{aligned} \tag{3.8}$$

We find :

$$E_a{}^{\underline{\mu}}E_{\underline{\mu}}{}^{\underline{\beta}} = -\frac{1}{2}\psi_a{}^{\underline{\alpha}}\{\delta_{\underline{\alpha}}{}^{\underline{\beta}} - E_{\underline{\alpha}}{}^m E_m{}^{\underline{\beta}}\}. \tag{3.9}$$

The bracket has an inverse and we can solve (3.9) for $\psi_a{}^{\underline{\alpha}}$

$$\frac{1}{2}\psi_a{}^{\underline{\alpha}} = -E_a{}^{\underline{\mu}}E_{\underline{\mu}}{}^{\underline{\beta}} \sum_{n=0}^{16} (X)^n{}_{\underline{\beta}}{}^{\underline{\alpha}} \tag{3.10}$$

where $X_{\underline{\beta}}{}^{\underline{\alpha}} = E_{\underline{\beta}}{}^m E_m{}^{\underline{\alpha}}$ is a nilpotent matrix. This is the reason why the infinite geometric series stops at X^{16}. Now it is easy to invert the first of the equs (3.7) :

$$e_a{}^m = E_a{}^m - \frac{1}{2}(E_a{}^{\underline{\mu}}E_{\underline{\mu}}{}^{\underline{\beta}}) \sum_{n=0}^{16} (X)^n{}_{\underline{\beta}}{}^{\underline{\alpha}}E_{\underline{\alpha}}{}^m. \tag{3.11}$$

Equs (3.7), (3.10) and (3.11) show that $E_a{}^M$ and $e_a{}^m$, $\psi_a{}^{\underline{\alpha}}$ are an equivalent set of superfields. If we identify the lowest component of $e_a{}^m$ and $\psi_a{}^{\underline{\alpha}}$ with the graviton and the gravitino respectively, then we have found those combinations which have a manifest Lorentz property for the gravitino.

$$\begin{aligned} e_a{}^m\Big|_{\theta=\bar{\theta}=0} &= e_a{}^m(x) \\ \psi_a{}^{\underline{\alpha}}\Big|_{\theta=\bar{\theta}=0} &= \psi_a{}^\alpha(x). \end{aligned} \tag{3.12}$$

To compute the transformation laws of these component fields we calculate the transformation of $\mathcal{D}_A V^B$, first by using its tensorial properties explicitly and secondly by varying $\mathcal{D}_A$ and V^B separately.

$$\delta(\mathcal{D}_A V^B) = -\xi^C \mathcal{D}_C \mathcal{D}_A V^B - L_A{}^C \mathcal{D}_C V^B + (\mathcal{D}_A V^C) L^B{}_C. \tag{3.13}$$

The second way to compute this variation is :

$$\delta(\mathcal{D}_A V^B) = (\delta\mathcal{D}_A) V^B + \mathcal{D}_A \delta V^B = (\delta\mathcal{D}_A) V^B + \mathcal{D}_A\{-\xi^C \mathcal{D}_C V^B + V^C L_C{}^B\}. \tag{3.14}$$

A comparison of the two expressions yields :

$$(\delta\mathcal{D}_A) V^B = -\xi^C \mathcal{D}_C \mathcal{D}_A V^B + \mathcal{D}_A(\xi^C \mathcal{D}_C V^B) - L_A{}^C \mathcal{D}_C V^B - (-)^{ac} V^C \mathcal{D}_A L_C{}^B. \tag{3.15}$$

We use equ. (2.8) and we obtain :

$$\begin{aligned}(\delta\mathcal{D}_A) V^B &= \{\xi^C T_{CA}{}^D + \mathcal{D}_A \xi^D - L_A{}^D\} \mathcal{D}_D V^B \\ &\quad - V^D\{(-)^{ab} \xi^C R_{CAD}{}^B + (-)^{ab} \mathcal{D}_A L_D{}^B\}.\end{aligned} \tag{3.16}$$

If we look at the definition of the covariant derivative (3.6) we would compute the above variation as follows :

$$\begin{aligned}(\delta\mathcal{D}_a) V^B &= (\delta e_a{}^m) \mathcal{D}_m V^B + e_a{}^m (\delta\mathcal{D}_m) V^B \\ &\quad - \tfrac{1}{2}(\delta\psi_a{}^{\underline{\alpha}}) \mathcal{D}_{\underline{\alpha}} V^B - \tfrac{1}{2}\psi_a{}^{\underline{\alpha}} (\delta\mathcal{D}_{\underline{\alpha}}) V^B = \\ &= (\delta e_a{}^m) \mathcal{D}_m V^B + e_a{}^m V^c \delta\phi_{mc}{}^B \\ &\quad - \tfrac{1}{2}(\delta\psi_a{}^{\underline{\alpha}}) \mathcal{D}_{\underline{\alpha}} V^B \\ &\quad - \tfrac{1}{2}\psi_a{}^{\underline{\alpha}} \{\xi^C T_{C\underline{\alpha}}{}^L + \mathcal{D}_{\underline{\alpha}} \xi^L - L_{\underline{\alpha}}{}^L\} \mathcal{D}_L V^B \\ &\quad + \tfrac{1}{2}\psi_a{}^{\underline{\alpha}} V^L \{(-)^b \xi^C R_{C\underline{\alpha}L}{}^B + (-)^{\ell} \mathcal{D}_{\underline{\alpha}} L_L{}^B\}.\end{aligned} \tag{3.17}$$

Here we have used the explicit definition of $\mathcal{D}_m V^B$ (2.13) as well as formula (3.16) to rewrite $(\delta\mathcal{D}_\alpha) V^B$ in equ. (3.17).

A direct comparison of (3.17) and (3.16) yields :

$$\begin{aligned}\delta e_a{}^m &= \{\mathcal{D}_a \xi^{\ell} - L_a{}^{\ell} + \xi^C T_{Ca}{}^{\ell} \\ &\quad + \tfrac{1}{2}\psi_a{}^{\underline{\alpha}}[\xi^C T_{C\underline{\alpha}}{}^{\ell} + \mathcal{D}_{\underline{\alpha}} \xi^{\ell}]\} e_{\ell}{}^m\end{aligned} \tag{3.18}$$

$$e_a{}^m \delta\phi_{mC}{}^B = -\mathcal{D}_a L_C{}^B - \xi^L R_{LaC}{}^B + (-)^{\ell} \tfrac{1}{2}\psi_a{}^{\underline{\alpha}}\{\xi^L R_{L\underline{\alpha}C}{}^B + \mathcal{D}_{\underline{\alpha}} L_C{}^B\}$$

$$
\begin{aligned}
-\frac{1}{2}\,\delta\psi_a{}^{\underline{\gamma}} &= \xi^C T_{Ca}{}^{\underline{\gamma}} + \mathcal{D}_a \xi^{\underline{\gamma}} \\
&- \frac{1}{2}\{\xi^C T_{Ca}{}^d + \mathcal{D}_a \xi^d - L_a{}^d\}\psi_d{}^{\underline{\gamma}} \\
&+ \frac{1}{2}\,\psi_a{}^{\underline{\alpha}}\{\xi^C T_{C\underline{\alpha}}{}^{\underline{\gamma}} + \mathcal{D}_{\underline{\alpha}}\xi^{\underline{\gamma}} - L_{\underline{\alpha}}{}^{\underline{\gamma}}\} \qquad (3.18)\\
&- \frac{1}{4}\,\psi_a{}^{\underline{\alpha}}\{\xi^C T_{C\underline{\alpha}}{}^d + \mathcal{D}_{\underline{\alpha}}\xi^d\}\psi_d{}^{\underline{\gamma}}
\end{aligned}
$$

The lowest component in θ of (3.18) gives the transformation law of the graviton and the gravitino. The interesting fact of (3.18) is that the graviton and gravitino transform into components of the torsion, curvature and into themselves. All the remaining components of the supergravity multiplet are contained in the torsion and curvature and, therefore, in the superfields R, R^+, $G_{\alpha\dot{\alpha}}$, T_α, $\overline{T}_{\dot{\alpha}}$, $W_{\alpha\beta\gamma}$ and $\overline{W}_{\dot{\alpha}\dot{\beta}\dot{\gamma}}$. Their transformation properties can be directly derived from (3.4). Due to the additional relations (2.34) there are exactly 24 bosonic and 24 fermionic component fields in the supergravity multiplet if we use the natural constraints. This multiplet, however, is further reducible.

There are two ways to reduce the reducible (24, 24) multiplet.

If we impose as a further constraint :

$$T_\alpha = \overline{T}_{\dot{\alpha}} = 0 \qquad (3.19)$$

then we reduce the (24, 24) multiplet to the well-known minimal supergravity multiplet. It has 12 bosonic and 12 fermionic degrees of freedom. It is minimal because it should accomodate the gravitino described by a spin 3/2 field $\psi_a{}^{\underline{\alpha}}$. Such a field has 16 components off-shell. The four gauged supersymmetry transformations reduce the number of independent fields to 12. The graviton is described by the vierbein $e_m{}^a$ with 16 components as well. The four gauged general coordinate transformations and the six gauged Lorentz transformations reduce this number to six. Six auxiliary fields have to be added. They are found to be the lowest components of R (complex) and $G_{\alpha\dot{\alpha}}$ (real).

Table 1

(12, 12) multiplet.

Dimension	Superfield	Lowest Component
0	$e_m{}^a$	$e_m{}^a$
1/2	$\psi_a{}^{\underline{\alpha}}$	$\psi_a{}^{\underline{\alpha}}$
1	R	$-\frac{1}{6}$ M
1	$G_{\alpha\dot{\alpha}}$	$-\frac{1}{3}\, b_{\alpha\dot{\alpha}}$

There is another way to reduce the (24, 24) multiplet by imposing the constraints

$$R = 0. \tag{3.20}$$

This leads for $n \neq 0$ to the non-minimal multiplet with 20 + 20 components :

Table 2

(20, 20) multiplet.

Dimension	Superfield	Lowest Component
0	$e_m{}^a$	$e_m{}^a$
1/2	$\psi_a{}^{\underline{\alpha}}$	$\psi_a{}^{\underline{\alpha}}$
	$T_{\underline{\alpha}}$	T_α
1	$G_{\alpha\dot{\alpha}}$	$-\frac{1}{3} b_{\alpha\dot{\alpha}}$
	S	S
	$\overline{\mathcal{D}}_{\dot{\alpha}} T_\alpha$	$c_{\alpha\dot{\alpha}} + i d_{\alpha\dot{\alpha}}$
3/2	$\overline{\mathcal{D}}_{\dot{\alpha}} S$	$\overline{\lambda}_{\dot{\alpha}}$

The superfield combination S has been listed in this table.

$$\begin{aligned} S &= \mathcal{D}^\alpha T_\alpha - (n+1)T^\alpha T_\alpha \\ \overline{S} &= \overline{\mathcal{D}}_{\dot{\alpha}} \overline{T}^{\dot{\alpha}} - (n+1)\overline{T}_{\dot{\alpha}} \overline{T}^{\dot{\alpha}}. \end{aligned} \tag{3.21}$$

As a consequence of (2.34) S has to satisfy the following equation :

$$\mathcal{D}_\alpha S = 8 T_\alpha R^+ \tag{3.22}$$

i.e. S is chiral if (3.20) is imposed.

For both multiplets, as they are listed in table one and two, the dynamics can be formulated by the following Lagrangian in superspace :

$$L = \int d^4x d^2\theta d^2\overline{\theta} \;\; \mathrm{Sdet}(E_M{}^A). \tag{3.23}$$

Here Sdet means the super determinant or Berezin-determinant. The Lagrangian is just the invariant volume in superspace. The variation of the vielbein has to be subject to the respective constraints to produce the right equations of motion.

For the (12, 12) multiplet the Lagrangian has a simple form in x-space :

$$\begin{aligned} \mathcal{L} = &- \frac{1}{2}(\det e_m{}^a)\{R_{ab}{}^{ab} + MM^* - b^a b_a \\ &+ \varepsilon^{abcd}(\psi_a \sigma_b \mathcal{D}_c \overline{\psi}_d - \overline{\psi}_a \overline{\sigma}_b \mathcal{D}_c \psi_c)\}. \end{aligned} \tag{3.24}$$

For the (20, 20) multiplet the Lagrangian is too lengthy to be written down, however, the two multiplets describe exactly the same dynamics on mass-shell. This is most likely true for the interaction of these multiplets with matter fields as well. Therefore, I will only deal with the minimal multiplets and its interaction with chiral matter fields in the next lecture. For the non-minimal multiplet I refer to the literature [6].

CHIRAL MATTER-COUPLED TO MINIMAL SUPERGRAVITY AND THE KÄHLER-POTENTIAL

The notion of chiral superfields is generalized to curved space through the condition

$$\bar{\mathcal{D}}_{\dot\alpha}\phi = 0. \tag{4.1}$$

A Lorentz-covariant treatment requires the following definition of the component fields :

$$A = \phi\Big|_{\theta=\bar\theta=0}\ , \quad \chi_\alpha = \frac{1}{\sqrt{2}}\mathcal{D}_\alpha\phi\Big|_{\theta=\bar\theta=0,} \quad F = -\frac{1}{4}\mathcal{D}^\alpha\mathcal{D}_\alpha\phi\Big|_{\theta=\bar\theta=0.} \tag{4.2}$$

These component fields can again be arranged to a superfield with new θ variables :

$$\phi = A(x) + \sqrt{2}\,\Theta^\alpha\chi_\alpha + \Theta^\alpha\Theta_\alpha F(x). \tag{4.3}$$

In this expression the Θ variables carry local Lorentz indices rather than Einstein indices. The transformation law of the chiral multiplet can be calculated from (3.4). It can be summarized with the help of the new Θ variables by the following formula :

$$\delta\phi = -\eta^m(x,\Theta)\frac{\partial}{\partial x^m}\phi - \eta^\alpha(x,\Theta)\frac{\partial}{\partial\Theta^\alpha}\phi. \tag{4.4}$$

The new transformation parameters η are functions of the old parameters θ and the supergravity multiplet. The exact form of these functions is of no importance for our discussion - it can be found in ref. [3]. It is, however, of importance that a chiral density exists with component fields that are functions of the supergravity multiplet.

$$\delta\mathcal{E} = -\frac{\partial}{\partial x^m}(\eta^m\mathcal{E}) + \frac{\partial}{\partial\Theta^\alpha}(\eta^\alpha\mathcal{E}). \tag{4.5}$$

The chiral density has the following component structure :

$$2\mathcal{E} = (\det e_m{}^a)\{1+i\Theta^\alpha\sigma^a_{\alpha\dot\alpha}\bar\psi_a^{\dot\alpha} - \Theta^2[M^* + \bar\psi_a\frac{1}{4}(\bar\sigma^a\sigma^b-\bar\sigma^b\sigma^a)\psi_b]\}. \tag{4.6}$$

The product of the chiral density with a chiral superfield is again a chiral density :

$$\delta(\mathcal{E}\phi) = -\frac{\partial}{\partial x^m}(\eta^m\mathcal{E}\phi) + \frac{\partial}{\partial\Theta^\alpha}(\eta^\alpha\mathcal{E}). \tag{4.7}$$

This fact allows us to construct invariant actions from chiral superfields.

$$\mathcal{L} = \int d^4x\, d^2\Theta\,\mathcal{E}F(\phi,\phi^+) \tag{4.8}$$

will be invariant if F satisfies the chirality condition

$$\bar{\mathcal{D}}_{\dot\alpha}F(\phi,\phi^+) = 0. \tag{4.9}$$

Fortunately, it is quite easy to project any superfield on a chiral superfield. The covariant projection operator is :

$$\Delta = (\bar{\mathcal{D}}_{\dot{\alpha}}\bar{\mathcal{D}}^{\dot{\alpha}} - 8R). \tag{4.10}$$

Applying this operator to any superfield which is a scalar under Lorentz transformations will yield a chiral superfield.

The most general Lagrangian that couples chiral matter fields to supergravity is of the form

$$\mathcal{L} = \int d^2\theta 2\mathcal{E}[\tfrac{3}{8}(\bar{\mathcal{D}}^2 - 8R)K(\phi^+,\phi) + g(\phi)] + hc. \tag{4.11}$$

Here $K(\phi,\phi^+)$ is an arbitrary function of ϕ^+, ϕ and $g(\phi)$ is an arbitrary function of ϕ. The action (4.11) includes the pure supergravity part as well :

$$\mathcal{L}_{SG} = - 6 \int d^2\theta\mathcal{E}R + hc \tag{4.12}$$

is equivalent to (3.24) componentwise. Thus if K is a polynomial which starts with one we have already included supergravity with the right normalization.

The Lagrangian (4.11) has to be expanded in component fields, then the auxiliary fields M, b_m and F have to be eliminated via there equation of motion. This yields a coupling of the curvature scalar to the scalar fields of the form

$$- \frac{1}{2} \det(e_m{}^a)K(A^+,A)(R_{ab}{}^{ab}(x)). \tag{4.13}$$

To eliminate this coupling we have to redefine the vierbein field :

$$\sqrt{K(A^+,A)}\, e_m{}^a \rightarrow e_m{}^a. \tag{4.14}$$

As a final result we obtain the following Lagrangian for the Bosonic fields :

$$\begin{aligned}\mathcal{L}_{BM} = (\det e_m{}^a)\{- \frac{1}{2} R_{ab}{}^{ab}(x) - G_k{}^{\ell}(A^+,A)\partial_m A^k \partial^m A^+{}_{\ell} \\ - e^{G(A^+,A)}[G^{\ell}G^{-1}{}_{\ell}{}^{k}G_k - 3]\},\end{aligned} \tag{4.15}$$

with the notation :

$$\begin{aligned}&G(A^+,A) = - 3 \ln K(A^+,A) + \ln|g(A)|^2 \\ &G_{\ell} = \frac{\partial G}{\partial A^{\ell}} \qquad G^k = \frac{\partial G}{\partial A^+{}_k} \qquad G_k{}^{\ell} = \frac{\partial^2 G}{\partial A_k \partial A^{+\ell}}.\end{aligned} \tag{4.16}$$

The Lagrangian (4.15) has first been derived by Cremmer et al. in the component formalism [7]. The interesting fact is that the scalar fields always enter via the Kähler potential $G(A^+,A)$ and its derivatives. Kähler invariance is manifest :

$$\begin{aligned}&K \rightarrow K|f(A)|^2 \\ &g \rightarrow gf^3\end{aligned} \tag{4.17}$$

where f is an arbitrary function of A. This invariance can also be traced to superspace [8], where it is a combined Kähler-Weyl invariance.

Different choices of $K(\phi^+,\phi)$ have been discussed in the literature. One obvious choice is :

$$K = e^{-\frac{1}{3} A^+ A} \tag{4.18}$$

which leads to a canonical kinetic energy for the A-fields in (4.15) :

$$G_i{}^j = \delta_i{}^j. \tag{4.19}$$

We shall call such a model canonical.

Another choice is

$$K = 1 - \frac{1}{3} A^+ A. \tag{4.20}$$

This model would be obtained by a minimal extension of a flat space model to curved space. Therefore, we shall call it a minimal model. It has interesting features [9]. The kinetic energy of the A-fields represents a $U(n,1)/U(n) \times U(1)$ $CP^{(n)}$-model. This can be seen as follows : First calculate the Kähler-metric $G_\ell{}^k$ from (4.19). It is

$$G_\ell{}^k = \frac{\delta^k{}_\ell}{1 - \frac{1}{3} AA^+} + \frac{1}{3} \frac{A_\ell^+ A^k}{(1 - \frac{1}{3} AA^+)^2}. \tag{4.21}$$

This metric should be compared with the metric which we obtain on the "sphere" :

$$\sum_{i=1}^{n} |Z^i|^2 - 3|Z^o|^2 = -3. \tag{4.22}$$

The stability group of this "sphere" is $U(n)$. The point $Z^i = 0$, $Z^o = 1$ e.g. is left invariant under the $U(n)$ group acting on the Z^i variables. The manifold of the sphere is $U(n,1)/U(n)$. To arrive at the $CP^{(n)}$ manifold $U(n,1)/U(n) \times U(1)$ we have to divide out the phase transformation $U(1)$ which changes the phase of all the Z^i variables. This can be done by the following parametrization :

$$Z^i = \frac{A^i}{\sqrt{1 - \frac{1}{3} A^\ell A_\ell^+}}, \qquad Z^o = \frac{1}{\sqrt{1 - \frac{1}{3} A^\ell A_\ell^*}}. \tag{4.23}$$

In this parametrization Z^o is always real. To derive the metric on the A-manifold we have to modify the metric (4.21) in such a way that two points of the sphere that can be transformed into each other by the U(1) phase transformation have a vanishing distance. The U(1) transformation should be only a gauge degree of freedom. A physicist can solve this problem in the language of gauge theories [10]. Consider the following Lagrangian :

$$\mathcal{L} = (\partial_m + iA_m)\overline{Z}^i(\partial_m - iA_m)Z_i - 3(\partial_m + iA_m)\overline{Z}^o(\partial_m - iA_m)Z_o. \tag{4.24}$$

The phase transformation on the Z^i fields has been gauged. The fields are also supposed to be restricted via the condition (4.22). This allows us to solve the Euler equation for A_m

$$A_m = \frac{i}{6}(\overline{Z}^i \overleftrightarrow{\partial}_m Z_i - 3\overline{Z}^o \overleftrightarrow{\partial}_m Z_o). \tag{4.25}$$

Substituting this back into (4.24) gives the following Lagrangian :

$$\mathcal{L}' = \partial_m \bar{Z}^i \partial_m Z_i - 3\partial_m \bar{Z}^o \partial_m Z^o$$
$$- \frac{1}{12}[\bar{Z}^i \overleftrightarrow{\partial}_m Z_i - 3\bar{Z}^o \overleftrightarrow{\partial}_m Z^o][\bar{Z}^\ell \overleftrightarrow{\partial}^m Z_\ell - 3\bar{Z}^o \overleftrightarrow{\partial}^m Z_o]. \tag{4.26}$$

From (4.26) we read off the following metric :

$$ds^2 = d\bar{Z}_j \eta_i{}^j dZ^i - \frac{1}{12}(\bar{Z}_j(\eta_i^j)dZ^i - d\bar{Z}_j \eta_i^j Z^i)^2$$
$$\eta_i^j = (1\ldots 1,-3) \tag{4.27}$$

or, in matrix form

$$ds^2 = (dZ^i, d\bar{Z}_i)$$

$$\begin{pmatrix} -\frac{1}{12}\bar{Z}_k \bar{Z}_\ell \eta_i^\ell \eta_j^k & \frac{1}{2}(\eta_i^j + \frac{1}{6}\bar{Z}_k Z^\ell \eta_i^k \eta_\ell^j) \\ \frac{1}{2}(\eta_j^i + \frac{1}{6}\bar{Z}_k Z^\ell \eta_\ell^i \eta_j^k) & -\frac{1}{12} Z^k Z_\ell \eta_k^i \eta_\ell^j \end{pmatrix} \begin{pmatrix} dZ^j \\ d\bar{Z}_j \end{pmatrix} \tag{4.28}$$

This matrix has a zero eigenvalue, the corresponding eigenvector is $(Z^j, -\bar{Z}_j)$. The infinitesimal change of the Z variables due to a phase transformation is :

$$dZ^i = id\varphi Z^i, \quad d\bar{Z}_i = -id\varphi \bar{Z}_i. \tag{4.29}$$

We find that ds^2 is zero for such differentials, the metric (4.27) has the desired properties for a $CP^{(n)}$ manifold.

To compute the metric on the A-manifold we substitute into (4.27) the differentials dZ expressed in terms of dA as they are given through (4.23) :

$$dZ^i = \frac{\partial Z^i}{\partial A^j} dA^j + \frac{\partial Z^i}{\partial A^*_j} d\bar{A}_j$$
$$d\bar{Z}^i = \frac{\partial \bar{Z}^i}{\partial A^j} dA^j + \frac{\partial \bar{Z}^i}{\partial \bar{A}_j} d\bar{A}_j. \tag{4.30}$$

The result is the $CP^{(n)}$ metric (4.21).

We now turn to the potential part in (4.15). We have to invert the metric (4.21) and to insert it in (4.15). The result is :

$$V = e^{G(A^+A)}[G^\ell G^{-1}{}_\ell{}^k G_k - 3] =$$
$$= \frac{1}{(1-\frac{1}{3}AA^*)^2}(\sum_i |\frac{\partial g}{\partial A^i}|^2 - 3|g - \frac{1}{3}A^i \frac{\partial g}{\partial A^i}|^2). \tag{4.31}$$

This potential has an interesting property. It can vanish without vanishing g(A). Take one simple field a(x) and a super potential

$$g(a) = \mu(1 + \frac{1}{\sqrt{3}} a)^3. \tag{4.32}$$

Here, μ is an arbitrary parameter. From (4.31) we compute

$$V(a,a^*) = 0. \tag{4.33}$$

We learn that the vacuum expectation value of the scalar field a is not determined by the potential. This is the mechanism of the sliding scale models.

The potential (4.32) can be combined with other fields which enter the potential in second or higher order. The minimum of the potential will be degenerate and it will not determine the vacuum expectation value of a. It could be determined by radiative corrections and, as a consequence, a new scale would be introduced into the model. Based on this idea, N. Dragon, M.G. Schmidt and U. Ellwanger have built a model where all the mass parameters used to define the model are of the order of the Planck mass. The sliding vacuum expectation value of the field a introduces a new scale which generates a hierarchy between the electroweak and the GUT scale. This model is in agreement with the standard model and our present day particle phenomenology.

It should also be noted that the potential (4.31) can have its minimum at zero energy thus being consistent with a vanishing cosmological constant.

If we have found a potential which has its minimum at zero energy :

$$\langle V \rangle = 0 \tag{4.34}$$

then there are simple mass relations for this model :

$$m_{3/2} = e^{\frac{1}{2}\langle K \rangle} . \tag{4.35}$$

The gravitino mass which is a scale for supersymmetry breaking is obtained from the Kähler potential. The fields have to be taken at their vacuum expectation value. For the super potential (4.32) we find :

$$m_{3/2} = \mu\left[\frac{1 + \langle a \rangle}{\sqrt{1 - |\langle a \rangle|^2}}\right]^3 \tag{4.36}$$

the gravitino mass slides.

There is also an interesting relation for the fermion and boson masses :

$$\sum_{\text{bosons}} m^2 - \sum_{\substack{\text{fermions}\\ \text{of spin } 1/2}} m^2 =$$

$$= 2m_{3/2}^2\{(N+1) - R^i_{\;j} G^{-1}{}_i{}^{\ell} G^{-1}{}^j{}_k \frac{\partial g}{\partial A^\ell} \frac{\partial g}{\partial A^*_k}\}. \tag{4.37}$$

N is the number of complex scalar fields in the model and $R^i_{\;j}$ is the Ricci tensor with the metric (4.16).

For the canonical model (4.18) which is Kähler flat we obtain from (4.37)

$$\sum_{\text{bosons}} m^2 - \sum_{\substack{\text{fermions}\\ \text{of spin } 1/2}} m^2 = 2m_{3/2}^2(N+1). \qquad (4.38)$$

For the minimal case (4.20) we find an Einstein metric :

$$R_i{}^j = \frac{1}{3}(n+1)G_i{}^j. \qquad (4.39)$$

As a consequence

$$\sum_{\text{bosons}} m^2 - \sum_{\substack{\text{fermions}\\ \text{of spin } 1/2}} m^2 = 0. \qquad (4.40)$$

This is a very desirable relation because most of the cancellations of divergences in supersymmetric theories are a consequence of (4.40).

REFERENCES

[1] H.E. Haber and G.L. Kane, Phys. Rep. 117 (1985) 75.
[2] L.D. Landau, E.M. Lifshitz, Theor. Phys. II.
[3] J. Wess and J. Bagger, Supersymmetry and Supergravity, Princeton University Press 1983.
[4] N. Dragon, Z. Phys. C2 (1979) 29.
[5] G. Girardi, R. Grimm, M. Müller, J. Wess, Z. Phys. C26 (1984) 123-140.
[6] G. Girardi, R. Grimm, M. Müller, J. Wess, Z. Phys. C26 (1984) 123; R. Grimm, M. Müller and J. Wess, Z. Phys. C26 (1984) 427.
[7] E. Cremmer, B. Julia, S. Scherk, S. Ferrara, Phys. Lett. 79B (1978) 231, and Nucl. Phys. B147 (1979) 105.
[8] J. Wess, KA-THEP-85-5, University of Karlsruhe preprint.
[9] N. Dragon, M.G. Schmidt and U. Ellwanger, Phys. Lett. 145B (1984) 192.
[10] L. Alvarez-Gaume and D.Z. Freedman, Proceedings of the Europhysics Conference, Erice, Italy, March 17-24 (1980), New York Plenum 1980, p. 41.

TOPICS IN COSMOLOGY AND PARTICLE PHYSICS

Keith A. Olive

Department of Physics - University of Minnesota

Minneapolis, Minnesota 55455

In these three lectures, I will review some of the major topics which concern both particle physics and cosmology. In the first lecture, I will give a general overview of the standard cosmological Big Bang model and briefly go over Big Bang nucleosynthesis as well as Big Bang baryosynthesis. In the second lecture, I will review the inflationary model with particular emphasis on supersymmetry. In the final lecture I will discuss the dark matter problems which confront cosmology today and their demands on elementary particle physics.

LECTURE I - THE STANDARD MODEL

The standard Big Bang model now spans some 61 orders of magnitude in temperature. That is from today at $t \sim 10^{17}$ s and $T \sim 10^{-13}$ GeV (3°K) back to what is called the Planck epoch at $t \sim 10^{-44}$ s and $T \sim 10^{18}$ GeV. In Table 1, the major cosmological events are summarized. To be sure,

Table 1
Big Bang Chronology

t(sec)	T(GeV)	Event
10^{-44}	10^{18}	Planck epoch : 1) Quantum Gravity ? 2) Kaluza-Klein ? 3) Supergravity-Superstrings ?
10^{-35}	10^{15}	Grand Unification/Baryon Generation/ Inflation/Resolution of Cosmological Problems
10^{-10}	10^{2}	Weak Symmetry Breaking
10^{-5}	10^{-1}	Confinement Transition
$1\text{-}10^{2}$	$10^{-3}\text{-}10^{-2}$	Big Bang Nucleosynthesis
10^{12}	10^{-9}	Recombination/Galaxy Formation
10^{17}	10^{-13}	Today

the Planck epoch is the least certain as it is defined to be the time at which gravitational interactions are as important as all others. Popular theories of gravitational interactions include quantum gravity [1], Kaluza-Klein theories [2] and supergravity theories [3].

The time scale for the Grand Unified Theory (GUT) epoch is basically set by the GUT mass M_x taken here to be $O(10^{15})$ GeV. The major cosmological application of GUTs is of course the generation of a net baryon asymmetry in the early Universe [4], and will be discussed shortly. Inflation [5] and the resolution of the problems in the standard Big Bang model will be discussed in the next lecture. Although the table lists no events taking place in the "desert" between 10^{15} - 10^2 GeV, new interactions, supersymmetry or composite models may play a role here. At 10^2 GeV, the Universe will have gone through the weak symmetry breaking transition, though it appears with little major impact on cosmology [6]. At still lower temperatures, $T \sim O(100)$ MeV, quarks and gluons will have combined to form hadrons [7] at about the same time as the chiral symmetry phase transition.

The light elements such as D, ^{3}He, ^{4}He and ^{7}Li were not formed until the Universe was a couple of minutes old during Big Bang nucleosynthesis [8,9]. This will also be discussed shortly in more detail. Familiar astrophysical objects such as galaxies did not begin to form until the epoch when electrons combined with protons to form neutral hydrogen when the Universe was about 10^5 yrs old.

Of course in order to believe any model which spans so many orders of magnitude in time, we must ask ourselves what is the evidence which allows for such an extrapolation. The most important piece of evidence which supports a hot Big Bang is the observation of the 3K microwave background radiation [10]. The temperature of the background is now known quite well 2.73 ± .05 [11] and the lack of observed anisotropies indicated by $\delta T/T \lesssim$ few $\times 10^{-5}$ [12] gives one confidence that the description of the early Universe is well fit by the isotropic and homogeneous Friedmann-Roberston Walker model which I am about to discuss. Since photons have been decoupled from themselves since the epoch of recombination, the background radiation gives a direct information back to temperatures ~ 1 eV.

The second major piece of evidence comes from both observational and theoretical considerations, i.e., Big Bang Nucleosynthesis and the abundances of the light elements. The abundances as we will see cover some 9 orders of magnitude and are all very well fit with the simplest model. This enables the extrapolation to go back to temperatures ~ 1 MeV. The third piece of evidence (and it might be premature to call it evidence) is purely theoretical and deals with the capability of GUTs in producing a net baryon asymmetry of the right order of magnitude. Hence extrapolations to 10^{15} GeV.

In the remainder of this lecture I will briefly review what is contained in the now standard Big Bang model. The starting point is, as I said, with assuming a homogeneous and isotropic model which gives us the Friedmann-Roberston-Walker metric

$$ds^2 = -dt^2 + R^2(t)\left[\frac{dr^2}{1-kr^2} + r^2(d\theta^2 + \sin^2\theta\, d\phi^2)\right] \qquad (1)$$

where R(t) is the R-W scale factor and k is the curvature constant describing the overall geometry (k=0,+1,-1 for a flat, closed or open Universe). Einstein's equations lead to the Friedmann equation which relates the expansion rate to the energy density and curvature,

$$H^2 \equiv (\dot{R}/R)^2 = 8\pi\rho/3M_p^2 - k/R^2 + \Lambda/3 \tag{2}$$

where H is the Hubble parameter, ρ is the total mass-energy density, Λ is the cosmological constant and $M_p = G_N^{-1/2} = 1.2 \times 10^{19}$ GeV is the Planck mass.

When combined with the equation for energy conservation

$$\dot{\rho} = -3\left(\frac{\dot{R}}{R}\right)(\rho + p) \tag{3}$$

where p is the isotropic pressure, the Friedmann equation (2) leads to several typical expansion stages in the early Universe. At early times, the Universe is thought to have been dominated by radiation so that the equation of state was just $p = \rho/3$ and at early times we can assume that the contribution to H from k and Λ were negligible, we have that $\dot{\rho} = -4\rho\, \dot{R}/R$ or $\rho \sim R^{-4}$ and hence $\dot{R} \sim R^{-1}$ so that

$$R \sim t^{1/2} \tag{4}$$

for a radiation dominated Universe. One then also finds the time temperature relation through

$$t = (3/32\pi G\rho)^{1/2} + \text{constant} \tag{5}$$

and realizing that $\rho \sim T^4$. Similarly for a matter dominated Universe (late times) we take $p = k = \Lambda = 0$ and find $\dot{\rho} = -3\rho\, \dot{R}/R$ or $\rho \sim R^{-3}$ and

$$R \sim t^{2/3}. \tag{6}$$

If we maintain that $\Lambda = 0$, we can define a critical energy density ρ_c such that $\rho = \rho_c$ for $k = 0$

$$\rho_c = \frac{3H^2}{8\pi G_N}. \tag{7}$$

In terms of the present value of the Hubble parameter

$$\rho_c = 1.88 \times 10^{-29}\, h_o^2 \text{ g cm}^{-3}, \tag{8}$$

where

$$h_o = H_o/(100 \text{ km Mpc}^{-1}\text{s}^{-1}) \tag{9}$$

is the present value of the Hubble parameter in units of 100 km $\text{Mpc}^{-1}\text{s}^{-1}$. The cosmological density parameter is then defined as the ratio of the present energy density to the critical density

$$\Omega \equiv \rho/\rho_c. \tag{10}$$

Furthermore, the value of Ω will determine the sign of k. For $\Omega > 0$ we have $k = +1$, $\Omega = 1$ corresponds to $k = 0$ and $\Omega < 0$ to $k = -1$. In terms of Ω the Friedmann equation can be rewritten as

$$(\Omega - 1)H_o^2 = \frac{k}{R^2}. \tag{11}$$

The observational limits on h_o and Ω are [13]

$$1/2 < h_o < 1 \tag{12a}$$

$$0.1 < \Omega < 4. \tag{12b}$$

The density of the Universe is generally determined by means of a mass-to-light ratio. The mass of a galaxy or gravitational system if in gravitational equilibrium can be computed via the virial theorem from measured rotational velocities. The total mass of the system is then compared with its absolute luminosity which is derived from the measured apparent luminosity. The total density ρ is then

$$\rho = (\frac{M}{L})\mathcal{L}, \tag{13}$$

where (M/L) is the above described mass-to-light ratio and $\mathcal{L}$ is the total luminosity density of the night sky

$$\mathcal{L} \approx 2 \times 10^8 \; h_o \; L_\Theta \; Mpc^{-3}, \tag{14}$$

where L_Θ is the solar luminosity $L_\Theta = 3.9 \times 10^{33}$ erg s^{-1}. We can now define a critical mass-to-light ratio

$$(M/L)_c = \rho_c/\mathcal{L} \approx 1200 \; h_o \tag{15}$$

and the cosmological density parameter is given by

$$\Omega = (M/L)/(M/L)_c. \tag{16}$$

In principle this could give us an accurate determination of Ω. The problem is that the derived value of Ω seems to depend on what scale we measure (M/L). For example, the following four systems all give different values [14] of Ω

1) solar neighborhood

$(M/L) \sim 2 \pm 1 \Rightarrow \Omega \sim (0.0016 \pm .0008)/h_o$

2) central parts of galaxies

$(M/L) \sim (10 - 20)h_o \Rightarrow \Omega \sim (0.008 - 0.017)$

3) binaries and small groups of galaxies

$(M/L) \sim (60 - 180)h_o \Rightarrow \Omega \sim (0.05 - 0.15)$

4) clusters of galaxies

$(M/L) \sim (300 - 1000)h_o \Rightarrow \Omega \sim (0.25 - 0.8)$.

The dependence on h_o of the last three mass-to-light ratios is due to the uncertainties in estimating the mass and absolute luminosities of distant objects. It is evident that as we look on larger and larger scales the value of Ω seems to be increasing. This is known as the missing mass problem. In particular, it seems to indicate that there is dark matter present in the Universe on large scales. This problem will be discussed in detail in the third lecture.

As I have indicated, at temperatures greater than ~ 1 eV the Universe is usually taken to be radiation dominated so that the content of the radiation plays a very important role. Today, the content of the microwave background consists of photons. We can calculate the energy density of photons by:

$$\rho_\gamma = \int E_\gamma dn_\gamma, \tag{17}$$

where the density of states is given by

$$dn_\gamma = \frac{g_\gamma}{2\pi^2} [\exp(E_\gamma/T)-1]^{-1} q^2 dq \tag{18}$$

and $g_\gamma = 2$ simply counts the number of degrees of freedom for photons, $E_\gamma = q$ is just the photon energy (momentum). (We are using units such that $\hbar = c = k_B = 1$ and will do so through the remainder of these lectures). On performing the integral in (17) we have that

$$\rho_\gamma = \frac{\pi^2}{15} T^4 \tag{19}$$

which is the familiar blackbody result.

In general, at very early times, at very high temperatures, other particle degrees of freedom join the radiation background when $T \sim m_i$ for each particle type i if that type is brought into thermal equilibrium through interactions. In equilibrium the energy density of a particle type i is given by

$$\rho_i = \int E_i dn_{q_i} \tag{20}$$

and

$$dn_{q_i} = \frac{g_i}{2\pi^2} [\exp[(E_{q_i} - \mu_i)/T]\pm 1]^{-1} q^2 dq, \tag{21}$$

where again g_i counts the total number of degrees of freedom for type i,

$$E_{q_i} = (m_i^2 + q_i^2)^{1/2}. \tag{22}$$

μ_i is the chemical potential if present and $\pm$ corresponds to either fermi or bose statistics.

In the limit that $T >> m_i$ the total energy density can be expressed by

$$\rho = (\Sigma_B g_B + \frac{7}{8} \Sigma_F g_F) \frac{\pi^2}{30} T^4 \equiv \frac{\pi^2}{30} N(T) T^4, \tag{23}$$

where $g_{B(F)}$ are the total number of boson (fermion) degrees of freedom and the sum runs over all boson (fermion) states with $m << T$. The factor of 7/8 is due to the difference between the fermi and bose integrals. Equation (23) defines N(T) by taking into account new particle degrees of freedom as the temperature is raised. We can also rewrite Eq. (5) giving us a relationship between the age of the Universe and its temperature

$$t = (90/32\pi^3 G_N N(T))^{1/2} T^{-2}. \tag{24}$$

Put into a more convenient form

$$t T_{MeV}^2 = 2.4 [N(T)]^{-1/2}, \tag{25}$$

where t is measured in seconds and T_{MeV} in units of MeV.

The value of N(T) at any given temperature depends on the particle physics model. In the standard SU(3) × SU(2) × U(1) model, we can specify N(T) up to temperatures of O(100)GeV. This is done in the following table.

Table 2

Temperature	New Particles	4N(T)
$T < m_e$	γ's + ν's	29
$m_e < T < m_\mu$	$e^\pm$	43
$m_\mu < T < m_\pi$	$\mu^\pm$	57
$m_\pi < T < T_c^*$	π's	69
$T_c < T < m_{strange}$	$-\pi$'s + u, $\bar{u}$, d, $\bar{d}$ + gluons	205
$m_s < T < m_{charm}$	$s\bar{s}$	247
$m_c < T < m_\tau$	$c\bar{c}$	289
$m_\tau < T < m_{bottom}$	$\tau^\pm$	303
$m_b < T < m_{top}$	$b\bar{b}$	345
$m_t < T < m_w$	$t\bar{t}$	387

* T_c corresponds to the confinement-deconfinement transition between quarks and hadrons.

At higher temperatures, N(T) will be model dependent. For example, in the minimal SU(5) model, one needs to add to N(T), 6 states coming from $W^\pm$, Z, 24 for the X and Y gauge bosons, another 24 from the adjoint Higgs, and another 10 from the 5. Hence for $T > M_x$ in minimal SU(5), N(T) = 160.75. In a supersymmetric model [15] this would at least double, with some changes possibly necessary in the table if the selectron (scalar partner of the electron) has a mass below M_w.

The origins of the current connection between particle physics and cosmology really began with the generation [16] of a small but finite baryon to entropy ratio using grand unified theories (GUTs) [4]. The problem in cosmology is basically that there is apparently very little antimatter in the Universe and the number of photons greatly exceeds the number of baryons. If we define

$$\eta = (n_B - n_{\bar{B}})/n_\gamma \tag{26}$$

where $n_{B,\bar{B},\gamma}$ is the number density of baryons, antibaryons and photons, we find that

$$\eta \approx n_B/n_\gamma \sim 10^{-10} - 10^{-9}. \tag{27}$$

In a standard model, the entropy density today is related to n_γ by

$$s \simeq 7n_\gamma \tag{28}$$

so that eq. (27) implies $n_B/s \sim 10^{-10} - 10^{-11}$. This ratio is conserved however and hence represents another undesirable initial condition, with its origin unknown.

Let us for the moment, assume that in fact $\eta = 0$. We can compute the final number density of nucleons left over after annihilations have frozen out. At very high temperatures (neglecting a quark-hadron transition) $T > 1$ GeV, nucleons were in thermal equilibrium with the photon background and $n_N = n_{\bar{N}} = 3/2 n_\gamma$ (a factor of 2 accounts for neutrons and protons and the factor 3/4 for the difference between fermi and bose statistics). As the temperature fell below m_N, annihilations kept the nucleon density at its equilibrium value $(n_N/n_\gamma) = (m_N/T)^{3/2} \exp(-m_N/T)$ until the annihilation rate $\Gamma_A \approx n_N m_\pi^{-2}$ fell below the expansion rate. This occurred at $T \approx 20$ MeV. However, at this time the nucleon number density has already dropped to

$$n_N/n_\gamma = n_{\bar{N}}/n_\gamma \approx 10^{-18}, \tag{29}$$

which is eight orders of magnitude too small [17] aside from the problem of having to separate the baryons from the antibaryons. If any separation did occur at higher temperatures (so that annihilations were as yet incomplete) the maximum distance scale on which separation could occur is the causal scale related to the age of the Universe at that time. At $T = 20$ MeV, the age of the Universe was only $t = 2 \times 10^{-3}$ sec. At that time, a causal region (with distance scale defined by 2ct) could only have contained $10^{-5}\ M_\odot$ which is very far from the galactic mass scales which we are asking for separations to occur, $10^{12}\ M_\odot$.

A final possibility might be statistical fluctuations, but in a region containing $10^{12}\ M_\odot$, there are $\sim 10^{80}$ photons so that one would only expect statistical fluctuations to produce an asymmetry $\eta \sim 10^{-40}$! Thus we are left with the problem as to the origin of a small non-zero value for η. We can assume that it was an initial condition to start off with and in a baryon number conserving theory it would remain nearly constant. [The production of entropy (photons) could cause it to fall]. In this case, however, we must still ask ourselves, why is it so small ? A more attractive possibility, however, is to suppose that the baryon asymmetry was in some way generated by the microphysics. Indeed, if one can show that a small non-zero value for η developed from $\eta = 0$ (or any other value) as a initial condition, we could consider the question solved. In the rest of this section, we will look at this second possibility for generating a non-zero value of η using GUTs [4].

There are three basic ingredients necessary [16] to generate a non-zero η. They are

1. baryon number violating interactions
2. C and CP violation
3. a departure from thermal equilibrium.

The first condition is rather obvious, unless there is some mechanism for violating baryon number conservation, baryon number will be conserved and an initial condition such as $\eta = 0$ will remain fixed. C and CP violation indicate a direction for the asymmetry. That is, should the baryon number violating interactions produce more baryons than antibaryons ? If C or CP were conserved, no such direction would exist and the net baryon number would remain at zero. The final ingredient is necessary in order to insure that not all processes are actually occurring at the same rate. For example, in equilibrium if every process which produced a positive baryon number was accompanied by an equivalent process which destroyed it, again no net baryon number would be produced.

The first two of these ingredients are contained in GUTs, the third in an expanding universe where it is not uncommon that interactions come in and out of equilibrium. In SU(5), the fact that quarks and leptons are in the same multiplets allows for baryon non-conserving interactions such as $e^- + d \leftrightarrow \bar{u} + \bar{u}$, etc., or decays of the supermassive gauge bosons X and Y such as $X \rightarrow e^- + d$, $\bar{u} + \bar{u}$. Although today these interactions are very ineffective because of the masses of the X and Y bosons, in the early Universe when $T \sim M_X \sim 10^{15}$ GeV these types of interactions should have been very important. C and CP violation is very model dependent. In the minimal SU(5) model, the magnitude of C and CP violation is too small to yield a useful value of η. The C and CP violation in general comes from the interference between tree level and first loop corrections.

The departure from equilibrium is very common in the early Universe when interaction rates cannot keep up with the expansion rate. In fact, the simplest (and most useful) scenario for baryon production makes use of the fact that a single decay rate goes out of equilibrium. It is commonly referred to as the out of equilibrium decay scenario [18]. The basic idea is that the gauge bosons X and Y (or Higgs bosons) may have a lifetime long enough to insure that the inverse decays have already ceased so that the baryon number is produced by their free decays.

More specifically, let us call X, either the gauge boson or Higgs boson, which produces the baryon asymmetry through decays. Let α be its coupling to fermions. For X a gauge boson, α will be the GUT fine structure constant, while for X a Higgs boson, $(4\pi\alpha)^{1/2}$ will be the Yukawa coupling to fermions. The decay rate for X will be

$$\Gamma_D \sim \alpha\, M_X. \tag{30}$$

However decays can only begin occurring when the age of the Universe is longer than the X lifetime Γ_D^{-1}, i.e., when $\Gamma_D > H$

$$\alpha\, M_X \gtrsim N(T)^{1/2}\, T^2/M_P \tag{31}$$

or at a temperature

$$T^2 \lesssim \alpha\, M_X M_P\, N(T)^{-1/2}. \tag{32}$$

Scatterings on the other hand proceed at a rate

$$\Gamma_S \sim \alpha^2 T^3/M_X^2 \tag{33}$$

at hence are not effective at lower temperatures. In equilibrium, therefore, decays must have been effective as T fell below M in order to track the equilibrium density of X's (and $\bar{X}$'s). Thus the condition for equilibrium is that at $T = M_X$, $\Gamma_D > H$ or

$$M_X \lesssim \alpha\, M_P(N(M_X))^{-1/2} \sim 10^{18}\ \alpha\ \text{GeV}. \tag{34}$$

In this case, we would expect no net baryon asymmetry to be produced.

For masses $M_X \gtrsim 10^{18}\ \alpha$ GeV, the lifetime of the X bosons is longer than the age of the Universe when $T \sim M_X$. Decays finally begin to occur when $T < M_X$, however, the density of X's is still comparable to photons $n_X/n_\gamma \sim 1$ whereas the equilibrium density at $T < M_X$ is $n_X/n_\gamma \sim (M_X/T)^{3/2} \times \exp[-M_X/T] \ll 1$. Hence, the decays are occurring out of equilibrium (inverse decays are not occurring), and we have the possibility for producing a net asymmetry.

Let us now look at what happens during the decay of an X, $\bar{X}$ pair. If we consider the example of the X gauge boson and its decays to $\bar{u}$, $\bar{u}$ with branching ratio r and net baryon number change $\Delta b_1 = -2/3$ and to e^-, d with branching ratio 1-r, and net baryon number change $\Delta b_2 = +1/3$.

$$X \underset{r}{\rightarrow} \bar{u} + \bar{u} \qquad \Delta b_1 = -2/3 \tag{35a}$$

$$X \xrightarrow[1-r]{} e^- + d \qquad \Delta b = +1/3. \tag{35b}$$

A similar set of decays will occur for $\bar{X}$

$$\bar{X} \underset{\bar{r}}{\rightarrow} u + u \qquad \Delta b_{\bar{1}} = +2/3 \tag{36a}$$

$$\bar{X} \xrightarrow[1-\bar{r}]{} e^+ + d \qquad \Delta b_{\bar{2}} = -1/3. \tag{36b}$$

If C and CP are violated then $r \neq \bar{r}$ and we can define the total net baryon number produced per decay of X and $\bar{X}$

$$\Delta B = (\Delta b_1)r + (\Delta b_2)(1-r) + (\Delta b_{\bar{1}})\bar{r} + (\Delta b_{\bar{2}})(1-\bar{r})$$

$$= \bar{r} - r. \tag{37}$$

The value of $\bar{r} - r$ will of course depend on the specific model for C and CP violation.

The total baryon density that will have been produced by the X, $\bar{X}$ pair [provided Eq. (34) is not satisfied] is

$$n_B \simeq (\Delta B)n_X \tag{38}$$

and since we also have $n_X = n_{\bar{X}} \approx n_\gamma$,

$$n_B \approx (\Delta B)n_\gamma. \tag{39}$$

Although the net baryon number is conserved during the subsequent evolution of the Universe, the photon number density is not. A more useful quantity just after baryon generation is the baryon-to-specific entropy ratio, n_B/s. The entropy density, is

$$s = \frac{2\pi^2}{45} N(T)T^3. \tag{40}$$

At $T \lesssim M_X \sim 10^{15}$ GeV, we expect $N(T) \gtrsim 0(100)$ so that $s \sim 0(100)n_\gamma$. Thus the baryon-to-entropy ratio we would expect to produce in the out-of-equilibrium decay scenario would be

$$n_B/s \sim 10^{-2}(\Delta B) \tag{41}$$

The value of n_B/s that we are looking for must be related to the limits on which will be discussed in what follows. η in the range $(3-10) \times 10^{-10}$ corresponds to a value of n_B/s in the range $(4.3 - 14) \times 10^{-11}$. Comparing this with the expected production, Eq. (41) gives us a lot of hope that GUTs may provide us with a viable mechanism for generating a small (but not too small) value for η.

As was noted in the introduction, the two most important pieces of evidence in support of the standard big bang model are the observation [10] of the 3°K microwave background radiation and the explanation [19] of the origin of the light elements and their abundances. Because of the initially high temperatures and densities and the large abundance of neutrons relative to protons, the chains of nuclear reactions similar to those occurring in stars might have occurred. Indeed in the simplest model of nucleosynthesis, one can compute the produced abundances of deuterium, ^{3}He, ^{4}He and ^{7}Li and one finds an amazing degree of agreement with the observed abundances. (The observations which must be compared with the big bang abundances must be from sources where little or no subsequent nucleosynthesis has taken place). Here I will briefly review the predictions of big bang nucleosynthesis and its cosmological consequences in terms of limits on particle physics.

The temperature region of interest is one typical of nuclear energies, i.e., $T \sim 1$ MeV. The initial conditions for the problem will therefore be set at $T >> 1$ MeV. Once again, because the asymmetry between baryons and antibaryons is so small and since we do not expect very different asymmetries among the leptons (standard GUT models even predict their similarity) we will take all chemical potentials to be zero. One of the chief quantities of interest will be the neutron-to-proton ratio (n/p). At very high temperatures ($T >> 1$ MeV), the weak interaction rates for the processes

$$\begin{aligned} n + \nu_e &\leftrightarrow p + e^- \\ n + e^+ &\leftrightarrow p + \bar{\nu}_e \\ n &\leftrightarrow p + e^- + \bar{\nu}_e \end{aligned} \quad (42)$$

were all in equilibrium, i.e., $\Gamma_W > H$. Thus we would expect that initially $(n/p) = 1$. Actually in equilibrium, the ratio is essentially controlled by the boltzmann factor so that

$$(n/p) = \exp(-\Delta m/T), \quad (43)$$

where $\Delta m = m_n - m_p$ is the neutron-proton mass difference. For $T >> \Delta m$, $(n/p) \simeq 1$.

At temperatures $T >> 1$ MeV, nucleosynthesis cannot begin to occur even though the rate for forming the first isotope, deuterium, is sufficiently rapid. To begin with, at $T \gtrsim 1$ MeV deuterium is photodissociated because $E_\gamma > 2.2$ MeV (the binding energy of deuterium; $E_\gamma = 2.7T$ for a blackbody). Furthermore, the density of photons is very high $n_\gamma/n_B \sim 10^{10}$. Thus the onset of nucleosynthesis will depend on the quantity

$$\eta^{-1} \exp[-2.2 \text{ MeV}/T] \quad (44)$$

where η is defined as before. When this quantity (44) becomes $\lesssim O(1)$, the rate for $p + n \rightarrow D + \gamma$ finally becomes greater than the rate for dissociation $D + \gamma \rightarrow p + n$. This occurs when $T \sim 0.1$ MeV or when the Universe is a little over 2 min. old.

Because nucleosynthesis begins when $T \sim 1$ MeV, the rates for processes which control (n/p) (42) as well as those which keep neutrinos in equilibrium are frozen out. Furthermore, because the rates for processes (42) also freeze out (at $T \sim 1$ MeV), the neutron to proton ratio must be adjusted from its equilibrium value. When freeze out occurs, the ratio (n/p) is relatively fixed at

$$(n/p) \sim 1/6. \quad (45)$$

This equilibrium value is adjusted by taking into account the free neutron decays up until the time at which nucleosynthesis begins. This reduces the ratio to

$$(n/p) \sim 1/7. \quad (46)$$

Since virtually all the neutrons available end up in deuterium which gets quickly converted to ^{4}He, we can estimate the ratio of the ^{4}He nuclei formed compared with the number of protons left over

$$X_4 \equiv (N_{^4He}/N_H) = 1/2(n/p)[1 - (n/p)] \tag{47}$$

or more importantly the ^{4}He mass fraction

$$Y_4 = 4X_4/(1 + 4X_4) = 2(n/p)/[1 + (n/p)]. \tag{48}$$

For $(n/p) \simeq 1/7$, we estimate that $Y_4 \simeq 0.25$ which is very close to the observed value.

The actual calculated value of Y_4 will depend on a numerical calculation which runs through the complete sequence of nuclear reactions [8,9]. The nuclear chain is temporarily halted because there are gaps at masses A = 5 and A = 8, i.e., there are no stable nuclei with those masses. There is some further production, however, which accounts for the abundances of ^{6}Li and ^{7}Li. Once again because of the gap at A = 8 there is very little subsequent nucleosynthesis in the big bang. A second chief factor in the ending of nucleosynthesis is that during this whole process the Universe continues to expand and cool. At lower temperatures it becomes exponentially difficult to overcome the Coulomb barriers in nuclear collisions. In spite of these effects, numerical calculations of the elemental abundance continue the chain up until Al.

There are three parameters which have a very strong effect on the results. They are

1) the baryon-to-photon ratio η;
2) the neutron half-life $\tau_{1/2}$;
3) the number of light particles or, in particular, the number of neutrino flavors N_ν.

As we have seen above, the value of η controls the onset of nucleosynthesis (44). Basically what happens is that for a larger baryon-to-photon ratio η the quantity (44) becomes smaller thus allowing nucleosynthesis to begin earlier at a higher temperature. Remember also that a key ingredient in determining the final mass fraction of ^{4}He, Y_4, was (n/p) [see Eq. (48)] and that the final value of (n/p) was determined by the time at which nucleosynthesis begins thus controlling the time available for free decays after freeze out. If nucleosynthesis begins earlier, this leaves less time for neutrons to decay and the value of (n/p) and hence Y_4 is increased.

The value of η cannot be determined directly from observations. If we break it up we find that

$$n_B = \rho_B/m_B = \Omega_B\rho_c/m_B = 1.13 \times 10^{-5}\ \Omega_B h_o^2\ \text{cm}^{-3}, \tag{49}$$

where ρ_B is the energy density in baryons, m_B is the nucleon mass, Ω_B is that part of Ω which is in the form of baryons and ρ_c is the critical energy density. The number density of photons is just

$$n_\gamma = 400(T_o/2.7)^3\ \text{cm}^{-3}, \tag{50}$$

where T_o is the present temperature of the microwave background radiation. Putting η back together we find

$$\eta = 2.81 \times 10^{-8}\ \Omega_B h_o(2.7/T_o)^3. \tag{51}$$

Thus we could determine η if we knew Ω_B, h_o and T_o.

The second parameter $\tau_{1/2}$, is important in that it also plays a role in determining the value of Y_4. Although we don't usually consider $\tau_{1/2}$ a parameter, the uncertainties in its measured value are significant from the point of view of nucleosynthesis. After all, it is this quantity which will control the weak interaction rates and hence determine the freeze-out temperature. The common value of $\tau_{1/2} = 10.6$ min. is actually uncertain by about two percent and this is enough to affect the production ^4He. The range we will consider is

$$10.4 \text{ min.} < \tau_{1/2} < 10.8 \text{ min.} \tag{52}$$

As in the case of η, increasing $\tau_{1/2}$ leads to a larger value of Y_4. We can see this by looking again at a comparison between the weak interaction rates and the expansion rate. If we parametrize the weak interaction rate by $\Gamma_w = AT^5$ and the expansion rate by $H = BT^2$ then the freeze-out temperature is determined by

$$H(T_D) = \Gamma_w(T_D) \tag{53}$$

or

$$T_d^3 = B/A. \tag{54}$$

If we now increase $\tau_{1/2}$, this corresponds to decreasing $\Gamma_w \sim \tau_{1/2}^{-1}$ or decreasing the value of A. This in turn gives a higher value for T_d. Now if T_d is larger, this will give a larger value of (n/p) at freeze-out via Eq. (43) and hence more ^4He via Eq. (48).

The final input parameter we said was the number of light particles. Specifically, what we mean is the number of degrees of freedom corresponding to particles which are still relativistic ($m \ll T$) when $T < O(1)$ MeV. In addition, we must require that these particles be relatively stable so that they will be present when freeze-out occurs, thus $\tau >$ few seconds. As we hinted to above, likely candidates for these particles are neutrinos and thus the number of neutrino flavors N_ν becomes important. Of course any other types of light particles such as Higgsinos or axions, etc., may also be important.

The number of neutrino flavors N_ν will also affect the primordial abundance of ^4He and like η and $\tau_{1/2}$, increasing N_ν increases Y_4. The expansion rate is proportional to $[N(T)]^{1/2}$. At $T \gtrsim 1$ MeV, N(T) is given by

$$N(T) = 2 + \frac{7}{2} + \frac{7}{4} N_\nu \tag{55}$$

which takes into account the contribution of γ's, $e^\pm$'s, and N_ν flavors of neutrinos. Thus increasing N_ν, increases B in the notation of Eq. (54) and again leads to higher value of T_d, with the same effect of producing more ^4He.

Let us now look at the observations [20] which tells us the abundances of the light elements. In particular, we will be interested in the abundances of D, ^3He, ^4He, and ^7Li. Deuterium is the most easily destroyed of the light elements. It is also very difficult to produce in astrophysical systems where it is not further processed to form ^3He. Therefore, any of the observed D is generally assumed to be primordial. Furthermore because deuterium is so easily destroyed (or burned) we must assume that the abundance of D produced in the big bang is greater than the observed value or

$$(D/H)_{BB} \geq (D/H)_{OBS} \gtrsim (1\text{-}2) \times 10^{-5} \tag{56}$$

where (D/H) is the ratio (by number) of deuterium to hydrogen.

Unlike deuterium, ^3He is very difficult to destroy in its entirety in stellar systems. One has the constraint therefore that

$$(D + {}^3He)/H|_{BB} \leq (D + {}^3He)/H|_{Pre\Theta} + O(3) \frac{{}^3He}{H}|_{Pre\Theta} \lesssim 10^{-4}. \tag{57}$$

In order to be consistent with the abundances of D and ^3He, η must lie in the range [9]

$$(3\text{-}4) \times 10^{-10} \leq \eta \leq (7\text{-}10) \times 10^{-10} \tag{58}$$

we note here only that the ^7Li abundance ($^7Li/H \lesssim 10^{-10}$) is consistent for $10^{-10} \leq \eta \leq 7 \times 10^{-10}$ in agreement with (58).

This brings us to ^4He which is probably the most important of the isotopes studied. The main reason ^4He is so important is that there is so much of it. Next to hydrogen it is the most abundant element around and its abundance is quite well known. Unlike the other light elements which have observational uncertainties of $\gtrsim 100$ %, the ^4He abundances are measured to within a few per cent. The main problem is that it is also produced in stars and care must be taken in trying to derive the "observed" primordial abundance.

To be sure, one can place an upper limit on the primordial abundance by $Y_{4BB} < Y_{4OBS}$ (Y_4, remember is the total ^4He mass fraction). However, in order to use big bang nucleosynthesis to set limits on particle physics (e.g., N_ν) a must more accurate determination of Y_{4BB} is needed. Spectral measurements [20] of galactic HII regions give very accurate values of Y_4, however, there they have been contamined with by-products of stellar processing. The observations of galaxies with low metal abundances could in principle yield an accurate value of Y_{4BB} but these measurements are difficult because these galaxies are typically very far away. It is not possible within the scope of these lectures to cover completely the discussion of Y_4. The best estimates consistent with the observations place Y_4 in the range

$$0.22 \leq Y_4 \leq 0.25. \tag{59}$$

If we restrict ourselves as before to $N_\nu = 3$, $\tau_{1/2} = 10.6$ min., the upper limit on Y_4 implies an upper limit on η from Fig. 1

$$\eta \leq 5 \times 10^{-10} \tag{60}$$

which is once again consistent with the previous limits Eq. (58). (The lower limit on Y_4 does not give an interesting bound on η).

Figure 1 actually contains significantly more information than just a limit on η. In Fig. 1, we see clearly the behavior of Y_4 with respect to all three parameters; η, $\tau_{1/2}$ and N_ν. It is clear how Y_4 increases with increasing values of any of the three parameters. It is also immediately clear that we can set a limit [21,9] on N_ν provided that we have a lower limit to η. Using $n \geq 3 \times 10^{-10}$ and $Y_4 \leq 0.25$, we find that $N_\nu \leq 4$ with the equality being at best marginal. This implies that at most one more generation is allowed, assuming that the neutrinos associated with each generation are light and stable.

The strong dependence of Y_4 on the three parameters requires great precision to strengthen the limits due to nucleosynthesis. Strictly speaking, $\eta \geq 3 \times 10^{-10}$ and $\tau_{1/2} > 10.4$ min. allows $N_\nu = 4$ only if

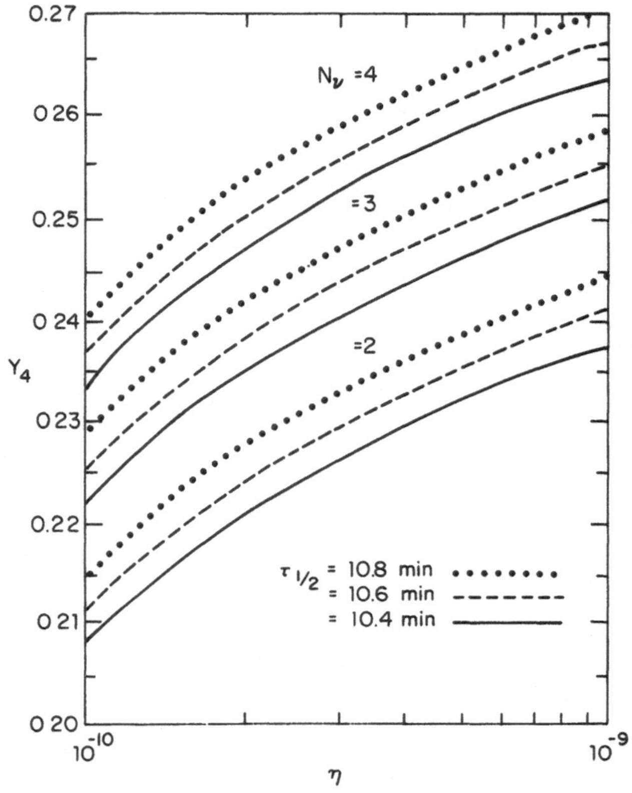

Fig. 1 : The abundance (by mass) of ^{4}He as a function of η for $N_\nu = 2$, 3 and 4 and for $\tau_{1/2} = 10.4$ min. (solid), 10.6 min. (dashed), and 10.8 (dotted).

$Y_4 \geq 0.253$; however, we are not yet in a position to believe the third decimal place. For $\tau_{1/2} > 10.4$ min., the limit Eq. (60) on η can be relaxed so that $N_\nu \leq 3$, $Y_4 \leq 0.25$ implies $\eta < 7 \times 10^{-10}$. We can also turn the limits around and set a lower limit to the helium abundance by assuming $\eta \geq 3 \times 10^{-10}$ and $N_\nu \geq 3$ then we have $Y_4 > 0.24$. If future observations actually yield $Y_4 < 0.24$, one would have to argue that perhaps ν_τ is heavy and/or unstable (the present limit is only $m_{\nu_\tau} < 160$ MeV). If we only assume $N_\nu \geq 2$, then the lower limit on Y_4 becomes $Y_4^\tau \geq 0.22$. Any observation of the primordial helium abundance less than 0.22 would indicate an inconsistency with the standard model.

There is still one more important consequence of the above limits, that is the limit on η can be convereted to a limit on the baryon density and Ω_B. If we turn around Eq. (51) we have

$$\Omega_B = 3.56 \times 10^7 \ \eta h_o^{-2} (T_o/2.7)^3, \tag{61}$$

and using the limits on η Eq. (58), h_o and T_o from (2.7 - 2.8)°K we find a range for Ω_B

$$0.01 \leq \Omega_B \leq 0.15. \tag{62}$$

Recall that for a closed Universe $\phi > 1$, thus from Eq. (62) we can conclude that the Universe is not closed by baryons. This does not exclude the possibility that other forms of matter (e.g., massive neutrinos, etc.) exist in large quantities to provide for a large Ω. In fact, if large clusters of galaxies were representative of Ω the limit from nucleosynthesis would indicate that some form of dark matter must exist.

LECTURE II - INFLATION

As there are already several reviews [5] about inflation [22], I will try to be brief here. However, any review of the very early Universe would be incomplete if it did not at least touch upon inflation. In short what is meant by inflation, is the effect of exponential expansion due to a supercooled phase transition in order to resolve several finetunings regarding the initial conditions in the standard big bang model.

As examples of these problems, I will briefly describe what is known as the horizon problem and the curvature problem. The horizon volume or causally connected volume today, is just related to the age of the Universe $V_o \propto t_o$. The microwave background radiation with the temperature $T_o \sim 3$°K has been decoupled from itself since the epoch of recombination at $T_d \sim 10^4$ K. The horizon volume at that time was $V_d \propto t_d$. Now the present horizon volume scaled back to the period of decoupling will be $V_o' = V_o(T_o/T_d)^3$ and the ratio of this volume to the horizon volume at decoupling is

$$V_o'/V_d \sim (V_o/V_d)(T_o/T_d)^3 \sim (t_o/t_d)^3 \ (T_o/T_d)^3 \sim 10^5, \tag{63}$$

where I have used $t_d \sim 3 \times 10^{12}$ sec and $t_o \sim 5 \times 10^{17}$ sec. The ratio (63) corresponds to the number of regions that were casually disconnected at recombination which grew into our present visible Universe. Because the anisotropy of the microwave background is so small, [12], $\delta T/T \lesssim$ few $\times 10^{-5}$, the horizon problem, therefore, is the lack of an explanation as to why 10^5causally disconnected regions at t_d all had the same temperature to within one part in 10^4!

The curvature problem (also known as the flatness or oldness problem) stems from the fact that although the Universe is very old, we still do not know whether it is open or closed. If we look at the Friedmann equation (2) for the expansion of the Universe and use the limits $\Omega < 4$ and $H_o < 100$ km s^{-1} Mpc^{-1} we can form a dimensionless constant

$$\hat{k} = k/R^2T^2 = (\Omega-1)H_o^2/T^2 \lesssim 3H_o^2/T_o^2 < 2 \times 10^{-58} \tag{64}$$

where I have used $T_o > 2.7°K$. In an adiabatically expanding Universe, $\hat{k}$ is absolutely constant ($R \sim T^{-1}$) and thus the limit (64) represents an initial condition which must be imposed so that the Universe will have lived this long looking still so flat.

A more natural initial condition might have been $\hat{k} \sim O(1)$. In this case the universe would have become curvature dominated at $T \sim 10^{-1} M_p$. For $k = +1$, this would signify the onset of recollapse. Even for $\hat{k}$ as small as $O(10^{-40})$ the Universe would have become curvature dominated when $T \sim 10$ MeV or when the age of the Universe was only $O(10^{-2})$ sec. Thus not only is (64) a very tight constraint, it must also be strictly obeyed. Of course, it is also possible that $k = 0$ and the Universe is actually spatially flat.

These are the two main problems that led Guth [22] to consider inflation. In the problems that were just discussed it was assumed that the Universe has always been expanding adiabatically. During a phase transition, however, this is not necessarily the case. If we look at a potential describing a phase transition from a symmetric false vacuum state $\langle\Sigma\rangle = 0$ to the broken true vacuum at $\langle\Sigma\rangle = v$ as in Fig. 2, and we suppose that because of the barrier separating the two minima the phase transition was a supercooled first-order transition. If in addition, the transition takes place at T_c such that $T_c^4 < V(0)$, the energy stored in the form of vacuum energy will be released. If released fast enough, it will produce radiation at a temperature $T_R^4 \sim V(0)$. In this reheating process entropy has been created and

$$(RT)_f \sim (T_R/T_c)(RT)_i \tag{65}$$

provided that T_c is not too low. Thus we see that during a phase transition the relation $RT \sim$ constant need not hold true and thus our dimensionless constant $\hat{k}$ may actually not have been constant.

The inflationary Universe scenario [22], is based on just such a situation. If during some phase transition, such as $SU(5) \to SU(3) \times SU(2) \times U(1)$ the value of RT changed by a factor of $O(10^{29})$, these two cosmological problems would be solved. The isotropy would in a sense be generated by the immense expansion; one small causal region could get blown up and hence our entire visible Universe would have been at one time in thermal contact. In addition, the parameter $\hat{k}$ could have started out $O(1)$ and have been driven small by the expansion.

If, in an extreme case, a barrier as in Fig. 2 caused a lot of supercooling such that $T_c^4 << V(0)$, the dynamics of the expansion would have greatly changed. In the example of Fig. 2 the energy density of the symmetric vacuum, V(0) acts a cosmological constant with

$$\Lambda = 8\pi\, V(0)M_p^2. \tag{66}$$

If the Universe is trapped inside the false vacuum with $\langle\Sigma\rangle = 0$, eventually the energy density due, to say, radiation will fall below the vacuum energy density, $\rho << V(0)$. When this happens, the expansion rate will be dominated by the constant V(0) and we will get the De Sitter-type expansion and from Eq. (2)

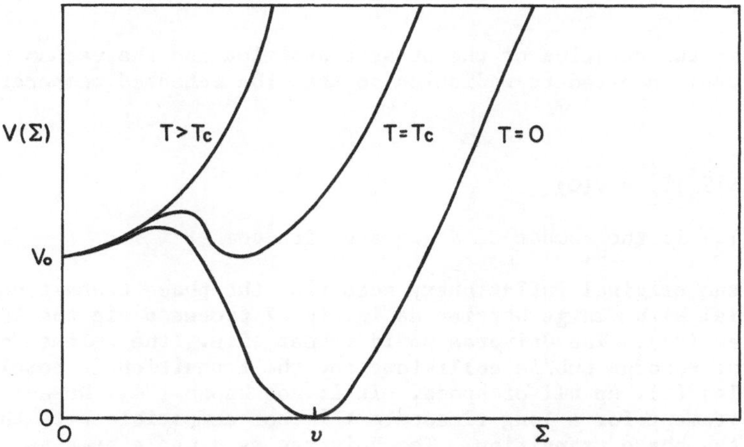

Fig. 2 : The scalar potential for a first order phase transition.

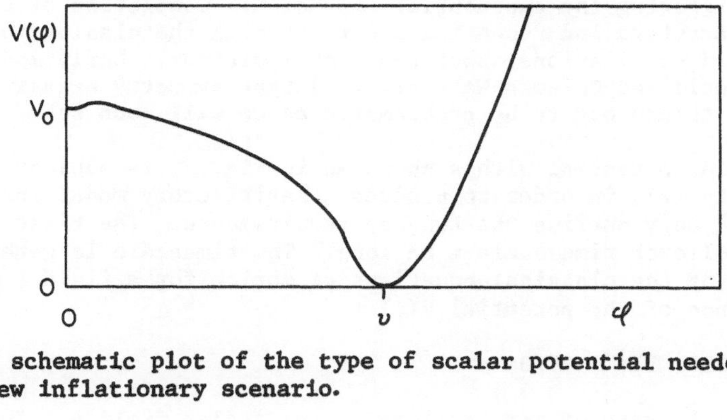

Fig. 3 : A schematic plot of the type of scalar potential needed for the new inflationary scenario.

$$R \sim \exp[Ht], \tag{67}$$

where

$$H^2 = \Lambda/3 = 8\pi\, V(0)/3M_P^2. \tag{68}$$

The cosmological problems could be solved if

$$H\tau \gtrsim 65, \tag{69}$$

where τ is the duration of the phase transition and the vacuum energy density was converted to radiation so that the reheated temperature is found by

$$\frac{\pi^2}{30} N(T_R)T_R^4 = V(0) \tag{70}$$

where $N(T_R)$ is the number of degrees of freedom at T_R.

In the original inflationary scenario, the phase transition given by a potential with a large barrier as in Fig. 2 proceeds via the formation of bubbles [23]. The Universe would reheat, i.e., the release of entropy must occur through bubble collisions and the transition is completed when the bubbles fill up all of space. It is now known [24], however, that the requirement for a long timescale τ is not compatible with the completion of the phase transition. The Universe as a whole remains trapped in the exponentially expanding phase containing only a few isolated bubbles of the broken SU(3) × SU(2) × U(1) phase.

The well-known solution to this dilemma is called the new inflationary scenario [25]. If the shape of the potential $V(\phi)$ resembles that of Fig. 3 rather than Fig. 2, the phase transition would proceed not through the formation of bubbles, but rather, by the long rollover in which ϕ picks up a vacuum expectation value. (ϕ may be the adjoint in the case of SU(5) or any other field with potential V). During the rollover, the vacuum energy density would remain essentially constant for a long period of time triggering the exponential expansion. Completion of the transition is thus guaranteed and reheating occurs through the dissipation of energy due to field oscillations about the global minimum. Early models for new inflation utilized Coleman-Weinberg [26] type symmetry breaking for SU(5). These too, turned out to be problamatic as we will soon see.

A scalar potential with a shape as in Fig. 3, is subject to several requirements [27] in order to produce a satisfactory model for inflation. Here I will only outline the two key requirements. The first is obviously that the rollover time scale τ be long. The timescale is generally determined by the classical equations of motion for a field ϕ moving under the influence of the potential $V(\phi)$

$$\ddot{\phi} + (3H + \Gamma)\dot{\phi} + \partial V/\partial\phi = 0 \tag{71}$$

where Γ is the rate of interactions of the scalar field ϕ. Initially we can neglect $\ddot{\phi}$ and the $\Gamma\dot{\phi}$ term is only relevant for $\Gamma \geq H$. Inflation (a slow rollover) can only occur if $\Gamma < H$ so that $3H\dot{\phi} + \partial V/\partial\phi = 0$ and

$$\tau^{-1} \approx \dot{\phi}/\phi \sim (\partial^2 V/\partial\phi^2)/3H. \tag{72}$$

Hence, near the origin $\phi < H << v$, we must require

$$\partial^2 V/\partial\phi^2\big|_{\phi\approx 0} < 3H^2/65 \tag{73}$$

i.e., we must have a flat potential near the origin.

A second key constraint concerns the production of density fluctuations during the phase transition. In general, there will be a time spread over which in certain regions of space, ϕ rolls down faster or slower than in others. Density perturbations have been calculated [28] in terms of this time spread

$$\frac{\delta\rho}{\rho} \propto H\,\delta\tau \tag{74}$$

where $\frac{\delta\rho}{\rho}$ is the magnitude of the perturbation as it enters the horizon and is calculated in terms of H and ϕ. Limits coming from the isotropy of the microwave background radiation imply that

$$\frac{\delta\rho}{\rho} \leq 10^{-4}. \tag{75}$$

In general the Coleman-Weinberg potential can be expressed as

$$V(\phi) = A\phi^4\left(\ln\frac{\phi^2}{v^2} - \frac{1}{2}\right) + D\phi^2 + \frac{1}{2}Av^4 \tag{76}$$

where the ϕ^4 coupling A is given by

$$A = \frac{1}{64\pi^2 v^4}\left[\Sigma\, g_B M_B^4 - \Sigma\, g_F M_F^4\right] \tag{77}$$

where $g_{B(F)}$ is the number of boson (fermion) helicity states of mass $M_{B(F)}$. The effective mass^2, D, takes into account all ϕ^2 terms such as bare masses, temperature dependent corrections, curvature corrections, etc. In standard SU(5), A is fixed

$$A = \frac{5625}{1024\pi^2} g^4 \approx 5 \times 10^{-2}. \tag{78}$$

The constraint (73) implies that $D < (10^{10}\ \mathrm{GeV})^2$ for $v \sim 10^{15}$ GeV, whereas one would expect $D \sim v^2$. The constraint (75) implies that $A \lesssim 10^{-10}$ (the above value (78) of A leads to $\frac{\delta\rho}{\rho} \sim 50$) in sharp contrast with Eq. 78. Along with other maladies [29,30] than those mentioned above, Coleman-Weinberg potentials appear not to be capable of producing an inflationary scenario.

A varient of the new inflationary scenario is called primordial inflation [31]. Primordial inflationary models are simply those in which the scalar field responsible for inflation is no longer associated with the 24, of SU(5), but rather with some other field ϕ, dubbed the inflation, which picks up a vacuum expectation value $\langle\phi\rangle \gg \langle\Sigma\rangle$, thus allowing for a longer rollover timescale.

Supersymmetry offers the possibility for keeping small those radiative corrections responsible for large scalar masses, i.e., supersymmetric theories naturally have flatter potentials [30,31]. As an example, if we look at the Coleman-Weinberg potential in an exactly supersymmetric model we find that $A \equiv 0$, i.e., there is no Coleman-Weinberg potential. The reason is that in a supersymmetric model $g_B = g_F$ and $M_B = M_F$. Of course supersymmetry must be broken in nature and we expect some splitting between fermion and boson masses,

$$M_B^2 - M_F^2 \sim M_S^2 \tag{79}$$

where M_S is the scale of supersymmetry breaking. Then for the SU(5) case above we have

$$A \approx \frac{75}{32\pi^2} g^2 (M_S^2/v^2) \quad (80)$$

where g is the GUT gauge coupling constant and the limit $A < 10^{-10}$ implies that $M_S \lesssim 10^{11}$ GeV (recall that M_S is really the mass splitting scale). Thus, it appears that supersymmetric models may produce a model satisfying the inflationary constraints.

Because of its capability of producing flat potentials, supersymmetric inflation has been of great interest lately. For example, attention has been focused on finding a theory in N = 1 supergravity which has inflationary capabilities [32]. In these theories one specifies a superpotential f, from which one derives [33] the scalar potential $V(\phi)$

$$V(\phi) = e^G \left[\frac{\partial G}{\partial \phi} \frac{\partial G}{\partial \phi *} \left(\frac{\partial^2 G}{\partial \phi \partial \phi *}\right)^{-1} - 3\right] \quad (81)$$

where

$$G = \phi\phi^* + \ln|f|^2 \quad (82)$$

is the Kähler potential in minimal N = 1 supergravity, f must then be chosen so that $V(\phi)$ complies with the conditions for inflation discussed earlier. The simplest example for f, satisfying the inflationary requirements, is [34]

$$f = m^2 \left(1 - \frac{\phi}{m_P}\right)^2 M_P. \quad (83)$$

The mass scale m is determined to be $O(10^{-4})$ from the calculation of $\delta\rho/\rho$.

Because the inflation is generally taken to be a gauge non-singlet, there is an additional concern regarding initial conditions. Normally, one expects that symmetries are restored at high temperatures so that a natural initial condition for $\langle\phi\rangle$ is $\langle\phi\rangle = 0$. Inflation then takes place as $\langle\phi\rangle$ moves from 0 to $v \approx M_P$. In these models "symmetry restoration" must be added as an additional constraint [35]. Namely, we must also require that at very high temperatures there exists a minimum at $\langle\phi\rangle = 0$. It has been shown [36], however, that no choice of a superpotential can simultaneously satisfy the inflationary constraints and the thermal constraint.

Another constraint on the initial conditions comes from the fact that at high temperatures ϕ is not localized near $\phi \sim 0$ but rather, $\phi \sim T$ and hence, as the Universe cools ϕ may fall directly to the global minimum at $\phi = v$ [37]. Recently, it has been shown that these difficulties are most easily avoided in models of primordial inflation [38].

A route to tackle the problem regarding the thermal constraint and finding a model with a suitable superpotential is to look at non-minimal supergravity models, i.e., those in which

$$\partial^2 G/\partial\phi\partial\phi^* = 1. \quad (84)$$

If, instead, one considered [39] a general form for G

$$G = g_\phi(\phi + \phi^*) + g_\phi^\phi(\phi\phi^*) + g_{\phi\phi}(\phi^2 + \phi^{*2}) + g_{\phi\phi}^\phi(\phi\phi^*)(\phi + \phi^*) \quad (85)$$

where g_ϕ, g_ϕ^ϕ etc. are couplings. It is then possible to derive relations between the g_ϕ's to satisfy the inflationary constraints as well as the thermal constraint. The extra degrees of freedom in G allowed the constraints to be satisfied without the inclusion of several scales.

There are also specific non-minimal models based on SU(n,1) supergravity [40] which satisfy all constraints [41]. In these models the Kähler potential has the form

$$G = -3 \ln(z + z^* - \phi\phi^*/3) + \ln|f|^2 \tag{86}$$

where z is the field which breaks supergravity (see references [42] for a discussion on the z field in minimal supergravity theories) ϕ is the inflation and f is the superpotential. In these theories it is possible to write down a superpotential as in Eq. (83) which in addition satisfies the thermal constraint [41]

$$f = m^2(\phi - \phi^4/4M_P^3). \tag{87}$$

I note here that SU(n,1) models have also been of interest in recent superstrings models [43]. Hence it appears that supergravity can supply some models for the inflationary scenario.

LECTURE III - DARK MATTER

In the first lecture, I indicated that the value of Ω (determined from mass to light ratios) increases with astrophysical scales. In addition, if inflation did occur then from Eq. (11) because of the exponential expansion in R, $\Omega = 1$ to a very good approximation. We will now look and the astrophysical and cosmological reasons for dark matter (DM) and the candidates in particle physics models.

Already on galactic scales, there is good evidence from rotation curves [14] of spiral galaxies for the presence of dark matter and a galactic halo. The rotation curve is a measure of the velocity as a function of distance from the center of the galaxy of a star as it revolves around the galaxy. If there were no DM, one would expect that at distances beyond the bulk of the luminous matter that $v^2 \sim 1/r$. Instead one finds flat rotation curves ($v^2 \sim$ constant) out to very large distances ($\gtrsim 50$ kpc). This implies that the mass of the galaxy must continue to increase $M \sim r$ beyond the luminous region.

In addition to the DM problem on the scale of galactic halos (which I will return to shortly) there are several other DM problems [44]. Even locally, in our galactic disk, there appears to be a DM problem [45]. From local stellar distributions there is evidence that some 50 % of the disk mass is non-luminous. However, this DM probably cannot be in the form of elementary particles because it must have undergone some dissipation in order to become flattened into a disk. Astrophysical arguments indicate that black holes are also a bad choice [46], while white dwarfs remain an interesting possibility [47].

A more subtle DM problem is the one in relation to the growth of density perturbations and galaxy formation. One of the features of the spectrum of density perturbations produced by inflation (which are preferred for galaxy formation) is that as the different fourier modes fall within the horizon scale (i.e., $\lambda \sim ct$) they all have the same magnitude $\frac{\delta\rho}{\rho}$ at that time. Once within the horizon, these modes cannot really begin to grow further as long as the Universe is radiation dominated. At a temperature of a few thousand degrees, the Universe becomes matter dominated and density perturbations begin growing as

$$\frac{\delta\rho}{\rho} \sim R(t) \sim \frac{1}{T}. \tag{88}$$

Now to reach non-linear growth we must have had $\frac{\delta\rho}{\rho} \sim 1$ at the time when the oldest galaxies and quasars were forming or at $T \sim 4\ T_o \sim 10°K$. This means that at the time of matter dominance $\frac{\delta\rho}{\rho} \gtrsim 10^{-3}$. However we know from limits on the anisotropy of the microwave background [11] that $\frac{\delta\rho}{\rho} <$ few $\times 10^{-5}$. We will see shortly that this is really a DM problem.

As a first guess as to the identity of the DM, one might pick baryons i.e., ordinary matter. As we saw from big bang nucleosynthesis, there are good limits (Eqs 58 and 62) on the value of Ω in the form of baryons. Recall, in the standard model, one finds good agreement for the predicted abundances of the light isotopes D, ^{3}He, ^{4}He, ^{7}Li only for a range in Ω_B between 0.01 and 0.15. For $\Omega_B < 0.01$, D and ^{3}He are overproduced while for $\Omega_B > 0.15$, ^{4}He is overproduced and D is underproduced. If inflation is correct and $\Omega = 1$, then at least some of the DM must be non-baryonic.

Returning to the growth of perturbations, if there exists some form of non-baryonic DM, the Universe may have become matter dominated earlier. For example in the case of massive neutrinos $T_{MD} \sim m_\nu/6$. Density perturbations could then begin to grow earlier at say $T \sim 10^4 - 10^5\ T_o$ while baryonic perturbations could not until decoupling at $T \sim 10^3\ T_o$. After decoupling, the baryons would fall into the perturbation already formed by the neutrinos. Hence the existence of DM could help enormously in the growth of perturbations for galaxy formation. For a complete review see [48].

Returning once more to the question of DM on the scale of galactic halos, although one needs $\Omega \geq 0.05$ and that is consistent with Ω in baryons, there are several arguments [49] against baryonic matter in halos. Put briefly, it is very difficult to have a large baryon density in such a way that it is unobservable. In the form of gas the baryons would heat up and emit X-rays in violation of observed limits. To put the baryons in non-nuclear burning stars (Jupiters) would require an extrapolation of the stellar mass distribution which is very different from what is observed. Dust or rocks along with dead remnants such as neutron stars or black holes would require a metal abundance in great excess of the galactic metallicity. Very massive ($\geq 100\ M_\odot$) black holes remain a possibility.

There are of course, many other candidates for the DM. Because of its important role in the formation of galaxies, DM has classified [50] into three types : hot, warm and cold DM. They are distinguished by their effective temperature at the time they decoupled from the thermal background. Examples of hot particles are neutrinos or very light Higgsinos with < 100 eV masses. These particle decouple at $T_d \sim 1$ MeV and are thus still relativistic at T_d. Warm particles decouple earlier and have higher masses (up to ~ 1 keV). Any superweakly interacting neutral particle is a warm candidate such as a right-handed neutrino. Cold particles are non-relativistic at temperatures relevant for galaxy formation and usually have masses $\gtrsim 1$ GeV. Examples of these include heavy neutrinos, photinos/Higgsinos, sneutrinos and axions. With this classification, the specific identity of the particle is no longer important. The benefits and problems associated with each type of DM with regards to galaxy formation has been nicely reviewed in [51] and the reader is referred there for further details.

Given the need for DM, we can ask what sort of constraints are there. The most common cosmological constraint is on the mass of a stable particle and is derived from the overall mass density of the Universe. The mass density of a particle X can be expressed as

$$\rho_X = m_X Y_X n_\gamma \leq \rho_c \simeq 10^{-5}\ h_o^2\ \mathrm{GeV/cm^3} \tag{89}$$

where $Y_x = n_x/n_\gamma$ is the density of x's relative to the density of photons $n_\gamma \simeq 400(T_\gamma/2.7°K)^4$ for $\Omega h_o^2 < 1$. Hot particles have limits characteristic to that of neutrinos. For neutrinos [52] $Y_\nu = 3/11$ and one finds

$$\Sigma(\frac{g_\nu}{2})m_\nu < 100 \text{ eV}(\Omega h_o^2) \tag{90}$$

where the sum runs over neutrino flavors and $g_\nu = 2$ for a Majorana mass neutrinos and g_ν for a Dirac mass neutrino. All hot particles with abundances Y similar to neutrinos will have mass limits as in Eq. (90).

Warm particle limits are derived from Eq. (89) as well. Warm particles have lower abundances than neutrinos and the corresponding mass limits are weaker. Recall that $Y_\nu = 3/11$ is derived from the conservation of entropy before and after $e^\pm$ annihilation. Neutrinos at this time are decoupled so that after the annihilations $(T_\nu/T_\gamma)^3 = 4/11$ and $Y_\nu = (3/4)(T_\nu/T_\gamma) = 3/11$. (The factor of 3/4 is due to the difference between Fermi and Bose statistics). If a particle x interacts more weakly than neutrinos then the ratio $(T_x/t_\gamma)^3$ will be lowered [53] due to other particle species' annihilations. Thus Y_ν is reduced allowing [54] for a larger value for m_x. If the particle x decouples around the GUT epoch, then Y_x could be as low as $O(10^{-2})$ and $m_x \lesssim O(1)$ keV.

For cold particles the analysis is somewhat different. The abundance is now a function of m_x and in most cases one finds a lower limit to m_x. The reason for this is that for large m_x, Y_x is controlled by the annihilations of x. When the annihilations freezeout, Y_x is fixed. The freezeout will then depend on the annihilations cross-section and roughly one finds $Y_x \sim (m_x\sigma_A)^{-1}$ and $\rho_x \sim 1/\sigma_A$. This situation was first analyzed for neutrinos [55]. The annihilation cross-section in this case is basically $\sigma_A \sim m_\nu^2/m_w^4$ so that $\rho_x \sim 1/m_\nu^2$ and yields [55,56] $m_\nu \gtrsim 2$ GeV for Dirac mass neutrinos and $m_\nu \gtrsim 6$ GeV for Majorana mass neutrinos [57,56].

Supersymmetric theories introduce several DM candidates. The reason is that if the R-parity (which distinguishes between "normal" matter and the supersymmetric partners) is unbroken then there is at least one supersymmetric particle which must be stable. Candidates for the stable particle include the photino, Higgsino, sneutrino, gravitino, and goldstino. If we assume for simplicity that all of the scalar quarks and leptons have equal masses then the photino annihilation cross-section can be expressed as [58,59,60]

$$\langle\sigma v\rangle_A \simeq \frac{8\pi\alpha^2}{m_{sf}^4} \sum_f q_f^4(1 - z_f^2)^{1/2}\, m_{\tilde\gamma}^2(z_f^2 + 2x(1 - 17z_f^2/8)) \tag{91}$$

where α is the fine structure constant, m_{sf} is scalar fermion mass, q_f the electric charge of the fermion f, $z_f = m_f/m_{\tilde\gamma}$ and $x = T/m_{\tilde\gamma}$. For $m_{sf} \simeq 40$ GeV, $m_{\tilde\gamma} > 1.8$ GeV. For Higgsinos [60], the annihilations are controlled by the fermion Yukawa couplings and the cosmological bound requires $m_{\tilde H} \gtrsim m_b$ or about 5 GeV.

Sneutrinos are an interesting example in that there is in general no cosmological limit on their mass [61]. In addition to the standard weak annihilations of sneutrinos, there is also the process $\tilde\nu + \tilde\nu \rightarrow \nu + \nu$ via zino exchange. In this case $\langle\sigma v\rangle_A \sim 1/M_{\tilde Z}^2$ and is independent of $m_{\tilde\nu}$. Thus $\rho_{\tilde\nu}$ is fixed by parameters other than $m_{\tilde\nu}$ making the sneutrino mass free from cosmological bounds. Before turning to the gravitino, I note that the goldstino arguments are essentially warm particle limits, and will depend on its specific interactions.

The remaining possible supersymmetric DM candidate is the gravitino. Although in most models the gravitino is not stable, there is nothing which prevents its stability. If stable, the gravitino mass limit would again be $m_{3/2} \lesssim O(1)$ keV as for a warm particle [62] assuming that its abundance $Y_{3/2}$ was determined by considering gravitino decoupling at the Planck time. Such an early decoupling will make $Y_{3/2}$ sensitive to the details of inflation [63]. For a more complete discussion on the role of gravitinos I refer the reader to [64].

To conclude this lecture, I would like to briefly mention some recent ideas concerning some signatures of cold particles in our galactic halo. One interesting suggestion [65] has been to use a very cold detector implanted with superconducting grains which would flip as the DM passes through. Another suggestion [66] is to examine carefully the spectrum of cosmic rays due to DM annihilation. For the case of photinos as the DM, although the γ ray flux produced in $\tilde{\gamma}$ annihilations is well below the established backgrounds, the predicted fluxes of positrons and antiprotons is very close to observed values. This may be significant for antiprotons as the low energy antiprotons observed [67] are otherwise difficult to explain. Another possible test of galactic halo DM is the neutrino spectrum of DM annihilations occurring in the sun [68] or in the earth [69]. DM passing through the sun will be trapped and if the DM is a massive Dirac neutrino or a sneutrino it will also be trapped in the earth. In many cases, the neutrino flux is comparable or exceeds the flux expected from atmospheric neutrinos.

In conclusion, there is a broad region of overlap between cosmology, astrophysics and particle physics with respect to DM. The need for DM comes from several sources; inflation; galaxy formation; galactic halos, etc. Supersymmetry is thus of great interest in that it most probably guarantees one stable new particle. Indeed combining theory with new experimental results seems to require [70] that $\Omega_{\tilde{\gamma}} h_o^2 \geq 0.0025$ (recall the limit for baryons is only $\Omega_B h_o^2 \gtrsim 0.01$). The consequences of a positive detection of DM in the halo are enormous. A single galaxy formation scenario may be singled out and if supersymmetric the identity of the lightest and stable SUSY partner would greatly narrow the choice of supersymmetric models.

REFERENCES

[1] For review see articles by B.S. DeWitt and S.W. Hawking in *General Relativity : An Einstein Centenary Survey*, S.W. Hawking and W. Israel (eds) (Cambridge Univ. Press, 1979).
[2] E. Witten, Nucl. Phys. B186 (1981) 412;
A. Salam and J. Strathdee, Ann. Phys. 141 (1982) 316; in the cosmological context see E.W. Kolb, Fermilab preprint 85-17 (1985).
[3] See reviews by : R. Arnowitt, A.H. Chamseddine and P. Nath, Northeastern Univ. Preprints 2597, 2600, 2613 (1983);
J. Ellis, CERN preprint Th. 3718 (1983);
D.V. Nanopoulos, CERN preprint Th. 3699 (1983);
H.-P. Nilles, Phys. Rep. 110C (1984) 1;
J. Polchinski, Harvard Univ. preprint. HUTP-83/A036 (1983).
[4] For a review see : A.W. Kolb and M.S. Turner, Ann. Rev. Nucl. Part. Sci. 33 (1983) 645.
[5] For reviews see : A.D. Linde, Rep. Prog. Phys. 47 (1984) 925;
R. Brandenberger, Rev. Mod. Phys. 57 (1985) 1.
[6] see e.g., M. Crawford, Ph.D. thesis, Univ. of Chicago (1984).
[7] E.V. Shuryak, Phys. Rep. 61 (1980) 71;
K.A. Olive, Nucl. Phys. B190 (1981) 483;
E. Suhonen, Phys. Lett. 119B (1982) 81;
T. De Grand and K. Kajantie, Phys. Lett. 147B (1984) 273.

[8] D.N. Schramm and R.V. Wagoner, Ann. Rev. Nucl. Part. Sci. 27 (1977) 37;
K.A. Olive, D.N. Schramm, G. Steigman, M.S. Turner and J. Yang, Ap. J. 246, 547 (1981);
A. Bosegard and G. Steigman, Ann. Rev. Astron. and Astrophys. 1985.
[9] J. Yang, M.S. Turner, G. Steigman, D.N. Schramm and K.A. Olive, Ap. J. 281 (1984) 493.
[10] A.A. Penzias and R.W. Wilson, Ap. J. 142 (1965) 419.
[11] G.F. Smoot et al., LBL preprint 18602 (1984).
[12] J. Uson and D. Wilkinson in *Inner Space Outer Space*, E. Kolb, M.S. Turner, K.A. Olive, D. Lindly and D. Seckel, eds. Univ. of Chicago Press, 1985.
[13] See e.g., G.A. Tammann, A. Sandage and A. Yahil, *Physical Cosmology*, R. Balian, J. Audouze and D.N. Schramm (eds.) (North Holland Pub. Co., Amsterdam, 1980).
[14] S.M. Faber and J.J. Gallagher, Ann. Rev. Astron. Astrophys. 17 (1979) 135.
[15] P. Fayet, these proceedings.
[16] A.D. Sakharov, Zk. Eksp. Teor. Fiz. Pisma. Red. 5 (1967) 32.
[17] G. Steigman, Ann. Rev. Astron. and Astrophys. 14 (1976) 339.
[18] S. Weinberg, Phys. Rev. Lett. 42 (1979) 850;
D. Toussaint, S.B. Treiman, F. Wilczek and A. Zee, Phys. Rev. D19 (1979) 1036.
[19] G. Gamow, Phys. Rev. 70 (1946) 572;
R.A. Alpher, H. Bethe and G. Gamow, Phys. Rev. 73 (1948) 803.
[20] For a recent compilation of observations see : the *Proceedings of the ESO Workshop on Primordial Helium*, Eds. P.A. Shaver, D. Kunth and K. Kajar, Garching, Germany, 1983.
[21] G. Steigman, D.N. Schramm, and J.E. Gunn, Phys. Lett. B86 (1977) 202;
J. Yang, D.N. Schramm, G. Steigman and R.T. Rood, Ap. J. 227 (1979) 697.
[22] A.H. Guth, Phys. Rev. D23 (1981) 347.
[23] S. Coleman, Phys. Rev. D15 (1977) 2929;
C. Callan and S. Coleman, Phys. Rev. D16 (1977) 1762.
[24] A.H. Guth and E. Weinberg, Phys. Rev. D23 (1981) 826; and Nucl. Phys. B212 (1983) 321.
[25] A.D. Linde, Phys. Lett. 108B (1982) 389;
A. Albrecht and P. Steinhardt, Phys. Rev. Lett. 48 (1982) 1220.
[26] S. Coleman and E. Weinberg, Phys. Rev. D7 (1973) 1888.
[27] P. Steinhardt and M.S. Turner, Phys. Rev. D29 (1984) 2105.
[28] W.H. Press, Phys. Scr. 21, 702;
S.W. Hawking, Phys. Lett. 115B (1982) 295;
A.H. Guth and S.Y. Pi, Phys. Rev. Lett. 49 (1982) 1110;
A.A. Starobinski, Phys. Lett. 117B (1982) 175;
J.M. Bardeen, P.J. Steinhardt, and M.S. Turner, Phys. Rev. D28 (1983) 679.
[29] A.D. Linde, Phys. Lett. 116B (1982) 335;
J. Breit, S. Gupta and A. Zaks, Phys. Rev. Lett. 51 (1983) 1007.
[30] J. Ellis, D. Nanopoulos, K.A. Olive and K. Tamvakis, Phys. Lett. 118B (1983) 335.
[31] J. Ellis, D.V. Nanopoulos, K.A. Olive and K. Tamvakis, Nucl. Phys. B221 (1983) 224.
[32] D.V. Nanopoulos, K.A. Olive, M. Srednicki and K. Tamvakis, Phys. Lett. 123B (1983) 41.
[33] E. Cremmer, B. Julia, J. Scherck, S. Ferrara, L. Girardello and P. Van Nieuwenhuizen, Phys. Lett. 79B (1978) 231; and Nucl. Phys. B147 (1979) 105;
E. Cremmer, S. Ferrara, L. Girardello and A. Van Proeyen, Phys. Lett. 116B (1982) 231; and Nucl. Phys. B212 (1983) 431.
[34] R. Holman, P. Ramond and G.G. Ross, Phys. Lett. 137B (1984) 343.
[35] G. Gelmini, D.V. Nanopoulos and K.A. Olive, Phys. Lett. 131B (1983) 53.

[36] B. Ovrut and P. Steinhardt, Phys. Lett. 133B (1983) 161.
[37] G. Mazenko, W. Unruh and R. Wald, Phys. Rev. D31 (1985) 273.
[38] A. Albrecht and R. Bradenberg, Phys. Rev. D31 (1985) 1225; G.D. Coughlan and G.G. Ross, Phys. Lett. 157B (1985) 151;
L. Jensen and K.A. Olive, Phys. Lett. 159B (1985) 99.
[39] L. Jensen and K.A. Olive, Nucl. Phys. B (1985).
[40] E. Cremmer, S. Ferrara, C. Kounnas and D.V. Nanopoulos, Phys. Lett. 133B (1983) 287;
J. Ellis, C. Kounnas, D.V. Nanopoulos, Nucl. Phys. B241 (1984) 406 and Phys. Lett. 143B (1984) 410 and CERN preprint Th. 3824.
[41] J. Ellis, K. Enquist, D.V. Nanopoulos, K.A. Olive and M. Srednicki, Phys. Lett. 152B (1985) 175.
[42] G.D. Coughlan, W. Fischler, E.W. Kolb, S. Raby and G.G. Ross, Phys. Lett. 131B (1983) 59;
M. Dine, W. Fischler and D. Nemeschansky, Phys. Lett. 136B (1984) 169;
G.D. Coughlan, R. Holman, P. Ramond and G.G. Ross, Phys. Lett. 140B (1984) 44.
[43] E. Witten, Princeton Univ. preprint 1985.
[44] D.N. Schramm, Nucl. Phys. B253 (1985) 53.
[45] J. Bahcall, Ap. J. 276 (1984) 169.
[46] J. Bahcall, P. Hut and S. Tremaine, Ap. J. 290 (1985) 15;
D. Heggi, E.W. Kolb and K.A. Olive, Ap. J. 300 (1986).
[47] R.B. Larson, Ap. J. (in press) 1985;
R. Mochkovitch, K.A. Olive and J. Silk, Ap. J. (submitted) 1985.
[48] J. Primack, SLAC preprint 3387 (1984).
[49] D. Hegyi and K.A. Olive, Phys. Lett. 126B (1983) 28; Ap. J. 1985.
[50] J.R. Bond and A. Szalay, Ap. J. 274 (1983) 443.
[51] M. Davis, Proceedings of the Sixth Workshop on Grand Unification, S. Rudaz and T. Walsh (eds) in press, 1985.
[52] R. Cowsik and J. McClelland, J. Phys. Rev. Lett. 29 (1972) 669;
A.S. Szalay, G. Marx, Astron. Astrophys. 49 (1976) 437.
[53] K.A. Olive, D.N. Schramm and G. Steigman, Nucl. Phys. B180 (1981) 497.
[54] K.A. Olive and M.S. Turner, Phys. Rev. D25 (1982) 213.
[55] P. Hut, Phys. Lett. 69B (1977) 85;
B.W. Lee and S. Weinberg, Phys. Rev. Lett. 39 (1977) 165.
[56] E.W. Kolb and K.A. Olive, Fermilab preprint 85/116 (1985).
[57] L.M. Krauss, Phys. Lett. 128B (1983) 37.
[58] H. Goldberg, Phys. Rev. Lett. 50 (1983) 1419.
[59] L.M. Krauss, Nucl. Phys. B227 (1983) 556.
[60] J. Ellis, J. Hagelin, D.V. Nanopoulos, K.A. Olive and M. Srednicki, Nucl. Phys. B238 (1984) 453.
[61] L.E. Ibanez, Phys. Lett. 137B (1984) 160;
J. Hagelin, G.L. Kane and S. Raby, Nucl. Phys. B241 (1984) 638.
[62] H.R. Pagels and J.R. Primack, Phys. Rev. Lett. 48 (1982) 223.
[63] J. Ellis, A.D. Linde and D.V. Nanopoulos, Phys. Lett. 118B (1982) 59.
[64] J. Ellis, D.V. Nanopoulos, S. Sarkar, Nucl. Phys. B259 (1985) 175;
K.A. Olive, Proceedings of the Sixth Workshop on Grand Unification, S. Rudaz and T. Walsh (eds) 1985.
[65] A. Drukier and L. Stodolosky, Phys. Rev. D30 (1984) 2795;
M. Goodman and E. Witten, Princeton Univ. preprint (1984).
[66] J. Silk and M. Srednicki, Phys. Rev. Lett. 53 (1984) 624;
J. Hagelin and G.L. Kane, Nucl. Phys. B (1985);
F.W. Stecker, S. Rudaz and T.F. Walsh, Phys. Rev. Lett. (1985).
[67] A. Buffington, S. Schindler and C. Pennypacker, Ap. J. 248 (1981) 1179;
R. Protheroe, Ap. J. 254 (1982) 391.
[68] J. Silk, K.A. Olive and M. Srednicki, Phys. Rev. Lett. 55 (1985) 257;
M. Srednicki, K.A. Olive and J. Silk, in preparation, 1985.

[69] K. Freese, CFA preprint 2180 (1985);
M. Srednicki, F. Wilczek and L.M. Krauss, Santa Barbara preprint NSF-ITP-85-58.

[70] J. Ellis, J. Hagelin and D.V. Nanopoulos, CERN preprint Th. 4157 (1985); see also J. Ellis, Proceedings of the Sixth Workshop on Grand Unification, S. Rudaz and T. Walsh (eds) 1985.

THE ELLIPTIC INTERPRETATION OF BLACK HOLES AND QUANTUM MECHANICS

G.W. Gibbons

D.A.M.T.P. - University of Cambridge
Silver Street
Cambridge CB3 9EW - U.K.

ABSTRACT

Preamble. The lectures as delivered contained an elementary introduction to the classical theory of black holes together with an account of Hawking's original derivation of the thermal emission from black holes in the quantum theory. I also described what is here called the "elliptic interpretation" partly because of its possible relevance to the lectures of Professor 't Hooft. It seemed to me that there was little point in repeating in the written version what is rather standard material so I have decided to give a rather more detailed account of the elliptic interpretation and refer the reader to the original literature [1, 2] for the elementary material.

INTRODUCTION (section 1)

Recently there has been renewed interest in the quantum properties of black holes [3, 4, 5, 6]. The original derivation by Hawking [2, 7] of the fact that black holes formed by gravitational collapse should radiate thermally with a temperature $T = 1/8\pi GM$ where M is the mass of the black hole was obtained by considering a background spacetime which contains the collapsing body producing the black hole. The spacetime is neither everywhere time independent nor is it symmetrical with respect to time reversals. There is a natural definition for the incoming no particle state $|0\rangle$ which is determined in terms of an appropriate and natural notion of positive frequency at past infinity. Similarly there is a natural definition of outgoing particles at future infinity. There is no natural definition of particles falling through the future horizon but Hawking showed that this did not affect the predicted outward flux of particles at infinity. Given these assumptions (and ignoring the back reaction of the geometry) Hawking's derivation follows unambiguously at least as far as non interacting fields are concerned. Inclusion of the back reaction must of course modify the result substantially at late times but is unlikely to make a great deal of difference at times much larger than the collapse time but short compared with the evaporation time. They would of course introduce non-thermal correlations so the outgoing flux would no longer be exactly thermal but the qualitative picture should remain unchanged.

Some authors [4, 5, 8] have however preferred to work not with the dynamic spacetime representing the formation of a black hole in gravitational collapse and the subsequent occurrence of a spacetime singularity inside a future event horizon but rather with the exact maximally extended vacuum solution. The geometry of this spacetime, M_K, is very different from that of the dynamic collapse. For example it is symmetric under time reversal

$$T : (U,V,\theta,\phi) \rightarrow (-V,-U,\theta,\phi) \tag{1.1}$$

in standard Kruskal null coordinates. This means that M_K contains both future and past horizons, and singularities both in the future and in the past. The extended manifold is also invariant under the spatial reflection

$$R : (U,V,\theta,\phi) \rightarrow (V,U,\theta,\phi). \tag{1.2}$$

This means that M_K contains not one but two asymptotically flat regions. Furthermore the behaviour of fields on M_K is not determined by initial data posed at either or both past infinities but requires also that data be specified on both past horizons.

The spacetime M_K is sometimes referred to as that of an "eternal black hole", however this description is not really accurate. It actually represents a black hole and a "white hole" in the same spacetime. [A "white hole" is usually defined as the time reversal of a black hole]. It is clear that the behaviour of quantum or classical fields on M_K is much less obvious than on the dynamical spacetime of gravitational collapse. Depending upon ones boundary conditions one could obtain virtually any result. Furthermore we have no good reason to believe that objects described by this spacetime actually will occur in nature. Purely as a model however it may provide interesting insights into how General Relativity and Quantum Mechanics may be combined.

I want to point out some consequences of a model in which the two asymptotic regions and the two singularities are regarded as *identical*. This model has something in common with the ideas of Professor 't Hooft described also in these proceedings but discussions at the School indicate that it is not the same as his. For the sake of a name I shall refer to it as the *elliptic interpretation.*

THE IDENTIFIED SPACETIME (section 2)

Attempts have been made in the past to identify the two asymptotic regions of M_K by identifying point x and Jx where J is an isometry of the Schwarzschild metric whose square is the identity (i.e. an involution). In order to obtain a non-singular quotient manifold $\widetilde{M}_K = M_K/J$ it is essential that J has no fixed points. This requirement is not insisted upon in all refs. in [9] and is not used explicitly in [4]. If we demand that J commute with the continuous symmetries of M_K, which will thus descend to $\widetilde{M}_K$, one finds that J must be given by

$$J : (U,V,\theta,\phi) \rightarrow (-U,-V,\pi-\theta,\phi+\pi). \tag{2.1}$$

[The identification $(U,V,\theta,\phi) \rightarrow (-U,-V,\theta,\phi)$ used in [9] would give a singularity at the "Boyer axis" $u = v = 0$].

One may check that J is space orientation preserving but time orientation reversing. Thus $\widetilde{M}_K$ is space orientable but not time orientable - there is no continuous assignment of future directed and past directed timelike vectors on $\widetilde{M}_K$. This ambiguity does not arise if one remains outside the horizon.

Even if one penetrates the horizon one can still *not* find closed smooth non-spacelike closed curves in $\tilde{M}_K$, for suppose one could : such a curve γ would correspond in M_K to a non closed curved joining two distinct points x and Jx. If one traverses γ twice in $\tilde{M}_K$ one gets a smooth closed curve Γ in M_K going from x to Jx to x. Since γ is everywhere non-spacelike Γ will be everywhere non-spacelike but M_K, unlike $\tilde{M}_K$, certainly contains no everywhere non-spacelike closed curves so γ can't exist.

This simple argument makes no essential use of the detailed geometry of the Schwarzschild metric. A simple calculation, which will not be repeated here, shows that $\tilde{M}_K$ *does* have closed timelike and null curves starting from a point x and going into the local future and returning to x from the local future. They are of course not smooth at x.

Because the fundamental group, $\pi_1(\tilde{M}_K)$ equals $\mathbb{Z}_2$ tensor fields on $\tilde{M}_K$ need not be single valued, they may be sections of a twisted bundle over $\tilde{M}_K$. Since we have a Lorentzian metric the appropriate structural group is the full Lorentz Group. This has 4 connected components related by P, T, and PT. Depending upon whether one represents these elements by ±1 we get ordinary tensors, time-pseudo tensors, space pseudo tensors and spacetime pseudo-tensors. These correspond to the 4 possible representations of the group $Z_2 \otimes Z_2$ generated by P and T. Since $\tilde{M}_K$ is neither time orientable nor spacetime orientable tensor fields of the second or fourth kind will be double valued, changing sign as one traverses a non-trivial loop in $\pi_1(\tilde{M}_K)$. On the covering space M_K they must be antisymmetric with respect to the involution J while ordinary tensors or space-pseudo tensors must be symmetric with respect to J.

In flat spacetime one usually regards the electromagnetic field strength $F_{\mu\nu}$ as a time pseudo tensor. This is because under time reversal the 4-current $J^\mu = (\rho,\underline{J}) \rightarrow (\rho,-\underline{J})$ and $F_{\mu\nu}$ is taken to have the same properties as J^μ under time reversal. If we maintain this convention in curved space we see that Maxwell fields on $\tilde{M}_K$ must be double valued or alternatively must be antisymmetric under J on the covering space M_K. This has an important consequence for electromagnetic black holes in the elliptic interpretation. Consider the Reissner-Nordstrom black holes. If one does not identify under the obvious generalization from Schwarzschild of the inversion J the electromagnetic field can have any "duality complexion", i.e. it may be pure electric, pure magnetic or any duality rotation of these. If one identifies however only the pure magnetic case is allowed since one may easily check that only in this case will $F_{\mu\nu}$ be odd with respect to J. Thus a black hole can't have an electric charge in the elliptic interpretation. It should be noted that in the case of extreme Reissner-Nordstrom black holes (which correspond in certain theories to solitons) the analogue of the inversion map J does not exist.

Since Weyl neutrinos transform in a well defined way under time reversal there is no problem in introducing them on $\tilde{M}_K$. This would not have been true if J had reversed parity. A reversal of parity is required in the Anti-de Sitter case and does indeed give rise to difficulties as will be discussed later.

Scalar propagators on $\tilde{M}_K$ may be obtained by taking any propagator on the covering space M_K, G(x,y) satisfying

$$(-\nabla_x^2 + m^2)G(x,y) = \delta(x,y) \tag{2.2}$$

and averaging with respect to both arguments using J. Thus $\tilde{G}(x,y)$ will be an appropriate propagator on $\tilde{M}_K$ if

$$\tilde{G}(x,y) = 1/2\{G(x,y) \pm G(Jx,y) \pm G(x,Jy) + G(Jx,Jy)\} \tag{2.3}$$

where the minus signs allow for twisted fields. If G(x,y) is invariant under J in the sense that

$$G(x,y) = G(Jx,Jy) \tag{2.4}$$

$\tilde{G}(x,y)$ is given by the simple formula

$$\tilde{G}(x,y) = G(x,y) \pm G(x,Jy) \tag{2.5}$$

so that we merely add an antipodal source in addition to the original one.

If G(x,y) is the usual retarded propagator on the Kruskal manifold which satisfies :

$$G^{ret}(Jx,Jy) = G^{adv}(x,y) \tag{2.6}$$

we find that the resulting propagator on the identified space is

$$1/2(G^{adv}(x,y) + G^{ret}(x,y) \pm G^{adv}(x,Jy) \pm G^{ret}(x,Jy)). \tag{2.7}$$

If x and y are both outside the horizon the 3rd and 4th terms will not influence them and so the net result of the elliptic interpretation in the exterior region is to force the use of half advanced plus half retarded Green's functions.

A different approach to field theory on $\tilde{M}_K$ is to pose Cauchy data on a t = constant hypersurface, e.g. U + V = 0. Call this hypersurface Σ. The involution J takes points on Σ to points on Σ. Call its restriction to Σ I. The metric restricted to Σ is

$$ds^2 = (1 + \frac{GM}{2\rho})^4 (dx^2 + dy^2 + dz^2) \tag{2.8}$$

where $\rho^2 = x^2 + y^2 + z^2$. The horizon is at $\rho = GM/2$, and Σ is all points such that $\rho > 0$, I is given by

$$I : x^i \to -(\frac{GM}{2})^2 \frac{x^i}{\rho^2} . \tag{2.9}$$

If the data on Σ are such that the fields on Σ are invariant under I and their time derivatives are odd under I they will evolve on the Kruskal manifold M_K to be invariant under J. This means that they are well defined on the identified space $\tilde{M}_K$. For twisted fields one merely interchanges even and odd in the above statements. It is clear from this that one may impose what ever Cauchy data one pleases outside the horizon and still be consistent with the elliptic interpretation. As far as the outside of the horizon is concerned it makes no difference.

Using an analogous construction for the initial data for Einstein's equations one can easily convince oneself that there is an entire class of time dependent spacetimes possessing an involutive isometry J (but no continuous isometries) in which the elliptic interpretation is possible. They represent as it were a different "sector" of the theory from what is usually considered in gravitational collapse. Although perhaps of limited physical or astrophysical interest it is certainly a mathematically consistent sector.

QUANTUM MECHANICS ON THE IDENTIFIED SPACE (section 3)

The question we wish to answer in this section is how to do quantum field theory on the identified space $\tilde{M}_K$. One approach might be to define

Feynman propagator using the averaging technique described in the last section. Thus one could take the Hartle-Hawking propagator [10] on M_K, $G^{H^2}(x,y)$ and use on $\widetilde{M}_K$,

$$\widetilde{G}^{H^2}(x,y) = 1/2(G^{H^2}(x,y) \pm G^{H^2}(x,Jy)). \tag{3.1}$$

This would come from analytically continuing from the Euclidean Schwarzschild metric

$$ds^2 = (1 - \frac{2GM}{r})d\tau^2 + \frac{dr^2}{1 - \frac{2GM}{r}} + r^2(d\theta^2 + \sin^2\theta\, d\phi^2). \tag{3.2}$$

The Euclidean Schwarzschild solution is periodic in the imaginary time τ with a period $8\pi GM$ [10, 11]. To ensure invariance under J one would further identify under the euclidean version of J, J_e.

$$J_e : (\tau,r,\theta,\phi) \to (\tau + 4\pi GM, r, \pi-\theta, \phi+\pi). \tag{3.3}$$

On the face of it this suggests that one gets finite temperature Green's functions with *twice* the usual Hawking temperature since the period is halved. However this is not correct. The Green's function $\widetilde{G}^{H^2}(x,y)$ does not arise as the expectation value of field operators in a Fock space. To see this we follow the analysis given in Allen [11] in the closely related de Sitter case. He wasn't considering the elliptic interpretation directly but his arguments are relevant here. Consider the symmetric scalar propagator $\widetilde{G}^{(1)}(x,y)$ corresponding to (3.1). If $p_i(x), n_i(x)$ are the positive and negative frequency functions defining the putative Fock vacuum we have

$$\widetilde{G}^{(1)}(x,y) = \sum_i p_i(x)p_i(y) + n_j(x)n_j(y). \tag{3.4}$$

By forming

$$\int_\Sigma \{\widetilde{G}^{(1)*}(x,y)\overleftrightarrow{\partial}_\mu(x)\}d\,\Sigma^\mu \tag{3.5}$$

where Σ is the t = constant surface used in section (2) and using the assumed completeness and orthogonality of $p_i(x)$ we obtain :

$$p_i(x) = \pm\, p_i(Jx). \tag{3.6}$$

But this means that

$$\int_\Sigma p_i^*(x)\overleftrightarrow{\partial}_\mu p_i(x)d\Sigma^\mu = 0. \tag{3.7}$$

Thus our putative positive frequency functions have zero norm. Our Green's function fails to satisfy the correct positivity properties. Usually one has an ambiguity in building up a Fock basis in curved space because one could choose any set of "positive frequency" functions $p_i(x)$ which can be normalized to unity. Any such choice determines a Fock basis. In the present case however insisting on invariance under J as in (3.6) implies that the norm of p(x) vanishes. From a mathematical point of view the real vector space of initial data for wave equation is endowed with a natural symplectic structure [13] given by :

$$(f,g) = \int_\Sigma (f\, \overleftrightarrow{\partial}_\mu\, g)d\,\Sigma^\mu. \tag{3.8}$$

Picking a positive frequency function gives the space of Cauchy data a complex structure [13] and enables us to construct a complex Hilbert space. Such a construction breaks down if we use a function satisfying (3.6) since the norm vanishes.

One may express this in more physical terms. J reverses the sense of time. Therefore in Quantum Mechanics we anticipate that J corresponds to an antiunitary operator $\hat{J}$.

$\hat{J}$ may be thought of as complex conjugation followed by the action of some unitary operator $\hat{U}$ say. Since $\hat{J}$ is involutary we have that on any state $|\psi>$

$$\hat{J}(\hat{J}|\psi>) = c|\psi> \tag{3.9}$$

whence

$$UU^*|\psi> = |\psi> \tag{3.10}$$

where * denotes complex conjugation. It follows that either $c = +1$ or $c = -1$. [In flat space, where $\hat{J}$ is replaced by CPT, $c = (-1)^F$ where F is the fermion number. This can be used in flat space to give the fermion superselection rate since if one superposes two states they must have the same value of c if CPT is to make sense].

Working on the identified space $\tilde{M}_K$ with Green's functions invariant under J means effectively that all physical measurements under J and so in effect one is imposing on ones states the condition

$$\hat{J}|\psi> = \alpha|\psi> \tag{3.11}$$

acting with J again shows that

$$|\alpha|^2 = +1 \tag{3.12}$$

so our states must have $c = \pm 1$ [presumably this means the fermion number must be zero mod. 2]. By multiplying $|\psi>$ by $\alpha^{-1/2}$ one can set $\alpha = 1$ with no loss of generality. The set of states $\{|\psi>\}$ satisfying (3.11) does not form a complex Hilbert space because $\hat{J}$ is antilinear and so does not commute with multiplication by a complex number. It does however constitute a *Real* Hilbert space.

Thus the elliptic interpretation seems to force us to use a Real Hilbert space. Since in any complex Hilbert space there is natural anti-linear isomorphism between kets and bras we see that the correspondence (3.1) using $\hat{J}$ provides a *linear* map between bras and kets which are in effect identified in a Real Hilbert space. Thus the elliptic interpretation defined by (3.11) makes contact with the ideas of Professor 't Hooft. The general principles of Quantum Mechanics are compatible with the restriction to real numbers though it makes it difficult to talk about Heisenberg's Uncertainty Principle and Heisenberg equations of motion. In the present case this is perhaps not surprising. Real Quantum Mechanics has some unusual features. For example continuous symmetries no longer give rise to observables. This is because continuous symmetries are effected by orthogonal operators. These are obtained by exponentiating skew symmetric operators which can't of course be diagonalized over the reals. Thus if we have global U(1) symmetry like the baryon number in pre GUT times, the elliptic boundary conditions tell us that the total charge Q corresponding to a conserved current must vanish

$$Q = \int_\Sigma J_\mu d\, \Sigma^\mu = 0. \tag{3.13}$$

We saw earlier that the total electric charge of a black hole must also vanish in the elliptic interpretation which is consistent.

MORE THAN ONE BLACK HOLE (section 4)

A natural question to ask is what happens if one has more than one black hole. Are there many asymptotic regions or just one ? We don't have exact solutions describing many black holes however some years ago Misner [14] described how one could construct initial data for many black holes, though in those days they weren't called black holes. In the simplest case one considers time symmetric data of the form

$$ds^2 = \phi^4 d\underline{x}^2 \tag{4.1}$$

where

$$\nabla^2\phi = 0 \tag{4.2}$$

where ∇^2 is the usual Laplacian in flat Euclidean 3-space. Setting

$$\phi = 1 + \frac{GM}{2|\underline{x}-\underline{x}_o|} \tag{4.3}$$

gives the Schwarzschild data (c.f. equation (2.8)). Let $\underline{g}$ be a conformal transformation of the flat metric dx^2 such that under g : $\underline{x} \to \underline{g}(\underline{x})$, $\underline{dx}^2 \to \Omega_g^4(x)\underline{dx}^2$. Since (4.2) is conformally invariant, if $\overline{\phi}(\underline{x})$ solves $\overline{(4.2)}$ so will

$$\Omega_g^{-1}(x)\phi(g(x)) = U(g)\phi. \tag{4.4}$$

I defined by (2.9) is such a conformal transformation and one has

$$U(I)1 = \frac{GM}{2|\underline{x} - \underline{x}_o|} \tag{4.5}$$

Consider now N - black holes at positions $\underline{x}_n$ and mass M_n, n = 1,2,...N. There are N inversion maps $I_1 \ldots I_N$. Consider the discrete Schottky like group G generated by $I_1 \ldots I_N$ subject to the obvious relations

$$I_n^2 = 1 \tag{4.6}$$

let g by a typical element. The conformal factor (assuming convergence)

$$\phi(x) = \sum_{g \in G} U(g)1 \tag{4.7}$$

is invariant under the discrete group G. A fundamental domain for G will be the exterior of the balls in $\mathbb{R}^3$ or the points $\underline{X}_n$ with radius $\frac{GM_n}{2}$. We assume these balls don't overlap. This domain D will be repeated infinitely often in the interior of the balls, i.e. the sets {gD}, g ∈ G will lie inside these balls. The time symmetric data using (4.7) in (4.1) now have infinitely many isometric regions related by elements of the group G. These may however all be identified if one pleases, since G acts without fixed points.

The data constructed in this way are similar to but different from the data constructed by Misner [14, see also 15] in which there are 2 isometric asymptotic regions even though there are N black holes. The isometry joining these 2 regions has fixed points on the horizon and so would be unsuitable for the identification proposed here. In fact Misner's construction is to use the maps

$$x^i \to + \left(\frac{GM}{2}\right)^2 \frac{x^i}{\rho^2}. \tag{4.8}$$

This has fixed points on the sphere of radius $\frac{GM}{2}$. It follows that the boundary of D, i.e. the surfaces of the N spheres is a totally geodesic submanifold. Thus Misner was able to take another copy of D, call it D^1 and join it to D across the spheres forming the "double". Misner constructed a 3-manifold for N black holes with 2 asymptotically flat regions. One cannot identify these regions however without getting singularities. The manifold with just two asymptotic regions is used by Professeur 't Hooft in his work. Finally I should mention that if one simply takes as data

$$\phi = 1 + \sum_{n=1}^{n=N} \frac{GM_n}{2|\underline{x} - \underline{x}_n|} \tag{4.9}$$

one has just N asymptotic regions which cannot be identified. Further possibilities ("wormholes") arise if two of the spheres have equal radius.

From the point of view of the elliptic interpretation I believe that the model with just one asymptotic region is the most natural. On the covering space one would presumably identify points to future in those regions obtained by an even number of inversions with points in the past in those regions obtained by an odd number of inversions. Thus we see that the Elliptic interpretation can deal satisfactorily with the case of more than one horizon.

DE SITTER AND ANTI-DE SITTER SPACETIME (section 5)

Historically the elliptic interpretation was first discussed in Cosmology. Eddington pointed out the possibility of identifying points which were *spatial* antipodes of one another on the Einstein-Static universe. The spatial cross-section would thus be $\mathbb{RP}^3$ or SO(3) rather than S^3 or SU(2). In fact other identifications are possible but these would not commute with the isometry group. If the cosmological model has a moment of time symmetry - e.g. a k = 1 model one may compose this with time reversal to get a fixed point free involutive isometry. This shares with our previous examples the property of reversing time orientation but not space orientation. A special case would be de Sitter space. The Robertson Walker form of the line element is

$$ds^2 = dt^2 - \cosh^2 \sqrt{\frac{\Lambda}{3}}\, t \,(dx^2 + \sin^2 x(d\theta^2 + \sin^2\theta\, d\phi^2)) \tag{5.2}$$

and we identify under

$$J : (t,x,\theta,\phi) \to (-t,\pi-x,\pi-\theta,\phi+\pi). \tag{5.2'}$$

Since de Sitter can be isometrically embedded as a hyperboloid in 5 dimensional Minkowski space M^5

$$(X^o)^2 - (X^1)^2 - (X^2)^2 - (X^3)^2 - (X^4)^2 = -3/\lambda. \tag{5.3}$$

J has the simple expression as inversion in M^5 :

$$J : X^A \to -X^A. \tag{5.4}$$

The identification under J was discussed by Schroedinger [16]. Since x and Jx are always spacelike separated he felt the identification to be rather natural but recognized the difficulties with time orientability.

Our previous discussion would apply. Classically we use half advanced plus half retarded propagators (the last 2 terms in (2.7) vanish). Quantum Mechanically we use Feynman propagators satisfying

$$G(x,Jy) = \pm G(x,y). \tag{5.5}$$

As Allen [12] has pointed out there is now no Fock Vacuum and our previous remarks about real quantum mechanics follow. In de Sitter space there is no unique de Sitter invariant Fock vacuum for scalar fields. This is essentially because we can always add antipodal source. There is a unique Green's function $G^e(x,y)$ which is regular on the euclidean section, S^4. Then there is a one-parameter family of vacua of the form

$$G^{\theta}(x,y) = \cosh\theta \ G^{\theta}(x,y) + \sinh\theta \ G^{\theta}(x,Jy) \tag{5.6}$$

the symmetric or antisymmetric Green's functions are not members of this family. Only the Euclidean propagator corresponds to a finite temperature propagator and is the one usually used in inflationary studies. In principle predictions using the elliptic interpretation might differ from the standard results. This has not, to my knowledge, been investigated. One might suspect that these effects should be small since the antipodal region would now be according to standard views almost infinitely far away from us.

A related case is Anti de Sitter space which arises in gauged supergravity. Quantum field theory in Anti-de Sitter space has received a lot of attention recently [17, 18]. Anti-de Sitter space is again a hyperboloid but now in a flat space with 2 timelike directions.

$$(X^o)^2 + (X^4)^2 - (X^1)^2 - (X^2)^2 - (X^3)^2 = -3/\Lambda. \tag{5.7}$$

The inversion J (still defined by 5.4) now preserves time orientation but reverses spatial parity. Since time is not reversed there is nothing funny about quantum mechanics in Anti-de Sitter space. However there is something funny about Parity. Avis, Isham and Storey [17] and Breiten-Iohner and Freedman [18] found that certain boundary conditions needed to be imposed on the fields in Anti-de Sitter space to give a well-defined quantum theory. One may express these boundary conditions in terms of the inversion map. They amount to demanding that the fields be even or odd under the action of J. Thus in a sense the elliptic interpretation is forced on one in Anti-de Sitter spacetime. Since J reverses parity this has the effect of ruling out chiral fermions in Anti-de Sitter space. [c.f. 19, 20].

The differences between the two cases may be understood in a more group theoretic way. In both cases J lies in the centre of the isometry group which is O(4,1) or O(3,2) respectively. Both groups have 4 connected components. In the case of O(3,2) J lies in the same component as time reversal. In the case of O(4,1) it lies in the same component as Parity.

CONCLUSION (section 6)

I hope I have convinced the reader that the Elliptic interpretation is mathematically consistent. As far as black holes formed in gravitational collapse is concerned it seems to me to be irrelevant. It is conceivable that primordial black holes should be described in this way and it may be important at a deeper level. The general idea that demanding symmetry under time reversal may require rethinking our ideas about quantum mechanics seems to be of some interest. This may be relevant to various speculations about the origin of time asymmetry in the universe

[21, 22]. Personally I find the most interesting aspect of work the possible relation between the global topology of spacetime and the structure of quantum mechanics. Topological effects concerned with *spatial* topology will never, it seems to me, lead to a substantial alteration of our ideas about quantum mechanics. If General Relativity is to cause any changes in the overall structure of Quantum theory it seems most likely that they will come about because of the modifications of our understanding of the nature of *time* that Einstein's theory brings about.

Finally let me conclude by thanking the organizers of the school for an extremely enjoyable stay in Cargese.

REFERENCES

[1] Les Houches Lectures 1972 "Black Holes" ed. B.S. DeWitt and C. de Witt.
[2] S.W. Hawking, Comm. Math. Phys. 248 (1974) 30.
[3] D. Gross, Nucl. Phys. B236 (1984) 349.
[4] G. 't Hooft, Journal of Geometry and Physics 1 (1984) 45; Nucl. Phys. B256 (1985) 727, (and these proceedings).
[5] K. Freese, C.T. Hill, M. Mueller, Nucl. Phys. B255 (1985) 693.
[6] T.D. Lee, Columbia preprint CU-TP 305 "Are Black Holes Black".
[7] S.W. Hawking, Nature (Lond) 248 (1974) 30.
[8] W. Israel, Phys. Lett. 57A (1976) 107.
[9] W. Rindler, Phys. Rev. Lett. 15 (1965) 1001;
W. Israel, Phys. Rev. 143 (1966) 1016;
W. Israel, Nature (Lond) 211 (1966) 466;
F.J. Belinfante, Phys. Lett. 20A (1966) 25;
J.L. Anderson and R. Gautreau, Phys. Lett. 20A (1966) 24.
[10] S.W. Hawking and J.B. Hartle, Phys. Rev. D13 (1976) 2188.
[11] G.W. Gibbons and M.J. Perry, Proc. Roy. Soc. (Lond) A358 (1978) 467.
[12] B. Allen, Santa Barbara preprint UCSB-TM3-1985 "Vacuum States in de Sitter Space".
[13] A. Ashtekar and A. Magnon, Proc. Roy. Soc. (Lond) A346 (1975) 375.
[14] C.W. Misner, Annals of Physics 24 (1963) 102-117.
[15] R.W. Lindquist, J. Math. Phys. 4 (1963) 938.
[16] E. Schrodinger, "Expanding Universes" Cambridge University Press.
[17] S. Avis, C.J. Isham and D. Storey, Phys. Rev. D18 (1978) 356.
[18] P. Breitenlohner and D.Z. Freedman, Ann. Phys. 144 (1982) 249.
[19] D.W. Dusenden and D.Z. Freedman, MIT preprint, CTP n° 1291 (1985).
[20] B. Allen and C.A. Lutken, private communication.
[21] A.D. Sakharov, J.E.T.P. 52 (1980) 348.
[22] S.W. Hawking, "The Arrow of Time in Cosmology", Phys. Rev. (in press).

LOW ENERGY PHYSICS FROM SUPERSTRINGS

Gino C. Segre

Department of Physics
University of Pennsylvania
Philadelphia, PA 19104

INTRODUCTION (chapter 1)

When I originally agreed to accept the kind invitation to lecture at the NATO Summer School in Cargese, my topic was to be the theory of CP violation. The developments of the past year have resulted in growing interest in the theory of superstrings, a subject which is on the one hand extraordinarily exciting in the promise it holds for solutions of many of the outstanding problems of particle physics and on the other hand rather forbidding in the amount of new knowledge which needs to be acquired by the average theorist to understand the papers that are now being published on the recent developments. These considerations have persuaded me "in extremis", to change the subject of my lectures.

These lectures are meant to ease the access of non-expert theorists to this field. They are of course no substitute for the reading of the papers, but hopefully will keep the non-initiated from throwing up his or her hands in frustration as they come across for the first time terms such as holonomy, Betti numbers, Chern classes, Calabi-Yau manifolds, etc. I will try to give the relevant references as we go along, though the list is necessarily incomplete in such a fast moving field. Of particular note is the set of papers by Witten and collaborators, a "must" for these new developments. This is not a review article so the references will be skimpy. In general, I will only provide a recent reference from which the reader can work backwards in time.

In a sense the term low energy superstrings is misleading : the work of the past fifteen years in string theory, culminating in last summer's stunning developments by Green and Schwartz have led theorists to believe a finite, consistent superstring theory can be formulated. An enormous amount of work is going on in this subject; we will not discuss it at all! Rather we will start with the premise that an effective field theory in ten space-time dimensions can be obtained from the superstring theory. Our lectures will cover this later stage, namely how does one proceed from the effective ten dimensional theory to an effective four dimensional theory, describing the world as we "see it". Even if the particular path we follow is incorrect, it seems likely that many of the techniques we will use may be needed in a more satisfactory formulation. They are, by and large, nothing more than some of the standard tools of differential geometry. We do wish to emphasize one more time that the route we have chosen whereby one goes from string theory to ten dimensional field theory to four dimensional field theory may not be the right one.

SUPERSTRINGS (chapter 2)

We just finished saying we would not discuss string theory, but at least a few words seems appropriate.

Consider a particle of mass m moving freely along a path $s(\tau)$ going from $s_i(\tau_i)$ to $s_f(\tau_f)$. For one dimensional motion the action is

$$S = m \int_{s_i}^{s_f} ds = m \int_{\tau_i}^{\tau_f} d\tau \sqrt{-\dot{x}(\tau)^2}. \tag{2.1}$$

Introducing the metric $\gamma_{\tau\tau} = \gamma$ one can imagine $x(\tau)$ is a scalar field in one dimension and rewrite S as

$$S = -\frac{1}{2} \int_{\tau_i}^{\tau_f} d\tau \sqrt{\gamma} \left\{ \frac{\dot{x}^2}{\gamma} + m^2 \right\} \tag{2.2}$$

(note one can let $m \to 0$ in 2.2).
Setting $\delta S/\delta\gamma = 0$, solving for γ and inserting back into (2.2) gives us (2.1). Equation (2.1) is easy to generalize to n dimensions, all functions of τ

$$S = m \int_{\tau_i}^{\tau_f} d\tau \frac{dx^\mu}{d\tau} \frac{dx_\mu}{d\tau}. \tag{2.3}$$

We see that we can re-parametrize τ, letting $\tau \to f(\tau)$.

This action is easily generalized to that of a string in which instead of a world-line, parametrized by τ, we have a world sheet parametrized by τ and σ

$$S = -\frac{1}{2} T \int_0^\pi d\sigma \int_{\tau_i}^{\tau_f} d\tau \sqrt{-\gamma}\, \gamma^{\alpha\beta}\, \partial_\alpha x^\mu\, \partial_\beta x_\mu \tag{2.4}$$

where $\gamma = \det\gamma$ and the space coordinate σ is taken to run from 0 to π. This action is invariant under

i) global Poincaré invariance

$$\delta x^\mu = \ell^\mu{}_\nu\, x^\nu + a^\nu \quad ; \quad \delta\gamma_{\alpha\beta} = 0 \tag{2.5a}$$

ii) local reparametrization invariance

$$\delta x^\mu = \xi^\alpha\, \partial_\alpha\, x^\mu$$

$$\delta\gamma_{\alpha\beta} = \xi^\lambda\, \partial_\lambda\, \gamma_{\alpha\beta} + \partial_\alpha\xi_\beta + \partial_\beta\xi_\alpha. \tag{2.5b}$$

iii) Weyl invariance

$$\delta\gamma_{\alpha\beta} = f(\sigma,\tau)\gamma_{\alpha\beta};\ \delta x^\mu = 0. \tag{2.5c}$$

Clearly Weyl-invariance is lost if we introduce the analogue of the m^2 term in (2.2). Note also that $\frac{1}{\gamma}$ appearing in (2.2) and $\gamma^{\alpha\beta}$ in (2.4) is consistent because $\gamma^{\alpha\beta}$ is the inverse of $\gamma_{\alpha\beta}$.

Equation (2.4) is the Lagrangian for the so-called bosonic string. Classically, we have eliminated γ as a variable by the field equation

$$\frac{\delta S}{\delta \gamma_{\alpha\beta}} = 0. \tag{2.6}$$

Classically, this is always possible : it corresponds to the vanishing of the energy momentum tensor for the string. Quantum mechanically, on the other hand, Weyl or conformal invariance only holds for n = 26, i.e., twenty six dimensional string [1,2].

The fermionic or supersymmetric string [3] is obtained by adding fermionic partners $\psi^\mu(\sigma,\tau)$ to the Lagrangian and

$$\frac{1}{2}\,\partial_\alpha x^\mu \partial_\beta x_\mu \rightarrow \frac{1}{2}\,\{\partial_\alpha x^\mu \partial_\beta x_\mu + i\psi^\mu \gamma_\alpha \partial_\beta \psi_\mu\} \tag{2.7}$$

In this case the critical dimension is ten, so $\mu = 1 \ldots 10$. To quantize the system we go to the so-called light-cone gauge in which we have eight transverse degrees of freedom. The gauge group is introduced by giving "quark" labels to the ends to the string so our modes on the string belong to $\bar{q} \times q$ representation, i.e. to the adjoint representation of the gauge group.

The S-matrix elements can be obtained from the action by the path integral formulation with appropriate vertex functions.
Finally, as we shall see in later sections, we will be interested in non-flat space-time. This alteration is introduced by having

$$\partial_\alpha x^\mu \partial_\beta x_\mu \rightarrow g_{\mu\nu}\, \partial_\alpha x^\mu \partial_\beta x^\nu \tag{2.8}$$

where $g_{\mu\nu}$ is the appropriate curved space metric.

So far we have described the open or type 1 string. Its lowest energy, i.e. zero mass excitations belong to the gauge group supermultiplet. A consistent string theory requires that we allow the ends of the string to close upon themselves to form the so-called closed string, whose lowest excitation is the gravity supermultiplet. The excited modes of the string typically have masses $\sim\sqrt{T}$, which in this theory is of order $M_{Planck} \sim 10^{19}$ Gev. We see thus that a consistent gauge theory has as excitations in the low energy world the graviton supermultiplet coupled to the gauge supermultiplet. As we shall see in this section on anomalies, the only allowed gauge groups are SO(32) and $E_8 \times E_8$. Of these two choices, only SO(32) can be introduced on the string by giving "quark" labels to the endpoints. $E_8 \times E_8$ requires a new and very interesting type of string theory, the heterotic string [4]. Both of course lead to an effective field theory of supergravity coupled to super Yang-Mills on a curved manifold in ten dimensions, which is the main thrust of these lectures.

As a final comment [5], let us count the number of dimensionless parameters in this theory. There are three dimensionfull parameters, given by the string tension T with dimension $(\text{length})^{-2}$, the gravitational coupling $\kappa \sim (\text{length})^4$ and the Yang-Mills coupling $g \sim (\text{length})^3$. This last statement may at first seem surprising, but remember we are in ten dimensions and though, as always, gA has the same dimensions as the derivative ∂_μ, A has dimensions of $(\text{length})^{-4}$. We would think that there would be two dimensionless parameters, but the consistency of the string, namely the relationship between closed and open strings requires

$$\kappa \sim g^2 T. \tag{2.9}$$

In addition there is at tree level a massless scalar in the supergravity multiplet (more about this later when we discuss ten dimensional supergravity) whose vacuum expectation value is undetermined. It turns out that shifting the vacuum expectation value rescales g and T keeping κ and g^2T fixed. Therefore the second dimensionless parameter labels, not a set of theories, but a set of vacua. If quantum mechanically the vacuum is fixed, so is the second dimensionless parameter g^4/κ^3.

MATHEMATICAL PRELIMINARIES (chapter 3)

In order to learn the necessary mathematical techniques we recommend study of a relatively elementary book such as Flanders [1] "Differential Forms" or, at a more advanced level, The Physics Report of Eguchi, Gilkey and Hanson [2]. We will nevertheless sketch the basic notions needed, sending the reader to texts for more details.

Differential Forms (section a)

Denote by U an open domain in an n dimensional space E^n. One forms at a point $x_1 \ldots x_n$ are written as

$$\sum_{i=1}^{n} a_i(x_1 \ldots x_n) dx^i \tag{3.1}$$

i.e. they are line elements in E^n. A zero form is just a smooth function in E^n and a two form is defined using the wedge product

$$dx^i \wedge dx^j = \frac{1}{2}(dx^i \otimes dx^j - dx^j \otimes dx^i). \tag{3.2}$$

The 2-form is

$$\Sigma\, b_{ij}(x_1 \ldots x_n) dx^i \wedge dx^j. \tag{3.3}$$

Similarly one may proceed to define 3-forms, 4-forms, etc. up to n-forms. A (n+1) form in n dimensions vanishes because the dx^i's are taken to all be anti-symmetrized.

$$\alpha_p = \alpha_{\mu_1 \ldots \mu_p} dx^{\mu_1} \wedge dx^{\mu_2} \ldots dx^{\mu_p}$$

is a p-form with the summation over repeated indices assumed. A p-form and a q-form may be multiplied together to obtain a p+q form.

$$\alpha_p \beta_q = \alpha_p \wedge \beta_q = (-1)^{pq} \beta_q \wedge \alpha_p. \tag{3.4}$$

Exterior derivation is an operation which takes a p-form α to a (p+1) form $d\alpha$

$$0 \to 1 \quad d\alpha = \frac{\partial \alpha}{\partial x^i} dx^i$$

$$1 \to 2 \quad d\alpha = \frac{\partial \alpha_j}{\partial x^i} dx^i \wedge dx^j$$

$$2 \rightarrow 3 \quad d\alpha = \frac{\partial \alpha_{jk}}{\partial x^i} dx^i \wedge dx^j \wedge dx^k \tag{3.5}$$

etc...

An important property of exterior derivation is that for any p form α

$$d(d\alpha) = 0. \tag{3.6}$$

For zero forms this follows for the equality of cross derivatives

$$dd\alpha = \frac{\partial^2 \alpha}{\partial_x{}^i \partial_x{}^j} dx^i \wedge dx^j. \tag{3.7}$$

A form α is said to be *closed* if $d\alpha = 0$ and *exact* if $\alpha = d\beta$. An exact form is obviously closed. A closed form is locally exact, but not necessarily globally.

Example

Three space p forms α_o, α_1, α_2, α_3

$$\alpha_o = f(x)$$

$$\alpha_1 = g_1 dx^1 + g_2 dx^2 + g_3 dx^3$$

$$\alpha_2 = h_1 dx^2 \wedge dx^3 + h_2 dx^3 \wedge dx^1 + h_3 dx^1 \wedge dx^2$$

$$\alpha_3 = j dx^1 \wedge dx^2 \wedge dx^3$$

$$d\alpha_2 = \left(\frac{\partial h_1}{\partial x^1} + \frac{\partial h_2}{\partial x^2} + \frac{\partial h_3}{\partial x^3}\right) dx^1 \wedge dx^2 \wedge dx^3$$

$$\alpha_1 \wedge \alpha_2 = (g_1 h_1 + g_2 h_2 + g_3 h_3) dx^1 \wedge dx^2 \wedge dx^3$$

etc...

We need to define also the Hodge star* or duality transformation which, in n dimensions takes a p form into an n-p form

$$* (dx^{i_1} \wedge \ldots dx^{i_p}) = \frac{1}{(n-1)!} \varepsilon_{i_1, i_2 \ldots i_{p+1} \ldots i_n} dx^{i_p+1} \wedge dx^{i_p+2} \ldots \wedge dx^{i_n}. \tag{3.8}$$

Repeating the * operation on a p form takes us back to the p form, i.e.

$$**\alpha_p = (-1)^{p(n-p)} \alpha_p. \tag{3.9}$$

We may also introduce the operator δ

$$\begin{aligned} \delta &= -*d* && \text{n even, all p} \\ \delta &= (-1)^p *d* && \text{n odd.} \end{aligned} \tag{3.10}$$

Whereas d takes p forms into p+1 forms, δ takes them into p-1 forms. By analogy to the operations with d, we say a p form α_p is *co-closed* if

$\delta\alpha_p = 0$ and *co-exact* if $\alpha_p = \delta\beta_{p+1}$.

We can also define an inner product of two p forms as

$$(\alpha_p,\beta_p) = \int_M \alpha_p \wedge {*\beta_p}. \tag{3.11}$$

The manifold M over which we are integrating is n dimensional and $\alpha_p \wedge *\beta_p$ is an n form since $*\beta_p$ is a (n-p) form. It is easy to show that

$$(\alpha_p, d\beta_{p-1}) = (\delta\alpha_p, \beta_{p-1}) \tag{3.12}$$

so we say that δ is the adjoint of d. The Laplacian (for flat metrics) is given by

$$\Delta = (d+\delta)^2 = d\delta + \delta d \tag{3.13}$$

since both dd and $\delta\delta$ give zero operating on a p form.

Since

$$\begin{aligned}(\alpha_p,\Delta\alpha_p) &= (\alpha_p, d\delta\alpha_p) + (\alpha_p,\delta d\alpha_p)\\ &= (\delta\alpha_p,\delta\alpha_p) + (d\alpha_p, d\alpha_p)\end{aligned} \tag{3.14}$$

we see that, for sufficiently well behaved forms α_p is *harmonic*, i.e.

$$\Delta\alpha_p = 0 \text{ if and only if } \delta\alpha_p = d\alpha_p = 0. \tag{3.15}$$

Stokes' theorem says that if M is an n dimensional manifold with boundary ∂M, with $\alpha_n = d\alpha_{n-1}$

$$\int_M \alpha_n = \int_M d\alpha_{n-1} = \int_{\partial M} \alpha_{n-1}. \tag{3.16}$$

This process cannot be repeated one more time with $\alpha_{n-1} = d\alpha_{n-2}$ because i) $\alpha_n = dd\alpha_{n-2} = 0$ and/or ii) $\partial\partial M = 0$ (the boundary of a boundary is always empty).

Example

Euclidean Maxwell's Equations

($\mu = 1,2,3,4;\ i = 1,2,3$).

Gauge potential 1-form : $A = A_\mu(x)dx^\mu$.

Field strength 2-form : $F = dA = \partial_\mu A_\nu\, dx^\mu \wedge dx^\nu = \frac{1}{2}(\partial_\mu A_\nu - \partial_\nu A_\mu)dx^\mu dx^\nu$.

Gauge transform : $A' = A + d\Lambda(x)$

$F' = F$ since $dd\Lambda = 0$.

Bianchi Identity : $dF = ddA = 0$.

δF is a 1-form which we call J, where of course

$J = J_\mu dx^\mu$.

We may now identify the ordinary vector $\vec{E}$ and $\vec{B}$ fields as

$$F = E_i \, dx^i \wedge dx^4 + \frac{1}{2} B_i \varepsilon_{ijk} \, dx^j \wedge dx^k. \tag{3.17}$$

Then

$$\begin{aligned} dF = 0 = & \frac{\partial}{\partial x^j} E_i \, dx^j \wedge dx^i \wedge dx^4 \\ & + \frac{1}{2} \frac{\partial}{\partial x^\ell} B_i \varepsilon_{ijk} \, dx^\ell \wedge dx^j \wedge dx^k \\ & + \frac{1}{2} \frac{\partial}{\partial x^4} B_i \varepsilon_{ijk} \, dx^4 \wedge dx^j \wedge dx^k \end{aligned} \tag{3.18}$$

$$= \nabla . \vec{B} \, dx^1 \wedge dx^2 \wedge dx^3 + \frac{1}{2}(\frac{\partial}{\partial x^4} B_i + (\nabla \times \vec{E})_i) \varepsilon_{ijk} \, dx^i \wedge dx^j \wedge dx^k.$$

Similarly, using (3.17) and (3.8)

$$\begin{aligned} {}^*F = & \frac{1}{2} \varepsilon_{i4jk} \, E_i \, dx^j \wedge dx^k \\ & + \frac{1}{4} B_i \varepsilon_{ijk} \varepsilon_{jk\ell 4} \, dx^\ell \wedge dx^4 \\ = & \frac{1}{2} \varepsilon_{ijk} \, E_i \, dx^j \wedge dx^k + B_i \, dx^i \wedge dx^4 \end{aligned} \tag{3.19}$$

so, under duality $F \leftrightarrow F^*$ and $E \leftrightarrow B$. We may now also construct δF :

$$\delta F = -{}^*d{}^*F = -\nabla . E \, dx^4 + (\frac{\partial \vec{E}}{\partial x^4} + \nabla \times \vec{B}) . d\vec{x} \tag{3.20}$$

so Maxwell's equations are $dF = 0$ and $\delta F = J$. If $J \equiv 0$, F is harmonic, i.e. $\Delta F = 0$. We seem to have miraculously proved $\nabla . B = 0$. What goes wrong if there is a magnetic monopole present. We wrote $F = dA$, but if a monopole is present A is not defined globally. We have to introduce two coordinate patches $U_\pm$ covering the $z > -\varepsilon$ and $z < \varepsilon$ regions so that $F = dA_\pm$ in $U_\pm$. A_+ and A_- differ on the equator by a gauge transformation so F is uniquely defined but $A_\pm$ are respectively singular in $U_\mp$.

So far we have only treated the gauge field 1-form for the case of electromagnetism. If we consider instead a so-called matrix valued 1 form

$$A = A_\mu dx^\mu = -iA_\mu{}^a \lambda_a dx^\mu \tag{3.21}$$

where the λ_a are the matrix generators of the gauge group the field strength F is

$$\begin{aligned} F &= dA + A \wedge A \\ &= \frac{1}{2}(\partial_\mu A_\nu - \partial_\nu A_\mu + [A_\mu, A_\nu]) dx^\mu \wedge dx^\nu. \end{aligned} \tag{3.22}$$

The equation $dF = 0$ is no longer valid because since dA is a 2-form and A a 1-form by (3.4) $A \wedge dA = dA \wedge A$. To recover a Bianchi identity we must define the covariant derivative DF by

$$DF = dF + [A,F] = 0. \tag{3.23}$$

This is obviously true since

$$dF = d(dA + A\wedge A) = dA\wedge A - A\wedge dA$$
$$= (dA + A\wedge A)\wedge A - A\wedge(dA + A\wedge A) = [F, A]. \tag{3.24}$$

In fact (3.23) can be generalized to any tensor valued p form

$$\Sigma^{a_1\ldots a_r}_{b_1\ldots b_\ell} = \Sigma^{a_1\ldots a_r}_{b_1\ldots b_\ell;\,\mu_1\ldots\mu_p}\, dx^{\mu_1}\ldots dx^{\mu_p} \tag{3.25}$$

with r upper and ℓ lower group indices

$$D\Sigma = d\Sigma + [A,\Sigma]$$
$$= d\Sigma + A^{a_1}_{a'}\ \Sigma^{a'\ldots a_r}_{b_1\ldots b} + \text{(all upper indices)} \tag{3.26}$$
$$- (-1)^p\ \{\Sigma^{a_1\ldots a_r}_{b'\ldots b}\ A^{b'}_{b_1} + \text{(all lower indices)}\}.$$

For a gauge transformation g we gave

$$A \to g^{-1}(A+d)g$$
$$F \to g^{-1}Fg \tag{3.27}$$
$$\Sigma^{a_1\ldots}_{b_1\ldots} \to (g^{-1})^{a_1}_{a'}\ \Sigma^{a'\ldots}_{b'\ldots}\ (g)^{b'}_{b_1}.$$

<u>Riemannian Geometry</u> (section b)

In Riemannian geometry the distance ds between two infinitesimally nearby points x^μ and $x^\mu + dx^\mu$ is given by

$$ds^2 = g_{\mu\nu}dx^\mu \otimes dx^\nu$$

where $\otimes$ means ordinary multiplication and $g_{\mu\nu}$ is the metric tensor of the D dimensional manifold M. The Greek indices μ, ν, $\lambda\ldots$ of a tensor correspond to coordinate frame indices, i.e. to motion along the manifold. Similarly we may erect at each point an orthonormal tangent frame for which the metric η_{ab} is flat (either (++++...+) for Euclidean space or (-++++...+) for Minkowski space) with indices a, b, c... . We go from one metric to the other by the so-called vielbeins [3] $e^a_\mu(x)$

$$g_{\mu\nu} = e^a_\mu e^b_\nu \eta_{ab}$$
$$\eta^{ab} = g^{\mu\nu}e^a_\mu e^b_\nu. \tag{3.28}$$

The inverse of the vielbein is

$$E^\mu_a = \eta_{ab}g^{\mu\nu}e^b_\nu \tag{3.29}$$

for which we have

$$E^{\mu}_{a}e^{b}_{\mu} = \eta_{ac}g^{\mu\nu}e^{c}_{\nu}e^{b}_{\mu} = \eta_{ac}\eta^{cb} = \delta_{a}^{b}$$

$$E^{\mu}_{a}e^{a}_{\nu} = \delta^{\mu}{}_{\nu}.$$

The 1-forms

$$e^{a} \equiv e^{a}_{\mu}dx^{\mu} \tag{3.30}$$

can be thought of as matrix valued 1-forms, acted on by orthogonal rotations in the orthonormal tangent frame

$$e^{a} \rightarrow L^{a}{}_{b}(x)e^{b} \tag{3.31}$$

where $L^{a}{}_{b}$ are rotations in either SO(D) or SO(1,D-1) depending on our choice of η_{ab}. We have

$$\eta_{cd}L^{c}{}_{a}L^{d}{}_{b} = \eta_{ab} \tag{3.32}$$

so

$$g = g_{\mu\nu}dx^{\mu} \otimes dx^{\nu} = e^{a} \otimes e^{b}\eta_{ab} \rightarrow e^{c} \otimes e^{d}L^{a}{}_{c}L^{b}{}_{d}\eta_{ab} = g \tag{3.33}$$

i.e. the metric is invariant. In other words the e^{a}, which can be taken as a basis for 1-forms, are defined up to local tangent frame rotations.

It is important to understand clearly the distinction between the tangent space and the underlying curved space. The group of local frame rotations generated by $L^{a}{}_{b}(x)$ act to "rotate" tensors in the tangent space. The local invariance group of the manifold (diffeomorphisms) is that of coordinate reparametrizations. Of course the two are intertwined and we shall display the formalism in both frames and how to go from one to another.

We need to start by introducing a connection which shows how to parallel transport objects in either frame. Later we will show the relationship between the two connections.

In the tangent frame we observe that a general p-form Lorentz vector (vector under either SO(D) or SO(1,D-1) depending on our metric)

$$S^{a} = S^{a}{}_{\mu_{1}\ldots\mu_{p}}dx^{\mu_{1}}\wedge\ldots dx^{\mu_{p}} \tag{3.34}$$

does not transform the same way as dS^{a}, i.e. dS^{a} is a p+1 form by definition, but not necessarily a Lorentz vector. The situation is entirely analogous to what we saw in the last section, where Σ and $d\Sigma$ transformed differently; we needed to define the covariant derivative $D\Sigma$. Similarly here we introduce a matrix valued 1-form

$$\omega^{a}{}_{b} = \omega^{i}{}_{\mu}(T^{i})^{a}{}_{b}dx^{\mu} \tag{3.35}$$

where T^{i} is the representation matrix of SO(D) or SO(1,D-1). It is defined so that

$$DS^{a} = dS^{a} + \omega^{a}{}_{b}\wedge S^{b} \tag{3.36}$$

is a Lorentz vector 2-form. We immediately generalize the covariant derivative as in (3.26) to

$$D = d + [\omega, \ ]. \tag{3.37}$$

We may then define the particular Lorentz vector 2-form

$$T^a = De^a = de^a + \omega^a{}_b \wedge e^b \tag{3.38}$$

which we call torsion. The Lorentz tensor 2-form is

$$R^a{}_b = d\omega^a{}_b + \omega^a{}_c \wedge \omega^c{}_b. \tag{3.39}$$

Equations (3.38) and (3.39) are called Cartan's structure equations. T^a is a Lorentz vector. $R^a{}_b$ transforms linearly

$$(R'^a{}_b) = L^a{}_c R^c{}_d (L^{-1})^d{}_b \tag{3.40}$$

but $\omega^a{}_b$ does not

$$\omega'^a{}_b = L^a{}_c \omega^c{}_d (L^{-1})^d{}_b + L^a{}_c d(L^{-1})^c{}_b \tag{3.41}$$

in a manner entirely similar to the field strength and gauge field in gauge theories. It is easy to verify that, for an arbitrary vector 1 form ζ^a

$$\begin{aligned} D^2\zeta^a &= d(D\zeta^a) + \omega^a{}_b \wedge D\zeta^b \\ &= R^a{}_b \wedge \zeta^b \end{aligned} \tag{3.42}$$

and that

$$\left.\begin{aligned} DR^a{}_b &= 0 \\ DT^a &= R^a{}_b \wedge e^b \end{aligned}\right\} \quad \text{Bianchi identities.} \tag{3.43}$$

The e^a form a basis of 1-forms so we can also expand R and T

$$T^a \equiv T^a_{bc} e^b \wedge e^c \tag{3.44a}$$

$$R^a{}_b \equiv \frac{1}{2} R^a_{bcd} e^c \wedge e^d. \tag{3.44b}$$

<u>Example</u>

Two sphere :

The metric is

$$\begin{aligned} ds^2 &= r^2 d^2\theta + r^2 \sin^2\theta \, d\phi^2 \\ &= (e^1)^2 + (e^2)^2 \end{aligned}$$

so $e^1 = r \, d\theta$ $\qquad$ $e^2 = r \sin\theta \ d\phi$.

Since $\omega_{ab} = -\omega_{ba}$ only ω_{12} need be calculated.

$$de^1 = \frac{\partial r}{\partial \phi} d\phi \wedge d\theta = 0 \qquad de^2 = \frac{\partial r \sin\theta}{\partial \theta} d\theta \wedge d\phi$$

$$= r \cos\theta \; d\theta \wedge d\phi.$$

We know that $De^1 = De^2 = 0$ so

$$\omega^1{}_2 \wedge e^2 = 0 \qquad -\omega^2{}_1 \wedge e_1 = r \cos\theta \; d\theta \wedge d\phi$$

i.e. $\omega^1{}_2 = -\cos\theta \; d\phi$

and the curvature

$$R^1{}_2 = d\omega^1{}_2 + \omega^1{}_2 \wedge \omega^2{}_1$$

$$= R^1{}_{212} e_1 \wedge e_2 = \frac{1}{r^2} e_1 \wedge e_2.$$

So far out treatment has focussed entirely on the tangent space. Under local frame rotations $e^a \to L^a{}_b(x) e^b$, we have demanded that D transform covariantly.

We may focus instead on the coordinate frame basis and define a connection 1-form

$$\Gamma^\alpha{}_\beta = \Gamma^\alpha{}_{\beta\gamma} dx^\gamma \tag{3.45}$$

so that

$$\nabla = d + [\Gamma, \;] \tag{3.46}$$

transforms as a covariant derivative acting on (coordinate) tensor valued forms $\Sigma^{\mu_1 \cdots}_{\nu_1 \cdots}$

The curvature two form can be defined in terms of the connection (ω is called the spin connection and Γ the Christoffel connection) [4].

$$R^\alpha{}_\beta = d\Gamma^\alpha{}_\beta + \Gamma^\alpha{}_\sigma \Gamma^\sigma{}_\beta = \frac{1}{2} R^\alpha_{\beta\sigma\tau} dx^\sigma \wedge dx^\tau \tag{3.47}$$

from which it follows that

$$\nabla \wedge \nabla = [R, \;] \tag{3.48}$$

i.e. parallel transport of a tensor around an infinitesimal area is equivalent to the commutator of the tensor with the curvature tensor (d gives infinitesimal shifts, but D or ∇ give infinitesimal shifts by parallel transport, i.e. the connection tells us how the frames are rotated when we move along a trajectory on the manifold). $\Gamma^\alpha{}_\beta$ and $\omega^a{}_b$ can be expressed in terms of each other using the vielbein

$$\Gamma^\alpha{}_\beta = E_a{}^\alpha \omega^a{}_b e^b{}_\beta + E_a{}^\alpha de^a{}_\beta. \tag{3.49}$$

The key point is that, for any Riemannian manifold, the connection is unique provided we require

i) the metric be covariantly constant and

ii) the connection have zero torsion, i.e.

$$0 = \nabla_\alpha g_{\mu\nu} = \partial_\alpha g_{\mu\nu} - g_{\lambda\nu} \Gamma^\lambda_{\alpha\mu} - g_{\mu\lambda} \Gamma^\lambda_{\alpha\nu} \tag{3.50a}$$

$$0 = T^\alpha = \frac{1}{2} T^\alpha_{\beta\gamma} dx^\beta \wedge dx^\gamma. \tag{3.50b}$$

The analogue of (3.50a) in tangent space is the antisymmetry of ω_{ab} which follows from the fact that η_{ab} should be left invariant as we move on the manifold.

We stated earlier that the curvature's commutator with a tensor was a measure of the tensor's change when parallel transported. If we parallel transport a vector around some closed loop on a manifold, it generally will not coincide with the original vector. The parallel transported vector is rotated by a linear transformation which we call holonomy. The group of such transformations is of course called the holonomy group. As an example, parallel transporting a vector in a flat plane leaves the vector unchanged while parallel transporting a vector on the surface of a sphere gives the original vector rotated by an angle which depends on the path. The holonomy group is O(2).

ANOMALIES (chapter 4)

Introduction (section a)

One of the key ingredients in superstring theory is the remarkable way in which anomalies cancel. Since this forms the basis for our selection of the viable gauge groups, we will enter here into a brief discussion of anomalies, referring the reader to lengthier articles for the needed fuller treatment. We have gauge field anomalies [1], gravitational anomalies [2] and mixed ones [3] . We begin by discussing the first of these three kinds.

We defined in (3.21) and (3.22) the matrix valued 1-form gauge field and the 2-form field strength F. Similarly we can define an infinitesimal matrix valued zero form gauge transformation v(x). The 1-form

$$Dv \equiv dv + [A,v] = -\delta_v A \tag{4.1}$$

is just the gauge transformation of A, which we call $\delta_v A$. $\delta_v F$ is then given by

$$\delta_v F = - [F,v]. \tag{4.2}$$

The axial $U_A(1)$ anomaly can be shown to be proportional, for 2n dimensions, to

$$\Omega_{2n}(A) \equiv tr(F^n) \tag{4.3}$$

where F^n really means F F...F n times (it is clear that F^n is a 2n-form since F is a 2-form). The form is dictated by invariance requirements; it has dimension 2n, is gauge invariant and odd under parity and time reversal. $\Omega_{2n}(A)$ is known as the n'th Chern form (sometimes also called the Chern character). It is a term in the expansion of the Chern character (or total Chern form) [4]

$$ch(F) \equiv tr\, e^{iF/2\pi}. \tag{4.4}$$

The Chern forms are local forms on the manifold M, constructed from field strength 2-forms, or analogously for gravity from curvature 2-forms, whose integrals over the manifold are sensitive to non-trivial topologies. $\Omega_{2n}(A)$ is gauge invariant as can be easily seen, and is a closed form

$$d\Omega_{2n} = d\ \mathrm{tr}\ F^n = n\mathrm{tr}\ (dFF^{n-1})$$

$$= n\ \mathrm{tr}\ \{(dF + AF - FA)F^{n-1}\} \tag{4.5}$$

$$= n\ \mathrm{tr}\ \{(DF)F^{n-1}\} = 0$$

where technically by trace we really mean supertrace Str

$$\mathrm{Str}(B_1 \ldots B_k) = \frac{1}{k!} \sum_{(i_1 \ldots i_k)} \mathrm{Tr}(B_1 \ldots B_k) \tag{4.6}$$

the sum being over all permutations $(i_1 \ldots i_k)$ of $(1 \ldots k)$. In (4.5) we have used the cyclic property of traces and the fact that F is an even form. As we saw, following equations (3.6) and (3.7), $d\Omega_{2n} = 0$ means that Ω_{2n} is locally exact (This is Poincaré's lemma).

$$\Omega_{2n}(A) = d\omega^{o}_{2n-1}(A) \tag{4.7}$$

where ω^{o}_{2n-1} is known as the Chern-Simons form. If (4.7) were true globally, the integral of Ω_{2n} over the manifold would, by Stokes' theorem equal the integral of $\omega^{o}_{2n-1}(A)$ over the boundary ∂M and hence would be zero if ∂M was empty. In general (4.7) is not true globally. A curved manifold requires a covering by several patches, each one of which can be deformed to R^{2n}. The gauge fields are defined on each patch with the rule that on the intersection of adjoining patches, they are related to one another by gauge transformations. The topological information is encoded in these gauge transformations. Consider for instance the case where $F \to 0$ on ∂M, as occurs with four dimensional theories with spatial infinity being the boundary. Then A goes to a pure gauge so we have

$$\begin{aligned} & A \to w = g^{-1}dg \\ & F \to 0 \Rightarrow dA \to -A^2 = -w^2 \\ & \omega^{o}_{2n-1} \to (\ldots)\mathrm{tr}\ w^{2n-1} \\ & \int_M \mathrm{tr}\ F^n \to (\ldots) \int_{\partial M} \mathrm{tr}\ w^{2n-1} \end{aligned} \tag{4.8}$$

where (...) denotes numerical factors and M is a 2n dimensional manifold so ∂M is a 2n-1 dimensional manifold.

More generally for g(x) elements of a simple lie group, $w = g^{-1}dg$ is a 1-form. On an n dimensional manifold tr w^n is trivially closed, but not necessarily globally exact. On $M = S_n$ the n dimensional sphere we define

$$Q = \int_{S_n} \mathrm{tr}\ w^n. \tag{4.9}$$

If Tr w^n is globally exact, i.e. Tr $w^n = d\rho$, then by Stokes' theorem $Q = \int_{\partial S_n} \rho$ and hence $Q = 0$. Since $\partial S_n = 0$, we then say the homotopy group $\pi_n(G)$ is trivial, i.e. the topology is basically trivial. As an example for U(1) we have $g = e^{i\theta}$, $w = \partial_\mu \theta\ dx^\mu$ and letting θ go from 0 to 2π, we see that w fails to be exact at the boundary

$$Q = \int_{S_1} \frac{\partial\theta}{\partial x}\, dx = 2\pi. \tag{4.10}$$

Non-Abelian Anomalies. (section b)

We have already seen that the Abelian $U_A(1)$ anomaly in 2n dimensions is proportional to $\Omega_{2n} = \mathrm{tr}\, F^n$. A sophisticated and non-trivial argument, which we only sketch [1], shows that the non-Abelian anomalies in 2n dimensions can be derived from Ω_{2n+2}. Of course since in 2n dimensions, we have at most 2n-forms, Ω_{2n+2} is defined only formally (no pun intended). The Bianchi identities ensure that Ω_{2n+2} is closed, eq. (4.5). We then have that locally

$$\Omega_{2n+2} = d\omega^{(o)}_{2n+1}(A). \tag{4.11}$$

Since Ω_{2n+2} is gauge invariant, we have

$$\delta_v[d\omega^{(o)}_{2n+1}] = d[\delta_v\omega^{(o)}_{2n+1}] = 0 \tag{4.12}$$

which implies the existence of a 2n-form linear in the infinitesimal quantity v

$$d\omega^{(1)}_{2n}(v,A) = \delta_v\omega^{(o)}_{2n+1}. \tag{4.13}$$

Now we assume there is no topological obstruction to extending our 2n dimensional gauge fields to a 2n+1 dimensional S_{2n+1} whose boundary is S_{2n}. We then have

$$\int_{S_{2n}} \omega^{(1)}_{2n}(A) = \int_{S_{2n+1}} \delta_v\omega^{(o)}_{2n+1} \tag{4.14}$$

and, after some work, we can show that the non-Abelian anomalies $G_i(x)$ are given by

$$G(v,A) \equiv \int v^i(x)G_i(x). \tag{4.15}$$

To summarize, the procedure we follow is to first find the Chern forms Ω_{2n}; using them we construct the Chern-Simons forms and finally find the non-Abelian anomalies.

As a final comment in this section, we point out that in differential forms the divergence of a current is obtained by applying the current 1-form J the operator d*; * is defined in (3.8) so e.g. in 2n dimensions

$$d{*}J = \partial_\lambda J^\lambda \frac{1}{2n!} \varepsilon_{\mu_1\ldots\mu_{2n}} dx^{\mu_1}\cdots dx^{\mu_{2n}}. \tag{4.16}$$

J is a 1-form; *J is a 2n-1 form and d*J is a 2n form. If J is the $U_A(1)$ Abelian current

$$d{*}J = (\text{constant})\Omega_{2n} \tag{4.17}$$

is the anomaly equation.

String Anomalies. (section c)

As we saw in Chapter 2, the fermionic string is formulated as a theory with ten dimensions. Hence we are forced to consider anomalies in a field theory formulated on a ten-dimensional manifold. Furthermore since anomalies are basically long wavelength phenomena, namely break-downs of gauge invariance and of local Lorentz invariance, the absence of anomalies in ten dimensions will ensure that the theory is anomaly free after compactification [6]. We also believe that the anomalies of the string are the same as those of the field theory so let us concentrate on the latter.

First of all, we observe that in 2n dimensions, anomalies can all be related to those of the n+1 polygon, i.e. triangles in four dimensions and hexagons in ten dimensions so we have diagrams of the type

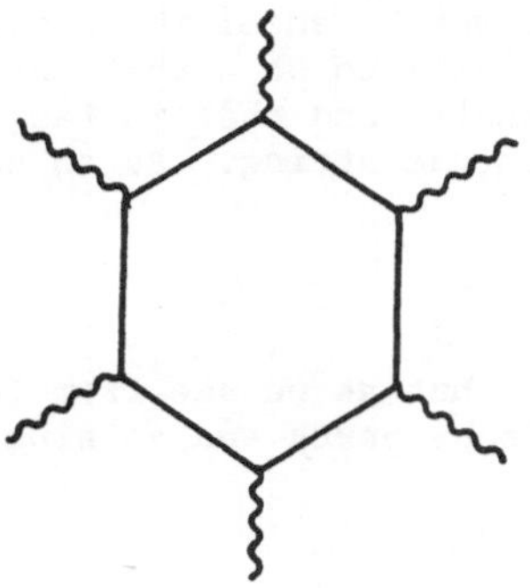

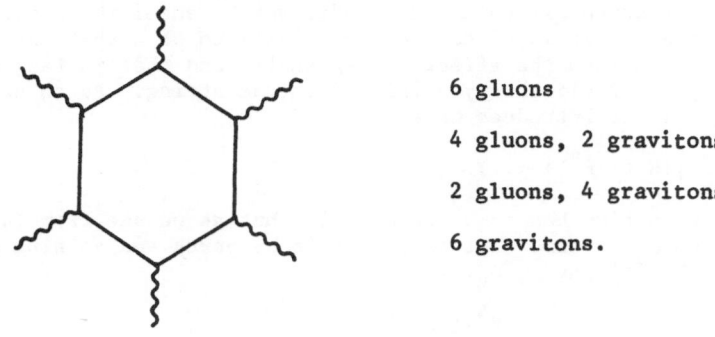

6 gluons

4 gluons, 2 gravitons

2 gluons, 4 gravitons

6 gravitons.

We saw in the previous section that the ten-dimensional anomalies lead us to consider formally in twelve dimensions the twelve-form

$$\Omega_{12} = \mathrm{TR}\, F^6. \tag{4.18}$$

Now the trace taken here is for the matrices in the adjoint representation of the group. We will almost always wish to evaluate traces in the fundamental representation, in which case we use the notation

$$\mathrm{Tr}\, F^n. \tag{4.19}$$

There are several keys to the anomaly cancellation [5]. One of them is the existence in the ten-dimensional field theory of an antisymmetric tensor 2-form field B (see next section), which is not gauge invariant.

It's field strength, H, is defined by

$$H = dB - \omega^o_{3,y} - \omega^o_{3,L} \tag{4.20a}$$

$$H = dB - \frac{1}{30}(\omega^o_{3,y_1} + \omega^o_{3,y_2}) - \omega^o_{3,L} \tag{4.20b}$$

where ω^o_3's are the Chern-Simons forms discussed in the previous section. The subscripts y refer to Yang-Mills fields and L to Lorentz fields.

$$\omega^o_{3y} = \mathrm{Tr}(AF - \frac{1}{3}A^3) \tag{4.21a}$$

$$\omega^o_{3L} = \mathrm{Tr}(\omega R - \frac{1}{3}\omega^3). \tag{4.21b}$$

Equation (4.20a) is the form for SO(N) and (4.20b) for $E_8 \times E_8$ (note in this case the adjoint and fundamental representations coincide. If we used matrices in the adjoint representation of SO(32), there would be a 1/30 in (4.20a) as well). In (4.21b) ω is the connection and R the curvature 2-form, as defined in (3.35) and (3.39).

The anomalies lead to effective interactions of the form

$$\begin{aligned}&\mathrm{tr}\,F^6,\ (\mathrm{tr}\,F^4)(\mathrm{tr}\,F^2),\ \mathrm{Tr}\,R^6\\&(\mathrm{tr}\,R^4)(\mathrm{tr}\,R^2),\ (\mathrm{tr}\,F^4)(\mathrm{tr}\,R^2),\ \text{etc.}\end{aligned} \tag{4.22}$$

Interactions which are of the form of a product of two traces can be cancelled by adding appropriate B dependent terms to the effective action. Basically these are pole diagrams in which a B particle is exchanged leading to effective six point amplitudes which cancel the anomalous hexagon diagrams. It is of course non-trivial to show that these necessary terms can be put into the effective Lagrangian and that in fact they are generated by the field theory arising from the string. As an example of what we need to introduce consider

$$\Delta S = c \int (B\ \mathrm{tr}\,F^4 + \ldots) \tag{4.23}$$

The effective action has an H^2 term in it, but as we see from (4.20), H^2 leads to a dB $\omega^o_{3,y}$ term. Integrating this by parts and joining the two pieces by a B propagator, D_B, we find

$$\mathrm{tr}\,F^4\, D_B\, d\omega^o_{3,y} = (\mathrm{tr}\,F^4)(\mathrm{tr}\,F^2)D_B, \tag{4.24}$$

i.e. we can generate the necessary interactions. This has been shown by Green and Schwartz [5] to occur for gauge groups SO(32) and $E_8 \times E_8$.

The tr F^6 and tr R^6 anomalies cannot however be cancelled in this manner, i.e. by pole diagrams. Something else has to happen and it does for SO(32) and $E_8 \times E_8$. For SO(N), tr F^6 appears with coefficient

$$\Omega_{12} = (N-32)\mathrm{tr}\,F^6 + 15\mathrm{tr}\,F^2\ \mathrm{tr}\,F^4 \tag{4.25}$$

so it vanishes for N = 32. The case of E_8 is even simpler; there are no linearly independent sixth order or fourth Casimir invariants. Ω_{12} is proportional to $(\mathrm{tr}\,F^2)^3$ and Ω_8 to $(\mathrm{tr}\,F^2)^2$.

The gravitational anomaly can be derived from the formal 12-form

$$\left\{\frac{1}{32} + \frac{n-496}{13824}\right\} (\mathrm{tr}\,R^2)^3 + \left\{\frac{1}{8} + \frac{n-496}{5760}\right\} \mathrm{tr}\,R^2\ \mathrm{tr}\,R^4 + \left\{\frac{n-496}{7560}\right\} \mathrm{tr}\,R^6 \tag{4.26}$$

where we have included one left handed spin 3/2 gravitino, one right handed spin one half field and n left handed spin 1/2 fields. These latter are the superpartners of the gauge fields in the gauge supermultiplet. As we already stated, the dangerous term in (4.26) is the last one. The others can be cancelled by adding additional local terms to the action.

These are local gauge variant terms which will cancel the anomalies. They supposedly originate from taking the low energy limit of the superstring. The leading term however cannot be cancelled in such a manner. It must vanish. In our case what is n ? It equals of course the number of gauge bosons; for SO(32) this is $\frac{1}{2}(32 \times 31) = 496$ and for E_8 it is 248 so for $E_8 \times E_8$ it is also 496. End of argument. All anomalies are products of traces.

We have not talked at all about global anomalies. We refer the reader to existing discussions [7].

Ten Dimensions. (section a)

We saw, however, briefly in section II, that a consistent superstring theory could be formulated in ten dimensions. It seems likely that a formulation in terms of an effective ten dimensional field theory should be possible. The massless degrees of freedom in such a field theory should be the supergravity multiplet, present as excitations of the closed string, the gauge super-multiplet from the open string and other possible fields as required. We are led therefore to consider ten-dimensional supergravity coupled to a Yang-Mills supermultiplet [1]. There will certainly be differences, i.e. additional interactions whose origin is "stringy" constrained in ways we cannot determine by the field theory. On the other hand, our intuition and understanding of field theory at this point is far greater than that of string theory so it seems to be a worthwhile guide. One main difference, of course, is that the string theory is finite while the field theory is not. We shall use the field theory not so much as a calculational tool, but rather as a guide.

Cremmer, Julia and Scherk [2] established the existence of a local N = 1 supergravity in eleven dimensions. Scherk [3] later showed that this theory could be truncated consistently to give an N = 1 supergravity in ten dimensions. Later, work by Cremmer and Julia [4], Chamseddine [5], Bergshoff et al. [6], Chapline and Manton [7] and others showed this could be consistently coupled to supersymmetric Yang-Mills.

We start therefore with local supergravity in eleven dimensions. The fields in the multiplet are an elfbein e_M^A (see section 3.b) where $A = 1,2,\dots 11$ and $M = 1,2,\dots 11$, a single gravitino since this is N = 1 supergravity ψ_M, which is however a 32 dimensional spinor (see later section on spinors) and an antisymmetric tensor A_{MNP}. The gravitino is a Majorana spinor for which

$$\bar{\psi}_M = \psi_M^T C \qquad C\Gamma_M C^{-1} = -\Gamma_M^T. \tag{5.1}$$

We can use the local SO(1,10) symmetry to write the elfbein as

$$e_M^A = \begin{pmatrix} \tilde{e}_m^a & \tilde{e}_m^{11} \\ & \\ 0 & \tilde{e}_{11}^{11} \end{pmatrix} \tag{5.2}$$

and similarly with ψ_M and A_{MNP} as

$$\begin{aligned} \psi_M &= (\psi_m, \psi_{11}) \\ A_{MNP} &= (A_{mnp}, A_{mn11}) \end{aligned} \tag{5.3}$$

with of course a, m, n, p = 1,2,...10. Special properties of ten dimensions allow us to impose simultaneously Majorana and Weyl conditions on spinors. We need to do this because if we keep all the fields in (5.3), we will have an N = 2 supergravity in ten dimensions (i.e. two gravitini). The consistent truncation is to impose

$$e_m^{11} = 0 \qquad A_{mnp} = 0 \qquad \dots \tag{5.4}$$

$$\psi_m^R \equiv (1 - \gamma_{11})\psi_m = 0$$

$$\ldots \quad \psi_{11}^L \equiv (1 + \gamma_{11})\psi_{11} = 0 \qquad \psi_{11}^R \equiv \lambda. \tag{5.4}$$

For consistency the constant spinor used in supersymmetry transformation ε, e.g.

$$\delta e_M^A = \frac{\varepsilon}{2}\Gamma^A\psi_M \tag{5.5}$$

must also be right handed since $\delta\psi_M^R = \partial_\mu\varepsilon^R + \ldots$ and $\delta\psi_M^R = 0$. In the above γ_{11} is the generalization of γ_5 in 4 dimensions. The gamma matrices in eleven dimensions will be labelled as Γ^M

$$i\gamma_{11} = \Gamma^{11} = -\Gamma_{11} \qquad (\gamma_{11})^2 = 1 \qquad (\Gamma_{11})^2 = -1 \tag{5.6}$$

(see later section on spinors for a discussion). After this truncation, we are left with a zehnbein (zehnbein means ten legs in German, vierbein four legs), a left handed gravitino, a right handed spinor ψ_{11}, an antisymmetric tensor and a scalar field ϕ.

An additional advantage of the choice (5.2) for the elfbein is that its determinant factorizes into

$$\tilde{e} = \det(\tilde{e}_m{}^a)\tilde{e}_{11}{}^{11}. \tag{5.7}$$

This allows us to rescale the kinetic terms for the Einstein action in ten dimensions so that they will have the canonical form by a Weyl rescaling of the metric. In particular, with $e_{11}{}^{11} = \phi$ we see that rescaling $\tilde{e}_m^a$ by

$$\tilde{e}_m^a = \hat{e}_m^a\,\phi^{-(1/d-2)}, \tag{5.8}$$

and hence $\tilde{g}^{mn}$ by $\phi^{2/d-2}$, has the desired effect, namely

$$\begin{aligned}\tilde{e}\tilde{g}^{mn}\partial_m\partial_n &\to \phi\phi^{-(d/d-2)}\phi^{2/d-2}\hat{e}\hat{g}^{mn}\partial_m\partial_n \\ &= \hat{e}\hat{g}^{mn}\partial_m\partial_n,\end{aligned} \tag{5.9}$$

for d dimensions; in our case d = 10. In addition to the canonical Einstein action, there is a kinetic energy term for

$$-\frac{1}{2}\left(\frac{d-1}{d-2}\right)\left(\frac{\partial_\mu\phi}{\phi}\right)^2. \tag{5.10}$$

One also rescales the gravitino spinor field by $\phi^{-(1/2(d-2))} = \phi^{-1/16}$ so the Rarita Schwinger action is canonical, and so on. Full details are presented in the earlier keys papers [2-7].

<u>Supergravity - Super Yang-Mills.</u> (section b)

After the Weyl rescaling of the metric the supergravity Lagrangian takes the form (see section 6 for a discussion of SO(1,9) spinors)

$$\frac{\mathcal{L}}{\hat{e}} = -\frac{1}{2}R - \frac{1}{2}i\ \bar{\psi}_m\Gamma^{mnp}D_n\psi_p + i/2\ \bar{\lambda}\Gamma^m D_m\lambda$$

$$+ \frac{g}{16}\left(\frac{\partial_\mu\phi}{\phi}\right)^2 + \frac{3}{4}\phi^{-3/2}\ H_{mnp}H^{mnp}$$

$$+ \frac{3}{8}\sqrt{2}\ \bar{\psi}_m \frac{\partial\!\!\!/\phi}{\phi}\Gamma^m\lambda - \tag{5.11}$$

$$-\frac{\sqrt{2}}{16}\phi^{-3/4}\ H_{mnp}(i\ \bar{\psi}_s\Gamma^{smnpr}\psi_r + 6i\ \bar{\psi}^m\Gamma^n\psi^p$$

$$+ \sqrt{2}\ \psi_s\ \Gamma^{mnp}\Gamma^s\lambda) + \text{four fermi interactions}$$

where λ is the right-handed spinor field ψ^R_{11} and H_{mnp} is the field strength of the antisymmetric tensor $B_{mn} = A_{mn11}$. Note that the H_{mnp} kinetic term does involve the ϕ field as well. We now couple in the super Yang-Mills fields $(A^\alpha_m, \chi^\alpha)$ where α is the gauge field index, A^α_m are gauge fields and χ^α left handed gauginos. The terms one adds to (5.11) are

$$-\frac{1}{4}\phi^{-3/4}\ F^\alpha_{mn}F^{\alpha mn} + \frac{i}{2}\bar{\chi}^\alpha\Gamma^m(D_m\chi)^\alpha$$

$$-\frac{i}{4}\kappa\phi^{-3/8}(\bar{\chi}^\alpha\Gamma^m\Gamma^{np}F^\alpha_{np})(\psi_m + \frac{i\sqrt{2}}{12}\Gamma_m\lambda) \tag{5.12}$$

$$+ \frac{i\sqrt{2}}{16}\phi^{-3/4}\ H_{mnp}\ \bar{\chi}^\alpha\Gamma^{mnp}\chi^\alpha.$$

In (5.11) and (5.12) D_m is of course the covariant derivative. In obtaining the proper scaling factors remember that H_{mnp} is really the field strength of a third rank antisymmetric tensor in eleven dimensions, i.e. B_{mn} equals A_{mn11} in equation (5.3). The field strength H has turned out to be the focus of much interest. Chapline and Manton [7] realized that H = dB, in the language of differential forms, had to be modified by introducing the Chern-Simons form, cf. equation (4.21a) :

$$H = dB - \omega^o_{3,y} \tag{5.13}$$

in order to preserve supersymmetry. Green and Schwartz [8] showed that for superstrings an additional term must be present in H, cf. eqs. (4.20a,b), (4.21a,b)

$$H = dB - \omega^o_{3,y} - \omega^o_{3,L} \tag{5.14}$$

in order to have cancellation of anomalies. Superficially, the addition of the Lorentz Chern-Simons form in H seems to violate supersymmetry. It is hoped/assumed that other terms present in the effective Lagrangian arising from the string will restore supersymmetry. Since the string itself is supersymmetric, it is plausible that the effective string theory should be as well. We will return in our final chapter to the discussion of supersymmetry breaking. In Chapter 7, we will see the importance of the modification of H.

Reduction to Four Dimensions. (section c)

Jumping ahead for a moment, we repeat the arguments of section a) to reduce from d = 10 to d = 4. Once again we have the local Lorentz invariance to bring the zehnbein to the form in which its determinant will factorize

$$\hat{e}_m^a = \begin{pmatrix} \hat{e}_\mu^\alpha & \hat{e}_\mu^s \\ & \\ 0 & \hat{e}_i^j \end{pmatrix} \qquad \begin{matrix} \mu,\alpha = 1,2,3,4 \\ \\ i,j = 5,6,\ldots 10 \end{matrix} \tag{5.15}$$

where $\hat{e}_\mu^\alpha$ is now the vierbein. Again we do a Weyl rescaling of the metric; if we put det $\hat{e}_i^j = \Delta^{1/2}$, the Weyl rescaling is by

$$(\Delta^{1/2})^{-\frac{1}{4-2}} = \Delta^{-1/4}$$
$$\hat{e}_\mu^\alpha = e_\mu^\alpha \, \Delta^{-1/4} \tag{5.16}$$

where e_μ^α is the vierbein we actually use from now on.

Now let us count the fields we are left with in four dimensions. What we mean by this is that we classify fields by their four dimensional indices $\mu,\nu,\lambda\ldots$. Additional indices in the six dimensional compactified space (technically speaking we haven't yet talked about compactification, but let's jump ahead) are seen simply as scalar indices in four dimensions. We see as an example that e_μ^α leads to the graviton, but e_μ^i gives six vector fields (i = 5,6,7,8,9,10) and e_i^j leads to twenty one scalars, namely $\frac{6\times 7}{2}$ since indices are symmetrized. To summarize from the boson fields, simply relabeling A_{mn11} as A_{mn}

$$e_m^a \to \begin{cases} e_\mu^\alpha & : \text{graviton spin 2} \\ e_\mu^i & : \text{6 vectors} \\ e_i^j & : \text{21 scalars} \end{cases}$$

$$B_{mn} \to \begin{cases} B_{\mu\nu} & : \text{one scalar} \\ B_\mu^i & : \text{6 vectors} \\ B_{ij} & : \text{15 scalars} \end{cases} \tag{5.17}$$

$$\phi \to \phi \; : \text{scalar.}$$

In (5.13), there are fifteen scalars B_{ij} since we antisymmetrize i and j. $B_{\mu\nu}$ is a four dimensional scalar; in fact as we shall see later it is an axion, coupling to $F\wedge F$. At this point we have not yet distinguished parities, so we call $B_{\mu\nu}$ a scalar though it really is a pseudoscalar.

How many fermions are there ? As we said earlier, in ten dimensions, our spinors are both chiral and Majorana. They are $2^5 = 32$ dimensional objects, restricted as we just said by the Majorana and Weyl conditions. Every 32 dimensional Majorana-Weyl spinor gives four independent spinors in four dimensions (see discussion in section 6). To (5.17) we then append the fermionic degrees of freedom

$$\psi_m^L \rightarrow \begin{cases} \psi^L_{\mu,a} & 4 \text{ gravitini} \\ \psi^L_{s,a} & 4 \times 6 \text{ (values of s) spin } 1/2 \end{cases}$$

$$\psi_{11}^R = \lambda^R \rightarrow \{ \lambda,a \quad 4 \text{ spin } 1/2$$

$$a = 1,2,3,4 \tag{5.18}$$

where the index just means we have four distinct spinors in four dimensions.

The fact that we have four distinct gravitini in four dimensions means that we have N = 4 supergravity in four dimensions. We want N = 1 supergravity! We also need a means of removing most of the degrees of freedom in the four dimensional supergravity multiplet from the visible spectrum. Both our goals will hopefully be met by the process of compactification. So far we have not said anything about how the tendimensional space-time shrinks down to Minkowski space M_4 × compact manifold K. As a warm-up for the more sophisticated examples, let us consider the baby case of compactifying ordinary four dimensional space by curling up the three spatial dimensions and demanding that the resulting fields only be functions of time

$$\begin{aligned} &A_{\mu_1,\mu_2\ldots\mu_N}(t,x) \\ &A_{\mu_1,\mu_2\ldots\mu_N}(t) \end{aligned} \tag{5.19}$$

so, for instance an ordinary four dimensional vector $A_\mu(x)$ would go over to four independent functions of time.

If, however we demand invariance under rotations in 3-D space, i.e. we say only the modes whose indices do not rotate are massless

$$A_\mu(x) \rightarrow A_o(t) \tag{5.20}$$

i.e. only a single function is left. If we demand invariance under an SO(2) rotation about the 3 axis $A_o(t)$ and $A_3(t)$ survive. For a symmetric tensor $\phi_{\mu\nu}$ and an antisymmetric tensor $\psi_{\mu\nu}$ the surviving modes are

$$\begin{aligned} &\phi_{00}(t), \phi_{03}(t), \phi_{33}(t),\ \phi_{xx} + \phi_{yy} \\ &\psi_{03}, \psi_{xy}. \end{aligned} \tag{5.21}$$

In fact we will see that of the four gravitini, only one survives at low energies. A naive way of seeing this is to say that the four gravitini belong to the 4 of SU(4) (or equivalently to spinor representation of SO(6)) but that the zero mass modes are those which do not rotate under an SU(3) subgroup of SU(4). Under this SU(3), 4 = 1 ⊕ 3 so only a singlet remain massless, and N = 1 supergravity is what we are left with.

Orthogonal Groups : SO(2N). (section a)

We will begin by discussing the orthogonal group SO(2N), emphasizing its relationship to SU(N).

Consider a set of N operators X_i (i = 1...N) and their Hermitean conjugates X_i satisfying the anti-commutation relations

$$\{X_i,X_j^\dagger\} = \delta_{ij}$$
$$\{X_i,X_j\} = 0. \qquad (6.1)$$

The operators

$$T_j^i = X_i^\dagger X_j \qquad (6.2)$$

are the generators of U(N)

$$[T_j^i,T_\ell^k] = \delta_j^k T_\ell^i - \delta_\ell^i T_j^k \qquad (6.3)$$

and the generators Σ_{jk} of SO(2N) satisfying the commutation relations

$$[\Sigma_{jk},\Sigma_{\ell m}] = i(-\delta_{jm}\Sigma_{k\ell} - \delta_{k\ell}\Sigma_{jm} + \delta_{j\ell}\Sigma_{km} + \delta_{km}\Sigma_{j\ell}) \qquad (6.4)$$

can be constructed also from the X_i's as follows [1]. Define a Clifford algebra of rank 2N, from the 2N Hermitean Γ_i's satisfying

$$\{\Gamma_i,\Gamma_j\} = 2\delta_{ij} \qquad (6.5)$$

The Σ's are related to the Γ's by

$$\Sigma_{ij} = \frac{1}{2i}[\Gamma_i,\Gamma_j] \qquad (6.6)$$

and finally the Γ_a's can also be constructed from the X_i's by

$$\Gamma_{2j-1} = -i(X_j - X_j^\dagger)$$
$$\Gamma_{2j} = (X_j + X_j^\dagger). \qquad (6.7)$$

We see from (6.4) that we can simultaneously diagonalize $\Sigma_{12},\Sigma_{34}, \ldots \Sigma_{2N-1,2N}$. Each Σ will have eigenvalues ± 1, so SO(2N) has a 2^N spinorial representation. This can be split into two 2^{N-1} representations using the generalization of γ_5

$$\Gamma = (i)^N \Gamma_1 \ldots \Gamma_{2N} \qquad (6.8)$$

which obviously anticommutes with all Γ_a's. This implies that Γ commutes with all the Σ_{ab}'s and hence is invariant under SO(2N) rotations. Our 2^N spinorial representation is reducible, the irreducible representations being labelled additionally by the eigenvalue of Γ. The spinorial representation can be defined in terms of the X_i's by defining a vacuum state $|0\rangle$ and then acting on it with creation operators $X_i^\dagger$. With a little work, we can show, introducing $n_j \equiv X_j^\dagger X_j$, that

$$\Gamma = \prod_{j=1}^{N} (1-2n_j) = \prod_{j=1}^{N} (-1)^{n_j}$$
$$= (-1)^n \qquad n = \sum_{j=1}^{N} n_j \tag{6.9}$$

so that $\Gamma_\pm$ is a chirality projection operator where

$$\Gamma_\pm = \frac{(1 \pm \Gamma)}{2} \tag{6.10}$$

$\Gamma_\pm$ projecting out states with an even or odd number of the X particles.

Ex. 1. SO(6)

We will use this example in the next section when we discuss compactifying the six dimensional manifold. SO(6) has an eight dimensional spinorial representation. The states can be constructed as follows :

$$|\Omega\rangle = |0\rangle \qquad i,j,k = 1,2,3$$
$$|\Omega_i\rangle = X_i^\dagger|0\rangle$$
$$|\overline{\Omega}^i\rangle = \frac{1}{2}\,\varepsilon^{ijk}\, X_j^\dagger X_k^\dagger|0\rangle \tag{6.11}$$
$$|\overline{\Omega}\rangle = \frac{1}{6}\,\varepsilon^{ijk}\, X_i^\dagger X_j^\dagger X_k^\dagger|0\rangle$$

with $|\Omega\rangle$ and $|\overline{\Omega}^i\rangle$ having one chirality and $|\overline{\Omega}\rangle$ and $|\Omega_i\rangle$ having the opposite chirality.

Constructing the generators of SU(3) using $X_i^\dagger$ and X_i as in (6.2), we see that $|\Omega\rangle$ and $|\overline{\Omega}\rangle$ are both singlet states while $|\Omega_i\rangle$ and $|\overline{\Omega}^i\rangle$ transform like the 3 and $\overline{3}$ of SU(3). $|\overline{\Omega}\rangle$ and $|\Omega^i\rangle$ have opposite chirality (eigenvalue of Γ) while $|\Omega\rangle$ and $|\overline{\Omega}_i\rangle$ have negative chirality. In fact, SO(6) is isomorphic to SU(4) and the positive chirality spinors are the fundamental 4 representation of SU(4) while the negative chirality spinors are the $\overline{4}$ representation.

In the previous example we saw how to construct representations for SU(N) using the generators of the Clifford algebra for SO(2N). The spinorial representations of SO(2N) can also be constructed directly in the basis where we simultaneously diagonalize $\Sigma_{12}, \Sigma_{34}, \Sigma_{56}\cdots\Sigma_{2N-1,2N}$. The 2^N dimension basis can be labelled as

$$|\varepsilon_1\ \varepsilon_2\ \varepsilon_3\cdots\varepsilon_N\rangle \tag{6.12}$$

where $\varepsilon_i = \pm 1$; alternatively we can view the space as a product of N two component spinors. For N = 1, we can choose [2]

$$\Gamma_1 = \tau_2 \qquad \Gamma_2 = \tau_1$$
$$\Sigma_{12} = \frac{1}{2i}[\tau_2,\tau_1] = -\tau_3 \tag{6.13}$$

where of course τ_3 has eigenvalues ± 1 in a basis where it is diagonal. For SO(2) we would then write the basis vector and Clifford algebra as

$$|\varepsilon_1\rangle \text{ and } \tau_2,\tau_1.$$

This can be extended to SO(2N) by the labeling (6.12) of the states and having

$$\Gamma_{2K-1} = \underbrace{1 \times 1 \times 1 \ldots \times}_{K-1 \text{ times}} \tau_2 \underbrace{\times \tau_3 \times \tau_3 \times \tau_3}_{N-K \text{ times}} \cdots$$

$$\Gamma_{2K} = \underbrace{1 \times 1 \times 1 \ldots \times}_{K-1 \text{ times}} \quad \tau_1 \underbrace{\times \tau_3 \times \tau_3 \times \tau_3}_{N-K \text{ times}} \cdots \tag{6.14}$$

so, e.g. for SO(6) we have

$$\begin{array}{lll} \Gamma_1 = \tau_2\tau_3\tau_3 & \Gamma_2 = \tau_1\tau_3\tau_3 & \Gamma_3 = 1 \times \tau_2 \times \tau_3 \\ \Gamma_4 = 1 \times \tau_1 \times \tau_3 & \Gamma_5 = 1 \times 1 \times \tau_2 & \Gamma_6 = 1 \times 1 \times \tau_1 . \end{array} \tag{6.15}$$

These satisfy by inspection the Clifford algebra. We have

$$\Sigma_{2K-1,2K} = \underbrace{1 \times 1 \times 1}_{K-1 \text{ times}} \ldots (-\tau_3) \underbrace{\times 1 \times 1 \ldots 1}_{N-K \text{ times}} \tag{6.16}$$

Chirality, Charge Conjugation and Parity (section b))

We defined in (6.8) an operator Γ. In the spinorial representation of the Γ matrices we see that

$$\Gamma = (i)^N \underbrace{(\tau_2\tau_1)\,(\tau_2\tau_1) \ldots (\tau_2\tau_1)}_{N \text{ times}}$$

$$= \underbrace{\tau_3 \times \tau_3 \times \ldots \times \tau_3}_{N\text{-times}} \tag{6.17}$$

We see therefore that Γ has eigenvalues

$$\prod_{i=1}^{N} \varepsilon_i = \pm 1 \tag{6.18}$$

where we defined our spinors as right handed if the eigenvalue is -1 and left handed if it is +1.

We see from the above that if we have a spinor of SO(2N+2M), we can write its transformation properties under SO(2N) × SO(2M) as

$$|\varepsilon_1 \ldots \varepsilon_N ;\ \varepsilon_{N+1} \ldots \varepsilon_{M+1}\rangle . \tag{6.19}$$

A spinor of SO(2N+2M) contains an equal number of left and right handed spinors of SO(2N) since

$$\prod_{i=1}^{N+M} \varepsilon_i = \prod_{i=1}^{N} \varepsilon_i \prod_{i=N+1}^{M} \varepsilon_i . \tag{6.20}$$

We saw earlier that a representation of SO(2N+2M) is 2^{N+M} dimensional. We also saw that it splits up into 2^{N+M-1} dimensional representations of chirality +1 and -1 since Γ ... commutes with all $\overline{\Gamma}$'s. If we take e.g., the +1 sign, we see that under SO(2N) × SO(2M)

$$2_+^{N+M-1} \to (2_+^{N-1}, 2_+^{M-1}) + (2_-^{N-1}, 2_-^{M-1}) \tag{6.21}$$

and similarly for the negative chirality states.

We can also define a group charge conjugation C so that $\phi^T C\phi$ is invariant by :

$$C^{-1}\Sigma_{ij}^T C = -\Sigma_{ij}. \tag{6.22}$$

Using the definition of Σ_{ij} in terms of Γ matrices we see that we must have

$$C^{-1} \Gamma_i^T C = \pm\Gamma_i \tag{6.23}$$

in order for (6.22) to be satisfied; we will choose the negative sign in (6.23). From (6.14) we see (since $\tau_2^T = -\tau_2$ and $\tau_{1,3}^T = \tau_{1,3}$)

$$\begin{aligned} \Gamma_{2K-1}^T &= -\Gamma_{2K-1}^T \\ \Gamma_{2K}^T &= \Gamma_{2K}^T \end{aligned} \tag{6.24}$$

and therefore we can define C by

$$\begin{aligned} C &= \prod_{i\ \mathrm{odd}} \Gamma_i \\ &= \tau_2(-i\tau_1)\tau_2(-i\tau_1)\dots . \end{aligned} \tag{6.25}$$

Since both τ_1 and τ_2 flip a state with $\varepsilon_i = \pm 1$ to one with $\varepsilon_i = \mp 1$, i.e. an eigenstate of τ_3 with eigenvalue ±1 goes into one with eigenvalue ∓1, we see that C reverses the sign of all ε_i. We have already seen that the spinorial representation of SO(2N) splits into two 2^{N-1} representations, which we call S^+ and S^- corresponding to the two eigenvalues of Γ. Thus C acting on a state in S^+ may or may not take it into a state of S^+ depending on the eigenvalue of Γ.

Parity, on the other hand, takes $S^\pm$ into $S^\mp$. In ordinary four dimensional space, under parity, $\vec{x} \to -\vec{x}$, $t \to t$ and spinors are transformed by γ_o. In 2N dimensional space, reflection about the K'th axis takes spinorial states into Γ_K times themselves.

$$P = \Gamma_K \qquad \Gamma_K\Gamma_i\Gamma_K = \begin{cases} -\Gamma_i & i \neq K \\ \Gamma_i & i = K. \end{cases} \tag{6.26}$$

Γ_K changes the eigenvalue ε_K into $-\varepsilon_K$, leaving all other ε's unchanged. Since, by (6.18) the eigenvalues of Γ are given by the product of all ε_i's, we see that under parity

$$S^+ \leftrightarrow S^-. \tag{6.27}$$

If we adjoin parity to SO(2N) to form O(2N), the irreducible spinorial representations are clearly $S^+ \oplus S^-$.

It is easy to generalize to $SO(2N-1)$: use the first $2N-1$ Γ matrices. The spinorial representation is 2^{N-1} dimensional and irreducible.

As a final comment, note we could interchange τ_2 and τ_1 in the definitions in (6.13), i.e. have

$$\Gamma_1 = \tau_1 \qquad \Gamma_2 = \tau_2$$
$$\Sigma_{12} = \frac{1}{2i}[\tau_1,\tau_2] = \tau_3. \tag{6.28}$$

For consistency, we would then need to change $(i)^N$ to $(-1)^N$ in (6.17) and have C be the product of all Γ_i with i even.

$\underline{E_6}$. (section c)).

We will say very little about E_6 [3]. There are numerous discussions of it as a group for phenomenology; we will give a few references to this in section 9. E_6 is a group of rank six with 78 generators. It has three maximal subgroups with rank six, which are $SU_3 \times SU_3 \times SU_3$, $SU_2 \times SU_6$ and $SO_{10} \times U_1$. We will not give here the Dynkin diagrams, postponing some of that material to section 9.

The adjoint representation of E_6 is of course 78 dimensional, and the fundamental representation is 27 dimension and complex, i.e. $27 \neq \overline{27}$. The Lie algebra of E_6 can be written conveniently in a SO(10) decomposition [4] involving the Σ_{ij} generators of SO(10), a U(1) generator Y and sixteen complex generators X_a with their conjugates $\overline{X}^a$ transforming like the spinorial representation of SO(10) [5].

CHIRAL FERMIONS (chapter 7)

Introduction (section a)

One of the triumphs of superstring theory has been to have a plausible theory with $d = 4$ chiral fermions. In the compactification of $d = 10$ theories, there are several no-go theorems on this subject, so the eluding of them, which we will discuss in sections seven and eight, is no mean feat. Much of the work in this subject arises from analyses of Kaluza-Klein theories. Some of the earlier papers are by Palla [1] and by Witten [2] trying to find zero mode solutions of the Dirac equation on higher dimension manifolds, by Chapline, Slansky and Manton [3] and by Witterich [4] working out many of the details and kinematics. Some models are proposed by Randjbar-Daemi, Salam and Strathdee [5]; we shall mainly be following a discussion of Witten's [6].

Let us begin by explaining just what me mean by chiral fermions in four dimensions. We will call fermions chiral if they cannot acquire a bare mass, that is the Lagrangian of the theory forbids a mass term for fermions by gauge invariance. We call such a theory chiral. If, on the other hand, bare fermion masses are allowed, the theory is said to be vector-like. Of course, it may happen in a vector-like theory that particles do not acquire mass because of a global symmetry, but that is another matter.

A four component field ψ can always be written as the sum of two component Weyl spinors

$$\psi = \psi_L + \psi_R = \frac{(1+\gamma_5)}{2}\psi + \frac{(1-\gamma_5)}{2}\psi$$
$$= P_L\psi + P_R\psi \tag{7.1}$$

where P_L, P_R are projection operators.

Under charge conjugation

$$\psi \underset{C}{\rightarrow} \psi^c \equiv C\,\overline{\psi}^T$$
$$\psi_{L,R} \underset{C}{\rightarrow} \psi^c_{L,R} \equiv P_{L,R}\,\psi^c = C\,\overline{\psi}^T_{R,L} \tag{7.2}$$

i.e., ψ_L annihilates a left particle or creates a right antiparticle while ψ^c_L annihilates a left antiparticle or creates a right particle; similarly for ψ_R and ψ^c_R. Q.C.D. is an example of a vector-like theory, that is the interaction between gauge bosons $A^\mu = \lambda/2.A^\mu$ and fermions is of the form

$$\overline{\psi}\gamma_\mu A^\mu\psi = \overline{\psi}_L\gamma_\mu A^\mu\psi_L + \overline{\psi}_R\gamma_\mu A^\mu\psi_R \tag{7.3}$$

so if we classify our fermions as left handed we see that quarks belong to the ψ_L and ψ^c_L representations, i.e. under color SU(3) they transform like a $3 \oplus \overline{3}$, that is the representation is equal to its complex conjugate, which transforms like $\overline{3} \oplus 3$. The gauge theory allows a bare mass term coupling the 3 to the $\overline{3}$ fields; there are three equivalent ways of writting these mass terms, depending on our choice of notation

$$\psi^T_L\, C\, \psi^c_L,\ \overline{\psi}_L\psi_R,\ \overline{\psi}\psi. \tag{7.4}$$

In the case of the Q.C.D. Lagrangian, there are in fact no such mass terms present. The Lagrangian has a global chiral U(3) symmetry for e.g. three quarks, namely the u, d and s. This global symmetry is then broken spontaneously, giving rise to pseudo-scalar Goldstone bosons, etc. We emphasize however that the gauge theory allows the bare mass terms to be present. Weak SU(2), on the other hand, is a chiral theory; left-handed quarks are in doublets and right-handed ones in SU(2) singlets so a term such as (7.4) is forbidden in the Lagrangian by a gauged symmetry, not a global one. Of course the gauged symmetry can be broken spontaneously by Higgs bosons with non-zero v.e.v.'s. The difference, we repeat, between chiral theories and vector-like ones is whether or not gauged symmetries forbid bare mass terms for fermions.

At present we believe fermions are chiral and fall into families, which are classified by their transformation properties under SU(3) × SU(2) × U(1); one family of left-handed quarks and leptons have labels of $(3,2,1/3) \oplus (\overline{3},1,-4/3) \oplus (\overline{3},1,2/3) \oplus (1,1,2) \oplus (1,2,-1)$. This classification is sufficient to ensure that fermions cannot have masses much larger than that of the breaking of SU(2) × U(1). Of course, we see that at least for the first two families, their masses are much smaller. We cannot exclude the presence of so-called "mirror" fermions, that is fermions with V + A couplings to the usual W and Z bosons and the same other quantum nos. as our families. In this case, our theory would be like Q.C.D. with L and R fermions as in (7.3). In fact such mirror fermions do exist in models where families are put in one gauge super-multiplet and gauge bosons exist which connect families [7]. These models have problems of course; it is difficult to avoid generating large masses for fermions.

One primary reason for believing fermions are chiral is the cancellation of anomalies. In theories with mirror fermions they automatically cancel since these theories are vector like, e.g. Q.C.D. In $SU(2) \times U(1)$ we see rather miraculously that the quark contributions to the anomalies are exactly cancelled by those of the leptons. Ins't nature telling us something ? Why invoke mirror fermions if you don't need them ?

Chiral Fermions in High Dimensions (section b))

We saw in chapter five that in order to formulate a d = 10 SUGRA (supergravity) we need to start with a left-handed supermultiplet in d = 11. It is clear that we will have to study the chirality of fermions in d = 4 that are derived from d = 10 fermions. For the sake of generality, let the number of dimensions be 4 + n, though we will quickly turn to the case where n = 6. We will discuss the group $O(1,3+n)$ with metric $(-1;1,1\ldots1)$ rather than $O(4+n)$ since we're interested in Minkowski space. We will also redefine $\overline{\Gamma}$, $\overline{\Gamma}^{(4)}$ and $\overline{\Gamma}^{(n)}$, following Witten [6] as

$$\begin{aligned} \overline{\Gamma} &= \Gamma_1\Gamma_2\ldots\Gamma_{4+n} \\ \overline{\Gamma}^{(4)} &= \Gamma_1\Gamma_2\Gamma_3\Gamma_4 \\ \overline{\Gamma}^{(n)} &= \Gamma_5\Gamma_6\ldots\Gamma_{4+n}. \end{aligned} \tag{7.5}$$

The change in metric from the previous chapter can be effected by simply redefining Γ_1 so that $\Gamma_1^2 = -1$. We see that

$$\overline{\Gamma} = \overline{\Gamma}^{(4)}\,\overline{\Gamma}^{(n)}. \tag{7.6}$$

The idea behind obtaining chiral zero mass fermions is the following. In 4+n dimensions, let us study a representation of $O(1,3+n)$ with a definite chirality, i.e. eigenvalue of $\overline{\Gamma}$. There is then a correlation between the eigenvalues of $\overline{\Gamma}^{(4)}$ and $\overline{\Gamma}^{(n)}$; their product must equal the eigenvalue of $\overline{\Gamma}$. Now let us turn to the Dirac equation

$$\begin{aligned} \not{D}\psi &= \sum_{i=1}^{4} \Gamma_i D^i\psi + \sum_{i=5}^{4+n} \Gamma_i D^i\psi \\ &= \not{D}^{(4)}\psi + \not{D}^{(n)}\psi. \end{aligned} \tag{7.7}$$

From the point of view of four dimensional space time

$$\sum_{i=5}^{4+n} \Gamma_i D^i\psi \sim M\psi \tag{7.8}$$

i.e. the higher dimensional terms look like a mass in four dimensions. If we start with a definite chirality in d = 4+n, we already saw that the chiralities in d=4 and d=n are correlated. What we are aiming for is a situation in which we will have zero mass in d=4, i.e. zero mode solutions in the internal space transforming like a complex representation S of some symmetry group G. We will then look for the case in which there is a net non-zero number of chiral fermions in S, i.e. $N_L(S) - N_R(S) \neq 0$ on the compact manifold. Since the d=4 chirality is coupled to the d=n chirality, we will have a left over chirality in d=4.

Let us assume for the moment that $\overline{\Gamma}$ has eigenvalue +1. From (7.6), we see then

$$\eta\,\overline{\Gamma}^{(4)} = \eta^{-1}\,\overline{\Gamma}^{(n)} = +1 \quad \text{or} \quad \eta\,\overline{\Gamma}^{(4)} = \eta^{-1}\,\overline{\Gamma}^{(n)} = -1 \tag{7.9}$$

where η is a phase factor, which we will determine in a moment. Clearly chiralities are correlated. It is also obvious, presumably, that one must start with a chiral theory in 4+n dimensions : if one allowed both eigenvalues of $\overline{\Gamma}$, i.e. ± 1, the zero mode solutions in d=4 would be present with both chiralities. We would have a vector-like d=4 theory.

In d = 4k + 2, i.e. d = 2,6,10,... $\overline{\Gamma}$ squared equals +1 (remember $(\Gamma_1)^2 = -1$) so we can choose $\overline{\Gamma} = \pm 1$. We limit ourselves to the eigenvalue choice $\overline{\Gamma} = +1$, i.e. start with fermions of one chirality. $\overline{\Gamma}^{(4)}$ squared equals -1 and $\overline{\Gamma}^{(n)}$ squared also equals -1. The difference in sign between $(\overline{\Gamma})^2$ and $(\overline{\Gamma}^{(n)})^2$, both of which are in 4k + 2 dimensions, k differing by one, is due to the fact that $(\Gamma_1)^2 = -1$. We see thus that to have $\overline{\Gamma} = +1$ we must have

$$\overline{\Gamma}^{(4)} = -\overline{\Gamma}^{(n)} = \pm i. \tag{7.10}$$

<u>Zero Modes</u> (section c)

Consider, instead of $\overline{\Gamma}^{(n)}$, $\hat{\Gamma}$ (we are following ref. [6] here)

$$\hat{\Gamma} = i\overline{\Gamma}^{(n)}. \tag{7.11}$$

Since $\overline{\Gamma}^{(n)}$ had eigenvalues $\pm i$, $\hat{\Gamma}$ has eigenvalues ± 1. Now define a "Hamiltonian" $H^{(n)} = (i\not{D}^{(n)})^2$; we saw in the last section that if $H\psi = 0$, ψ will certainly be a zero mode of the Dirac equation. Now

$$[\hat{\Gamma}, H^{(n)}] = 0 \tag{7.12}$$

so we can simultaneously diagonalize $H^{(n)}$ and $\hat{\Gamma}$, but

$$\begin{aligned} H^{(n)}\psi = E\psi \Rightarrow H^{(n)} i\not{D}^{(n)}\psi &= i\not{D}^{(n)} H\psi = i\not{D}^{(n)} E\psi \\ &= E i\not{D}^{(n)}\psi \end{aligned} \tag{7.13}$$

so ψ and $i\not{D}^{(n)}\psi$ are degenerate in energy. On the other hand

$$\{\not{D}^{(n)}, \hat{\Gamma}\} = 0 \tag{7.14}$$

so $\not{D}^{(n)}\psi$ and ψ have opposite eigenvalues of $\hat{\Gamma}$. This implies that $E \neq 0$ eigenstates come in pairs, degenerate in energy, but with opposite chiralities. The argument does not hold for the zero energy eigenstates, i.e. zero eigenvalues need not be paired. The difference in number between zero eigenvalues of opposite chirality is called the index of $\not{D}^{(n)}$.

The index of $\not{D}$ is a topological invariant; any smooth deformation of the manifold will not disturb the pairing off of energy states, sending one of them to zero. The index depends on the topology of the manifold, not on the local structure.

What we actually want in our case is a slightly different topological invariant, the character valued index of $\not{D}^{(n)}$, which corresponds not to the difference in opposite chirality zero modes itself, but to the difference in opposite chirality modes in a given representation S. To see the reason for this, let us go to a Majorana representation in which the n Γ matrices are real. $\overline{\Gamma}^{(n)}$ is real, but it has eigenvalues $\pm i$ as we have seen. Now take a solution of $H^{(n)}\psi = E\psi$ with $\overline{\Gamma}^{(n)}$ eigenvalue i.

$$\overline{\Gamma}^{(n)}\psi = i\psi \Rightarrow \overline{\Gamma}^{(n)}\psi^* = -i\psi^* \tag{7.15}$$

Since $\overline{\Gamma}^{(n)}$ is real; $H^{(n)}$ is also real so we see that ψ and ψ^* are degenerate with opposite chiralities. The index of $\not{D}^{(n)}$ therefore vanishes. On the other hand the character valued index does not if S is a complex representation : ψ belongs to the S representation and ψ^* to the $\overline{S}$ representation. Our statement then becomes that the number of left zero modes in the S equals the number of right zero modes in the $\overline{S}$.

Gauge Fields and Topology (section d)

In attempting to understand why there are fermion zero modes on the internal manifold, we are necesarily led to the coupling of fermions to background gauge fields (i.e. gauge fields with nonzero v.e.v.'s) with non trivial topological properties. The character valued index's vanishing when there are no elementary gauge fields is a theorem due to Atiyah and Hirzebruch [8,9,6]; as discussed in ref. [6], this deals a "death blow" to attempts to generate chiral fermion theories in Kaluza-Klein like theories. It is clear also that if we have background gauge fields which can be continuously deformed to zero, the character valued index will vanish as well. We thus see the necessity of having background fields with a non-trivial topology.

The simplest example of such a singularity is a magnetic monopole, as we discussed following equation (3.20). Consider [5,6] a spin 1/2 particle constrained to move on the surface S^2 of a sphere. Define the helicity at a point x on the surface of the sphere as $\vec{J}.\hat{x}$, where $\vec{J}$ is the angular momentum of the particle. The chirality of the particle is $\pm 1/2$, the total angular momentum can take on the values J = 1/2, 3/2, 5/2 . . . . It turns out in this case there are no zero modes and states with $\Gamma = +1$ and $\Gamma = -1$ are paired off at values of J = 1/2, 3/2, 5/2 .... Now turn on a magnetic monopole of strength eg = n/2, for some integer n. This gives a contribution to the angular momentum of $\vec{J}_{mon} = eg\,\hat{x}$. The helicity now takes on the values eg + 1/2 and eg − 1/2 so that the allowed values of angular momentum are

$$\begin{array}{lll} \hat{\Gamma} = +1 & J = eg + 1/2, & eg + 3/2, \ldots \\ \hat{\Gamma} = -1 & J = eg - 1/2, & eg + 1/2,\ eg + 3/2, \end{array} \tag{7.16}$$

so we see that we have paired states at eg + 1/2 + k, k an integer. which must be degenerate. We also have a single $\hat{\Gamma} = -1$, J = eg − 1/2 state, which constitutes the spectrum of zero modes. It is amusing to see the details of how the Dirac equation has a zero mode in this monopole case.

To see that this is really non-trivial, consider the Dirac operator squared

$$\begin{aligned} (i\not{D})^2 &= -(\Gamma_i\Gamma_j\, D^iD^j) \\ &= -(\tfrac{1}{2}\{\Gamma_i,\Gamma_j\} + \tfrac{1}{2}[\Gamma_i,\Gamma_j])D^iD^j \end{aligned} \tag{7.17}$$

or, using the anticommutation relations it equals (remember Eq. 3.48)

$$(-D_iD^i = \tfrac{1}{4}[\Gamma_i,\Gamma_j][D^i,D^j])$$

$$= -D_iD^i - \tfrac{1}{32}[\Gamma_i,\Gamma_j][\Gamma_k,\Gamma_\ell]R^{ijk\ell} \qquad \ldots (7.18)$$

$$\cdots = -D_i D^i + \frac{1}{4} R \tag{7.18}$$

so, since $-D_i D^i$ is a non-negative operator we find that for $R > 0$ there are no zero mode solutions.

For a $d = 2$ sphere, the curvature is proportional to r^{-2} where r is the radius of the sphere. The positive curvature contribution is exactly cancelled by a negative monopole contribution to the electromagnetic field. If we did not know to be the case because of the index theorem, we would regard it as a miracle.

Another example is the instanton configuration for gauge fields in four dimensions [10], where one proves the sum rule

$$n_- - n_+ = \nu. \tag{7.19}$$

Here $n_\pm$ are the number of fermion zero modes of opposite chirality and ν is the winding number of the gauge field configuration, a topological invariant.

Index Theorem (section e)

The two examples we have just given are specific cases of the index theorem. This theorem, or rather class of theorems, tells us the number of zero modes on a Riemannian manifold with background gauge fields. Some appropriate references are given [11], but we clearly cannot present the technical background. We will instead jump right away to the result and then say a few words about it (see also next chapter). The difference between the number of positive chirality and negative chirality zero modes of the Dirac operator on a Riemannian manifold M is

$$n_+ - n_- = \int_M \hat{A}(M) \wedge ch.(V) \tag{7.20}$$

where the Chern character ch(V) is defined by

$$\begin{aligned} ch(V) &= \mathrm{Tr}\, \exp(\frac{i}{2\pi} V) = \sum_k \frac{1}{k!} \mathrm{Tr}(\frac{iV}{2\pi})^k \\ &= k + c_1(V) + \frac{1}{2}(c_1^2 - 2c_2) + \ldots \end{aligned} \tag{7.21}$$

with iV being given in this case by the two form field strength F. $\hat{A}(M)$, known as the A roof genus (see Eguchi, Gilkey and Hanson, ref. [11] p. 332)

$$\hat{A}(M) = 1 - \frac{1}{24} p_1 + \frac{1}{5760} (7p_1^2 - 4p_2) + \ldots \tag{7.22}$$

with p_k being defined as the expansion terms in the Pontrjagin class of a bundle with curvature 2 form R (EGH, ref. [11], p. 311)

$$\mathrm{Det}(1 - \frac{R}{2\pi}) = 1 + p_1 + p_2 + \ldots . \tag{7.23}$$

A simplified expression of the $\hat{A}$ is given by Zumino [12]; it equals

$$\sqrt{\det \frac{R/4\pi i}{\sinh R/4\pi i}} = 1 + \ldots \tag{7.24}$$

where R is the curvature 2 form of (3.47). Note that there is no term linear in R in the above expression. This means that on a d=2 manifold

$A(M)$ equals one and $n_+ - n_-$ is simply given by the integral

$$n_+ - n_- = \int_M ch(V) = \int_M \frac{iV}{2\pi} \tag{7.25}$$

which is the monopole charge since $V = iF = \frac{i}{2} F_{\mu\nu}\, dx^\mu \wedge dx^\nu$. For an instanton we have to evaluate the first term in the expansion of A. On S^4 we have however $Tr(R\wedge R) = 0$, so again $\hat{A}(M)$ is just equal one. The spinors are in the spin 1/2 representation of the SU(2) group and gauge fields are three 2×2 matrices; they are traceless so c_1 in (7.21) vanishes and

$$n_+ - n_- = -\int_M c_2(V) = \frac{1}{8\pi^2} \int_M Tr(\Omega\times\Omega) \tag{7.26}$$

where

$$\Omega = \frac{1}{2}(\lambda^a/2i)\, F^a_{\mu\nu}\, dx^\mu \wedge dx^\nu. \tag{7.27}$$

COMPACTIFICATION (chapter 8)

<u>SUSY in d = 4</u> (section a)

Starting from the field theory limit of the superstrings, we need now to proceed with the compactification. The dynamics of the ten dimensional theory should lead us to a stable vacuum in a space of the form $M_4 \times K$, where M_4 is Minkowski space time and K is a compact manifold characterized by a scale $R \sim (M_{Planck})^{-1}$. The fields in question in ten dimensions are those introduced in section 5, namely the supergravity multiplet, and the gauge super-multiplet belonging to SO(32) or $E_8 \times E_8{}'$, as discussed in chapter 5. One significant deviation from the ten-dimensional supergravity model of Chapline and Manton [1] is the need to introduce in the antisymmetric tensor's field strength a Lorentz Chern-Simons form (see eqs. 4.20a,b). This may require additional terms for the preservation of supersymmetry, but we will not pursue further this last point.

Starting then with the supergravity super Yang-Mills multiplets in d = 10, CHSW [2] formulate the requirements for an acceptable compactification. They are :

i) Geometry be of the form $M_4 \times K$, where M_4 is maximally symmetric (this means homogeneous and isotropic) and K be compact.

ii) An unbroken N = 1 SUSY survives in four dimensions.

iii) Particle spectrum be realistic, in particular we be left with an acceptable gauge theory which includes at least $SU(3)_c \times SU(2) \times U(1)$ at low energies and fermions be in chiral families.

We will sketch how these conditions are sufficiently restrictive to lead us to K being a Kähler manifold with SU(3) holonomy (see chapter 6 for an intuitive version of this argument and chapter 3 for a discussion of holonomy). Our fields then, as introduced in chapter 5 (with apologies to readers however, because now we will let M, N, P = 1,...10 and m,n = 5,6,...10 whereas earlier M ran from 1 to 11 and m from 1 to 10) are

$$g_{MN},\ \psi_M,\ H_{MNP},\ \lambda, \phi \tag{8.1}$$

and the gauge supermultiplet

$$F^\alpha_{MN},\ \chi^\alpha \tag{8.2}$$

We start by examining the condition for unbroken N = 1 SUSY in d = 4. We will expand around bose and fermion background fields. Since M_4 is maximally symmetric, we require all fermion background fields to vanish (i.e. no preferred orientation). Since in SUSY the variation of a bose field b is proportional to a background fermion field, we immediately have

$$\delta_{SUSY}\, b = 0. \tag{8.3}$$

For fermions, we calculate the variation under a SUSY transformation generated by a parameter ε. The variations of ψ_μ, ψ_m (μ = 1,2,3,4, m = 5,6,7,8,9,10), λ and χ^a are therefore all proportional to ε; they can involve as well background values for F_{mn}, H_{pqr}, and ϕ, all of which are of course seen in d = 4 as scalar fields. We will use a notation $\otimes$ for the $M_4 \otimes K$ product space so, e.g., the gamma matrices in ten dimensions can be written as

$$\begin{aligned} &\gamma^\mu \otimes 1 \\ &\gamma_5 \otimes \Gamma_i \end{aligned} \tag{8.4}$$

where the Γ_i form the Clifford algebra of SO(6). We will also need the sechsbein (or six-bein, see ch. 3, sec. b) to go from the tangent frame to K so e.g., we may want to write

$$\Gamma_m = e^i_m\, \Gamma_i \tag{8.5}$$

i.e., i,j,k will be tangent space indices and m,n,p curved space indices. Γ_{ij}, Γ_{ijk} will be products of completely antisymmetrized SO(6) matrices and Γ_{mn}, Γ_{mnp} their curved space counterparts. The spin connection, defined in (3.35) will be written as a^i_j or ω^{ij} (we use now i,j rather than a,b) or as a 1-form $\omega^{ij}_m\, dx^m$ in K space. The covariant derivative in K is defined as

$$\nabla_m = \partial_m + \frac{1}{4}\Gamma_{ij}\,\omega^{ij}_m \tag{8.6}$$

and similarly the d = 4 covariant derivative is ∇_μ. We will also define

$$\widetilde{H} = H_{pqr}\,\Gamma^{pqr} \qquad \widetilde{H}_m = H_{mqr}\,\Gamma^{qr}. \tag{8.7}$$

With this as notation, we then write down the variations under a SUSY transformation of the fermion fields. We obtain

$$\begin{aligned} \delta\psi_\mu &= \nabla_\mu\varepsilon + \frac{\sqrt{2}}{32}\, e^{2\phi}(\gamma_\mu\gamma_5 \otimes \widetilde{H})\varepsilon \\ \delta\psi_m &= \nabla_m\varepsilon + \frac{\sqrt{2}}{32}\, e^{2\phi}(\Gamma_m\widetilde{H} - 12\widetilde{H}_m)\varepsilon \\ \delta\lambda &= \sqrt{2}\,(\Gamma_m\,\nabla^m\phi)\varepsilon + \frac{1}{8}\, e^{2\phi}\widetilde{H}\varepsilon \\ \delta\chi^\alpha &= -\frac{1}{4}\, e^\phi\, F^{(\alpha)mn}\,\Gamma_{mn}\varepsilon. \end{aligned} \tag{8.8}$$

The attentive reader may be bewildered by the appearance of e^ϕ factors in (8.8) since up to this point the dilaton has been a multiplicative factor; with apologies to our readers we have switched notation once again to conform to CHSW. The ϕ here is the logarithm of the ϕ in chapter 5;

one of the hardest things about a quickly moving field is the lack of consistent notation! Of course in what follows in this chapter it will make no difference.

Now let us begin to analyze the results of setting (8.8) equal to zero. From $\delta\psi_\mu = 0$ we see

$$\nabla_\mu \varepsilon = -\frac{\sqrt{2}}{32} e^{2\phi}(\Gamma_\mu \gamma_5 \otimes \Gamma_{pqr})H^{pqr}\varepsilon \tag{8.9}$$

and hence the integrability condition

$$[\nabla_\mu, \nabla_\nu]\varepsilon = \frac{1}{(16)^2} e^{4\phi}(\Gamma_\mu \Gamma_\nu \otimes \tilde{H}^2)\varepsilon. \tag{8.10}$$

In deriving (8.10) we have used the fact that the background fields do not depend on χ_μ and are scalars w.r.t. to the first four spatial components. If we assume now that M_4 is maximally symmetric

$$R_{\mu\nu\rho\sigma} = \kappa(g_{\mu\rho}g_{\nu\sigma} - g_{\mu\sigma}g_{\nu\rho}) \tag{8.11}$$

and using the well known relation for $[\nabla_\mu, \nabla_\nu]$ cf. (3.48) we find

$$\tilde{H}^2\varepsilon = \frac{(16)^2}{2}\frac{\kappa}{2} e^{-4\phi}\varepsilon. \tag{8.12}$$

Similarly $\delta\lambda = 0$ leads to

$$\nabla_m\phi\ \nabla^m\phi = -\kappa \tag{8.13}$$

so $\nabla_m\phi$ is constant. The dilaton field ϕ only depends on the internal compact manifold, as we already said; it must therefore assume a maximum value somewhere, at which point $\nabla_m\phi$ will vanish. Since $\nabla_m\phi$ is a constant, however, it vanishes everywhere and hence κ equals zero.

At this point we are merely parroting CHSW so let us jump ahead to the conclusions which are that we must have

i) κ in (8.11) equals zero, i.e. M_4 is flat Minkowski space;

ii) ε satisfies the following equations

$$(\nabla_m - \beta\Gamma_{ij}\ H_m^{ij})\varepsilon = 0 \tag{8.14a}$$

$$(\Gamma_{mnp}\ H^{mnp})\varepsilon = 0 \tag{8.14b}$$

$$(\Gamma_{mn}\ F^{(\alpha)mn})\varepsilon = 0 \tag{8.14c}$$

where, in (8.14a)

$$\beta = \frac{3}{8}\sqrt{2}\ e^{2\phi}. \tag{8.15}$$

An important modification from the formalism of Chapline and Manton [1] is the need to modify the three-form H to include a Yang-Mills Chern-Simons form. We discussed the reasons briefly at the end of chapter four, to wit the cancellation of anomalies in string theory. The needed change is to replace $H = dB - \omega_{3,y.m.}$ by the H in equation in (4.20b). We see from there that

$$dH = tr\ R\wedge R - \frac{1}{30}\ Tr\ F\wedge F. \tag{8.16}$$

Though it may not be necessary [3], the simplest solution is to set H_{pqr} equal to zero, so (8.14b) is satisfied; we then identify the background gauge fields and spin connections in such a way that

$$tr\ R\wedge R = \frac{1}{30}\ Tr\ F\wedge F \tag{8.17}$$

holds. CHSW give general arguments as to why solutions $H_{pqr} \neq 0$ would be unreliable; they are related to the fact that the two sides of (8.16) scale differently so $H_{pqr} \neq 0$ would fix the scale at some length L presumably of order $(M_{Planck})^{-1}$. At that scale our field theoretic perturbation like calculations are unreliable. These scale problems will come up time and time again; see section 11c for a somewhat lengthier discussion. In any case, from here on we assume $H_{pqr} = 0$.

Let us turn now to the analysis of (8.14a) and (8.14c). The spinor ε in d = 10 can be decomposed into products of spinors ξ in M_4 times spinors η in K so that

$$\varepsilon = \sum_{\Lambda} \xi_{\Lambda} \otimes \eta_{\Lambda} \tag{8.18}$$

where we take η's to commute and ξ's to anticommute. In addition ε is subject to the Weyl constraint in d = 10 (see chapter 5). ε and η can be taken to be real.

SO(6) is locally isomorphic to SU(4) se we can write, instead of the real eight dimensional (2^3) Majorana spinor, a complex four dimensional spinor belonging to the fundamental representation of SU(4). It will turn out to be more convenient to work in this SU(4) basis. We have already seen how the second of the three (8.14) equations is solved. Let us now turn to the first : choose a spinor η which does not depend on the coordinates x_m. Eq. (8.14a) then becomes

$$\Gamma_{ij}\left\{\frac{\omega_m^{ij} - 4\ \beta H_m^{ij}}{4}\right\}\eta = 0 \tag{8.19}$$

or, more succintly if we define a one-form A^{ij} as the quantity in the brackets in (8.19) times dx^m, we rewrite (8.19) as

$$\Gamma_{ij}\ A^{ij}\eta = 0. \tag{8.20}$$

The eight dimensional SO(6) Majorana spinor can be decomposed, as we saw in chapter six into two four-dimensional spinors of opposite chirality. Since $i\Gamma_{ij}$ is a generator of SO(6), and the generators commute with chirality, i.e. the representation is reducible, (8.20) can be written as

$$\begin{pmatrix} A & 0 \\ \hline 0 & A^* \end{pmatrix} \begin{pmatrix} \eta \\ \hline \eta^* \end{pmatrix} = 0 \tag{8.21}$$

where A is a four by four antihermitian matrix representing $\Gamma_{ij}\ A^{ij}$. It is clear that the eight dimensional SO(6) spinorial representation corresponds to the $4 + \overline{4}$ representation of SU(4); the 4 and $\overline{4}$ representations of SU(4) have opposite chirality so A is an SU(4) transformation acting on a spinor in SU(4). If we make a gauge rotation so that η is real and lies along the four direction, we see that the most general solution to (8.21) is for A to equal

$$A = \begin{pmatrix} \tilde{A} & 0 \\ 0 & 0 \end{pmatrix} \quad (8.22)$$

where $\tilde{A}$ is a 3 × 3 antihermitian traceless matrix. In other words the spin connection is an SU(3) matrix; this means that the holonomy group is SU(3) (see end of chapters three and five for earlier discussions).

CHSW also show that the manifold K is a Kähler manifold. The latter is defined as follows : consider a Hermitian metric on K

$$ds^2 = g_{ab}\, dz^a\, d\bar{z}^{\bar{b}} \quad (8.23)$$

where g_{ab} is a Hermitian metric. The Kähler form $\tilde{K}$ is given by

$$\tilde{K} = 1/2\; g_{a\bar{b}}\, dz^a \wedge d\bar{z}^{\bar{b}} \quad (8.24)$$

where we use $\tilde{K}$ so as not to confuse the form with the manifold K (it probably would have been better to use e.g. M_6 for the manifold, but we are trying to conform with the notation of CHSW). A manifold is said to be Kähler if it admits a Kähler metric, and a metric is said to be Kähler if $\tilde{K}$ is closed, i.e.

$$d\tilde{K} = 0. \quad (8.25)$$

It is also shown in CHSW that K is Ricci flat, i.e. that the once contracted Riemann tensor vanishes. This is seen by observing that

$$\nabla_m \eta = 0 \qquad [\nabla_m, \nabla_n]\eta = R_{mnpq}[\Gamma^p, \Gamma^q]\eta = 0. \quad (8.26)$$

After some work this implies that

$$R_{mp} = 0$$

i.e. the manifold is Ricci flat, a quality which appears to be necessary in order to have a consistent string theory.

In conclusion we have seen a great deal. The requirements of unbroken N = 1 SUSY and a geometry of the form M_4 × K have led us to Minkowski space for M_4 and a Ricci flat Kähler manifold with SU(3) holonomy for K.

Calabi-Yau Manifolds (section b)

Our six-dimensional manifold K has SU(3) holonomy. Consider first a manifold with U(3) holonomy. The coordinates on the manifold transform of course as the six-dimensional vector representation of O(6), but as we have already seen, this forms the $3 + \bar{3}$ representation of U(3). (At the beginning of chapter six we saw how one could build Γ's out of X and $X^\dagger$; this is equivalent to building the vector 2N representation of O(2N) out of N and $\bar{N}$ of SU(N)). This means that an O(6) coordinate differential

$$c_m dx^m = c_a dz^a + c_{\bar{b}} d\bar{z}^{\bar{b}} \quad (8.27)$$

where dz^a and $d\bar{z}^{\bar{b}}$ transform as the 3 and $\bar{3}$ of U(3) and of course $a, \bar{b} = 1,2,3$. As we stated at the end of the last section, K is a Kähler manifold. The spin connection on the Kähler manifold is a U(3) or SU(3) × U(1) gauge field. The metric on a Kähler manifold

$$ds^2 = \frac{\partial^2 \tilde{K}}{\partial z^a \partial \bar{z}^{\bar{b}}} dz^a \, d\bar{z}^{\bar{b}} \tag{8.28}$$

is not unique. One can always add to K functions of $\bar{z}^{\bar{a}}$ and functions of z^a only without changing the metric. The problem then in having SU(3) holonomy is to find a manifold for which the U(1) part of the U(3) spin connection can be choosen to be zero.

If we call $\Omega^{(1)}$ the field strength of the U(1) spin connection, we need to have $\Omega^{(1)mn}_{mn}$ vanish. Calabi showed that there is in general a topological obstruction to having

$$\int ds^{mn} \, \Omega^{(1)}_{mn} \tag{8.29}$$

vanish and conjectured that a Kähler manifold with vanishing first Chern class would admit a Kähler metric with SU(3) holonomy (Chern classes are briefly discussed in chapters 4 and 8; remember N^{th} Chern class is proportional to the product of $\Omega\Lambda\Omega$ N times). Yau proved this conjecture; these spaces are therefore known as Calabi-Yau spaces [4]. We will not present here examples of such spaces; see instead CHSW, where various polynomial constructions are given (one example will be given at the beginning of the next chapter).

We will now turn back to the third of our (8.14) conditions for N = 1 SUSY. It is easiest to see how to solve this using the SU(3) decomposition of SO(6) that we gave at the beginning of chapter six. Using (6.1) and (6.7) we see that the Γ matrices can be written in terms of X_i and $X^{i\dagger}$. In chapter six we simply called these operators X_i and $X_i^\dagger$; here we will switch to subscript-superscript notation to underline the fact that X_i transforms like the 3 of SU(3) and $X^{i\dagger}$ like the $\bar{3}$.

We have already seen that the coordinates x^m on the manifold split up into three complex coordinates; the a indices are known as holomorphic and the $\bar{a}$ as anti-holomorphic. Similarly to the sechsbein e^i_m which takes us from tangent space (i = 1...6) to base space (m = 1,...6) we must now define the analogue which takes us from the complex space parametrized by z^a to the tangent space. The one forms are

$$e^i = e^i_a dz^a + \bar{e}^i_{\bar{b}} d\bar{z}^{\bar{b}} \tag{8.30a}$$

$$\bar{e}_i = \bar{e}_{ia} dz^a + e_{i\bar{b}} d\bar{z}^{\bar{b}} \tag{8.30b}$$

where $(e^i_a)^* = e_{i\bar{a}}$. We assume we have a Hermitian complex space so that $\bar{e}_{ia} = \bar{e}^i_{\bar{b}} = 0$. Equations (8.30) then simplify to

$$e^i = e^i_a dz^a \qquad \bar{e}_i = e_{i\bar{b}} d\bar{z}^{\bar{b}} \tag{8.31}$$

and the metric is

$$g_{a\bar{b}} = e^i_a \bar{e}_{i\bar{b}} \qquad g^{a\bar{b}} = e^a_i \bar{e}^{i\bar{b}} \tag{8.32}$$

with $g_{ab} = g_{\bar{a}\bar{b}} = 0$.

We now turn back to (8.14c), the third remaining condition for N = 1 SUSY that has to be satisfied. The vacuum state is $|\Omega\rangle + |\bar{\Omega}\rangle$, using the notation of equation (6.11) (with $X^{i\dagger}$ instead of $X_i^\dagger$ for clarity as we already stated). We have

$$\Gamma_{mn} F^{(\alpha)mn} = e^i_m e^j_n \Gamma_{ij} F^{mn} \tag{8.33}$$

where i,j = 1,2...6 and m,n = 1...6. Going over to the complex basis where now a,b = 1,2,3 and i,j = 1,2,3 we rewrite (8.33) as

$$\begin{aligned}\Gamma_{mn} F^{(\alpha)mn} &= e^i_a e^j_b F^{(\alpha)ab} [X_i, X_j] \\ &+ \bar{e}_{i\bar{a}} \bar{e}_{j\bar{b}} [X^i, X^{j\dagger}] F^{(\alpha)\bar{a}\bar{b}} \\ &+ \bar{e}_{i\bar{a}} e^j_b [X^{i\dagger}, X_j] F^{(\alpha)\bar{a}b} \\ &+ e^i_a \bar{e}_{j\bar{b}} [X_i, X^{j\dagger}] F^{(\alpha)a\bar{b}}.\end{aligned} \tag{8.34}$$

Applying this to $|\Omega\rangle + |\bar{\Omega}\rangle$ we see that the condition (8.14c) implies

$$F^{(\alpha)ab} = F^{(\alpha)\bar{a}\bar{b}} = 0 \tag{8.35a}$$

$$g_{a\bar{b}} F^{(\alpha)a\bar{b}} = 0. \tag{8.35b}$$

Equations (8.35a) are easily solved by letting the gauge fields $A_a^{(\alpha)}$ and $A_{\bar{a}}^{(\alpha)}$ (multiplied by a representation matrix $t^{(\alpha)}$ for SO(32) or $E_8 \times E_8$) satisfy

$$A_a = V^{-1}\partial_a V \qquad A_{\bar{a}} = V^{-1}\partial_{\bar{a}} V \tag{8.36}$$

where V is a unimodular group element in SO(32) or $E_8 \times E_8$. Equation (8.35b), on the other hand, is a complicated non-linear partial differential equation [5]. The solutions can be classified topologically.

<u>Models</u> (section c)

We must now satisfy the final condition, (8.17). The spin connection is, as we already stated, an SU(3) gauge field for manifolds with SU(3) holonomy. E_8 has an SU(3) × E_6 maximal subgroup so we may identify the SU(3) spin connection with the SU(3) gauge fields of SU(3) × E_6. When we take the traces of R∧R and F∧F in the fundamental representations, we miraculously obtain the factor of thirty in (8.17). The fundamental representation for R is the six dimensional one of O(6), equivalent to the 3 + $\bar{3}$ of SU(3), while for E_8 the fundamental and adjoint representations coincide and are the 248 dimensional representation of E_8. This decomposes under SU(3) × E_6 into

$$248 \rightarrow (8,1) + (3,27) + (\bar{3},\overline{27}) + (1,78). \tag{8.37}$$

The (8,1) form the vector supermultiplet of SU(3) and the (1,78) the supermultiplet of E_6, i.e. the adjoint representation of E_6 is seventy eight dimensional, as we said at the end of chapter six (see also following chapter nine for more on E_6). As compared to one 3 + $\bar{3}$ in the six of O(6) we clearly see there are twenty-seven 3 + $\bar{3}$'s in the 248 of (8.37). The additional factor of three needed to make up 3 + 27 = 30 in (8.17) comes from the (8,1) representation in (8.37). We see thus that we have satisfied the requirements i) and ii) from the beginning of the chapter, namely geometry is M_4 × K and we have an unbroken N = 1 SUSY. Let us now turn to the third requirement, a realistic spectrum of chiral fermions.

Let us recapitulate. Our gauge group is $E_8 \times E_8$: one of the E_8's is unbroken. We will discuss in later chapters how it may act as a hidden sector for the breaking of SUSY. The other E_8 is broken by fields in its SU(3) subgroup taking on v.e.v.'s (of course the fields that do this are F_{mn} components; not $F_{\mu\nu}$'s since we are not breaking Lorentz invariance). These background gauge fields are identified with the SU(3) spin connection gauge fields. This would seem to leave an unbroken E_6, but we shall see in the next chapter how in fact E_6 may be broken dynamically to a subgroup of E_6 which includes SU(3) × SU(2) × U(1), the known group of particle physics.

Now consider the spectrum of massless particles that arises from the 248 ten dimensional vector superfields. The (1,78) includes $A_\mu^{(\alpha)}$ vector superfields, the gauge bosons and gauginos of E_6 and fields $A_m^{(\alpha)}$ where α is of course the E_6 index. Since $A_m^{(\alpha)}$ transform like the six of O(6) or $3 + \overline{3}$ under SU(3) holonomy, they are presumably very massive, i.e. $M \sim R^{-1}_{\text{compactification}}$. Similarly the (8,1) fields are presumably also all super-massive. That leaves us with the (3,27) and $(\overline{3},\overline{27})$ multiplets; this is the only place for chiral fermions since the multiplets transform non-trivially under E_6 and under the SU(3) in E_8 which will be identified with the holonomy. What happens is that the $A_m^{(3,27)}$ or $A_m^{(\overline{3},\overline{27})}$ fields can be overall singlets of holonomy by having the 3 or $\overline{3}$ spatial indices combine with the 3 or $\overline{3}$ group indices to form singlets (remember that effectively both the group and spatial indices denote holonomy because of the identification of $R\wedge R$ with $1/30\ F\wedge F$). These singlets can then be massless particles since only holonomy singlets do not acquire masses of order R^{-1}_{comp}. The notion of chiral fermions in 27 and $\overline{27}$ representations is extremely attractive since there seems to be a good possibility for developing an adequate phenomenology, as we shall see in the next chapter.

Now we ask how many massless 27 and $\overline{27}$'s are there. Naively we would say there is an equal number because we started with a (3,27) and a $(\overline{3},\overline{27})$; this conclusion would lead us to a vector-like theory. The reason this isn't true is of course that we have background fields with non-trivial topological properties, as we explained in chapter seven. To calculate the number of massless 27's, we must use the character valued index described in chapter seven, sections c, d and e. The procedure is as follows : we start with the Dirac equation in ten dimensions, as written, e.g. in (7.7). The ten dimensional space factorizes into $M_4 \times K$ and the ten dimensional spinor ψ decomposes into the sum of products of four dimensional spinors $\psi_i^{(4)}(x)$ and six dimensional spinors $\psi_i^{(6)}(y)$ where x are the coordinates of M_4 and y those of K.

$$\psi^{(10)}(x,y) = \sum_i \psi_i^{(4)}(x)\psi_i^{(6)}(y). \tag{8.38}$$

We use the decomposition of the gamma matrices given in equation (8.4) to write the Dirac equation, following (7.7),

$$\begin{aligned} \not{D}\psi^{(10)}(x,y) &= \sum_i \not{D}^{(4)}\psi_i^{(4)}(x) \otimes \psi_i^{(6)}(y) \\ &+ \sum_i \gamma_5\psi_i^{(4)}(x) \otimes \not{D}^{(6)}\psi_i^{(6)}(y) \end{aligned} \tag{8.39}$$

where $\not{D}$ is the full covariant derivative, including gauge field connection and the spin connection. If we have a solution of the Dirac equation in ten dimensions, the four dimensional Dirac equation is

$$D^{(4)}\psi_i^{(4)}(x) = -\gamma_5\psi_i^{(4)}(x) \otimes \frac{(\psi_i(y),\not{D}^{(6)}\psi_i(y))}{(\psi_i(y),\psi_i(y))} \tag{8.40}$$

where the inner product notation is hopefully fairly obvious. We see now that the zero modes in four dimensions correspond to the solutions of

$$\not{D}^{(6)}\psi_i(y) = 0. \tag{8.41}$$

From (8.37) we see the decomposition of the fields we are looking at : in ten dimensions we have one left handed 248 dimensional vector superfield. We saw already in chapter seven that in order to have chiral fields in four dimensions, we need to have background gauge fields with non-trivial topological properties, connected to the fermions. These will of course be the SU(3) gauge bosons in the decomposition of E_8 into the maximal subgroup SU(3) × E_6, which we identify with the spin connection, à la (8.17).

The background SU(3) gauge bosons break the E_8 gauge symmetry, naively leaving a E_6 symmetry (we shall see in the next chapter how for dynamical reasons this E_6 symmetry is also broken), and in any case leaving fields which can be described as 27 dimensional multiplets. Referring back to sections 7b, 7c, we are specifically looking at the character valued index in the 3 representation of SU(3) or rather in the (3,27) of SU(3) × E_6. This will give us

$$N = n_{27}^{L} - n_{27}^{R} \tag{8.42}$$

where N is the difference of left handed and right handed 27 zero modes on K arising from the 3 of SU(3). This will in turn give us N for four dimensional zero modes because of (8.40). We evaluate N using the formulae in chapter 7e.

<u>Chiral 27 Multiplets</u> (section d)

The index N is given by

$$N = \int_K \hat{A}(K)\mathrm{ch}\,(\frac{iF}{2\pi}) \tag{8.43}$$

where the traces in evaluating $\mathrm{ch}(\frac{iF}{2\pi})$ are taken in the 3 representation of SU(3). On the six dimensional manifold in question

$$\hat{A} = 1 + \frac{1}{24}c_2(R) \tag{8.44}$$

and N is given by

$$N = \frac{1}{6}\int \frac{\mathrm{tr}(iF^3)}{(2\pi)^3} + \frac{1}{24}\int \frac{\mathrm{tr}(iF)}{2\pi}c_2. \tag{8.45}$$

If we identify the spin connection with the gauge fields, we see that N can be determined entirely in terms of geometric quantities [6]. In fact N is given by one half the Euler characteristic $\chi(K)$. A simple (or relatively simple explanation is given in CHSW, namely the positive and negative chirality spinors on O(6) transform like the 4 and $\overline{4}$ of SU(4) (since O(6) is isomorphic to SU(4)). The Euler characteristic on K is

$$\chi(K) = \mathrm{index}\ (4) - \mathrm{index}\ (\overline{4}) \tag{8.46}$$

but under SU(3) $4 \approx 1 \oplus 3$ and $\overline{4} \approx \overline{1} \oplus \overline{3}$. The indices in the one cancel out and the difference of indices in 3 and $\overline{3}$ is just twice the index in the 3 representation, i.e. the character valued index we are calculating. We thus have $N = \frac{1}{2}\chi(K)$, or since left and right are a matter of defintion, $|N| = \frac{1}{2}|X(K)|$. CHSW evaluate X on six dimensional manifolds with SU(3) holonomy and find as possible values of X, 72, -200, -176, -144, -128. They then go on to show how this number can be reduced by having discrete symmetries on the manifold, the subject we will turn to in the next chapter.

In the meantime we wish to consider N further. In particular, we would like to know not just N, but n^L_{27} and n^R_{27} (or equivalently $n^L_{\overline{27}}$ since a right handed 27 field is the same as a left handed $\overline{27}$). This issue is discussed briefly in CHSW and more thoroughly in later works by Witten [7,8]. The skeleton of the argument goes as follows : we have a Kähler manifold with three complex coordinates, cf. eq. (8.27), so we characterize our space by three "holomorphic" indices ~ 3 and three "antiholomorphic" indices $\sim \overline{3}$. Differential forms are then of type (p,q) with p holomorphic indices and q anti-holomorphic indices $\psi_{a_1 \ldots a_p;\, \bar{a}_1 \ldots \bar{a}_q}$, so we talk about (p,q) forms rather than p+q forms. Spinors, as we saw in (6.11) may be regarded as (0,0), (0,2) forms of positive chirality and (0,1), (0,3) forms of negative chirality. The Dirac equation can be decomposed as well into

$$\not{D} = \sum_{a=1}^{3} (X^a\, D/Dz^a + X^{\bar{a}\dagger}\, D/D\bar{z}^{\bar{a}}) = \partial + \partial^*. \tag{8.47}$$

The number of linearly independent k forms ψ satisfying $d\psi = d^*\psi = 0$ (see chapter 3) is a topological invariant known as b_k, the k'th Betti number. On a Kähler manifold we have in addition that the individual solutions of $d\psi = d^*\psi = 0$ with p,q fixed adding up to k is a topological invariant $b_{p,q}$ known as the Hodge number. Furthermore on a Kähler manifold

$$dd^* + d^*d = 2(\partial\partial^* + \partial^*\partial) \tag{8.48}$$

so solutions of $\partial\psi = \partial^*\psi$ equal to zero are equivalent to solutions of $d\psi = d\psi^* = 0$. We conclude that zero mode spinors on a manifold with SU(3) holonomy are (0,q) forms characterized by Hodge numbers $b_{o,q}$.

We furthermore have, using (7.18) and the fact that the scalar curvature of our manifold is zero, since it is Ricci flat, that $i\not{D}\psi = 0$ implies $(i\not{D})^2\psi = 0$ which implies $D_m\psi = 0$. On a manifold of SU(3) holonomy, the only two solutions of this equation are forms of type (0,0) and (0,3). We now make use of a variety of identities among the Hodge numbers : by complex conjugation $b_{p,q} = b_{q,p}$. Poincaré duality says $b_{p,q} = b_{3-p,3-q}$. To summarize we have $b_{0,0} = b_{0,3} = b_{3,0} = b_{3,3} = 1$ and $b_{0,1} = b_{1,0} = b_{2,0} = b_{0,2} = b_{3,1} = b_{1,3} = b_{2,3} = b_{3,2} = 0$.

We have discussed in these topological considerations spinors coupled only to gravity. A spinor transforming however like the (3,27) of $SU(3) \times E_6$ is of course also a 27 dimensional representation transforming like the 3 of SU(3); this corresponds to an additional triplet holomorphic index, i.e. it is like X^a acting on the vacuum, because we have identified the SU(3) gauge group with the holonomy. This does not alter the chirality of spinors so we see we have positive chirality spinors whose numbers are fixed by $b_{1,0}$ and $b_{1,2}$ and negative chirality spinors with numbers fixed by $b_{1,1}$ and $b_{1,3}$. We just saw however that $b_{1,0}$ and $b_{1,3}$ were equal to zero. The upshot is that we have $b_{1,2}$ positive chirality spinors, i.e. left handed 27's and $b_{1,1}$ right handed 27's or, equivalently, left handed $\overline{27}$'s. N in (8.42) is given by $N = b_{1,2} - b_{1,1}$. All Kähler manifolds have $b_{1,1} \geq 1$ because the Kähler form defined in

(8.24) is a (1,1) form on the manifold which is covariantly constant and hence clearly satisfies $d\tilde{K} = d{*}\tilde{K} = 0$. We now assume [9] that in fact $\tilde{K}$ is the only zero mode (1,1) form on K and hence $b_{1,2} = N + 1$.

To summarize, we have seen that the chiral modes on K which transform non-trivially under the internal gauge group E_6 consist of N + 1 27's and one $\overline{27}$ (all left handed) and their anti-particles; $|N|$ is fixed to be one half the Euler characteristic $|\chi(K)|$. These are the candidates for families of chiral fermions. See also chapter 10, section b for a more pedestrian approach to some of these questions.

DISCRETE SYMMETRIES (chapter 9)

Discrete Symmetries on Manifolds (section a)

We saw in the previous sections that our compactification procedure had two major flaws : we had too many chiral fermion families and we were left with an E_6 invariant theory, without any visible means of breaking this symmetry further while maintaining N = 1 SUSY. Both these problems can be solved using discrete symmetries as we now show.

Let us concentrate on the simplest Calabi-Yau manifold [1], the solutions of the quintic polynomial

$$\sum_{i=1}^{5} z_i^5 = 0. \tag{9.1}$$

The Euler characteristic of this manifold is 200 so we have 100 complete families, far too many. We note however that we can instead study the much smaller manifold

$$K \to \tilde{K} = K/G \tag{9.2}$$

obtained by imposing a discrete symmetry G acting freely on K. An example is to have G isomorphic to $Z_5 \times Z_5$ where Z_5 are five element discrete groups : the discrete symmetries A and B act on K as follows :

$$\begin{aligned} A(z_1,z_2,z_3,z_4,z_5) &\to (z_5,z_1,z_2,z_3,z_4) \\ &\quad + \text{permutations} \\ B(z_1,z_2,z_3,z_4,z_5) &\to (\alpha z,\alpha^2 z_2,\alpha^3 z_3,\alpha^4 z_4,\alpha^5 z_5) \\ &\quad \alpha^5 = 1. \end{aligned} \tag{9.3}$$

Clearly A and B leave (9.1) unchanged. They also take us to a much smaller manifold, in fact one which has Euler characteristic

$$\chi(\tilde{K}) = \frac{\chi(K)}{N(G)} \tag{9.4}$$

where N(G) equals the number of elements in G, in this case 25. It is hopefully clear from the section on chiral fermions how this kind of "chopping up" of the manifold allows for fewer zero mass states. We see therefore that the discrete symmetry can take us to e.g. a four family model since from (9.4) $\frac{1}{2}\chi(\tilde{K}) = 100/25 = 4$.

Breaking of E_6 (section b)

A second more subtle advantage of acting on K with G is that it provides a means of breaking dynamically the residual gauge symmetry E_6.

To see this let us follow Witten's argument [2] : a function ψ acting on K is equivalent to a function on K which satisfies

$$\psi(g(x)) = \psi(x) \ \forall g \in G \tag{9.5}$$

for all points x on $\tilde{K}$, i.e. points x and g(x) are identified for all the g operations in G. In fact we do not require (9.5) because our Lagrangian is still E_6 invariant so really all we need to have is that acting with g on x is equivalent to a fixed rotation U_g of E_6 i.e.

$$\psi(g(x)) = U_g \psi(x). \tag{9.6}$$

Technically this is a homorphism of G into E_6 since $U_g U_{g'} = U_{gg'}$. Since G has N(G) elements we are selecting N(G) E_6 rotations. Some may be equal and so we have at most N(G) distinct E_6 rotations. Under a gauge transformation V in E_6 $\psi \to V\psi$; this is only compatible with (9.6) if

$$[V, U_g] = 0 \tag{9.7}$$

so we conclude E_6 is broken to J, consisting of the group of transformations which commute with U_g, $\forall g$.

To see this somewhat more physically, consider a Wilson loop defined on a contour γ in K

$$W = \exp\{-i \int_\gamma T^a A^a_m dx^m\} \qquad a = 1\ldots78 \qquad m = 5,6,7\ldots10. \tag{9.8}$$

The W's are clearly elements of E_6; from the point of view of d = 4 space $I^a = \int_\gamma A^a_m dx^m$ is a spin zero field, i.e. a "Higgs boson" in the adjoint representation. If $I^a \neq 0$, the "Higgs boson" has a non-zero v.e.v. In the vacuum the E_6 gauge field strength F_{mn} vanishes, but on a multiply connected manifold a loop may not necessarily be contractible so that we cannot use Stokes' theorem to relate $\int A.d\ell$ to $\int F\, ds$. Since $A^a \neq 0$ necessarily, $I^a \neq 0$ and hence the symmetry is broken to the group J consisting of all transformations in E_6 which commute with all $W \neq 1$. The W's represent U_g. Every inequivalent γ such that $W \neq 1$ is another element of U_g.

A simple example of this procedure is the following. Consider gauge fields belonging to SU_3 defined on $K_0 = S_2$ the two sphere. Let us now act with a discrete Z_2 on S_2 so $K = S_2/Z_2$, the real projective plane. Since antipodal points are identified, a contour which goes from the north to the south pole of K is closed, but clearly non-contractible. On the other hand, the contour in which we cover this path twice is contractible

$$\pi_1(S_2/Z_2) = Z_2, \tag{9.9}$$

so W has in general two non-zero elements $U_1 = 1$ and U_2 not necessarily equal to one. It depends on how we embed Z_2 in SU(3). The possibilities are

$$\text{a)} \begin{pmatrix} 1 & & \\ & 1 & \\ & & 1 \end{pmatrix} \quad \text{b)} \begin{pmatrix} 1 & & \\ & -1 & \\ & & -1 \end{pmatrix} \quad \text{c)} \begin{pmatrix} -1 & & \\ & -1 & \\ & & 1 \end{pmatrix} \quad \text{d)} \begin{pmatrix} -1 & & \\ & 1 & \\ & & 1 \end{pmatrix} \tag{9.10}$$

In case a) SU(3) is unbroken, i.e. J = SU(3) while in b), c), d),

$J = SU(2) \times U(1)$. Performing the path twice corresponds to a contractible loop, i.e. going from the north pole to the south pole, reappearing at the north pole and going back down to the south pole is contractible. Similarly, for $G = Z_5$, any contour γ when traced out five times leads to a contractible contour, i.e. a $W = 1$. For $G = Z_5 \times Z_5$ we thus have at most twenty five non unit W's, which of course satisfy the group properties of $Z_5 \times Z_5$.

The analogous procedure of finding the embeddings of G in E_6 was examined by Witten [2], by Breit, Ovrut and Segrè, and by Sen [3]. The unbroken subgroup J must contain at least $SU_3 \times SU_2 \times U_1$ for phenomenological reasons. In fact, since the embedding leaves the rank of the group unchanged, J must be at least as large as $SU_3 \times SU_2 \times U_1 \times U_1 \times U_1$.

If we classify E_6 by its $SU_3 \times SU_3 \times SU_3$ subgroups and let the first SU_3 be the color group, which must remain unbroken, as example of a possible embedding of $Z_5 \times Z_5$ in E_6 is

$$U = \begin{pmatrix} 1 & & \\ & 1 & \\ & & 1 \end{pmatrix} \times \begin{pmatrix} \alpha^j & & \\ & \alpha^j & \\ & & \alpha^{-2j} \end{pmatrix} \times \begin{pmatrix} \beta^k & & \\ & \beta^k & \\ & & \beta^{-2k} \end{pmatrix} \tag{9.11}$$

$$\alpha^5 = 1 \quad \beta^5 = 1 \qquad j,k = 1,2,3,4,5.$$

It remains of course to be proven that this mechanism for E_6 dynamical breaking actually takes place, i.e. that the physical vacuum corresponds to $W \neq 1$. So far, all we know is that it is possible; analogous calculations for a far simpler case have been carried out by Hosotani [4].

In Ref. [3], this problem was studied using the method of Weyl weights. The E_6 U's in which G is embedded are written generally as

$$U = \{\exp i\lambda_i H_i\} \qquad i = 1,2\ldots6 \tag{9.12}$$

where the H's are elements of the Cartan subalgebra of E_6 (maximal commuting set of generators) and λ are six dimensional real parameters $\lambda = (a_1 \ldots a_6)$. The λ's must have no projection along the direction of color SU_3 or weak SU_2; this fixes λ as

$$\lambda = (-c,c,a,b,c,0). \tag{9.13}$$

We can then write U as 27 dimensional diagonal matrix that depends on three real parameters a, b and c [5]. The diagonal elements are given in Table 1, together with the transformation properties under $SU(3)^c \times SU(2)^W$, under SO(10) and under SU(5) of the elements of the 27 plet on which they act. Another useful piece of input data is that, if $\vec{\alpha}$ is a root of E_6 corresponding to a generator E_α, the mass matrix of the corresponding gauge boson is given by $M_{\alpha\beta}$ where

$$M^2_{\alpha\beta} \propto g^2\, \delta_{\alpha\beta}(\alpha,\lambda)^2. \tag{9.14}$$

The non-zero E_6 roots are given in Table 2.

As an example, let us find all the embeddings of $G = Z_5 \times Z_5$ that break E_6 to $SU_4 \times SU_2^W \times U_1 \times U_1$ (Pati-Salam) [6]. From Table 1 we see that if the (e,ν) and (u,d) are to lie in a 4 of SU_4 we must have $b = -c$. (We have used the fact that the (e,ν) corresponds to the weak doublet in the $[16,\bar{5}]$ and the (u,d) to the doublet in the $[16,10]$). We can then either let a be independent or fix a = 0,c,2c,3c, or 4c. (Because of the Z_5 symmetries $5c \equiv 0$, $6c \equiv c$, etc.). For a independent we consult table 2 and find that (λ,α) is zero only for $\alpha = \pm(000001)$, $\pm(0100\text{-}10)$, $\pm(0\text{-}10011)$, $\pm(10001\text{-}1)$, $\pm(\text{-}100100)$, $\pm(\text{-}1101\text{-}1\text{-}1)$, $\pm(\text{-}10010\text{-}1)$, and the six roots (000000). So we have 20 massless vector bosons and these roots span a representation of $SU_4 \times SU_2^w \times U_1 \times U_1$. Similarly for a = 0 we have 46 massless vector bosons and the gauge symmetry is $SO_{10} \times U_1$; for a = c we have $SU_4 \times SU_2^w \times U_1 \times U_1$; for a = 2c, $SU_5 \times SU_2 \times U_1$; for a = 3c, $SU_5 \times SU_2 \times U_1$; and for a = 4c, $SU_4 \times SU_2^w \times U_1 \times U_1$. We break to the Pati-Salam group in three cases; (i) an independent, $G = Z_5 \times Z_5$, $e^{ic} = \alpha^j$, $e^{ib} = \alpha^{-j}$, $e^{ia} = \beta^k$; (ii) a = c, $G = Z_5$, $e^{ic} = \alpha^j$, $e^{ib} = \alpha^{-j}$, $e^{ia} = \alpha^j$; (iii) a = -c, $G = Z_5$, $e^{ic} = \alpha^j$, $e^{ib} = \alpha^{-j}$, $e^{ia} = \alpha^{-j}$.

By examining all the values for a, b and c allowed by the discrete symmetry, one can exhaust all the symmetry breaking induced by G. In particular we recover the embeddings found by the first method. In addition, we can find for which symmetries the effective Higgs VEVs are zero in some directions. We use this method in the next section to generate naturally light Higgs doublets.

Table 1.

Diagonal elements of the Wilson loops U.

U^{diag}	$SU_2^w \times SU_3^c$	SO_{10}	SU_5
exp{i(b-3c)}	(1,1)	1	1
exp{i(c+a-b)}	(2,1)	10	5
exp{2ic}	(1,3)		
exp{i(2c-a}	(2,1)	10	$\overline{5}$
exp{i(c-b)}	$(1,\overline{3})$		
exp{i(a+c)}	(1,1)	16	1
exp{ib}	(2,1)	16	$\overline{5}$
exp{i(a-c)}	$(1,\overline{3})$		
exp{i(a-b-2c)}	(1,1)	16	10
exp{i(b-a)}	$(1,\overline{3})$		
exp{-ic)}	(2,3)		

<u>Naturally Light Higgs Doublet Problem</u> (section c)

We would like to have light Higgs doublets to set the electroweak scale and give masses through Yukawa couplings to the ordinary fermions. In particular, as we shall see in the next section, it is necessary that at least one of the light doublets be in the $[10,\overline{5}]$ or [10,5] representations under $[SO_{10},SU_5]$. At the same time, supersymmetric E_6 theories contain extra color triplets than can mediate nucleon decay via dimension 4 or 5 baryon number violating, J invariant, operators. These triplets must be given very large masses. It is very difficult, even with fine tuning, to keep the doublet light while making the triplet heavy. Fortunately, the same mechanism that breaks the gauge symmetry gives us a natural method for splitting the doublets from the triplets. Although the method does not depend on which discrete symmetry we use, for simplicity we assume that $G = Z_5 \times Z_5$.

The nontrivial gauge fields that give rise to Wilson loops different from unity can also lead to the disappearence of chiral superfields from the zero mass spectrum. In other words, one of the 27's can "couple" to the $\overline{27}$ through a term involving an effective Higgs 78 VEV ($27 \times \overline{27} \times 78$ contains a singlet). These fields thus acquire a mass of order the inverse radius of the compactified dimensions, presumably the Planck mass, while the other four 27's remain massless. (Note that we cannot have a $27 \times \overline{27}$ bare mass term since, until E_6 is broken, the chiral superfields are all massless zero modes). Let us call the 27 and $\overline{27}$ that pair off χ and $\overline{\chi}$, and the remaining 27's ψ. We then expect all the components of χ and $\overline{\chi}$ to gain huge masses and disappear from the spectrum. As we now show, however, it is possible for some of these components to remain naturally massless. This occurs when the diagonal entries of the U's that multiply these components are unity. That is, the associated effective Higgs VEVs are zero in these directions.

Table 2.

Nonzero E_6 Roots.

Root $\vec{\alpha}$	(λ,α)	Root $\vec{\alpha}$	(λ,α)
(000001)	0	(0100-10)	0
(0-10011)	0	(10001-1)	0
(-11001-1)	3c	(-210000)	3c
(0-111-1-1)	a+b-2c	(00-12-10)	2b-a-c
(0-12-10-1)	2a-b-c	(1-11-110)	a-b-c
(101-10-1)	a-b-c	(1-11-11-1)	a-b-c
(0-11-101)	a-b-c	(001-1-10)	a-b-c
(0-11-100)	a-b-c	(00100-1)	a
(0-1101-1)	a	(00100-2)	a
(-1010-10)	a	(-1-11000)	a
(-1010-1-1)	a	(-100100)	b+c
(-1101-1-1)	b+c	(-10010-1)	b+c
(010-110)	2c-b	(000-120)	2c-b
(010-11-1)	2c-b	(-110-101)	2c-b
(-100-111)	2c-b	(-110-100)	2c-b
(-101-110)	a-b+2c	(-111-10-1)	a-b+2c
(-101-11-1)	a-b+2c	(-11-1011)	3c-a
(-12-1000)	3c-a	(-11-1010)	3c-a

As an example, let us use Table 1 to find which components of χ and $\overline{\chi}$ are left massless for the breaking to $SU_4 \times SU_2^w \times U_1 \times U_1$ given in section 2 : (i) for $b = -c$, a independent, none of the U^{diag} is one, the effective Higgs VEV then has no zeros, and all the components of χ and $\overline{\chi}$ are massive; (ii) for $b = -c$, $a = c$, $U^{diag} = 1$ for the color triplet in the $[16,\overline{5}]$ and the singlet in the $[16,10]$, and hence those components of χ and $\overline{\chi}$ remain massless; (iii) for $b = -c$, $a = -c$, the color triplet weak singlet in the $[16,10]$ and the $[16,1]$ remain massless.

Using Tables 1 and 2, we can find for which values of the parameters a, b and c we obtain light doublets and what the resulting gauge symmetries are. The light doublets in χ can be used as Higgs fields to break $SU_2^w \times U_1$ and to give masses to quarks and leptons. The corresponding light doublets in $\overline{\chi}$ cannot couple to ordinary matter, and hence we ignore them. We list below all the cases in which at least one weak doublet in χ is light while the color triplets are all heavy :

(i) $b = a + c$, a and c arbitrary; massless : doublet in $[10,5]$; gauge symmetry : $SU_3^c \times SU_2^w \times U_1 \times U_1 \times U_1$.

(ii) $b = 4c$, $a = 3c$; massless : doublet in $[10,5]$; gauge symmetry : $SU_5 \times SU_2 \times U_1$.

(iii) $b = 0$, $a = 4c$; massless : doublets in $[10,5]$ and $[16,\overline{5}]$, the singlet $[16,1]$; gauge symmetry : $SU_3^c \times SU_2^w \times SU_2 \times U_1 \times U_1$.

(iv) $a = 2c$, b arbitrary; massless : doublet in $[10,\overline{5}]$; gauge symmetry : $SU_3^c \times SU_2^w \times U_1 \times U_1 \times U_1$.

(v) $a = 2c$, $b = 0$; massless : doublets in $[10,\overline{5}]$ and $[16,\overline{5}]$, singlet in $[16,10]$; gauge symmetry : $SU_3^c \times SU_2^w \times SU_2 \times U_1 \times U_1$.

(vi) $a = 2c$, $b = 3c$; massless; doublets in $[10,5]$ and $[10,\overline{5}]$, the singlet $[1,1]$; gauge symmetry : $SU_3^c \times SU_2^w \times SU_2 \times U_1 \times U_1$.

(vii) $a = 2c$, $b = 4c$; massless : doublet in $[10,\overline{5}]$; gauge symmetry : $SU_5 \times SU_2 \times U_1$.

(viii) $b = 0$, a and c arbitrary; massless : doublet in $[16,\overline{5}]$; gauge symmetry : $SU_5 \times SU_2 \times U_1$.

The embeddings of G that give rise to these values for a, b and c can be readily found. For example, in case (i) let $e^{ia} = \alpha^j$, $e^{ic} = \alpha^k$, and $e^{ib} = \alpha^{j+k}$, $\alpha^5 = 1$. This corresponds to $G = Z_5 \times Z_5$. In case (ii) let $e^{ic} = \alpha^j$, $e^{ia} = \alpha^{3j}$, $e^{ib} = \alpha^{4j}$, which corresponds to $G = Z_5$.

It is worth re-emphasizing that our light Higgs doublets were not obtained by fine tuning. Setting, for example, $b = 3c$, $a = 2c$ as in case (vi) is a choice of parameters (optimistically a minimum for the vacuum configuration), but not a fine tuning, since e^{ia}, e^{ib} and e^{ic} are restricted by the discrete symmetry to be fifth roots of unity. These Higgs doublets are totally massless until supersymmetry is broken spontaneously.

Yukawa Couplings. (section d)

The relevant E_6 chiral superfields are (1) a 27 and $\overline{27}$, denoted by χ and $\overline{\chi}$, some of whose components can be naturally light and (2) four massless 27's which we call ψ_i, $i=1\ldots4$. The ψ's must couple to one another, and to χ, so as to produce the known low energy spectrum and interactions. As a guide we use the E_6 invariant coupling of three 27's, which we call respectively ρ, λ and ϕ. We label the states by their SO_{10} and SU_5 transformation properties. This coupling is

$$
\begin{aligned}
\mathcal{L}_{27 \times 27 \times 27} = {} & \rho[1,1]\{\lambda[10,5]\phi[10,\bar{5}]\} \\
& + \rho[16,1]\{\lambda[10,5]\phi[16,\bar{5}]\} \\
& + \rho[16,\bar{5}]\{\lambda[10,\bar{5}]\phi[16,10] + \lambda[10,5]\phi[16,1]\} \\
& + \rho[10,\bar{5}]\{\lambda[16,10]\phi[16,\bar{5}] + \lambda[10,5]\phi[1,1]\} \\
& + \rho[10,5]\{\lambda[16,10]\phi[16,10] + \lambda[16,1]\phi[16,\bar{5}] \\
& + \lambda[1,1]\phi[10,\bar{5}]\} + \rho[16,10]\{\lambda[16,10]\phi[10,5] \\
& + \lambda[10,\bar{5}]\phi[16,\bar{5}]\} + (\phi \leftrightarrow \lambda).
\end{aligned}
\tag{9.15}
$$

The relative coefficients will change under renormalization from the Planck scale to the GeV scale, but eq. (9.15) specifies the allowed couplings. We are faced with two immediate phenomenological issues, nucleon decay and the fermion mass spectrum. Each term is an E_6 invariant coupling of the form of Eq. (9.15). We specify the $[SO_{10}, SU_5]$ assignments of the fields by the suggestive labels

$$
\begin{aligned}
& [1,1] = N_0, \quad [16,1] = \nu^c, \\
& [10,5] = \begin{pmatrix} N^c \\ E^c \\ D \end{pmatrix}, \quad [10,\bar{5}] = \begin{pmatrix} N \\ E^- \\ D^c \end{pmatrix} \\
& [16,\bar{5}] = \begin{pmatrix} \nu \\ e^- \\ d^c \end{pmatrix}, \quad [16,10] = \begin{pmatrix} & u & \\ e^c & & u^c \\ & d & \end{pmatrix}
\end{aligned}
\tag{9.16}
$$

anticipating placing the ordinary fermions primarily in the 16 representation of SO_{10}. Our fermion spectrum consists of four conventional families, four additional charge $-1/3$ quarks, D_i, and extra leptons in ψ_i, χ and $\bar{\chi}$. We also have the scalar superpartners of each of these fields.

The coupling in (9.15) is E_6 invariant. At first sight, it would seem that all Yukawa couplings are E_6 invariant, leading to problems with e.g. the fermion mass spectrum. This is analogous to ordinary electroweak theory in which the Yukawa couplings are $SU(2) \times U_1$ invariant; the symmetry is broken by non-zero v.e.v.'s for the Higgs doublets.

The situation here is different, as Witten [2] has pointed out. It is once again the discrete symmetry which comes to the rescue. If there were no discrete symmetries on the manifold, the Yukawa couplings would be E_6 invariant, but there would also be e.g. 100 families. With discrete symmetries present, the massless modes are those invariant under $G + J$. Since quarks and leptons transform differently under J, their combinations which remain massless will also transform differently under G. The mass matrix appears to come from only J invariant Yukawa interactions, not E_6 invariant ones.

Let us give a schematic example of how this works : consider a superpotential which is invariant under some symmetry group S with the fields written as

$$\Omega_i = \begin{pmatrix} \phi_i^\alpha \\ \\ \zeta_i \end{pmatrix} \tag{9.17}$$

$i = 1 \ldots N$ being a family index and α representing some additional degree of freedom of the ϕ_i fields. The superpotential will have the form

$$\mathcal{L}_{Yuk} = \sum_{i,j,k=1}^{N} h_{ijk}\, \Omega_i \Omega_j \Omega_k \tag{9.18}$$

where h_{ijk} are family couplings. We have written explicitly the appropriate couplings for S, but $\mathcal{L}_{yuk}$ is implicitly S invariant (for example in E_6, $\Omega_i \sim 27$ and there is an invariant $27 \times 27 \times 27$ coupling). We are only interested however in the zero modes. These fields $\tilde{\phi}$ and $\tilde{\zeta}$ are linear combinations of ϕ and ζ

$$\tilde{\phi}_a = c^\alpha_{ai}\, \phi_{i,\alpha}$$

$$\tilde{\zeta}_a = b_{ai}\, \zeta_i \qquad a = 1 \ldots M \qquad M \leq N \tag{9.19}$$

where $M \leq N$ since not all fields remain massless. We rewrite $\mathcal{L}_{yuk}$, keeping only the terms involving $\tilde{\phi}_a$ and $\tilde{\zeta}_a$ to obtain $\tilde{\mathcal{L}}_{yuk}$, the effective zero mode superpotential. Since the c and b matrices are not equal, $\tilde{\mathcal{L}}_{yuk}$ is no longer S invariant. In our case $\tilde{\mathcal{L}}_{yuk}$ is J invariant, not E_6 invariant. This is an important and desirable result since it is well known that E_6 invariant Yukawa couplings with Higgs bosons only in the 27 representation cannot reproduce the desired mass spectrum [7].

The gauge boson couplings, on the other hand, are E_6 invariant. Their value at low energies is determined by a renormalization group analysis starting with their equality at the unification scale. The reason for this equality is that the gauge bosons are holonomy singlets, coming from the (1,78) representation of $SU(3) \times E_6$. In terms of our simplified example we would say the gauge boson coupling is

$$\Omega_i^\dagger\, \gamma^\mu\, \Omega_i\, A_\mu \tag{9.20}$$

so in fact the zero modes $\tilde{\phi}_a$ and $\tilde{\zeta}_a$ couple the same way as ϕ and ζ do. A detailed analysis of the renormalization group and of Yukawa couplings for E_6 models arising from superstrings has been performed by Dine et al. [8], to which we refer for details. They show which models are viable and which aren't.

A good deal of work has been going on in the study of manifolds and discrete symmetries, but it is outside the purview of these lectures [9]. At the present time there does not appear to be a fully acceptable phenomenological model [10].

TRUNCATED LAGRANGIAN (chapter 10)

Introduction (section a)

We saw in the last four chapters how one proceeds from a modified ten dimensional Lagrangian describing supergravity coupled to super Yang-Mills to a theory compactified on a Calabi-Yau manifold with discrete symmetries.

To get the "correct" four dimensional theory resulting from compactification is a Herculean task; not only must we know all the massless modes and their interactions with the massive modes, but these must then be integrated out to yield effective interactions.

Fortunately an illustrative example has been proposed by Witten [1], which hopefully contains many of the key features of compactification. In particular it may be a useful guide for analyzing how supersymmetry is broken. This chapter is basically a restatement of the contents of his paper.

The Four Dimensional Lagrangian (section b)

We are trying to find a Lagrangian with the essential features of our compactification. We start by assuming our fields depend only on the first four coordinates, i.e., for an arbitrary field ψ

$$\frac{\partial}{\partial x^i} \psi(x) = 0 \qquad i = 5,6,\ldots 10. \tag{10.1}$$

As we saw already in chapter 5c), the supergravity Lagrangian reduces in $d = 4$ to an $N = 4$ supersymmetry, whose generators we call Q^A. The group of rotations of the compact manifold in O(6), under which the six coordinates obviously transform as the six dimensional representation and the Q^A as the four dimensional spinorial one. As sketched in 5c), if we demand invariance under an SU(3) subgroup U of O(6) (remember O(6) is isomorphic to SU(4)), only $N = 1$ supersymmetry remains since the Q^A transform like the $3 + 1$ of SU(3). We are then demanding that the fields be invariant not only under translations of the extra coordinates (condition (10.1)) but under rotations as well.

Scalars and Pseudoscalars (section c)

Turning back to (5.17), we see that the vectors e^i_μ and $B_{\mu i}$ transform like the $3 + \overline{3}$ of SU(3) so they don't survive the test of invariance under SU(3) rotations. The graviton clearly does, as does the scalar field ϕ. From $B_{\mu\nu}$ we form the field strength $H_{\mu\nu\lambda}$ which in four dimensions is dual to a pseudoscalar field. To see this, let us write using differential forms, the equation of motion and the Bianchi identity for H. They are respectively

$$d{*H} = 0 \tag{10.2a}$$

$$dH = - Tr(F^2) + Tr(R^2). \tag{10.2b}$$

H is a 3-form so $*H$ is a 1-form in $d = 4$, which we call Y. Introducing a mass M for dimensional purposes, we have

$$dY = 0 \Rightarrow Y = Md\theta \tag{10.3}$$

and the Bianchi identity

$$\begin{aligned} dH &= d{**H} = d{*Y} = Md{*d\theta} \\ &= M \,\Box\, \theta = -Tr(F^2) + Tr(R^2). \end{aligned} \tag{10.4}$$

Equation (10.4) is the equation for an axion θ (remember $F^2 = F\wedge F \propto F_{\mu\nu}\tilde{F}^{\mu\nu}$). The $H_{\mu\nu\lambda} H^{\mu\nu\lambda}$ term in the Lagrangian will thus lead to the equations of motion for a pseudoscalar [2] θ coupled to $F\tilde{F}$ (we will see later that R^2 does not contribute for this simple compactification). The second pseudoscalar ζ also comes from the antisymmetric tensor B_{mn}; we see that we can form an SU(3) invariant

$$B_{ij} = \varepsilon_{ij}\,\zeta \tag{10.5}$$

where ε_{ij} is the SU(3) invariant tensor $\varepsilon_{5,6} = \varepsilon_{7,8} = \varepsilon_{9,10} = 1$, $\varepsilon_{ij} = -\varepsilon_{ji}$.

The two scalars both come from the metric tensor; the first is simply the dilation field ϕ described in chapter 5. The second is a mode σ that arises from compactification of the metric in ten dimensions.

$$g^{(10)}_{ij} = \delta_{ij}e^{\sigma} \qquad i,j = 5,6,\ldots 10. \tag{10.6}$$

We then do a Weyl-rescaling of $g_{\mu\nu}$: from (5.16) we see that the ten dimensional metric $g^{(10)}_{\mu\nu}$ is rescaled as

$$g^{(10)}_{\mu\nu} = g_{\mu\nu}\, e^{-3\sigma} \tag{10.7}$$

and hence the kinetic terms get rescaled as well. This completes the discussion of the supergravity multiplet. Two further remarks of interest should probably be made : the first is that we expect both ϕ and σ to be massless. ϕ is simply a dilational mode associated with scale invariance and therefore is massless, while σ is massless because we are compactifying on a manifold with no intrinsic size, i.e. no scalar curvature. Since the theory is supersymmetric we expect θ and ζ to pair up with ϕ and σ to form two massless complex scalar fields; of course they may well acquire a mass when supersymmetry is broken setting a new scale. The second remark is that in this simplified compactification our manifold corresponds topologically to $S_1 \times S_1 \times S_1 \times S_1 \times S_1 \times S_1$, i.e. it has no curvature and in fact has vanishing connection ω. The Lorentz Chern-Simons form is therefore absent.

Let us now turn our attention to the gauge fields A^{α}_m. The ten components of A^{α}_m are four dimensional vector fields A^{α}_{μ} and six four dimensional scalars A^{α}_i $i = 5,6,\ldots 10$ belonging to the adjoint representation of our symmetry group. If we demand invariance under rotations of the SU(3) group U, the A^{α}_i disappear from the low energy spectrum since they transform as $3 + \overline{3}$ under U. A more interesting procedure is to proceed as we did in the previous section and demand invariance not under U, but under the direct sum of $U + \tilde{U}$, where $\tilde{U}$ is an SU(3) subgroup of our symmetry group E_8. Clearly the $\tilde{U}$ will be the SU(3) in the SU(3) $\times$ E_6 decomposition of E_8. We already saw that the 248 A_m transform under $\tilde{SU}(3) \times E_6$ as

$$(8,1) + (3,27) + (\overline{3},\overline{27}) + (1,78). \tag{10.8}$$

In $d = 4$ the first four components of (1,78) survive to form the E_6 vector gauge fields since they are singlets under both U and $\tilde{U}$ while none of the (8,1) fields survive since they are octets under $\tilde{U}$ and either singlets or triplets under U. On the other hand, the $d = 4$ scalar components $(A_{5,6,7,8,9,10})$ of (3,27)) and $(\overline{3},\overline{27})$ can survive since they are 3's or $\overline{3}$'s, under both U and $\tilde{U}$. A general form for such an invariant is constructed using the generators T^a_x of E_8 belonging to the (3,27) representation normalized so that

$$\mathrm{Tr}(T^a_x (T^b_y)^*) = \delta^a_b \delta^x_y$$

$$\mathrm{Tr}(T^a_x T^b_y T^c_z) = \varepsilon^{abc} d_{xyz} \qquad (10.9)$$

where the trace is taken in the adjoint representation. In (10.9) ε^{abc} is just the usual antisymmetric tensor with a,b,c = 1,2,3 and d_{xyz} is the invariant coupling of three 27's in E_6. The form of the invariant under $U + \tilde{U}$ is

$$L^a = \sum_x T^a_x C_x \qquad (10.10)$$

where C_x are scalar fields in d = 4. This is analogous to writing $P = \tau_i \pi_i$ in SU(2) or $M = \lambda_i \pi_i$ in SU(3) i.e. writing mesons as components of a matrix. The index a = 1,2,3 and

$$L^1 = \frac{A^5 + iA^6}{\sqrt{2}} \qquad L^2 = \frac{A^7 + iA^8}{\sqrt{2}} \qquad L^3 = \frac{A^3 + iA^{10}}{\sqrt{2}} \qquad (10.11)$$

i.e. L^a transforms like the 3 of SU(3).

It may be redundant, but it is probably worth reiterating how it is that we end up with chiral fermions in this since this is often a source of confusion. Let us go back to fields A^α_m

$$A^\alpha_m = (A^\alpha_\mu, A^\alpha_a, A^\alpha_{\bar{a}})$$

$$\alpha = (8,1) + (3,27) + (\overline{3},\overline{27}) + (1,78). \qquad (10.12)$$

We saw that the only zero modes are those invariant under $U + \tilde{U}$, $A^{(78)}_\mu$ and C^x and its chiral anti-partner $C^{\bar{x}}$, where x is a 27 representation; note of course that $A^{(78)}_\mu$ is its own anti-partner. This is because with e.g. $A_{\bar{a}}$ we can form an invariant of $\bar{a}$ and the 3 in (3,27), but not of $\bar{a}$ with the $\overline{3}$ in $(\overline{3},\overline{27})$. The upshot is that we end up with one chiral left handed 27 and its anti-partner. If a left handed $\overline{27}$ and anti-partner had survived, we would have had a vector-like theory, i.e. one containing left 27 and left $\overline{27}$ (equivalent to right 27) and anti-partners.

The Bosonic Lagrangian (section d)

The terms in that need to be reduced to four dimensions are read off from (5.11) and (5.12). Consider e.g. $F^\alpha_{mn} F^{\alpha mn}$; we must evaluate $\mathrm{Tr}(F_{\mu i} F^{\mu i})$ and $\mathrm{Tr}(F_{ij} F^{ij})$, where i,j run over the compact dimensions. We then keep what we find in addition to the standard $\mathrm{Tr}(F_{\mu\nu} F^{\mu\nu})$.

$$\mathrm{Tr}(F_{\mu i} F^{\mu i}) = 6\, e^{2\sigma} D_\mu C^*_x D^\mu C_x$$

$$\mathrm{Tr}(F_{ij} F^{ij}) = 2e^{-2\sigma} \sum_{a,b} \mathrm{Tr}\{[L^a,L^b][L^*_b,L^*_a] + [L^a,L^*_b][L^b,L^*_a]\} \qquad (10.13)$$

$$\mathrm{Tr}(F_{\mu\nu} F^{\mu\nu}) = e^{6\sigma}\, \mathrm{Tr}(F_{\mu\nu} F^{\mu\nu})$$

where on the right the contraction is done with a d = 4 metric tensor. This is the origin of the various e^σ factors and follows directly from (10.6) and (10.7), e.g. we have an $e^{3\sigma}$ for $g^{(10)}_{\mu\nu}$ and an e^σ for $g^{(10)}_{ij}$. Witten [1], also evaluates

$$\sum_{a,b} \mathrm{Tr}[L^a,L^b][L^*_b,L^*_a] = 24|d_{xyz}\, C^y C^z|^2$$
$$\sum_{a,b} \mathrm{Tr}[L^a,L^*_a][L^b,L^*_b] = \frac{g}{f}\sum_\alpha (C^*,\lambda^\alpha C)^2 \tag{10.14}$$

where λ^α are the generators of E_6 normalized to $\mathrm{Tr}(\lambda^\alpha\lambda^\beta) = \delta^{\alpha\beta}$, with the trace taken in the adjoint representation of E_6. If instead we take the trace in the adjoint representation of E_8 we find $\mathrm{Tr}(\lambda^\alpha\lambda^\beta) = f\delta^{\alpha\beta}$. We must similarly evaluate $H_{mnp}H^{mnp}$, remembering the Chern-Simons term in H_{mnp}. Putting all the pieces together, Witten [1] gives an expression for the bosonic pieces of the Lagrangian, as read off from (5.11) and (5.12) after compactification. They are

$$\begin{aligned} e^{-1}\mathcal{L}_{\text{Bosonic}} = &-\frac{1}{2}R^{(4)} - 3\partial_\mu\sigma\partial^\mu\sigma - \frac{9}{16}\frac{\partial_\mu\phi\partial^\mu\phi}{\phi^2} \\ &-\frac{3}{2}e^{-2\sigma}\phi^{-3/2}(\partial_\mu\zeta - \frac{i\kappa}{2}C^*_x D_\mu C^x)^2 - \frac{1}{4}\phi^{-3/2}e^{-6\sigma}(\partial_\mu\theta)^2 - \frac{1}{4}\theta\,\mathrm{Tr}(F_{\mu\nu}\tilde{F}^{\mu\nu}) \\ &-\frac{1}{4}\phi^{-3/4}e^{3\sigma} f\,\mathrm{Tr}(F_{\mu\nu}F^{\mu\nu}) - 3e^{-\sigma}\phi^{-3/4}D_\mu C^*_x D^\mu C_x \\ &-\frac{8}{3}g^2\phi^{-3/4}e^{-5\sigma}\left|\frac{\partial W}{\partial C}\right|^2 - \frac{9}{2}\frac{g^2}{f}\phi^{-3/4}e^{-5\sigma}\sum_\alpha (C^*\lambda^\alpha C)^2 \\ &-8\kappa^2 g^2\phi^{-3/2}e^{-6\sigma}|W|^2 \end{aligned} \tag{10.15}$$

where

$$W = \frac{8}{\sqrt{2}}\, g\, d_{xyz}\, C^x C^y C^z \tag{10.16}$$

is the cubic superpotential for the fields C.

This is very nice, but it is far from clear how to recast this Lagrangian in the form of chiral superfields coupled to supergravity. The rules [3,4] for this state that there exists in the space of the chiral superfields, which we call z^i, a line element

$$ds^2 = \frac{\partial^2 K}{z^i z^*_j}\, dz^i dz^*_j = g_i^{j*} dz^i dz^*_j \tag{10.17}$$

where K is the Kähler potential. The superpotential is a function of z^i and the potential energy is

$$\begin{aligned} V(z^i,z^*_j) = {} & e^{K/M^2}\{g_i^{j*}\frac{DW}{Dz^i}\frac{DW}{Dz^*_j} - \frac{3}{M^2}|W|^2\} \\ & + \frac{1}{2}\sum_\alpha (D^\alpha)^2. \end{aligned} \tag{10.18}$$

Here $\frac{1}{M^2} \equiv \frac{1}{M^2_{\text{Planck}}}$ = Newton's constant and $(D^\alpha)^2$ is the so-called D-term which we will specify shortly. The metric g_i^{j*} is defined in (10.17) as $K''_i{}^{j*}$ where" means the second derivative and

$$\frac{DW}{Dz^i} \equiv D_{z^i}W = W'_i + \frac{K'_i W}{M^2}. \tag{10.19}$$

In practice, it is convenient to introduce G, a function of the fields, defined by

$$G = \frac{K}{M^2} + \ln \frac{|W|^2}{M^6} \tag{10.20}$$

whose derivatives are given by

$$G_i' = \frac{K_i'}{M^2} + \frac{W_i'}{W} = \frac{D_{z^i}W}{W} \tag{10.21a}$$

$$G_i''^{j*} = \frac{1}{M^2} g_i^{j*}. \tag{10.21b}$$

In deriving these expressions, remember that W is a function of z^i, W^* of z_i^* and $|W|^2 = WW^*$. The potential, or rather the bosonic part of the interaction Lagrangian is given by,

$$\begin{aligned} e^{-1} \mathcal{L}_{B,int.} &= -M^4 e^G\{G_i'(G''^{-1})^i_{j*} G'^{j*} - 3\} \\ &\quad - \frac{1}{2} g^2 \operatorname{Re} f^{-1}_{\alpha\beta}(K_i'(T^\alpha)^i_j z^j)(K_k'(T^\beta)^k{}_\ell z^\ell). \end{aligned} \tag{10.22}$$

The second piece in (10.22) being the so-called D term with $f_{\alpha\beta}$ a field dependent matrix function.

Fortunately, (10.14) can be recast in the form (10.22). Witten [1] shows how this occurs by introducing the two complex fields S and T defined as

$$\begin{aligned} S &= e^{3\sigma} \phi^{-3/4} + i\theta \\ T &= e^{\sigma} \phi^{3/4} - i/\sqrt{2}\,\zeta + C_x^* C^x. \end{aligned} \tag{10.23}$$

The corresponding Kähler potential is [5,6]

$$K = M^2\{- \ln(\frac{S+S^*}{M}) - 3 \ln(\frac{T+T^*}{M} - \frac{2C^*C}{M^2})\} \tag{10.24}$$

so (10.23) gives not only the kinetic terms for σ, ϕ, θ and ζ but for C and C^* as well. A lengthier discussion of this procedure is given in the recent article of Derendinger et al. [6], showing how one can deduce what S and T are from the fermion kinetic terms.

How good is this truncation procedure ? It reproduces some of the attractive features of Calabi-Yau manifolds in giving rise to chiral fermions, but clearly is unsatisfactory in the sense that it only has one 27 family and no $\overline{27}$'s. It clearly cannot represent the known fermions, but it may serve as a guide to the mechanisms for SUSY breaking. It is also useful to have a simplified version of the fuller theory for studying such phenomena as the breaking of $\mathcal{J}$.

Some of these topics will be treated in the next chapter.

BREAKING OF LOW ENERGY SYMMETRIES (chapter 11)

Breaking of SUSY (section a)

In the previous chapter we encountered the Kähler potential for the truncated Lagrangian. The superpotential dependended only on the C fields and there was no SUSY breaking, i.e. the gravitino $(\text{mass})^2$

$$m_{3/2}^2 = \langle e^{K/M^2} |W|^2/M^4 \rangle = 0. \tag{11.1}$$

The question of SUSY breaking is still an open one. The most attractive scenarios are those in which SUSY breaking occurs in the shadow world of the second E_8', which acts as a kind of hidden sector. Derendinger et al. [1] proposed that this occurs by giving a v.e.v. to the antisymmetric tensor field strength H_{ijk} of the supergravity multiplet. Dine et al. [2] observed that H_{ijk} and the E_8' gaugino condensate term $\bar{\chi}\Gamma_{ijk}\chi$ appear in the Lagrangian in the form of a perfect square so that one may give v.e.v.'s to both, break SUSY, but keep the cosmological constant equal to zero. The subject is not at all clear since the d = 10 field theory one starts with has to be modified to cancel anomalies by additional terms that are not supersymmetric themselves. The consistency of this procedure is in doubt. For a discussion of these points see the paper of Derendinger et al. [3].

We will follow the path suggested by Dine et al. [2] nevertheless. The idea goes as follows : the coupling of E_8' gauginos to supergravity introduces a term

$$(S + \bar{S})^2 \, \bar{\chi}\chi\bar{\chi}\chi \tag{11.2}$$

in the Lagrangian. As the E_8' becomes strong the gauginos condense and E_8' is broken to some subgroup $\mathcal{J}'$. This induces a superpotential for S

$$W(S) = M^3 \, h e^{-3S/2b_oM} \tag{11.3}$$

where M is the Planck mass, h is a constant and b_o is the coefficient of the one loop β function of $\mathcal{J}'$. They [2] also include another term in the superpotential which is independent of S coming from the v.e.v. of H_{ijk}

$$H_{ijk} \sim c \, \varepsilon_{ijk} \, M^3 \tag{11.4}$$

with c a constant [4]. The full superpotential is then

$$W = M^3(c + h \, e^{-3S/2b_oM}) + W(C) \tag{11.5}$$

where W(C), given in (10.16), is cubic in the matter fields C. We assume that W(S) is such that the associated potential energy has a minimum for finite S and that W(S) is non-zero at this minimum. We would hope then the SUSY breaking is communicated to the matter fields C^x and that the vacuum is stable.

The C^x Tree Level Potential (section b)

The machinery we need is in place; we use formulae (10.20), (10.21) and (10.22) together with the Kähler potential (10.24). The metric of (10.21b) is expressed as a 3 × 3 matrix in the fields, S, T and C^x (we use only one entry for the generic C^x),

$$g = \begin{pmatrix} M^2/\hat{S}^2 & 0 & 0 \\ 0 & 3M^2/\hat{Q}^2 & \frac{-6M^2C^y}{\hat{Q}^2} \\ 0 & \frac{-6M\overline{C}_x}{\hat{Q}^2} & \frac{6M}{\hat{Q}}\left(\delta_x^y + \frac{2\overline{C}_x C^y}{M\hat{Q}}\right) \end{pmatrix} \tag{11.6}$$

where

$$\hat{S} = S + S^* \qquad \hat{T} = T + T^*$$
$$\hat{Q} = \hat{T} - 2\,\frac{\overline{C}_x C^x}{M}. \tag{11.7}$$

The inverse metric is

$$g^{-1} = \begin{pmatrix} \hat{S}^2/M^2 & 0 & 0 \\ 0 & \frac{\hat{Q}T}{3M^2} & \frac{\hat{Q}C^y}{3M^2} \\ 0 & \frac{\hat{Q}\overline{C}_x}{3M^2} & \frac{\hat{Q}\delta_x{}^y}{6M} \end{pmatrix} \tag{11.8}$$

Finally we need to know, as derived in Ref. [5], that in equation (10.22)

$$f_{\alpha\beta} = \delta_{\alpha\beta}(S/M) \tag{11.9}$$

in order to construct the potential for our truncated Lagrangian. The result is

$$V = \mathcal{L}_B/e = \frac{M^4}{\hat{S}\hat{Q}^3}\left\{\frac{\hat{S}^2}{M^2}\,|D_sW|^2 + \frac{\hat{Q}}{6M}\left|\frac{\partial W}{\partial C^x}\right|^2\right\}$$
$$+ 18\,\frac{g^2M^3}{\hat{Q}^2}\,\mathrm{Re}\,\frac{1}{S}\,(\overline{C}_x(T^\alpha)^x_y\,C^y)^2. \tag{11.10}$$

The tree level vacuum state is obtained by minimizing V and is given by

$$\langle D_SW\rangle = 0$$
$$\langle C^x\rangle = 0. \tag{11.11}$$

We assume that the minimum of $\langle S\rangle$ determined by the condition (11.11) and the form of W is at some nontrivial value; this is clearly only possible for $c \neq 0$ in (11.5) since otherwise S runs off to infinity. The field S also acquires a tree level mass, whereas the C^x field does not, a point we will come back to shortly. The gravitino mass is

$$m_{3/2}^2 = \frac{1}{\langle\hat{S}\rangle\langle\hat{T}\rangle^3}\,|\langle W_s\rangle|^2 \tag{11.12}$$

and is non-zero because $\langle W_s\rangle \neq 0$ at the minimum of the potential. The two key points to notice are that minimum of <V> is zero, i.e. we have no cosmological constant and that the tree level potential does not fix <T> and hence $m^2_{3/2}$ is undetermined. Such models, i.e. with Kähler potentials like (10.24) have been discussed previously in the literature [6] and dubbed as "no scale models" for the reasons listed above.

The most striking feature of the tree level potential is that it doesn't fix <T>. The second is that it doesn't induce any $C\bar{C}$ terms, i.e. doesn't generate a C mass term. In fact by appropriate re-scalings

$$\tilde{C}^x = \sqrt{\frac{6M}{\langle\hat{T}\rangle}}\, C^x \qquad \tilde{W}_c = \frac{1}{6}\sqrt{\frac{M}{\langle\hat{S}\rangle 6}}\, W_c$$
$$g = \sqrt{\frac{M}{\langle\hat{S}\rangle}}\, g \tag{11.13}$$

we can [7] rewrite the C part of the potential, in terms of the re-scaled fields as

$$V = \left|\frac{\partial\tilde{W}}{\partial C^x}\right|^2 + \tilde{g}^2(\tilde{\bar{C}}_x (T^\alpha)^x_y\, \tilde{C}^y)^2. \tag{11.14}$$

Since W and hence $\tilde{W}$ is cubic in the C fields we see that V is quartic in the C fields and has no quadratic or cubic terms present. Furthermore radiative corrections, a la Coleman-Weinberg [8], do not change this result since they are supersymmetric and hence do not alter the couplings.

The difference between this model and ordinary hidden sector models can be traced back to the appearance of the matter fields in the logarithm of the Kähler potential. Neither <T> nor $\langle\bar{C}C\rangle$ is fixed; the C^4 term in W(C) then drives <C> to zero. At a mechanical level we see that the only place in which the C fields appear at a power less than quartic is through the Q dependence in

$$\frac{M^2}{\hat{S}\hat{\tilde{Q}}^3}\;\frac{\hat{S}^2}{M^2}\,|D_s W|^2. \tag{11.15}$$

In the more standard hidden sector [9] models in which the Kähler potential is quadratic in matter fields ϕ we would have, using (10.21)

$$\begin{aligned} K &= \phi^+\phi + \dots \\ G_\phi &= \frac{\phi^+}{M^2} + \frac{1}{W}\frac{\partial W}{\partial \phi} \\ G'' &= \frac{1}{M^2} \end{aligned} \tag{11.16}$$

and hence end up having, in the potential a term of the form

$$\begin{aligned} V_{\phi,\text{mass}} &\sim \phi^+\phi\; e^{K/M^2}\frac{|W|^2}{M^4} \\ &\sim m^2_{3/2}\,\phi^+\phi \end{aligned} \tag{11.17}$$

i.e. a ϕ mass^2 which is $\sim m^2_{3/2}$.

The One Loop Potential (section c)

The calculation we displayed is of course only a tree level calculation. One might hope to do better by constructing the one loop effective potential [8], including in it the SUSY breaking terms. This procedure involves in general the evaluation of

$$V(\phi_c) = - \sum_n \frac{1}{n!} \Gamma^{(n)}(0,0\ldots0)[\phi_c(x)]^n \tag{11.18}$$

where $\Gamma^{(n)}$ is the sum of all n'th order one particle irreducible Feynman diagrams and ϕ_c is the v.e.v. of the field ϕ. Usually one obtains divergences which are then re-absorbed into the definition of the original parameters in the Lagrangian. For example, if we take

$$\mathcal{L} = \frac{1}{2}(\partial_\mu\phi)^2 - \frac{1}{2}\mu^2\phi^2 - \lambda/4!\ \phi^4 \tag{11.19}$$

we find terms in $V(\phi_c)$ that are quadratically and logarithmically divergent. To remove these terms one adds to the potential terms which have the same structure as those in the original Lagrangian so that

$$V_{\text{renormalized}}(\phi_c) = V(\phi_c) + \frac{a\phi_c^2}{2} + \frac{b}{4!}\phi_c^4 \tag{11.20}$$

where a and b are constants which depend on a cutoff Λ, e.g. $a \sim \Lambda^2$, and are such that $V_{\text{ren.}}(\phi_c)$ is finite.

In our case we follow the same procedure, evaluating the Feynman diagrams with a cutoff Λ. As in the ϕ^4 theory, we find a quadratically divergent term, due to loops involving the graviton, the gravitino [10] and the S fields. Explicitly it is given by

$$V_{1,\text{loop},\Lambda^2} = - \frac{\Lambda^2}{4\pi^2}\left\{\frac{1}{\langle S\rangle\hat{Q}^3}(\langle W_s\rangle + W_c)^2\right\} \tag{11.21}$$

where we have dropped the tildas ($\sim$) in the definition of the fields. Expanding $(\hat{Q})^{-3}$ about $\langle T\rangle$ we find in fact a term proportional to $\overline{C}C$ in $V_{1,\text{loop}}$

$$V_{1,\text{loop},\Lambda^2}\Big|_{\overline{C}C} \sim \Lambda^2\, m_{3/2}^2 \frac{\overline{C}C}{M^2}. \tag{11.22}$$

This would need to be interpreted differently from the conventional calculation in e.g. a ϕ^4 theory. We cannot absorb the Λ^2 term into a counter term since in fact the original Lagrangian has no $\overline{C}C$ terms; rather we would say that the theory must include a cutoff

$$\Lambda \lesssim \Lambda_{SUSY} < \Lambda_{\text{compactification}} \lesssim M \tag{11.23}$$

and that one should treat this model as an effective theory valid only for scales $\lesssim \Lambda$, where the parameter Λ is to be fixed by a fuller theory. We will come back to this point in the next section where we discuss scales of symmetry breaking.

Let us proceed, for the time being, with an analysis of (11.22). It looks like an interesting solution since we can imagine a scenario in which $\Lambda \lesssim m_{3/2} \sim 10^{11}$ Gev and obtain then a C mass of order $\lesssim 1$ Tev. This is however an incorrect conclusion; the correct deduction is that m_c

equals zero. To see this remember that the general definition of the $(\text{mass})^2$ is

$$(m_c)^2 = \left.\frac{\partial^2 V}{\partial C \partial \bar{C}}\right|_{T = \langle T\rangle,\ C = \langle C\rangle,\ S = \langle S\rangle} . \tag{11.24}$$

We have already found, at tree level, the values of $\langle S\rangle$ and $\langle C\rangle$, cf. eq. (11.11). We assume that these values are only shifted slightly at one loop. The main purpose of our evaluating $V_{1,\text{loop}}$ is to fix $\langle T\rangle$; we will therefore keep only graphs with external T legs with one exception, namely we will also keep graphs with one external C^x and one $\bar{C}_x$ in order to see if we generate a non-zero $(m_c)^2$ at one loop level. This greatly simplifies the calculations and [7] we may proceed as follows : i) observe that the only fields that run around the loop are S, the SUSY partners of S and T and the gravitino. We can rescale these fields by factors of Q to get canonical kinetic energies. Next (ii) we notice that our graphs must explicitly contain a factor of $m_{3/2}^2$ since they otherwise cancel by supersymmetry, e.g. a $\bar{C}C$ term from $\left|\frac{\partial W}{\partial C}\right|^2$ is cancelled by a $\bar{C}C\,\bar{\chi}_c\chi_c\bar{\chi}_c\chi_c$ term. In fact, since only S, χ_s, χ_T and ψ_μ run around the loop we may set W(C) equal to zero in our calculation. We then find that $V_{1,\text{loop}}$ only depends on T and C,$\bar{C}$ through the combination $Q = T - 2\bar{C}C/M$. The minimization of the potential to fix $\langle T\rangle$

$$\left.\frac{\partial V}{\partial T}\right|_{T = \langle T\rangle,\ C = \langle C\rangle = 0} = 0 \tag{11.25}$$

is then equivalent to a zero mass condition for C since

$$\begin{aligned} m_c^2 &= \left.\frac{\partial^2 V}{\partial C\, \partial \bar{C}}\right|_{T = \langle T\rangle,\ C = \langle C\rangle} \\ &= -\frac{2}{M}\left.\frac{\partial V}{\partial T}\right|_{T = \langle T\rangle,\ C = \langle C\rangle} = 0. \end{aligned} \tag{11.26}$$

We can regard therefore $V_{1,\text{loop}}$ as the potential which fixes $\langle T\rangle$. The simplifications we alluded to no longer hold at two loops, in particular we need to keep contributions due to W(C). This might be an interesting scenario for setting up a hierarchy of sorts in which $\langle T\rangle$ is fixed at a non-trivial value at one loop level and m_c^2 at a two loop level. Unfortunately this does not happen either.

The quadratically divergent one loop potential is given in (11.21). It is clear that its minimum is at

$$\langle T\rangle = 0 \qquad \langle C\rangle = 0 \tag{11.27}$$

and the potential is unbounded from below. This corresponds to an unstable point of our calculation, namely the radius of compactification going to zero and the mass of the T scalar going to infinity, since $\left.\frac{\partial^2 V}{\partial T^2}\right|_{\langle T\rangle = 0} \to \infty$. We are violating the assumptions of (11.23) on which we based our calculation.

It was speculated in ref. [7] that the next terms in the one loop potential might balance the quadratic term for some finite Λ, thereby generating a non-trivial $\langle T\rangle$. Binetruy and Gaillard [11] reached the same conclusion and in fact, evaluated the logarithmic divergent term in $V_{1\text{ loop}}$ displaying that this in fact could happen. Recently, (Ahn and Breit[12]) have shown, however, that the full one loop potential is in fact a monotonic function, whose minimum is at $\langle T\rangle = 0$, and therefore the correct interpretation is the one we just gave, namely we are at an unstable point of our calculation. The issue here is whether or not one may drop $\frac{1}{\Lambda^2}$, $\frac{1}{\Lambda^4}$ etc. terms and keep only Λ^2 and $\log \Lambda^2$. The answer is presumably no since the cutoff is just a parameter, small compared to the other scales of the theory given in eq. (11.23).

SCALES OF SYMMETRY BREAKING

We have just seen how our loop potential pushed us to a minimum where $\langle T\rangle \to 0$. Going back to (10.23), $\langle T\rangle \to 0$ with fixed $\langle S\rangle$ corresponds to $\sigma \to -\infty$, $\phi \to 0$. Since, by (10.6) e^{σ} sets the scale for compactification, this corresponds to an ultracompactified theory and since ϕ scales the coupling, it is also an ultrastrongly coupled theory, i.e. it is nonsense.

Let us step back for a moment and discuss generally the scales of symmetry breaking in our theory. Let us call the characteristic string mass^2 $M_s^2 = 8\pi T$, with T of course the string tension. We will follow some recent discussions in the literature [13,14]. We already said at the very beginning, namely ch. 2, what the dimensions of the coupling constants were in d = 10, namely $g_{10} \sim \frac{1}{M^3}$ and $\kappa \sim \frac{1}{M^4}$. The mass scale is M_s and so the notion that we can have a sensible semi-classical approximation, expanding perturbatively, is equivalent to

$$g_{10} \lesssim \frac{1}{M_s^3}, \quad \kappa_{10} \lesssim \frac{1}{M_s^4}. \tag{11.28}$$

The four dimensional coupling constant g_4 and κ_4 are related to the ten dimensional coupling constants by the scale factor of the volume of compactification which goes like M_c^{-6}.

$$g_4 = g_{10} M_c^3 \qquad\qquad \kappa_4 = \kappa_{10} M_c^3. \tag{11.29}$$

Since we expect the d = 4 gauge coupling constant g_4 to be O(1) and $\kappa_4 \sim 1/M_{pl}$ we see that (11.28) and (11.29) together imply

$$M_c \sim M_s \sim M_{pl} \tag{11.30}$$

i.e. there is no gap between the string scale and the compactification scale. The latter is also the scale at which the GUT is broken dynamically by the non-trivial Wilson loops, as we saw in section nine. This coalescing of scales of course casts serious doubts on our whole procedure of using ten dimensional field theory.

We will not discuss here the breaking of SUSY in the hidden sector, but it may in practice turn out to be hard to have the SUSY breaking scale appreciably below the compactification scale. The analysis depends on the details of how the coupling constants run, which in turn depends on how the shadow E_8 is broken dynamically.

CONCLUSIONS (chapter 12)

A certain amount of the unbounded optimism of early 1985 has been tempered. It is clear that a good deal of work remains before string theories can be decisively accepted or rejected. At this present they are certainly an extraordinarily attractive candidate for the ultimate theory. There has been a good deal of interesting work on low energy phenomenology connected in some way to superstring models [1], but there is a growing feeling that the theory itself has not been refined to the point where it makes sharply testable predictions in the low energy world. In fact, as we have repeatedly emphasized, the route proceeding for d = 10 string theory → d = 10 field theory → d = 4 field theory may not be a valid one. We do nevertheless believe that many of the qualitative tools that one needs to develop to proceed along this path will be more generally useful, even if the path changes. It seems likely that the ideas of compactification, of topological invariants, of discrete symmetries on manifolds etc. will need to be part of the tools of the trade. Chapters 10 and 11 are less likely to be directly relevant, but we thought it useful to pursue one simple example.

We regret that we have not had time or space to cover many of the new developments; among these we might mention the new works on axions in cosmology [2], or vacuum degeneracy [4]. There have also been new promising technical developments such as the study of orbifolds [5], of monopoles [6] and other new topological phenomena [7].

In conclusion, we hope these notes may facilitate the reader's access to the special literature.

We would like to express our gratitude to the organizers of this Cargese school for the delightful atmosphere. In addition, we would like to thank John Breit and Burt Ovrut for working with us on some of the problems discussed in the notes. We would also like to thank John Breit for countless illuminating discussions on all aspects of these notes during their writing.

REFERENCES

Chapter 2

[1] Some recent reviews on string theory are listed below :
J.H. Schwarz, Phys. Rep. 89 (1982) 223;
M.B. Green, Surveys in High Energy Physics 3 (1982) 127;
L. Brink "Superstrings", CERN TH 4006/84.

[2] The World Scientific Press (Singapore) has announced publication in December of three books which should be helpful in bringing the reader up to date. They are :
i) J.H. Schwarz, Superstrings;
ii) Unified String Theories, Proceedings of the Santa Barbara Workshop, Aug. 1985, eds. M.B. Green and D.J. Gross;
iii) Yale Summer School on High Energy Physics, June 1985, eds. F. Gursey and M. Bowick.
This is not the place to try to give references to the new literature on string theories.

[3] P. Ramond, Phys. Rev. D3 (1971) 2415;
A. Neveu and J.H. Schwarz, Nucl. Phys. B31 (1971) 86; Phys. Rev. D4 (1971) 1109.

[4] D. Gross, J. Harvey, E. Martinec and R. Rohm, Phys. Lett. 54 (1985) 46; Nucl. Phys. B256 (1985) 253 and Princeton preprint;
See also P.G.O. Freund, Phys. Lett. 151B (1985) 387;
P. Goddard and D. Olive, "Conference on Vertex Operators...", J. Lepowsky ed. (Springer-Verlag, 1984).

[5] see e.g. E. Witten, Phys. Lett. 149B (1984) 351.

Chapter 3

[1] H. Flanders, "Differential Forms" New York, Academic Press (publ.) 1963.
[2] T. Eguchi, P. Gilkey and A.J. Hanson, Phys. Rep. 66 (1980) 213.
[3] See also references in ch. 4 to gravitational anomalies for general introduction to Riemannian geometry as well as ref. [2] cited above.
[4] For an old, but good introduction to curvature and general relativity, cf. L. Landau and E. Lifshitz, "The Classical Theory of Fields", Addison-Wesley, publ., see also S. Weinberg, "Gravitation and Cosmology" J. Wiley, publ.

Chapter 4

[1] Three recent references on gauge field anomalies are :
B. Zumino "Chiral Anomalies and Differential Geometry" in Relativity Groups and Topology (B.S. De Witt and R. Stora, eds.) North Holland, 1983;
B. Zumino, T.S. Wu and A. Zee, Nucl. Phys. B239 (1984) 477;
P.H. Frampton and T.W. Kephart, Phys. Rev. Lett. 50B (1983) 1343, 1437.
[2] L. Alvarez-Gaume and E. Witten, Nucl. Phys. B234 (1984) 269.
[3] L. Alvarez-Gaume and P. Ginsparg, Nucl. Phys. B243 (1984) 439; ibid Annals of Physics 161 (1985) 423.
[4] See also ref. [2] of ch. 3.
[5] M.B. Green and J.H. Schwarz, Phys. Lett. 149B (1984) 117.
[6] E. Witten, Phys. Lett. 149B (1984) 351.
[7] E. Witten, "Global Gravitational Anomalies", "Global Anomalies in String Theory", Princeton University Preprints (1985).

Chapter 5

[1] A good general reference is J. Wess and J. Bagger, "*Supersymmetry and Supergravity*", Princeton University Press (1983).
[2] E. Cremmer, J. Scherk and S. Ferrara, Phys. Lett. 74B (1978) 61.
[3] J. Scherk in Recent Developments in Gravitation, Cargese, Plenum Press (1978) eds. M. Levy and S. Deser.
[4] E. Cremmer and B. Julia, Nucl. Phys. B159 (1979) 141.
[5] A.H. Chamseddine, Nucl. Phys. B185 (1981) 403.
[6] E. Bergshoff, M. De Roo, B. De Wit, and P. Van Nieuwenhuizen, Nucl. Phys. B195 (1982) 97.
[7] G.F. Chapline and N.S. Manton, Phys. Lett. B120 (1983) 105.
[8] See Chapter 4, reference [5].

Chapter 6

[1] We are following here R.N. Mohapatra and B. Sakita, Phys. Rev. D21 (1980) 1062.
[2] Clearly it is awkward to define $\Gamma_{1,2} = \tau_{2,1}$ rather than $\Gamma_{1,2} = \tau_{1,2}$ but we wish to display an explicit form for the Γ matrices keeping the notation of ref. [1]. It would have been preferable to exchange Γ_{2i-1} and Γ_{2i} in (6.7).
[3] A quick summary of E_6 branching rules, etc. is contained in R. Slansky, Phys. Rep. 79 (1981) 1.
[4] See e.g.,
Y. Achiman, S. Aoyama and J.W. Van Holten, "Gauged Supersymmetric σ-Models and E_6 / SO(10) × U(1)", Wuppertal preprint, Jan. 1985.
[5] For an interesting discussion of exceptional groups, cf. F. Gursey in "To Fulfill a Vision" Jerusalem Einstein Centennial Celebr., Y. Neeman, ed. Addison-Wesley (publ. 1981).

Chapter 7

[1] Z. Horvath and L. Palla, Nucl. Phys. B142 (1978) 327;
L. Palla, Proceedings of 1978 Tokyo Conf. on High Energy Physics, p. 629.
[2] E. Witten, Nucl. Phys. B186 (1981) 412.
[3] N.S. Manton, Nucl. Phys. B193 (1981), 331;
G. Chapline and N. Manton, Nucl. Phys. B184 (1981) 391;
G. Chapline and R. Slansky, Nucl. Phys. B209 (1982) 461.
[4] C. Wetterich, Nucl. Phys. B211 (1983) 177.
[5] S. Randjbar-Daemi, A. Salam and A. Strathdee, Nucl. Phys. B214 (1983) 491; Phys. Lett. B124 (1983) 345.
[6] E. Witten, Proceedings of Shelter Island II, p. 227 (published by M.I.T. Press, 1985).
[7] Some representative references are :
M. Gell-Mann, P. Ramond and R. Slansky in "Supergravity" ed. P. Van Nieuwenhuizen and D.Z. Freedman (North Holland, 1979);
F. Wilczek and A. Zee, Phys. Rev. D25 (1982) 553;
H. Georgi, Nucl. Phys. B156 (1979) 126;
J. Kim, Phys. Rev. Lett. 45 (1980) 1916;
H. Sato, Phys. Rev. Lett. 45 (1980) 1997;
A. Davidson et al., Phys. Rev. Lett. 45 (1980) 1335;
J. Malampi and K. Engvist, Phys. Lett. 97B (1980) 217.
[8] M. Atiyah and F. Hirzebruch, "Essays on Topology and Related Topics" A. Haefliger and R. Narasimahan eds. (Springer-Verlag, New York 1970).
[9] E. Witten, J. Diff. Geom. 17 (1982) 661.
[10] See for a review S. Coleman, "The Uses of Instantons" in "Lectures on Field Theory", Oxford University press (1985).
[11] See ref. [2], ch. 3 for a detailed discussion of index theorem. Other good references are Luis Alvarez-Gaume, "Supersymmetry and Index Theory", Lectures at Bonn Summer School 1984, to be published;
J. Manes and B. Zumino, LBL-20234 preprint 1985;
L. Alvarez-Gaume, Math. Phys. 90 (1983) 161;
D. Friedan and P. Windey, Nucl. Phys. B235 (1984) 295.
[12] B. Zumino, Proceedings of Shelter Island II, p. 79 (published by M.I.T. Press, 1985).

Chapter 8

[1] Ch. 5, reference [7].
[2] CHSW stands for P. Candelas, G. Horowitz, A. Strominger and E. Witten, Nucl. Phys. B258 (1985) 46.
[3] Alternative schemes have been considered, following CHSW. For instance, manifolds with non-vanishing H_{mnp} have been considered by I. Bars, Phys. Rev. D33 (1986) 383;
I. Bars, D. Nemeschansky and S. Yankielowicz, "Torsion in Superstrings" SLAC-Pub 3758 (1985).
[4] E. Calabi in "Algebraic Geometry and Topology : A Symposium in Honor of S. Lefshetz (Princeton University Press) 1957;
S.T. Yau, Proc. Nat. Acad. Sci. 74 (1978) 177.
[5] K. Uhlenbeck and S.T. Yau, to be published, presents an analysis of this equation.
[6] For a detailed discussion which does not assume identification of the spin connection and SU(3) gauge fields, see I. Bars and M. Visser, "Number of Massless Fermion Fields in Superstring Theory", Univ. of So. Calif. preprint 85/016 (1985). See also R. Nepomechie, Y.S. Wu and A. Zee, Phys. Lett. 158B (1975) 311.
[7] E. Witten, Nucl. Phys. B258 (1985) 75.
[8] E. Witten, "New Issues in Manifolds of SU(3) Holonomy", Princeton preprint 1985.

[9] A. Strominger and E. Witten, "New Manifolds for Superstring Compactification", Princeton preprint.

Chapter 9

[1] We are following here again P. Candelas et al., ref. [2] of chapter 8.
[2] E. Witten, Nucl. Phys. B258 (1985) 75.
[3] J.D. Breit, B.A. Ovrut and G. Segrè, Phys. Lett. 158B (1985) 33; A. Sen, Phys. Rev. Lett. 55 (1985) 33.
[4] Y. Hosotani, Phys. Lett. 126B (1983) 303.
[5] For a detailed discussion of the group theoretic techniques involved see e.g., R. Slansky, Phys. Rep. 79 (1981) 1.
[6] J. Pati and A. Salam, Phys. Rev. D10 (1974) 275.
[7] See e.g., F. Gursey and M. Serdaroglu, Nuovo Cimento 65A (1981) 337; F. Gursey, P. Ramond and P. Sikivie, Phys. Lett. 60B (1976) 77.
[8] M. Dine, V. Kaplunovsky, M. Mangano, C. Nappi and N. Seiberg, "Superstring Model Building", Nucl. Phys. B259 (1985) 543.
See also : S. Ceccotti, J.P. Derendinger, S. Ferrara, L. Girardello and M. Roncadelli, Phys. Lett. 156B (1985) 318.
[9] See for example, A. Strominger and E. Witten, "New Manifolds for Superstring Compactification", Princeton (1985) to be published in Nucl. Phys. A. Strominger "Yukawa Couplings in Superstring Compactification" ITP (Santa Barbara) preprint (1985).
[10] For a good recent discussion of the overall picture see : J.P. Derendinger, L.E. Ibanez and H.P. Nilles, "On the Low Energy Limit of Superstring Theories", CERN Th 4228 preprint (1985).

Chapter 10

[1] E. Witten, "Dimensional Reduction of Superstring models", Princeton preprint. Phys. Lett. 155B (1985) 151.
[2] E. Witten, Ref. [5], section 2.
[3] E. Cremmer, S. Ferrara, L. Girardello and A. Von Proeyen, Phys. Lett. 116B (1982) 231; Nucl. Phys. B212 (1983) 412;
E. Cremmer, B. Julia, J. Scherk, S. Ferrara, L. Girardello and P. Van Nieuwenhuizen, Nucl. Phys. B147 (1979) 105.
[4] The formulation in terms of Kähler geometry is given in :
J. Bagger and E. Witten, Phys. Lett. 115B (1982) 202; Phys. Lett. 118B (1982) 303;
J. Bagger, Nucl. Phys. B211 (1983) 302.
[5] We have followed reference [1] so far. This form of Kähler potential was independently shown to arise in the analysis of so-called 16/16 supergravity models, cf. W. Lang, J. Louis and B. Ovrut, Univ. of Penn., preprint UPR-0280T (1985) and Karlsruhe preprint KA-THE P 85-2 (1985).
See also earlier references to G. Girardi, R. Grimm, M. Muller and J. Wess, Z. Phys. C26 (1986) 123; C26 (1984) 427; Phys. Lett. 147B (1984) 81. It is interesting to speculate on which of these two paths is a more realistic approximation to superstring phenomena.
[6] J.P. Derendinger, L.I. Ibanez and H.P. Nilles, ref. [10], section 9.

Chapter 11

[1] J.P. Derendinger, L.E. Ibanez and H.P. Nilles, Phys. Lett. 155B (1985) 65.
[2] M. Dine, R. Rohm, N. Seiberg and E. Witten, Phys. Lett. 156B (1985) 55.
[3] J.P. Derendinger, L.I. Ibanez and H.P. Nilles, "On the Low Energy Limit of Superstring Theories", CERN - TH. 4228/85 preprint.
[4] For a recent discussion see R. Rohm and E. Witten, "The Antisymmetric Tensor Field in Superstring Theory" Princeton University preprint (Nov. 1985).

See also R. Nepomechie, Y.S. Wu and A. Zee, Phys. Lett. 158B (1985) 311.

[5] E. Witten, ref. [1], ch. 10.

[6] J. Ellis, C. Kounnas and D.V. Nanopoulos, Nucl. Phys. B247 (1984) 373;
N.P. Chang, S. Ouvry and X. Wu, Phys. Rev. Lett. 51 (1983) 327;
E. Cremmer, S. Ferrara, C. Kounnas and D.V. Nanopoulos, Phys. Lett. 133B (1983) 61. See also ref. [5], ch. 10;
R. Barbieri, S. Ferrara, D.V. Nanopoulos and K. Stelle, Phys. Lett. 113B (1982) 219. See also M. Mangano, Z. Phys. C28 (1985) 613.

[7] This section is a condensed version of J.D. Breit, B.A. Ovrut and G. Segrè, Phys. Lett. 162B (1985) 303.

[8] S. Coleman and E. Weinberg, Phys. Rev. D7 (1983) 2369.

[9] For complete references, see e.g. H.P. Nilles, Phys. Rep. 110 (1984) 1. Some papers are : J. Polonyi, Budapest preprint K-FK1-38 (1977);
A.H. Chamseddine, R. Arnowitt and P. Nath, Phys. Rev. Lett. 49 (1982) 970;
L. Alvarez-Gaume, J. Polchinski and M. Wise, Nucl. Phys. B221 (1983) 435;
B. Ovrut and J. Wess, Phys. Lett. 112B (1982) 347;
L.I. Ibanez, Nucl. Phys. B218 (1983) 514;
L. Hall, J. Lykken and S. Weinberg, Phys. Rev. D27 (1983) 2369;
H.P. Nilles, M. Srednicki and D. Wyler, Phys. Lett. 120B (1983) 346;
J. Ellis, D.V. Nanopoulos and K. Tamvakis, Phys. Lett. 121B (1983) 1293;
For recent examples, see e.g. M. Cvetic and C. Preitschopf, SLAC-PUB-3685 (1985);
R. Barbieri and E. Cremmer, CERN TH 4177 (1985).

[10] R. Barbieri and S. Ceccotti, Z. Phys. C17 (1983) 183;
M. Srednicki and S. Theisen, UCSB preprint.

[11] P. Binetruy and M.K. Gaillard, "Radiative Corrections in Compactified Superstring Models" LBL-19972 preprint (1985).

[12] J. Ahn and J. Breit, "On One Loop Effective Lagrangians and Compactified Superstrings", Univ. of Pennsylvania preprint (1985).

[13] M. Dine and N. Seiberg, Phys. Rev. Lett. 55 (1985) 366.

[14] V.S. Kaplunovsky, Phys. Rev. Lett. 55 (1985) 1036.

Chapter 12

[1] For early references, see ch. 9, refs. [8-10]. Among more recent works, a sample is given by V. Kaplunosky and C. Nappi, "Phenomenological Implications of Superstring Theory", Princeton preprint 1985;
P. Binetruy, S. Dawson, I. Hinchcliffe and M. Sher, "Phenomenologically Viable Models from Superstrings". E. Cohen, J. Ellis; K. Enqvist and D.V. Nanopoulos, CERN TH preprints 4222/85, 4195/85;
R. Holman and D.B. Reiss, "Fermion Masses in $E_8 \times E_8$ Superstring Theories", Fermilab. preprint 1985;
T. Hubsch, H. Nishino and J.G. Pati, "Do Superstrings Lead to Preons", to appear in Phys. Lett. B.;
C.P. Burgess, A. Foni and F. Quevedo, "Low Energy Effective Action for the Superstring", Univ. of Texas Preprint;
R.N. Mohapatra, "A mechanism for Understanding Small Neutrino Masses in SUSY Theories", Univ. of Maryland preprint (1985);
S. Nandi and U. Sarkar, Univ. of Texas Preprint (1985).

[2] K. Choi and J.E. Kim, Phys. Lett. 154B (1985) 393;
S.M. Barr, Phys. Lett. 154B (1985) 397;
K. Yamamoto, Univ. of No. Carolina preprint (1985);
K. Choi and J.E. Kim, Seoul preprint SNUHE 85/10;
M. Dine and N. Seiberg, "String Theory and the Strong CP Problem", CUNY preprint 1985;
X.G. Wen and E. Witten, "World Sheet Instantons and the Peccei-Quinn Symmetry", Princeton preprint.

[3] M.J. Bowick, L. Smolin and L.C.R. Widjewardhana, Phys. Rev. Lett. 56 (1986) 424;
J. Lazarides, G. Panagiotakopoulos and Q. Shafi, Phys. Rev. Lett. 56 (1986) 432 and Bartol preprint BA-85 "Baryogenesis and the Gravitino Problem in Superstring Theories";
J.A. Stein-Schabes and M. Gleiser, "Low Energy Superstring Cosmology", Fermilab preprint 1985;
R. Holman and T. Kephart, "Axion Cosmology in Automatic $E_6 \times U(1)$ Models", Fermilab pub. 85/117;
E.W. Kolb, D. Seckel and M.S. Turner, "The Shadow World", Fermilab preprint.

[4] See e.g., M. Evans and B.A. Ovrut, "Splitting the Superstring Vacuum Degeneracy", Rockefeller preprint 1985.

[5] L. Dixon, J.A. Harvey, C. Vafa and E. Witten, "Strings on Orbifolds", Princeton preprint;
E. Witten, "Twistor-Like Transform in Ten Dimensions", Princeton preprint.

[6] X.G. Wen and E. Witten, "Electric and Magnetic Charges in Superstring Models", to appear in Nucl. Phys. B.

[7] For a general discussion see E. Witten, "Topological Tools in Ten Dimensional Physics", and also E. Witten, ref. [8], ch. 8.

AN INTRODUCTION TO QCD SUM RULES

J. Weyers

Institut de Physique Théorique
Université Catholique de Louvain
Louvain-la-Neuve, Belgium

INTRODUCTION

Even if QCD is the correct theory of strong interactions, we are still a long way off from being able to actually calculate hadronic properties, such as the masses and couplings of observed particles, from the fundamental lagrangian of the theory. Since the determination of the spectrum of QCD is a problem which presumably will never be solved exactly one must of course introduce either some approximation scheme or some model (usually both). Several such schemes or models have been developed over the years. Let me mention, in particular

- Monte Carlo simulations on a lattice
- bag models
- potential models
- QCD sum rules.

Since all these approaches have claimed some successes in hadron spectroscopy it is perhaps useful to discuss briefly what can be learned from them as well as what they can teach us about QCD itself.

Monte Carlo simulations on a lattice

This is undoubtedly the most promising and powerful approach to study the hadron spectrum in QCD. Adding one (controllable) parameter to the theory, namely the lattice spacing, one may hope, eventually, to calculate all spectroscopic quantities of interest. Since the pioneering work of Wilson a huge amount of practical calculations are presently being performed by a large number of physicists. For the moment the accuracy of lattice calculations is still very limited by the size of the lattices which can practically be used. In addition to this technological problem, there may be more fundamental difficulties related to fermions. Still, the lattice approach is the only one which, if all goes well, will put QCD to the ultimate test for a theory : it will tell us whether it is right or wrong!

Obviously one of the more interesting problems in QCD concerns the spectroscopic role of gluons : are there gluon degrees of freedom in the hadron spectrum or are gluons "amorphous" from the spectroscopic point of view ? In other words do there exist hadrons where gluons play a "constituent" role ? Such hadrons would, of course, not correspond to the $q\bar{q}$ or qqq states of the naïve quark model and could be either glueballs (GG... i.e. states with only constituent gluons) or hybrids ($q\bar{q}$G... i.e. states with constituent gluons and constituent quarks). From the experimental point of view there is still no compelling evidence for gluonic degrees of freedom although many glueball or hybrid candidates exist.

The lattice approach definitely predicts the existence of glueballs and, as far as I know, no final conclusion can yet be drawn concerning hybrids.

Bag Models

Bag models have an immediate intuitive appeal. Essentially they are based on an assumed two-phase structure of the vacuum : hadrons are viewed as bubbles of perturbative vacuum in a sea of strong vacuum. Colour singlet configurations of quarks and gluons propagate and interact inside the bubble of perturbative vacuum which they stabilize. From this picture, one clearly expects that any colour singlet collection of quarks and gluons can materialize as a physical hadrons : glueballs, hybrids, multiquark states (such as $qq\bar{q}\bar{q}$) are expected in these models together with the standard $q\bar{q}$ or qqq hadrons!

But, at least at the present state of the art it seems to me that, as a spectroscopic tool, bag models are, at best, extremely unreliable :

- chiral symmetry which we know to play a vital role in hadron spectroscopy cannot be included in bag models in any satisfactory way;
- independent oscillations of quarks and gluons in a bag give a spectrum of hadrons, in the most naive approximation which is much too rich;
- the vibrations of the cavity which confines the quarks and gluons and which should play a role in the spectroscopy of excited hadrons cannot be calculated and, if they could be, they would make too rich a spectrum even richer!

Potential models

Potential models have a glorious history in hadron physics starting with the quark model itself. At present they are mainly used as a tool to unravel the detailed spectroscopy of -onium systems ($c\bar{c}$, $b\bar{b}$ ...) and as such they have been and continue to be quite useful. From a conceptual point of view, it is of course unclear what potential models can actually teach us about QCD! Parametrizing the forces between a heavy quark and antiquark by a Coulomb plus confining potential is one thing but to relate the parameters of the potential to the fundamental lagrangian of QCD is quite another. Eventually lattice calculations in the static limit should solve this problem. On the other hand potential models cannot give any information on "unknown" sectors of the theory such as glueballs or hybrids and they are still of limited use for light quark mesons or for three body systems such as baryons.

QCD sum rules

The QCD sum rules [1,2] are a semi-phenomenological method to extract spectroscopic information from the QCD lagrangian. The basic idea of the sum rules is to extrapolate correlations functions from *large* Q^2 (short distances) where asymptotic freedom allows perturbative calculations to *low* Q^2 (large distances) where one knows that non-perturbative effects are at work and that the correlation functions are dominated by resonances. The non perturbative effects arise as power corrections (in $\frac{1}{Q^2}$) to the asymptotic freedom regime and they are introduced as *phenomenological parameters* through various "expectation values" such as

- the gluon condensate $<0|\frac{\alpha_s}{\pi} G^a_{\mu\nu} G^a_{\mu\nu}|0> \simeq (360 \text{ Mev})^4$
- the "light quark" condensate $<0|\bar{q}q|0> \simeq (-250 \text{ Mev})^3$.

The values of these parameters are determined from the spectrum i.e. phenomenologically : they are an *input* to the QCD sum rules. Here also one expects that lattice calculations, where these parameters can be determined, will confirm their phenomenological values and hence the consistency of the sum rule approach.

Provided there is some "overlap region" where both the large and low Q^2 representations are valid, one can then *compute* resonance parameters (masses and couplings) from

- the fundamental parameters of the QCD lagrangian (m_q, α_s),
- the behaviour of the theory in the asymptotic freedom régime,
- a few "universal" (non-perturbative) parameters which characterize the QCD vacuum.

Let us describe the strategy of the QCD sum rules somewhat more precisely.

(A) The large Q^2 on short distance behaviour of the theory is studied via the operator product expansion (OPE) [3]. Consider for example a mesonic current

$$j_\Gamma(x) = \bar{q}_i(x)\Gamma q_j(x). \tag{1}$$

By an appropriate choice of the flavour indices (i,j) and the tensor structure (Γ) this current will create from the vacuum, a meson with specific J^{PC} and internal quantum numbers.

The OPE, valid at short distances, reads

$$i\int d^4x\, e^{iqx}\, T(j_\Gamma(x) j_\Gamma(o)) = C_1^\Gamma(q^2)\, \mathbb{1} + \sum_n C_n^\Gamma(q^2) O_n \tag{2}$$

where $\mathbb{1}$ is the identity operator and O_n are operators constructed from quark and gluon fields while the $C(q^2)_n$ are, c-number functions of q^2, are the so-called Wilson coefficients. In this OPE, the operators O_n are ordered by their dimensions and the corresponding $C_n^\Gamma(q^2)$ fall off by the appropriate power of q^2.

The validity of Eq. (2) is a fundamental ingredient of the QCD sum rule approach. For a current of the type (1) which is of dimension 3, the Wilson coefficients will behave as follows

$$C_1^\Gamma(q^2) \sim q^2 \ln(-q^2/\mu^2) \tag{3a}$$

$$C_n^\Gamma(q^2) \sim (q^2)^{2-d_n} \tag{3b}$$

where d_n is the dimension of the operator O_n.
Eq. (2) will be used, when sandwiched between vacuum states. Hence we can limit the expansion to O_n operators which are gauge invariant Lorentz scalars. These are easy to list (up to dimension 6)

$$
\begin{array}{ll}
1 & d = 0 \\
m\bar{q}q & d = 4 \\
G^a_{\mu\nu}G^a_{\mu\nu} & d = 4 \\
\bar{q}\Gamma_i q\ \bar{q}\Gamma_i q & d = 6 \\
m\bar{q}\sigma_{\mu\nu}T^a G^{a\mu\nu}q & d = 6 \\
f_{abc}G^a_{\mu\nu}G^b_{\nu\lambda}G^c_{\lambda\mu} & d = 6.
\end{array}
\tag{4}
$$

The matrix elements or vacuum expectation values of these operators reflect the structure of the QCD vacuum. They are *parameters* of the QCD sum rules. Note that they are universal parameters : they will enter in *all* two-point function expansions since they are independent of the flavour or tensor structure of the currents (Eq 1). More precisely, the same matrix elements $<o|O_n|o>$ will appear in the expansion of all two point functions (or n-point functions, for that matter)

$$i\int d^4xe^{iqx} <o|T(A(x)B(o))|o> = C_1^{AB}(q^2) + \sum_n C_n^{AB}(q^2) <o|O_n|o> . \tag{5}$$

In other words, the short distance behaviour of the theory is described by the Wilson coefficients of the OPE while the matrix elements of the O_n operators reflect the dynamical structure of the QCD vacuum. Because of asymptotic freedom, the short distance behaviour is "perturbative" and one can thus calculate the Wilson coefficients as series in the coupling constant. The matrix elements $<o|O_n|o>$, on the other hand, are essentially non perturbative : indeed all these matrix elements vanish in the perturbative vacuum, to all orders in perturbation theory.

Hence, the meaning which is given to the OPE is that it allows a separation between perturbative (essentially short distance) and non-perturbative (essentially large distance) effects. That such a separation would be meaningful in a theory like QCD is by no means obvious. A remarkably lucid discussion of this point is given in reference 4.

Expanding the vacuum matrix element of Eq (2) in terms of spectral invariant functions i.e.

$$i\int d^4xe^{iqx} <o|T\{j_\Gamma(x)j_\Gamma(o)\}|o> = \sum_k T^{(k)}_{\Gamma\Gamma}\Pi^{(k)}(Q^2) \tag{6}$$

where $Q^2 = -q^2$ and the $T^{(k)}_{\Gamma\Gamma}$ are various tensor structures constructed out of $g_{\mu\nu}$, $E_{\mu\lambda\rho\sigma}$, q_μ ... From Eq (6) we can extract *expressions for the invariant functions* $\Pi^{(k)}(Q^2)$ which depend on the fundamental parameters of the theory (α_s, m_q ...), on the vacuum structure parameters $<o|O_n|o>$ and on the Wilson coefficients which can be calculated perturbatively. We will denote these expressions by $\Pi^{(k)}_{QCD}(Q^2)$.

(B) On the other hand, from general analyticity arguments, one knows that the invariant functions $\Pi^{(k)}(Q^2)$ do satisfy dispersion relations

$$\Pi^{(k)}(Q^2) = \frac{1}{\pi}\int \frac{\text{Im}\ \Pi^{(k)}(s)}{s+Q^2}\, ds \tag{7}$$

(eventually with some subtractions which will not concern us for the moment). The imaginary part $\text{Im}\ \Pi^{(k)}(s)$ is a quantity about which we have some *phenomenological knowledge.*

In the best possible case, which occurs for the vector current, $\text{Im}\,\Pi^{(k)}(s)$ is directly related to the cross-section ($e^+ + e^- \to$ hadrons). If we restrict ourselves to u,d,s quarks for a moment, $\text{Im}\,\Pi^{(k)}(s)$ is then dominated by resonances ($\rho,\rho';\omega,\omega';\varphi,\varphi'$ ...) and continuum states. In less favorable cases where $\text{Im}\,\Pi^{(k)}(s)$ is not directly measurable we can still parametrize it by some resonant contribution states together with a continuum.

Let us denote such a parametrization by $\text{Im}\,\Pi^{(k)}_{R+C}(s)$ to emphasize that the imaginary part is saturated by resonances (R) and a continuum (C).

Ⓒ The QCD sum rules are now obtained by equating the two parametrizations which we just described

$$\Pi^{(k)}_{QCD}(Q^2) \approx \frac{1}{\pi} \int \frac{\text{Im}\,\Pi^{(k)}_{R+C}(s)}{s+Q^2}\, ds. \tag{8}$$

Let me insist once more on the fact that the left hand side of Eq (8) is calculated from the OPE while the right hand side is a parametrization in terms of physical resonances together with some continuum.

In most applications of the QCD sum rules, so far, the aim has been to extract from Eq (8) information about the lowest lying resonance which couples to the channel one is studying. For this purpose, two methods have been widely used : the moments method and the Borel transform.

The moments method consists in taking successive derivatives of Eq (8)

$$M_n(Q_o^2) = \frac{1}{n!}\left(-\frac{d}{dQ^2}\right)^n \Pi^{(k)}(Q^2)\Big|_{Q^2=Q_o^2} \approx \frac{1}{\pi} \int \frac{\text{Im}\,\Pi^{(k)}_{R+C}(s)}{(s+Q_o^2)^{n+1}}\, ds. \tag{9}$$

Clearly if one assumes the dominance of a single resonance

$$\text{Im}\,\Pi^{(k)}_{R+C}(s) \propto \delta(s - m_R^2) \tag{10}$$

one finds

$$M_n(Q_o^2) \propto \frac{1}{(m_R^2+Q_o^2)^{n+1}}$$

and the ratio of two successive moments immediately gives the mass of the resonance. It is fairly easy to convince oneself that in a more realistic parametrization of the imaginary part, the lowest lying resonance will eventually dominate in the moments $M_n(Q_o^2)$ for n sufficiently large. The problem with moments is usually on the theoretical side of the QCD sum rules since the moments of the coefficients of higher dimensional operator contributions increase much faster with n that the moment of the coefficient of the operator 1. Luckily the behaviour is opposite for increasing Q_o^2. In practice a rather subtle analysis is needed to guarantee

- the validity of the approximation made in ignoring higher dimensional operators in the OPE
- the reliability of the perturbative calculation of the Wilson coefficients

- the dominance of the lowest lying resonance on the "physical" side of the sum rule.

We will not discuss these points here. The interested reader will find a detailed analysis in reference [2] and references quoted therein.

As it turns out, the moments method works best when there is an extra scale in the problem (like a heavy quark mass). For massless quarks one uses the Borel transform instead. It is defined as follows

$$L_M = \lim_{\substack{Q^2, n \to \infty \\ Q^2/n = M^2}} \frac{1}{(n-1)!} (Q^2)^n \left(-\frac{d}{dQ^2}\right)^n . \qquad (11)$$

Applying this transformation to Eq (8) one obtains

$$L_M \, \Pi^{(k)}_{QCD}(Q^2) = \frac{1}{\pi M^2} \int ds \; e^{-s/M^2} \; \mathrm{Im}\, \Pi^{(k)}_{R+C}(s). \qquad (12)$$

Note that by taking the Borel transform one does not have to worry about subtractions in dispersion relations.

Intuitively the Borel transform is a good compromise between conflicting requirements :

- to extract information about the lowest lying resonance in a given channel one should enhance as much as possible the low Q^2 region where the resonance is supposed to dominate (this is done by taking more and more derivatives);
- since the left hand side is a series in $\frac{1}{Q^2}$ (up to logs) one cannot emphasize the low Q^2 region too much without endangering the validity to the OPE calculations (one lets Q^2 go to infinity).

As sketched above, the QCD sum rule approach has met with many remarkable successes [1,2] in hadron spectroscopy. Let me remind you, for example, that it correctly predicted [1] the mass of the η_c at a time when the experimental indications were misleading. There are several "questions of principle" concerning the theoretical foundations of the method, like the OPE expansion [3], the separation between perturbative and non perturbative effects [4]. To my mind all these questions have been satisfactorily answered. The methods of analysis of the sum rules, on the other hand, are still quite clumsy (narrow resonances ets...) but it is not easy to do significantly better.

Since the sum rules have turned out to be a reliable tool for investigating the "known" hadron spectrum it is of course tempting to apply them to predict properties of unusual objects like hybrids and hence to provide real spectroscopic tests of QCD as a theory of the hadrons. We will discuss this point later on in somewhat more detail.

My main purpose in these lectures is to introduce the reader to the vast literature on QCD sum rules rather than give a review of the subject. They will be organized as follows : in Section I, I will describe the external field method to calculate Wilson coefficients; in Section II, I will briefly indicate how one projects two-point functions on specific J^{PC} states and in Section III, I will write explicitly a few sum rules. Section IV is a very short and biased summary of some of the main results obtained so far. It includes explicit expressions for the invariant functions of all quark bilinear two point functions.

THE BACKGROUND FIELD METHOD - SECTION I [5]

From the QCD lagrangian for one flavor

$$\mathcal{L} = -\frac{1}{4} F^a_{\mu\nu} F^{\mu\nu}_a + \frac{1}{2} i[\bar{\psi}\gamma^\mu D_\mu \psi - (D_\mu\bar{\psi})\gamma^\mu \psi] - m\bar{\psi}\psi \tag{13}$$

one derives the equations of motion

$$i\gamma^\mu D_\mu \psi = m\psi \tag{14a}$$

$$i(\overline{D_\mu \psi})\gamma^\mu = -m\bar{\psi} \tag{14b}$$

$$D^\nu_{ab} F^b_{\mu\nu} = g\bar{\psi}\gamma^\mu T_a \psi . \tag{14c}$$

In these equations A^a_μ and ψ are the gluon and quark fields, $F^a_{\mu\nu}$ is the gluon field strength

$$F^a_{\mu\nu} = \partial_\mu A^a_\nu - \partial_\nu A^a_\mu + g f^{abc} A^b_\mu A^c_\nu . \tag{15}$$

T_a are the colour matrices and the gauge covariant derivatives are

$$D_\mu = \partial_\mu - igA^a_\mu T^a \tag{16}$$

$$D^{ab}_\mu = \partial_\mu \delta^{ab} - g f^{abc} A^c_\mu \tag{17}$$

in the fundamental and adjoint representations respectively.

As already mentioned, the fundamental assumption of the QCD sum rules is the validity of the operator product expansion *in the physical vacuum* (Eq. 5).

The problem is now to compute the various Wilson coefficients. Basically two methods are available : one is to sandwich the OPE between various (simple) states selected in such a way as to isolate the corresponding Wilson coefficient and the other is the "external field" method which we now describe in some detail.

The idea of the method is to represent the vacuum fluctuations by classical background fields which satisfy the equations of motion and around which all fields are expanded as quantum fluctuations : in other words one shifts *all* fields by a classical background field

$$A^a_\mu(x) \rightarrow A^a_\mu(x) + \phi^a_\mu(x) \tag{18a}$$

$$\psi(x) \rightarrow \psi(x) + \eta(x) \tag{18b}$$

where $A^a_\mu(x)$ and $\psi(x)$ on the right hand side are the classical background fields which satisfy Eqs (14) while $\phi^a_\mu(x)$ and $\eta(x)$ are the quantum fluctuations over which the functional integral is performed. With the substitutions (18), the left hand side of Eq (5) can now be computed as an expansion in the background fields and, in the simpler cases, each term in that expansion which contains a scalar gauge invariant functional O_k of the external fields corresponds to the term $\langle o|O_k|o\rangle$ of the real OPE, while all other terms are discarded since they are either functionals of the quantum fluctuations (whose matrix elements vanish) or non scalar or non gauge invariant functionals of the external field (like the external field itself).

To summarize : in the background field method the true OPE expansion is replaced by an expansion in scalar gauge invariant functionals in the background fields. The coefficients of these functionals which can now be calculated by ordinary diagrammatic techniques are then easily related to the Wilson coefficients. The background fields have thus no physical significance : they are a technical device which simplifies the calculation of the Wilson coefficients in the OPE.

With the substitutions (18), the lagrangian (13) becomes

$$\begin{aligned}\mathcal{L}_{eff} &= \mathcal{L}(A,\psi,\bar{\psi}) + \mathcal{L}(\text{ghost}) + \bar{\eta}[i\gamma^{\mu}D_{\mu} - m]\eta \\ &+ \frac{1}{2}\phi^{a}_{\mu}[g^{\mu\nu}D^{2}_{ac} - (1-\frac{1}{\alpha})(D^{\mu}D^{\nu})_{ac} + 2gf_{abc}G^{\mu\nu}_{b}]\phi^{c}_{\nu} \\ &+ g(\bar{\psi}\gamma^{\mu}\phi^{a}_{\mu}T^{a}\eta + \bar{\eta}\gamma^{\mu}\phi^{a}_{\mu}T^{a}\psi) - g^{2}f^{abc}f_{adf}A^{b}_{\mu}\phi^{c}_{\nu}\phi^{\mu}_{d}\phi^{\nu}_{f} \\ - gf_{abc}&(\partial^{\mu}\phi^{\nu}_{a})\phi^{b}_{\mu}\phi^{c}_{\nu} - \frac{1}{4}g^{2}f^{abc}f_{adf}\phi^{b}_{\mu}\phi^{c}_{\nu}\phi^{d}_{\mu}\phi^{f}_{\nu} + g\bar{\eta}\gamma^{\mu}\phi^{a}_{\mu}T^{a}\eta\end{aligned} \tag{19}$$

where $\mathcal{L}(A,\psi,\bar{\psi})$ is the QCD lagrangian evaluated for the background fields, $G^{a}_{\mu\nu} = \partial_{\mu}A^{a}_{\nu} - \partial_{\nu}A^{a}_{\mu} + gf^{abc}A^{b}_{\mu}A^{c}_{\nu}$ is the background gluon field strength and D_{μ} or D^{ab}_{μ} are the background gauge covariant derivatives i.e. they depend on A^{a}_{μ} but not on ϕ^{a}_{μ}.

In writing Eq (19) we have used the background gauge for the gluon fluctuations i.e.

$$D^{ab}_{\mu}\phi^{\mu}_{b} = 0 \tag{20}$$

by adding the gauge fixing term $-\frac{1}{2\alpha}(D^{\mu}_{ab}\phi^{b}_{\mu})^{2}$ and $\mathcal{L}$(ghost) is the ghost contribution for that choice of gauge.

It is interesting to note that several a priori possible terms, such as $g\bar{\psi}\gamma^{\mu}\phi^{a}_{\mu}T^{a}\psi$, do not appear in the effective lagrangian. This follows from the use of the equations of motion for the background fields.

From the quadratic terms in η and ϕ^{a}_{μ} of the effective lagrangian (19) we obtain the propagators $S_{F}(x,y)$ and $S^{ab}_{\mu\nu}(x,y)$ of the quark and gluon fluctuations in the background fields. Taking the gauge parameter $\alpha = 1$, they are defined as follows

$$(i\gamma^{\mu}D_{\mu} - m)S_{F}(x,y) = i\delta^{4}(x-y) \tag{21}$$

$$[g^{\mu\nu}D^{2}_{ac} + 2gf_{abc}G^{\mu\nu}_{b}]S^{cd}_{\nu\rho}(x,y) = i\delta^{\mu}_{\rho}\delta^{d}_{a}\delta^{(4)}(x-y). \tag{22}$$

To solve these equations it turns out to be very convenient to fix the gauge freedom of the background field $A^{a}_{\mu}(x)$ by using the so-called Schwinger or fixed-point gauge

$$x^{\mu}A^{a}_{\mu} = 0 \tag{23}$$

since, in this gauge,

$$A_\mu^a(x) = \int_0^1 d\rho\rho x^\nu G^a_{\nu\mu}(\rho x) = \sum_{n=0}^{\infty} \frac{1}{(n+2)n!} x^{\alpha_1}\ldots x^{\alpha_n} x^\mu (D_{\alpha_1}\ldots D_{\alpha_n} G_{\nu\mu})^a(o). \tag{24}$$

Using this expression in the OPE will lead at once to gauge covariant results. Substituting Eq (24) in Eq (21) we obtain

$$[i\not\partial - m + \sum_{n=0}^{\infty} x^{\alpha_1}\ldots x^{\alpha_n} x^\nu \frac{1}{(n+2)} \frac{1}{n!} g(D_{\alpha_1}\ldots D_{\alpha_n} G_{\nu\mu})^a(o) T^a \gamma^\mu] S_F(x,y) = i\delta^4(x-y). \tag{25}$$

Taking the double Fourier transform, i.e.

$$S_F(x,y) = \int \frac{d^4p d^4q}{(2\pi)^8} e^{-ipx} e^{-iqy} S_F(p,q),$$

Eq (25) becomes

$$[\not p - m + \sum_{n=0}^{\infty} (-i)^{n+1} \frac{1}{(n+2)n!} g(D_{\alpha_1}\ldots D_{\alpha_n} G_{\nu\mu})^a(o) T^a \gamma^\mu \frac{\partial}{\partial p_{\alpha_1}} \ldots \frac{\partial}{\partial p_{\alpha_n}} \frac{\partial}{\partial p_\nu}] S_F(p,q) = (2\pi)^4 i\delta^4(p+q) \tag{26}$$

which we now solve as a series in the *external field* (not in g!)

$$S_F(p,q) = \sum_{i=0}^{\infty} S_F^{(i)}(p,q)$$

$$S_F^{(o)}(p,q) = \frac{i}{\not p - m} (2\pi)^4 \delta^4(p+q)$$

$$S_F^{(i+1)}(p,q) = (-1) \sum_{n=0}^{\infty} (-i)^{n+1} \frac{1}{(n+2)n!} g(D_{\alpha_1}\ldots D_{\alpha_n} G_{\nu\mu})^a(o) T^a \frac{1}{\not p - m} \gamma^\mu \frac{\partial}{\partial p_{\alpha_1}} \ldots \frac{\partial}{\partial p_{\alpha_n}} \frac{\partial}{\partial p_\nu} S_F^{(i)}(p,q).$$

Explicitly, this gives for the first few terms

$$S_F^{(1)}(p,q) = (2\pi)^4 g \left\{ -\frac{1}{2} G^a_{\nu\mu}(o) T^a \frac{1}{\not p - m}\gamma^\mu + \frac{i}{3}(D_\alpha G_{\nu\mu})^a(o) T^a \frac{1}{\not p - m}\gamma^\mu \frac{\partial}{\partial p_\alpha} + \frac{1}{8}(D_\alpha D_\beta G_{\nu\mu})^a(o) T^a \frac{1}{\not p - m}\gamma^\mu \frac{\partial}{\partial p_\alpha}\frac{\partial}{\partial p_\beta} + \ldots \right\} \frac{\partial}{\partial p_\nu}\left(\frac{1}{\not p - m}\delta^4(p+q)\right) \tag{27a}$$

$$S_F^{(2)}(p,q) = g\{\frac{i}{2} G^a_{\nu\mu} T^a \frac{1}{\not p - m} \gamma^\mu \frac{\partial}{\partial p_\nu} + \frac{1}{3}(D_\alpha G_{\nu\mu})^a(o) T^a \frac{1}{\not p - m} \gamma^\mu \frac{\partial}{\partial p_\alpha} \frac{\partial}{\partial p_\nu}$$

$$- \frac{i}{8}(D_\alpha D_\beta G_{\nu\mu})^a(o) T^a \frac{1}{\not p - m} \gamma^\mu \frac{\partial}{\partial p_\alpha} \frac{\partial}{\partial p_\beta} \frac{\partial}{\partial p_\nu}\} S_F^{(1)}(p,q) \quad (27b)$$

$$+ \ldots$$

Similar expressions can be derived for the gluon propagator and one obtains in configuration space (with $Z = x-y$)

$$S^{ab}_{\mu\nu}(x,y) = \frac{1}{4\pi^2} \frac{g_{\mu\nu}\delta^{ab}}{Z^2} + \frac{1}{8\pi^2} g f^{abc} G^c_{\mu\nu}(o) \ln(-Z^2)$$
$$- \frac{1}{8\pi^2} g f^{abc} g_{\mu\nu} G^c_{\lambda\sigma}(o) \frac{x^\lambda y^\sigma}{Z^2} + \ldots \quad (28)$$

It is important to realize that the choice of the Schwinger gauge for the external fields *does break* translation invariance. Of course, translation invariance will be restored at the end of the calculation but some care is needed in the intermediate steps.

We are now ready to compute any term in the operator product expansion of any n-point function by proceeding as follows :

- make the external field substitution in all operators of the n point function;
- calculate all terms by usual perturbation techniques.

Obviously one will be left with functionals of the external fields only and it is then a simple matter to extract the coefficients of the Lorentz scalar gauge invariant functionals and to relate them to the Wilson coefficients of the OPE.

Consider, for example, the dimension four operators in the Wilson expansion for some two-point function

$$C^W_{G^2} \langle o|\frac{\alpha_s}{\pi} G^2|o\rangle + C^W_m \langle o|m\bar{q}q|o\rangle. \quad (29)$$

The external field substitution gives

$$C^W_{G^2} \langle o|\frac{\alpha_s}{\pi} G^2_{Ext}|o\rangle + C^W_m \langle o|m\bar{\psi}_{Ext}\psi_{Ext}|o\rangle + C^W_m \langle o|m\bar{\eta}\eta|o\rangle. \quad (30)$$

To first order in α_s, the last term will give a contribution corresponding to the diagram

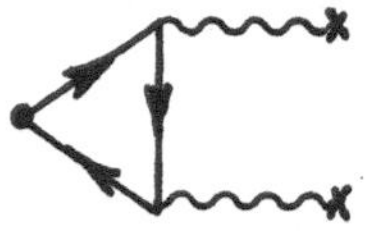

which we write as $c \langle o|\frac{\alpha_s}{\pi} G^2_{Ext}|o\rangle$.

Hence, by identifying the coefficients $C^{Ext}_{G^2}$ and C^{Ext}_m of the functionals $\frac{\alpha_s}{\pi} G^2_{Ext}$ and $m\bar{\psi}_{Ext}\psi_{Ext}$ we obtain

$$C^W_{G^2} + cC^W_m = C^{Ext}_{G^2} \qquad (31a)$$

$$C^W_m = C^{Ext}_m . \qquad (31b)$$

Note that in the massless quark limit C_m vanishes.

QUANTUM NUMBERS AND TENSOR STRUCTURE

The two-point functions which can be studied to analyse the meson spectrum are built out of mesonic currents which we can order according to their (scale) dimension and for which the accessible quantum numbers are listed below (current conservation and equations of motions have been ignored and C is listed for identical flavors).

Currents of dimension three

Type	J^{PC}
$\bar{q}q$	0^{++}
$\bar{q}\gamma_\mu q$	1^{--}, 0^{+-}
$\bar{q}\sigma_{\mu\nu}q$	1^{--}, 1^{+-}
$\bar{q}\gamma_\mu\gamma_5 q$	1^{++}, 0^{-+}
$\bar{q}\gamma_5 q$	0^{-+}

Currents of dimension four

$(\overleftrightarrow{D}_\mu = \overrightarrow{D}_\mu - \overleftarrow{D}_\mu)$

Type	J^{PC}
$\bar{q}\overleftrightarrow{D}_\mu q$	1^{--}, $\underline{0^{+-}}$
$\bar{q}\gamma_\mu \overleftrightarrow{D}_\nu q$	2^{++}, 1^{-+}, 0^{++}, 0^{++}; 1^{-+}, 1^{++}
$\bar{q}\gamma_5 \overleftrightarrow{D}_\mu q$	1^{+-}, $\underline{0^{--}}$
	etc...

Currents of dimension five

Type	J^{PC}
$g\bar{q}T^aG^a_{\mu\nu}q$	1^{+-}, 1^{--}
$g\bar{q}T^aG^a_{\mu\nu}\gamma^\mu q$	1^{-+}, 0^{++}
$g\bar{q}T^aG^a_{\mu\nu}\sigma^\nu_\alpha q$	2^{++}, 1^{-+}, 0^{++}, 0^{++}; 1^{++}, 1^{-+}
$g\bar{q}T^aG^a_{\mu\nu}\gamma^\mu\gamma_5 q$	1^{+-}, 0^{--}
$g\bar{q}T^aG^a_{\mu\nu}\gamma_5 q$	1^{+-}, 1^{--}
	etc...

For two-point functions built out of these currents, we want to isolate contributions due to states with specific J^{PC}. To this end we must construct "projection operators" for the various J^{PC} which can contribute as intermediate states. The simplest way to proceed is as follows : consider the matrix element of a given current between the vacuum and a state with given J^{PC} and polarization λ,

$$\langle n;\lambda | J^{(i)}_{\dots} | o \rangle = f^{(i),\lambda}_n T^{(i)(\lambda)}_n$$

where $f_n^{(i),\lambda}$ is a coupling constant and $T_n^{(i),\lambda}$ a tensor structure which is easily built out of the four momentum and the polarization tensor of the state n. For example

Lorentz structure of the current	$J^{PC}(n)$	$T^{(i)(\lambda)}_{J^{PC}}$
no Lorentz index	0^+	1
one Lorentz index	0^+	q_μ
	1^-	$e_\mu^{(\lambda)}$ (polarization vector)
two Lorentz indices (symmetric traceless)	0^+	$(4\frac{q_\mu q_\nu}{q^2} - g_{\mu\nu})$
	1^-	$i(q_\mu e_\nu^{(\lambda)} + q_\nu e_\mu^{(\lambda)})$
	2^+	$e_{\mu\nu}^{(\lambda)}$ (polarization tensor)
(antisymmetric)	1^-	$i(q_\mu e_\nu^{(\lambda)} - q_\nu e_\mu^{(\lambda)})$
	1^+	$i\ \varepsilon_{\mu\nu\rho\sigma} q^\rho e^{\sigma(\lambda)}$
(trace)	0^+	$g_{\mu\nu}$

With these expressions it is fairly straightforward to construct projectors on specific J^{PC} invariant functions

$$P^{(i)(j)}_{J^{PC}} = \sum_\lambda T^{(i)(\lambda)*}_{J^{PC}} T^{(j)(\lambda)}_{J^{PC}} .$$

For example, up to normalization factors, we have for

spin 0 $\quad P_o^{\alpha\beta} = \frac{q^\alpha q^\beta}{q^2}$

$$P_o^{\mu\nu,\rho\sigma} = g^{\mu\nu} g^{\rho\sigma} \text{ (trace, trace part)}$$

$$P_o'^{\mu\nu,\rho\sigma} = (4\frac{q^\mu q^\nu}{q^2} - g^{\mu\nu})(4\frac{q^\rho q^\sigma}{q^2} - g^{\rho\sigma}) \text{ (symmetric, symmetric part)}$$

$$P_o''^{\mu\nu,\rho\sigma} = (4\frac{q^\mu q^\nu}{q^2} - g^{\mu\nu}) g^{\rho\sigma} \text{ (symmetric, trace part)}$$

etc...

and, similarly,

for spin 1 $P_1^{\alpha,\beta} = \frac{q^\alpha q^\beta}{q^2} - g^{\alpha\beta}$

$$P_{1^+}^{\mu\nu,\rho\sigma} = \frac{1}{2}[(\frac{q^\mu q^\rho}{q^2} - g^{\mu\rho})(\frac{q^\nu q^\sigma}{q^2} - g^{\nu\sigma}) - (\frac{q^\mu q^\sigma}{q^2} - g^{\mu\sigma})(\frac{q^\nu q^\rho}{q^2} - g^{\nu\rho})]$$

$$P_{1^-}^{\mu\nu,\rho\sigma} = -\frac{1}{2} \boxed{\frac{\mu}{\nu}} \boxed{\frac{\rho}{\sigma}} g^{\rho\mu} \frac{q^\nu q^\sigma}{q^2}$$

where $\boxed{\frac{\alpha}{\beta}}$ means antisymmetrization in α and β.

While for spin 2 $P_2^{\mu\nu,\rho\sigma} = \frac{1}{2}(\eta_{\rho\mu}\eta_{\sigma\nu} + \eta_{\rho\nu}\eta_{\sigma\mu}) - \frac{1}{3}\eta_{\mu\nu}\eta_{\rho\sigma}$

where $\eta_{\mu\nu} = \frac{q_\mu q_\nu}{q^2} - g_{\mu\nu}$.

It now follows easily that for non conserved vector currents, the two-point function has the structure

$$i\int d^4x e^{iqx} \langle o|T\{J_\mu(x),J_\nu(o)\}|o\rangle = (\frac{q_\mu q_\nu}{q^2} - g_{\mu\nu})\Pi_V(q^2) + \frac{q_\mu q_\nu}{q^2}\Pi_s(q^2)$$

while, e.g. for an antisymmetric tensor current

$$i\int d^4x e^{iqx} \langle o|T\{\bar{q}(x)\sigma_{\mu\nu}q(x),\bar{q}(o)\sigma_{\alpha\beta}q(o)\}|o\rangle$$

$$= \frac{1}{2}(g^{\mu\beta}g^{\nu\alpha} - g^{\mu\alpha}g^{\nu\beta} + g^{\mu\alpha}\frac{q^\nu q^\beta}{q^2} + g^{\nu\beta}\frac{q^\mu q^\alpha}{q^2} - g^{\mu\beta}\frac{q^\nu q^\alpha}{q^2} - g^{\nu\alpha}\frac{q^\mu q^\beta}{q^2})\Pi_{1^+}(q^2)$$

$$+ \frac{1}{2}(g^{\mu\alpha}\frac{q^\nu q^\beta}{q^2} + g^{\nu\beta}\frac{q^\mu q^\alpha}{q^2} - g^{\mu\beta}\frac{q^\nu q^\alpha}{q^2} - g^{\nu\alpha}\frac{q^\mu q^\beta}{q^2})\Pi_{1^-}(q^2)$$

where the invariant functions Π_V, Π_S, $\Pi_{1^\pm}$ receive contributions from vector, scalar, and $1^\pm$ intermediate states only.

SOME EXPLICIT SUM RULES

A. Calculation of the Wilson coefficients

As a first example of how the calculations of the coefficients in the Wilson expansion are performed, let us consider the following currents

$$J_1(x) = \bar{q}_1(x)\Gamma_1 q_2(x) \tag{32a}$$

$$J_2^+(x) = \bar{q}_2(x)\bar{\Gamma}_2 q_1(x) \tag{32b}$$

where Γ_1, Γ_2 are any Dirac matrix and $\bar{\Gamma}_2 = \gamma^o\Gamma_2^+\gamma^o$. The two point function

$$\Pi_{\Gamma_1\Gamma_2}(q) = i\int d^4x e^{iqx} \langle o|T\{J_1(x)J_2^+(o)\}|o\rangle \tag{33}$$

becomes, after substitution of the background fields, a functional of these fields and it is straightforward to isolate the coefficients of the various polynomials in these fields.

Operator 1

First of all we have a term independent of the background fields which corresponds to the diagram (to lowest order in α_s)

and whose value is given by

$$- 3i\int d^4x e^{iqx} \mathrm{Tr}\, S^{(o)}_{F,2}(x)\overline{\Gamma}_2 S^{(o)}_{F,1}(-x)\Gamma_1$$

where the factor 3 comes from colour. Using the explicit expression of the quark propagator to 0th order in the external field, we have

$$\frac{3i}{(2\pi)^8}\int d^4x d^4k d^4k_2\, e^{i(q-k).x}\, \mathrm{Tr}\, \frac{\not{k}+\not{k}_2 + m_1}{(k+k_2)^2 - m_1^2}\overline{\Gamma}_2 \frac{\not{k}_2 + m_2}{k_2^2 - m_2^2}\Gamma_1 .$$

If there is an even number of γ matrices in $\Gamma_1\overline{\Gamma}_2$, this becomes

$$\frac{3i}{(2\pi)^4}\int\frac{d^4k_2}{[(q+k_2)^2 - m_1^2][k_2^2-m_2^2]}\{m_1m_2\, \mathrm{Tr}\,\overline{\Gamma}_2\Gamma_1 + (q+k_2)^\alpha k_2^\beta\, \mathrm{Tr}\gamma_\alpha\overline{\Gamma}_2\gamma_\beta\Gamma_1\}$$

with the help of a Feynman parameter y and a dimensional regularisation of the integral we are finally left with the expression

$$\frac{3}{16\pi^2}\int_0^1 dy[\{\tfrac{1}{2}\, g^{\alpha\beta}L - q^\alpha q^\beta y(1-y)\}\mathrm{Tr}(\gamma_\alpha\overline{\Gamma}_2\gamma_\beta\Gamma_1)\ln L + m_1m_2\, \mathrm{Tr}\,\overline{\Gamma}_2\Gamma_1\, \ln L] \tag{34a}$$

where $L = -q^2y(1-y) + m_1^2(1-y) + m_2^2y$.
Similarly, for an odd number of γ matrices in $\overline{\Gamma}_2\Gamma_1$, the same manipulations lead to

$$\frac{3}{16\pi^2}\int_0^1 dy\{m_1(1-y)\mathrm{Tr}\,\not{q}\,\overline{\Gamma}_2\Gamma_1 - m_2y\, \mathrm{Tr}\,\overline{\Gamma}_2\,\not{q}\,\Gamma_1\}\ln L. \tag{34b}$$

The expressions (34a,b) are to lowest order in α_s the coefficients of the operator 1 in the Wilson expansion.

Operator G^2

The next coefficient to calculate in the one multiplying quadratic expressions in the external gluon field $G_{\alpha\beta}$ i.e. all diagrams containing a factor $G_{\alpha\beta}G_{\gamma\delta}$. In general we will obtain an expression of the form

$$C^{\alpha\beta\gamma\delta}\,\frac{\alpha_s}{\pi}\, G_{\alpha\beta}G_{\gamma\delta}. \tag{35}$$

Using $\frac{\alpha_s}{\pi} \langle G_{\alpha\beta} G_{\gamma\delta} \rangle = \frac{1}{12}(g_{\alpha\gamma} g_{\beta\delta} - g_{\alpha\delta} g_{\beta\gamma}) \langle \frac{\alpha_s}{\pi} G^2 \rangle$ (36)

we identify $\frac{1}{12}(g_{\alpha\gamma} g_{\beta\delta} - g_{\alpha\delta} g_{\beta\gamma}) C^{\alpha\beta\gamma\delta} = C^W_{G^2}$, i.e. the Wilson coefficient of the operator $\langle \frac{\alpha_s}{\pi} G^2 \rangle$ in the massless limit. [See the discussion leading to Eq (31)].

To lowest order in α_s, we have three diagrams to calculate

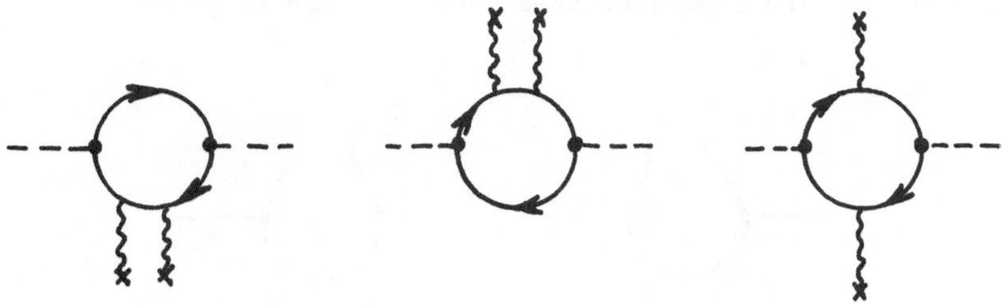

With the same "technology" (Feynman parameter and dimensional regularisation) one obtains after some work, for an even number of γ matrices in $\Gamma_1 \overline{\Gamma}_2$

$$
\begin{aligned}
C_{G^2} = &- \frac{\alpha_s}{48\pi} \int_0^1 dy \, \frac{1}{L^3} \Big\{ \frac{1}{4}[m_1^2(1-y)^3 + m_2^2 y^3] L \, \mathrm{Tr}\, \gamma^\alpha \overline{\Gamma}_2 \gamma_\alpha \Gamma_1 + m_1 m_2 [(-q^2 y + m_1^2(1-y) \\
&+ m_2^2 y)(1-y)^3 + (-q^2(1-y) + m_1^2(1-y) + m_2^2 y) y^3] \mathrm{Tr}\, \overline{\Gamma}_2 \Gamma_1 \\
&+ y(1-y)[m_1^2(1-y)^3 + m_2^2 y^3] \mathrm{Tr}\, \not{q} \overline{\Gamma}_2 \not{q} \Gamma_1 \Big\} \\
&- \frac{\alpha_s}{48\pi} \int_0^1 dy \, \frac{y(1-y)}{L^2} \frac{1}{8} \Big\{ L(10 \, \mathrm{Tr}\, \gamma^\alpha \overline{\Gamma}_2 \gamma_\alpha \Gamma_1 - \mathrm{Tr}\, \gamma^\alpha \gamma^\beta \gamma^\gamma \overline{\Gamma}_2 \gamma_\gamma \gamma_\beta \gamma_\alpha \Gamma_1) \\
&+ y(1-y)[4q^2 \, \mathrm{Tr}\, \gamma^\alpha \overline{\Gamma}_2 \gamma_\alpha \Gamma_1 + 4 \, \mathrm{Tr}\, \not{q} \overline{\Gamma}_2 \not{q} \Gamma_1 - 2 \, \mathrm{Tr}\, \gamma^\alpha \not{q} \gamma^\beta \overline{\Gamma}_2 \\
&\gamma_\beta \not{q} \gamma_\alpha \Gamma_1] - m_1 m_2 [8 \, \mathrm{Tr}\, \overline{\Gamma}_2 \Gamma_1 - 2 \, \mathrm{Tr}\, \gamma^\alpha \gamma^\beta \overline{\Gamma}_2 \gamma_\beta \gamma_\alpha \Gamma_1]
\end{aligned}
$$

and a similar expression for an odd number of γ matrices in $\Gamma_1 \overline{\Gamma}_2$ which we will not give here [6].

Operator $\overline{\psi}\psi$

It is easy to argue, first of all, that a quark condensate term for bosonic currents will in fact be multiplied by a quark mass. Secondly one can also show by the so-called heavy quark mass expansion [7] that condensates of heavy quarks can in fact be neglected.

For simplicity let us restrict our discussion to the case of the light quarks (u,d). The terms in the external field expansion which concern us are clearly

$$i\int d^4x e^{iqx}\, \bar{\psi}_1(x)\Gamma_1\, \eta_2(x)\bar{\eta}_2(o)\bar{\Gamma}_2\, \psi_1(o)' + i\int d^4x e^{iqx}\, \bar{\eta}_1(x)\Gamma_1\psi_2(x)\bar{\psi}_2(o)\bar{\Gamma}_2\, \eta_1(o). \tag{37}$$

Note that the background fields $\bar{\psi}_1(x)$ or $\psi_2(x)$ still have to be expanded around x = 0. Hence, in general, for each of the expressions in Eq (37) there will be two types of contributions to the quark condensate : one will come from the mass term in the propagator $S_F(x)$ and the other from, say, $\partial_\alpha\psi_2(o)$ in the Taylor series of the external field around x = 0. We can represent these terms by the diagrams

and similarly

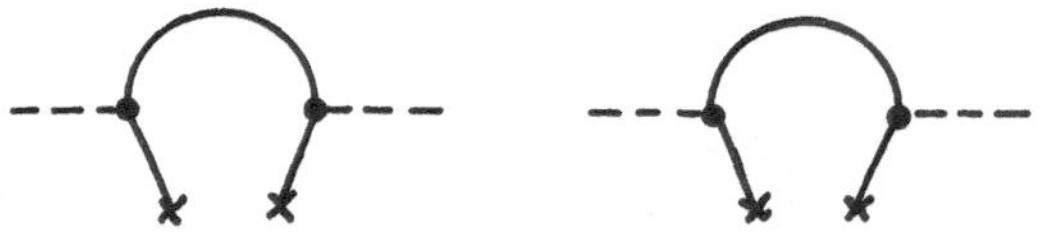

The terms with a derivative of the external fields will end up, through the equations of motion, as terms of the form "$m\bar{\psi}(o)\psi(o)$".

What about higher terms in the expansion of $\psi(x)$ e.g.

$$\bar{\psi}(o)\partial_\alpha\partial_\beta\partial_\gamma\, \psi(o). \tag{38}$$

Such an expression is of (naive) dimension 6. It can be rewritten in a more covariant looking way by replacing the ∂_α by covariant derivatives D_α (in the Schwinger gauge!). Through the use of the equations of motion, Eq (38) will contribute to dimension 6 operators but will also give a "correction term"

$$\frac{m^2}{q^2}\, m\bar{\psi}(o)\psi(o) \tag{39}$$

to the quark condensate.

This illustrates how tricky it becomes to calculate the full coefficient of the quark condensate. Luckily, in practice, the problem is greatly simplified since for light quarks (u,d) $\frac{m^2}{q^2} \ll 1$ while for heavy quarks, the heavy condensate can practically be ignored in the heavy quark mass expansion. For the strange quark, terms like Eq (39) have to be kept and turn out to be relatively small.

Anyway, it should be clear by now how to proceed for the higher dimensional operators.

If we work with $\bar{q}qG$ currents instead, the calculations proceed along similar lines and the coefficient of the operator $\mathbb{1}$ (still to lowest order in α_s) can be calculated from the diagram

while for the gluon condensate we have the diagram

and so on.

Clearly this technique can be extended to "radiative correction" diagrams which, for the operator $\mathbb{1}$ look as follows

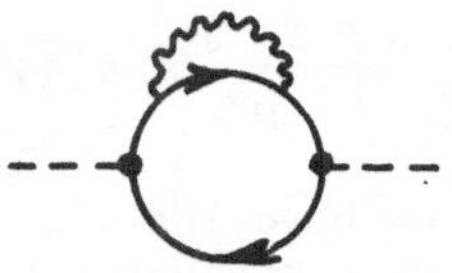

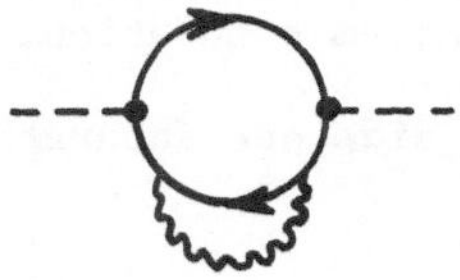

We will not discuss these diagrams any further. Let us remark, however, that there is a general "disagreement" on radiative corrections in the published literature, at least for the baryon sum rules. It seems to me that a careful calculation of these terms is still very much worthwhile.

B. Setting up the sum rules

For the various invariant spectral functions $\Pi(q^2)$, we can now write a dispersion relation with, in general, a certain number of subtractions

$$\mathrm{Re}\,\Pi(q^2) = \frac{1}{\pi}\int_{s_o}^{\infty} ds\left(\frac{q^2}{s}\right)^n \frac{\mathrm{Im}\,\Pi(s)}{s-q^2} + \sum_{k=1}^{n-1} a_k (q^2)^k. \qquad (40)$$

As repeatedly mentioned before one now takes either moments of this expression or one goes to its Borel transform Eq (11).

Obviously the Borel transform of a polynomial vanishes, while one can easily check that

$$L_M (q^2)^n \ln -q^2 = -n!(M^2)^n$$

$$L_M \left(\frac{1}{q^2}\right)^n = \frac{(-)^n}{(n-1)!}\left(\frac{1}{M^2}\right)^n \qquad \text{etc...}$$

Taking the Borel transform of Eq (40) leads to

$$L_M \, \mathrm{Re}\, \Pi(q^2) = \frac{1}{\pi M^2}\int_{s_o}^{\infty} ds\, e^{-s/M^2}\, \mathrm{Im}\, \Pi(s) \tag{41}$$

which is the generic form of the QCD sum rules : we equate $L_M \,\mathrm{Re}\,\Pi(q^2)$ calculated from the OPE (theoretical side of the sum rule) to $L_M \,\mathrm{Re}\,\Pi(q^2)$ obtained from phenomenological knowledge or modeling of $\mathrm{Im}\,\Pi(s)$ (usually saturated by one resonance plus a continuum).

Explicitly, for the current $j_\mu(x) = \bar{q}(x)\gamma_\mu q(x)$, one finds [1,2]

$$\int e^{-s/M^2}\, \mathrm{Im}\,\Pi(s)ds = \frac{1}{8\pi} M^2[1 + \frac{\alpha_s(M)}{\pi} + \frac{8\pi^2}{M^4}\langle o|m\bar{q}q|o\rangle + \frac{\pi^2}{3M^4}\langle o|\frac{\alpha_s}{\pi} G^2|o\rangle$$
$$- \frac{448}{81}\frac{\pi^3\alpha_s}{M^6}[\langle o|\bar{q}q|o\rangle]^2 + \ldots] \tag{42a}$$

where on the right hand side we have included the bare loop and its radiative correction, the quark condensate, the gluon condensate and the four quark operators (with vacuum saturation, see the next section).

On the phenomenological side one saturates with the ρ meson and a continuum with threshold S_C

$$g_\rho^2\, e^{-m_\rho^2/M^2} = -\int_{S_C}^{\infty} e^{-s/M^2}\, \mathrm{Im}\,\Pi_C(s) + \frac{1}{8\pi}M^2[1 + \frac{\alpha_s(M)}{\pi} + \frac{8\pi^2}{M^2}\langle o|m\bar{q}q|o\rangle + \text{other terms}] \tag{42b}$$

taking the ratio of this equation with its first derivative with respect to $\frac{1}{M^2}$ and using the numerical values for the various operators i.e.

$$\langle\bar{u}u\rangle = \langle\bar{d}d\rangle = (-.25 \text{ Gev})^3$$
$$\langle\frac{\alpha_s}{\pi}G^2\rangle = (.36 \text{ Gev})^4$$

gives us, finally, an equation for m_ρ^2

$$m_\rho^2 = M^2 \frac{[(1 + \frac{\alpha_s}{\pi})\{1 - (1 + \frac{S_C}{M^2})e^{-S_C/M^2}\} - \frac{.05}{M^4} + \frac{.06}{M^6}]}{(1 + \frac{\alpha_s}{\pi})\{1 - e^{-S_C/M^2}\} + \frac{.05}{M^4} - \frac{.03}{M^6}}. \tag{42c}$$

From this equation, with $S_C \simeq 1.5$ Gev^2, one obtains $m_\rho^2 \simeq (750 \text{ Mev})^2$. A detailed analysis of how m_ρ^2 is extracted from Eq (42) is given in reference [2].

C. The parameters

In the previous paragraph, we have used numerical values for the vacuum expectation values of various operators appearing in the OPE. Let me briefly summarize the common lore on this matter [2].

Gluon condensates

a) the dimension four gluon condensate

$$<o|\frac{\alpha_s}{\pi} G^a_{\mu\nu} G^a_{\mu\nu}|o> = (360 \pm 20 \text{ Mev})^4, \tag{43}$$

the value quoted corresponds to a fit to the charmonium spectrum. Other estimates based on lattice simulations or on theoretical models for the QCD vacuum (instanton gas) qualitatively agree.

b) gluon operators of dimension six

A priori one has the following possibilities

$$G^a_{\mu\nu} G^b_{\rho\sigma} G^c_{\alpha\beta}, \quad (D_\alpha G_{\mu\nu})^a (D_\beta G_{\rho\sigma})^b \text{ and } (D_\alpha D_\beta G_{\mu\nu})^a G^b_{\rho\sigma}.$$

However, using on the one hand the equations of motion

$$(D_\mu G_{\nu\mu})^a = g \sum_{\text{flavours}} \bar{q}(x)\gamma_\nu T^a q(x) = g j^a_\nu(x)$$

and, on the other hand, the Bianchi identities

$$D_\alpha G_{\beta\gamma} + D_\beta G_{\gamma\alpha} + D_\gamma G_{\alpha\beta} = 0$$

as well as the commutation relations of the covariant derivatives one readily convinces oneself that one is left with only *two* independent vacuum matrix elements, namely

$$<o|g^3 f^{abc} G^a_{\alpha\beta} G^b_{\beta\gamma} G^c_{\gamma\alpha}|o> \text{ and } <o|g^4 j^a_\nu j^a_\nu|o>.$$

Estimates of the $<G^3>$ condensate differ widely : for example, lattice and instanton-type calculations lead to opposite signs for this term! The $<G^3>$ condensate plays *no* role in sum rules for ordinary $q\bar{q}$ mesons. It could however be of essential importance as far as hybrid spectroscopy is concerned and is of course relevant in baryon spectroscopy. We will come back to the second vacuum matrix element when we discuss quark condensates.

c) higher dimension operators

It can be shown that there are seven independent vacuum matrix elements of operators of dimension 8. Existing estimates are usually based on a factorization plus vacuum saturation assumption. It seems to me that such a procedure is highly suspect.

Quark condensates

a) From PCAC, one has the relation

$$(m_u + m_d) <o|\bar{u}u + \bar{d}d|o> = - m_\pi^2 f_\pi^2. \tag{44}$$

Using the estimates of Gasser and Leutwyler [8] for the quark masses leads to the following value of the light quark condensates

$$<o|\bar{u}u|o> = (-225 \pm 25 \text{ Mev})^3 \quad (45)$$

$$<o|\bar{d}d|o> \cong <o|\bar{u}u|o>.$$

For the strange quark, both its mass and the value of its condensate are much more open to discussion. It is generally agreed that

$$<o|\bar{s}s|o> \simeq .8 \text{ tc } 1 <o|\bar{u}u|o>, \quad (46)$$

while for m_s, estimates range from 100 to 200 Mev.

For heavier quarks, the condensates do not play any role but once again estimates of the masses differ considerably.

b) The dimension 6 condensates $<o|\bar{\psi}\Gamma\psi\bar{\psi}\Gamma\psi|o>$ are usually "vacuum saturated" i.e. taken proportional to $(\bar{\psi}\psi)^2$. In my opinion, this is at best an order of magnitude estimate which could be off by as much as a factor of 2 or 3 as I will argue in the next section.

Mixed condensates

The dimension 5 condensate $<o|gG^a_{\mu\nu}\bar{\psi}\sigma^{\mu\nu}T^a\psi|o>$ has been parametrized by Novikov et al. as follows

$$<o|gG^a_{\mu\nu}\bar{\psi}\sigma^{\mu\nu}T^a\psi|o> = m_o^2 <\bar{\psi}\psi> \quad (47)$$

and they estimate $m_o^2 \simeq .5$ to 1 Gev^2.

Masses and couplings

These are of course taken as "running". A sample of representative values is the following

$$m_c(p^2 = -m_c^2) = 1.26 \pm .02 \text{ Gev}$$

$$m_b(p^2 = -m_b^2) = 4.23 \pm .05 \text{ Gev}$$

while

$$\alpha_s(4m_c^2) = .2 \pm .05 \qquad (\Lambda_{QCD} = 150 \pm 50 \text{ Mev})$$

but there is definitely no unanimity on these values. Other parameters which we do not discuss include the anomalous dimensions of the various operators considered.

RESULTS AND COMMENTS

I would like to end these lectures with a very selective and short summary of what I consider as the more remarkable results derived from the QCD sum rules for hadron masses.

Charmonium

The charmonium spectrum remains one of the triumphs of the sum rule method. Let me recall that Shifman, Vainshtein and Zakharov actually *predicted* the η_c at around 3 Gev. The present values for the lowest lying levels with different J^{PC} are the following [2]

^1S_o = 3.0± .02 Gev $\quad$ ^3P_o = 3.40 ± .01

$^3S_1$ = 3.09 ± .02 Gev $\quad$ $^3P_1$ = 3.5 ± .02

$^3P_2$ = 3.57 ± .02

$^1P_1$ = 3.51 ± .01 (predicted)

and the agreement with experiment is spectacular. Since one is studying a heavy quark system, it is appropriate to use the "moment method" and following Reinders et al. [2] one improves the dominance of the lowest lying resonance by taking moments at non zero momentum transfer squared.

It seems to me that one could do significantly better on the radial excitations of the charmonium system. In my opinion, one should consider a *set* of currents coupled to the same quantum numbers, with increasing dimension, and saturate *all* sum rules, i.e. diagonal and non diagonal ones, which one can derive from this set : I am convinced that one would then be able to find "windows" (in the moment number) where a specific radial excitations would dominate some subset of the sum rules. However, the calculations of radiative corrections are a serious obstacle for the realization of this program.

Bottomium

Here the physics is drastically different from the charmonium case and except for very gross features of the spectrum one does not expect the sum rule method to produce reliable information *unless* multi-order α_s corrections can be taken into account. The reason for all this is quite clear and well known : compared to the charmonium system, the gluon condensate in the $b\bar{b}$ spectrum is scaled by a factor $(\frac{m_c}{m_b})^4 \sim \frac{1}{100}$. Hence the radiative corrections will completely overshadow the non perturbative effects. In other words, the $b\bar{b}$ system is much more "Coulombic" than the $c\bar{c}$ system. It is precisely the interplay between "radiative corrections" and "non perturbative effects" (condensates) which lies at the heart of the successes of the sum rules in charmonium. Hence one should not expect (nor does it happen) that the same method will work as precisely in the two cases [2].

Light quark systems

Since in this case no natural scale is present, one is better off with a Borel transform than with moments.

I believe some progress has recently [9] been made concerning the L = 1 mesons which couple to dimension 3 currents, i.e. the $\delta(0^{++})A_1(1^{++})$ and $B(1^{+-})$. What makes these mesons interesting is the following : because of the subtraction in the corresponding dispersion relation, the Borel transforms Eq (42a) start as M^4 rather than M^2. This implies that the dimension four condensates do *not* contribute to the derivative sum rule. Hence dimension six operators play an essential role in stabilizing the sum rules. Hence, the L = 1 meson spectrum provides a unique window on dimension six operators. Of these, the $\langle G^3 \rangle$ term is known to vanish and one is left with operators

$$\bar{\psi}\Gamma\psi\bar{\psi}\Gamma\psi.$$

It is interesting to point out that a joint analysis of the δ, A, and B leads to an extremely satisfactory agreement with experiment *provided*, the vacuum dominance assumption is broken by a factor two or three for the scalar-scalar term

$$<o|\bar{\psi}\psi\bar{\psi}\psi|o>.$$

The consistency of this pattern can be checked for the strange mesons Q_A and Q_B.

For completeness, let us give the full expressions of the invariant functions for light quark bilinears (including radiative corrections to the coefficient of the operator 1)

$$0^{\pm +}\ \Pi(q^2) = -\frac{3}{8\pi^2}q^2\{\ln(-q^2/\mu^2) + \frac{\alpha_s}{\pi}[\frac{17}{3}\ln - q^2/\mu^2 - \ln^2(-q^2/\mu^2) + cte\}$$

$$+ \frac{1}{q^2}[-\frac{1}{8}<\frac{\alpha_s}{\pi}G^2> + (-\frac{1}{2}m_1 \mp m_2)<\bar{q}_1q_1> + (-\frac{1}{2}m_2 \mp m_1)<\bar{q}_2q_2>]$$

$$+ \frac{1}{(q^2)^2}[(\pm\frac{1}{2}m_2)<g\bar{q}_1G^a_{\alpha\beta}T^a\sigma_{\alpha\beta}q_1> + (\pm\frac{1}{2}m_1)<g\bar{q}_2G^a_{\alpha\beta}T^a\sigma_{\alpha\beta}q_2>$$

$$+ \frac{4\pi\alpha_s}{3}<o|[\bar{q}_1\gamma_\mu T^a q_1 + \bar{q}_2\gamma_\mu T^a q_2][\sum_q \bar{q}\gamma_\mu T^a q]|o>$$

$$\pm 4\pi\alpha_s <o|\bar{q}_1\sigma_{\alpha\beta}\begin{Bmatrix}1\\ \gamma_5\end{Bmatrix}T^a q_2\bar{q}_2\sigma_{\alpha\beta}\begin{Bmatrix}1\\ \gamma_5\end{Bmatrix}T^a q_1|o>]$$

$$1^{\mp\mp}\ \Pi(q^2) = -\frac{1}{4\pi^2}q^2\{\ln(-q^2/\mu^2)[1+\frac{\alpha_s}{\pi}] + cte\}$$

$$+ \frac{1}{q^2}[\frac{1}{12}<\frac{\alpha_s}{\pi}G^2> + (\pm m_2)<\bar{q}_1q_1> + (\pm m_1)<\bar{q}_2q_2>]$$

$$+ \frac{1}{(q^2)^2}[\frac{8\pi\alpha_s}{9}<o|[\bar{q}_1\gamma_\mu T^a q_1 + \bar{q}_2\gamma_\mu T^a q_2][\sum_q \bar{q}\gamma_\mu T_a q]|o>$$

$$+ 8\pi\alpha_s <\bar{q}_1\gamma_\alpha\gamma_5\begin{Bmatrix}1\\ \gamma_5\end{Bmatrix}T^a q_2\bar{q}_2\gamma_\alpha\gamma_5\begin{Bmatrix}1\\ \gamma_5\end{Bmatrix}T^a q_1>]$$

$1^{\mp -}$ (coming from the current $\bar{\psi}\sigma_{\mu\nu}\psi$ and including mass corrections)

$$\Pi(1^{\mp};q^2) = -\frac{1}{8\pi^2}q^2\{[1+6\frac{(\pm m_1m_2)}{q^2}]\ln - q^2/\mu^2 + \frac{\alpha_s}{3\pi}[\frac{7}{3}\ln(-q^2/\mu^2) + \ln^2(-q^2/\mu^2)] + cte\}$$

$$+ \frac{1}{q^2}[-\frac{1}{24}<\frac{\alpha_s}{\pi}G^2> + (\frac{1}{2}m_1 \pm m_2)<\bar{q}_1q_1> + (\frac{1}{2}m_2 \pm m_1)<\bar{q}_2q_2>]$$

$$+ \frac{1}{(q^2)^2}[\frac{1}{6}(2m_1 \pm m_2)<g\bar{q}_1G^a_{\alpha\beta}T^a\sigma_{\alpha\beta}q_1> + \frac{1}{6}(2m_2 \pm m_1)<g\bar{q}_2G^a_{\alpha\beta}T^a\sigma_{\alpha\beta}q_2>$$

$$+ \left(-\frac{8\pi\alpha_s}{9}\right) <o|[\bar{q}_1\gamma_\mu T^a q_1 + \bar{q}_2\gamma_\mu T^a q_2][\sum_q \bar{q}\gamma_\mu T^a q]|o>$$

$$+ (\mp 16\pi\alpha_s) <o|\bar{q}_1\begin{Bmatrix}\gamma_5\\1\end{Bmatrix}T^a q_2\bar{q}_2\begin{Bmatrix}\gamma_5\\1\end{Bmatrix}T^a q_1|o>].$$

With a breaking of the vacuum dominance assumption in the scalar four quark operator i.e.

$$<o|\bar{q}_1 T_a q_2 \bar{q}_2 T_a q_1> = (1+\alpha)(-\tfrac{1}{9}) <\bar{q}_1 q_1><\bar{q}_2 q_2>$$

one finds, for $\alpha = 1.7$, $m_B = 1235$ Mev ($s_o = 2.4$ Gev^2) while the other meson masses are essentially unchanged, i.e.

$m_\delta = 985$ Mev ($s_o = 1.5$ Gev^2)

$m_{A_1} = 1.27$ Gev ($s_o = 2.4$ Gev^2).

For more details, see reference [9].

Baryons

Undoubtedly one of the most remarkable results of the QCD sum rules is the one obtained by Ioffe [10] as a first approximation to the nucleon mass

$$m_N \cong [-8\pi^2 <\bar{q}q>]^{1/3} \simeq 1 \text{ Gev}.$$

Essentially this implies that the full nucleon mass is a result of chiral symmetry breaking!

The analysis of the octet and decuplet states gives a perfect agreement, at the precision one can hope for, between theory and experiment [2].

No systematics of higher multiplets exist so far.

Hybrids

For 0^{--} states, which are "exotic", there is no doubt that the QCD sum rules stabilize and hence that they predict the existence of hybrid hadrons. This is true for massless [11] as well as for massive quarks [12]. This prediction is conceptually extremely important, however it will not be easy to confirm it experimentally.

For 1^{-+} hybrids, there is disagreement [11,12,13] on
- whether or not the sum rules stabilize!
- where the mass of the resonance is, if there is a resonance!

In the case of massless quarks, the discrepancy comes from $<G^3>$ contributions and especially the next to leading order coefficient of this operator. It is however interesting to note that for *heavy quark* hybrids ($c\bar{c}g$ or $b\bar{b}g$) looking at *all* sum rules (diagonal and mixed) which can be written for $q\bar{q}g$ currents one finds that *not a single* 1^{-+} sum rule stabilizes [14]. Of course, the calculation is done without the d = 6 operators but it is highly unlikely that these operators would significantly affect the $b\bar{b}g$ spectrum!

For hybrids with "normal" quantum numbers, there is stabilization of the sum rules but the resonances which one obtains are always much higher than the normal $q\bar{q}$ state for the corresponding quantum number. This suggests the possible identification of some $q\bar{q}G$ states with radial excitations of the normal $q\bar{q}$ spectrum [12].

Clearly the dynamics of $q\bar{q}g$ states has to be better understood and, as I suggested before, quite a lot could be learned from coupled sum rules derived from currents with different scale dimensions.

3-point functions

The applications [2] which have been considered up to now in the literature are, essentially

- decays of hadrons into 2γ
- couplings of two hadrons to a real γ
- couplings of three hadrons.

The general procedure is quite similar to the 2-point function case but of course the analysis becomes much more complicated (one has to take double moments or a double borelization). Most of the results depend heavily on "pole dominance" assumptions which seem to work best for "Goldstone bosons".

One rather remarkable result [15] has been obtained for the pion nucleon coupling constant : it reads

$$g_{\pi N\bar{N}} = \frac{4m_q m_N^2\sqrt{2}}{f_\pi m_\pi^2} \tag{48}$$

while, the Goldberger Truman relation gives

$$g_{\pi N\bar{N}} = \frac{m_N\sqrt{2}}{f_\pi}. \tag{49}$$

If Eq (48) is not accidental (many approximations are needed to derive it) it is quite an impressive result since it relates $g_{\pi NN}$ to "chiral" parameters in a different way than (49).

This should certainly make a qualitative analysis of four point functions more than worthwhile : for the $\pi\pi \rightarrow \pi\pi$ amplitudes, for example, could one really expect relations which go beyond current algebra and PCAC ?

ACKNOWLEDGEMENTS

I have greatly benefited from many discussions with J. Govaerts and L. Reinders. The help of J. Govaerts in preparing these notes is also gratefully acknowledged.

REFERENCES

[1] M.A. Shifman, A.I. Vainshtein and V.I. Zakharov, Nucl. Phys. B147 (1979) 385, 448.
[2] Since the original work of reference [1], many excellent reviews have been written on various applications of the sum rules. Let me mention, in particular
L.J. Reinders, H. Rubinstein and S. Yazaki, Phys. Rep. 127 (1985) 1 where a complete list of references can be found.
[3] K.G. Wilson, Phys. Rev. 179 (1969) 1499.
[4] V.A. Novikov, M.A. Shifman, A.I. Vainshtein and V.I. Zakharov, Nucl. Phys. B249 (1985) 445.

[5] See e.g. M.A. Shifman, Z. Phys. C9 (1981) 347;
S.N. Nikolaev and A.V. Radyushkin, Nucl. Phys. B213 (1983) 285;
W. Hubschmid and S. Mallik, Nucl. Phys. B207 (1982) 29;
J. Govaerts, F. de Viron, D. Gusbin and J. Weyers, Nucl. Phys. B248 (1984) 1.

[6] S. Mallik, Nucl. Phys. B234 (1984) 45.

[7] D.J. Broadhurst, Phys. Lett. 101B (1981) 423.

[8] J. Gasser and H. Leutwyler, Phys. Rep. 87 (1982) 77.

[9] J. Govaerts, L.J. Reinders, F. de Viron and J. Weyers (to be published).

[10] B.L. Ioffe, Nucl. Phys. B188 (1981) 317.

[11] J. Govaerts et al., reference [5];
J.I. Latorre, S. Narison, P. Pascual and R. Tarrach, Phys. Lett. 147B (1984) 169.

[12] J. Govaerts, L.J. Reinders, H.R. Rubinstein and J. Weyers, Nucl. Phys. B258 (1985) 215;
J. Govaerts, L.J. Reinders and J. Weyers, Nucl. Phys. B262 (1985) 575.

[13] I.I. Balitsky, D.I. Dyakonov and A.V. Yung, Phys. Lett. 112B (1982) 71; Sov. J. Nucl. Phys. 35 (1982) 761.
I.I. Balitsky, D.I. Dyakonov and A.V. Yung (preprint, December (1985)).

[14] J. Govaerts, L.J. Reinders, P. Francken, X. Gonze and J. Weyers (to be published).

[15] L.J. Reinders, H.R. Rubinstein and S. Yazaki, Nucl. Phys. B213 (1983) 109.

INTRODUCTION TO EXCLUSIVE PROCESSES IN PERTURBATIVE QCD

E. Maina

Ist. di Fisica Teorica
Università di Torino
I.N.F.N. Sezione di Torino

INTRODUCTION

Every physics student by now knows that QCD [1] is the theory of strong interactions and that it is asymptotically free. It is also well known that the theory explains jets behaviour, massive lepton pair production in hadronic collisions, e^+e^- total cross section data, deep inelastic phenomena and many other processes, justifying the successes of the parton model and furthermore quantitatively explaining the differences between parton model predictions and experimental results. It is also usually familiar that confinement, while still not completely under control, seems to be supported by numerical experiments on the lattice. Less widely known are the difficulties the theory encounters in trying to connect the parton level phenomena to the hadrons detected in the final state.

In these notes I will try to present only one aspect of this problem, namely the description of exclusive processes.

THEORY

Attempts to interpret exclusive hadronic reactions in the framework of QCD began very early on. In 1973 [2] Brodsky and Farrar showed that the scaling with energy of large momentum transfer form factors and meson and/or baryon elastic scatterings, can be predicted if quark-quark interactions are scale invariant. Under this assumption, elastic scattering cross sections behave like :

$$\frac{d\sigma}{dt} = \frac{1}{S^{n-2}} f(\frac{t}{S})$$

where n is the number of partons participating in the process, while the electromagnetic form factor of an object made of m partons is predicted to scale as :

$$F_m(Q^2) = \text{const.x} \frac{1}{(Q^2)^{m-1}}$$

These result are in very good agreement with experiment for reactions going from $\gamma p \to \gamma p$ [3] to $pp \to pp$ and from the pion to the ^4He form factors [4] for t larger than about 5 GeV^2 as shown in Fig. 1 and 2.

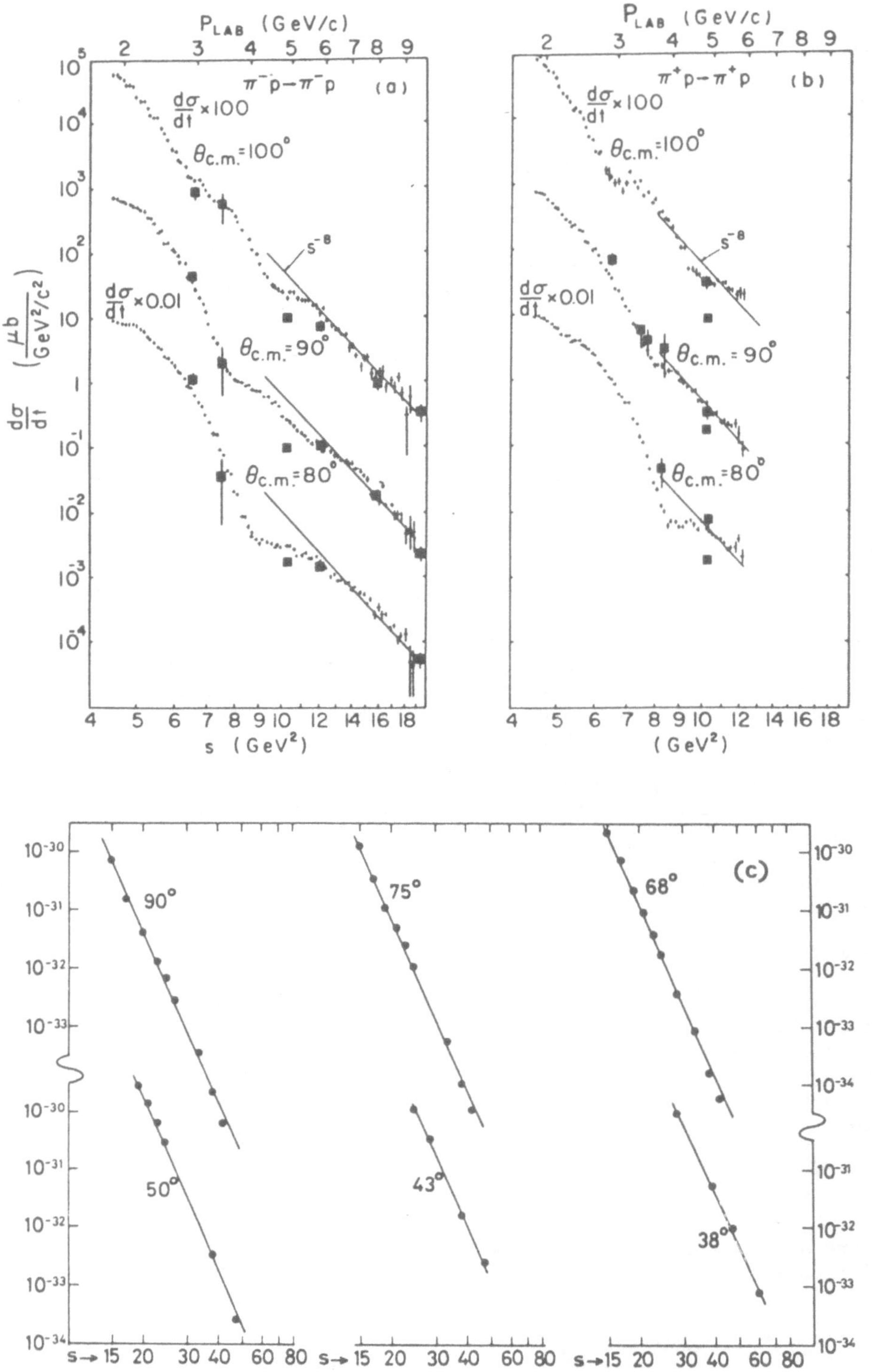

Fig. 1 : a,b) Plot of dσ/dt for πp elastic scattering at various center-of-mass scattering angles [3c].
c) Plot of dσ/dt for pp elastic scattering at various center-of-mass scattering angles [3b].

The successes of QCD applied to inclusive phenomena are largely due to factorization. There one can rigourously separate the long distance part of the interaction from the short distance one, which can then be treated perturbatively. In order to apply perturbative QCD to exclusive processes with the same confidence we feel in the inclusive case, we should be able to separate the presumably short distance physics that describes the hard scattering among constituents from the long distance

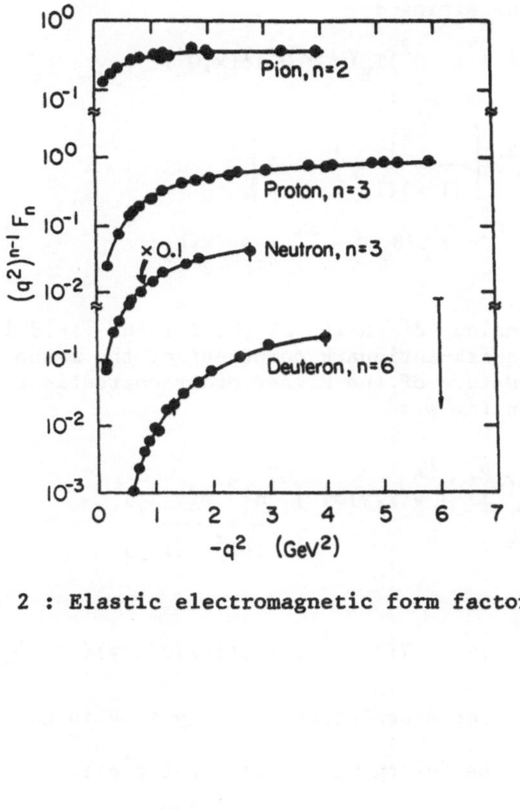

Fig. 2 : Elastic electromagnetic form factors [4].

interactions that keep them bound inside the hadrons. The amplitude A for an elastic scattering, for example, would then be written as :

$$A(Q^2,\theta) = \int dx dx' dy dy' \; \phi^*(x,Q^2) \; \phi^*(x',Q^2) T_H(x,x',y,y',\theta) \phi(y,Q^2) \phi(y',Q^2)$$

where the wavefunction ϕ describes the momentum distribution of bound quarks while T_H describe the hard momentum transfer [5].
The simplest and up to now only processes which can be fully analyzed this way is the meson form factor [6].
One of the four graphs describing $\gamma^*\pi \to \pi$ in lowest order is shown in

Fig. 3. Transverse momenta in the hard amplitude will be neglected, since for $p_t^2 \geq Q^2$, where Q^2 is the scale of the hard momentum transferred in the reaction, the diagrams are suppressed by large denominators, and will be integrated over in the wave function for $p_t^2 \leq Q^2$. The hadronic wavefunction will be taken to be proportional to $\gamma_5 \phi(x) \not{p}$ and all calculations will be performed in the light cone gauge n.A = 0 and n.n = 0. It is easy to show then that the four diagrams give :

$$T = \frac{4g^2 C_F}{Q^2} \left(\frac{1}{(1-x)(1-y)} e_1 + \frac{1}{xy} e_2 \right)$$

where e_1 and e_2 are the charges of the two quarks while $C_F = 4/3$. Furthermore in this gauge, the dominant diagrams in the leading log approximation are the ladder ones with vertex and self energy insertions. Vertex corrections are associated with the appropriate vertex and the self energies with the outermost rung of each loop. Since the two diagrams of Fig. 4 cancel because $Z_1 = Z_2$, in order to expand in α_s, one has to divide by the square of the self energy obtaining :

$$F_\pi(Q^2) = \int_0^1 dxdy\ \phi^*(y,Q^2) T_H(x,y,Q^2) \phi(x,Q^2)$$

with

$$T_H = \frac{16\Pi C_F}{Q^2} \alpha(Q^2) \left(\frac{e_1}{(1-x)(1-y)} + \frac{e_2}{xy} \right)$$

$$\phi(x,Q^2) = (\log Q^2/\Lambda^2)^{\gamma_F/\beta_o} \int^{Q^2} \frac{dk_\perp^2}{16\Pi^2} \psi(x,k_\perp)$$

where γ_F is the anomalous dimension of the fermion field in the light cone gauge and ψ is the quark-antiquark component of the meson wavefunction. From the ladder structure of the higher order corrections one can deduce an integral equation for ψ :

$$\frac{\psi(x,q_\perp)}{16\Pi^2} = \frac{C_F}{4\Pi} \frac{\alpha(q_\perp^2)}{q_\perp^2} \int_0^1 \hat{V}(x,y)dy \int^{q_\perp^2} \frac{dl^2}{16\Pi^2} \frac{\psi(y,l_\perp)}{y(1-y)}$$

where :

$$\hat{V}(x,y) = 2[x(1-y)\theta(y-x)(1+\frac{1}{y-x}) + y(1-x)\theta(x-y)(1+\frac{1}{x-y})]$$

It can be shown that the singularity at x = y in V is canceled by a similar term in γ_F.
The final result is the following expression for ϕ :

$$\phi(x,Q^2) = x(1-x) \sum_{n=0}^{\infty} a_n C_n^{3/2}(1-2x)\ \alpha(Q^2)^{\gamma_n}$$

where the $C_n^{3/2}$ are Gegenbauer polynomials and the anomalous dimensions γ_n are given by :

$$\gamma_n = \frac{C_F}{\beta_o}\left(1+4 \sum_{2}^{n+1} \frac{1}{k} - \frac{2}{(n+1)(n+2)}\right)$$

Since the γ_n are monotonously increasing with n one gets asymptotically :

$$\phi(x,Q^2) = \text{const}.x(1-x)$$

An analogous treatment for the baryon wavefunction would yield the asymptotic form :

$$\phi(x,y,z,Q^2) = \text{const} \; . \; xyz\delta(1-x-y-z) \; (\log Q^2/\Lambda^2)^{-2/3\beta_0}$$

In the case of baryon form factors though there are contributions coming from the exchange of soft gluons between almost on shell quarks which could spoil the proof of factorization. It has been suggested that Sudakov corrections, which will be discussed shortly, may suppress these contributions so that baryon form factors are indeed dominated by hard scattering diagrams and renormalization group equations.

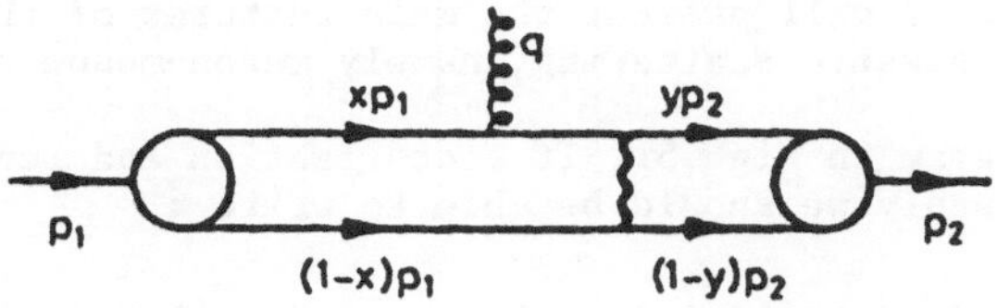

Fig. 3 : One of the four graphs describing $\gamma^*\pi \rightarrow \pi$ in lowest order.

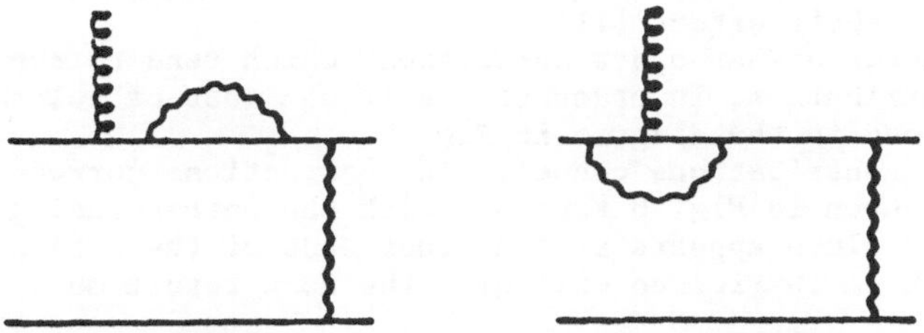

Fig. 4 : Higher order corrections to the diagram in Fig. 3.

A completely different approach to the determination of the appropriate form for the hadron wavefunction has been advocated in [7] by Chernyak and Zhitnitsky. They observe that present exclusive experiments are run at fairly low energies and that the asymptotic regime is approached only logarithmically so that we could actually be nowhere near the asymptotic region. Hence they propose to determine the hadron wavefunction from QCD sum rules [8]. The most interesting case is the baryon wavefunction for which they suggest the following form :

$$|P\uparrow\rangle = \text{const} \times \int_0^1 d_3x[\tfrac{1}{2}(V(x)-A(x))\,|u\uparrow(x_1)u\downarrow(x_2)d\uparrow(x_3)\rangle$$
$$+ \tfrac{1}{2}(V(x) + A(x))\,|u\downarrow(x_1)u\uparrow(x_2)d\uparrow(x_3)\rangle - T(x)\,|u\uparrow(x_1)u\uparrow(x_2)d\downarrow(x_3)\rangle]$$

$$V(x_1x_2x_3) = 120x_1x_2x_3(11.35(x_1^2 + x_2^2) + 8.82x_3^2 - 1.68x_3 - 2.94)$$

$$A(x_1x_2x_3) = 120x_1x_2x_3(6.72(x_2^2 - x_1^2))$$

$$T(x_1x_2x_3) = 120x_1x_2x_3(13.44(x_1^2 + x_2^2) + 4.62x_3^2 + 0.84x_3 - 3.78)$$

Recently Farrar made the observation [9] that at large momentum transfer, neglecting quark masses, a large number of different reactions depend on a much smaller set of hard quark amplitudes since asymptotically the flavour and spin hadron wavefunctions tend to their SU(6) expression. It is then possible to test whether we are in the asymptotic regime comparing the cross sections in various independent but similar processes. It must be noted that this argument does not depend on our ability to actually compute such amplitudes.

Sudakov effects [10] turn out to be crucially important in various areas of QCD [1]. I will present the main features of the subject in the simplest case of elastic scattering, namely meson-meson scattering, following Mueller.

Consider the diagram in Fig. 5. If factorization and renormalization group equations apply we should be able to write :

$$\frac{d\sigma}{dt} = \frac{1}{s^2} \left| \int_{-1}^{1} dx_1 dx_2 dx_1' dx_2' \phi(x_1,s)\phi(x_2,s)\phi(x_1',s)\phi(x_2',s)H(p,p',x,x') + \ldots \right|^2$$

If no denominator enhancement takes place then $d\sigma/dt = O(1/s^6)$. In fact such enhancements occur. For some range of k and k' three denominators vanish. Reinserting the fermion masses in the denominators in order to regulate the infinity one can evaluate the amplitude finding an additional factor of $(s/m^2)^{1/2}$. Thus $d\sigma/dt = O(1/s^5)$ rather than $O(1/s^6)$. This is called the Landshoff effect [11].

There are however higher order corrections which tend to suppress the Landshoff contribution. In order to see it one must calculate all double logs corrections to the diagram in Fig. 5.

All but eight contributions cancel. The corrections correspond to the two diagrams shown in Fig. 6 together with the corresponding ones where the extra soft gluon appears at the other ends of the original hard gluons. The diagram shown in Fig. 6a will give the Born result multiplied by the factor :

$$\frac{-g^2}{8\pi^2} \ln^2 s/m^2$$

while in the case of the diagram in Fig. 6b the factor is :

$$g^2/8\pi^2 \left\{ \ln\frac{Xs}{m^2} \ln\frac{s}{m^2} - \frac{1}{2}\ln^2\frac{Xs}{m^2} + \frac{1}{2}\ln^2\frac{s}{m^2} \right\}$$

where Xs is the minimum of the three denominators corresponding to a,b and c in Fig. 5. The sum of the two terms gives :

$$-\frac{g^2}{16\pi^2} \ln^2 \frac{1}{X}$$

times the original tree level diagram. If one sums the leading double logarithmic corrections at all orders the factor multiplying the tree level diagram is found to be :

$$\exp\left(-\frac{g^2}{4\pi^2}\ln^2\frac{1}{X}\right)$$

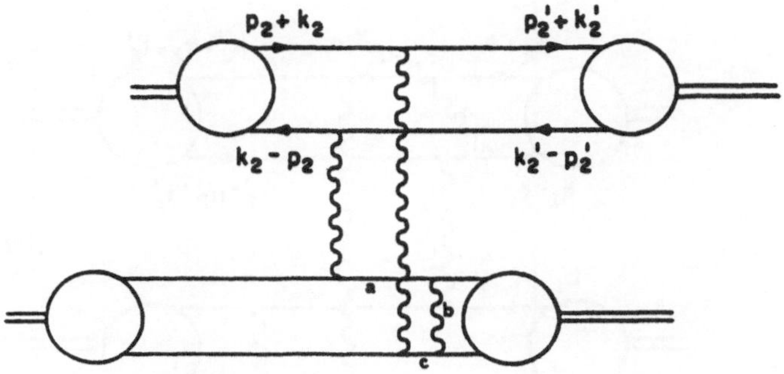

Fig. 5 : One of the "hard" diagrams in $\pi\pi$ elastic scattering.

In asymptotically free QCD this expression becomes :

$$\exp\left\{-\frac{2C_F}{11-2n_f/3}\left[\ln s(\ln\ln s - \ln\ln Xs) - \ln\frac{1}{X}\right]\right\}$$

This additional exponientiated factor suppresses the Landshoff contribution as $s \to \infty$, but also modifies the naive dimensional counting rules, leading to a behaviour intermediate between the two predictions. Kanwal [12] computed $\pi\pi$ elastic scattering taking Sukadov factors into account.

No treatment to all orders of Sukadov factors, beyond the leading double-logarithmic approximation, has been given to date, and without this ingredient it is impossible to prove rigorously that for large angle elastic scattering holds. Nonetheless it is worthwhile considering the scheme just presented as an adequate description of exclusive processes

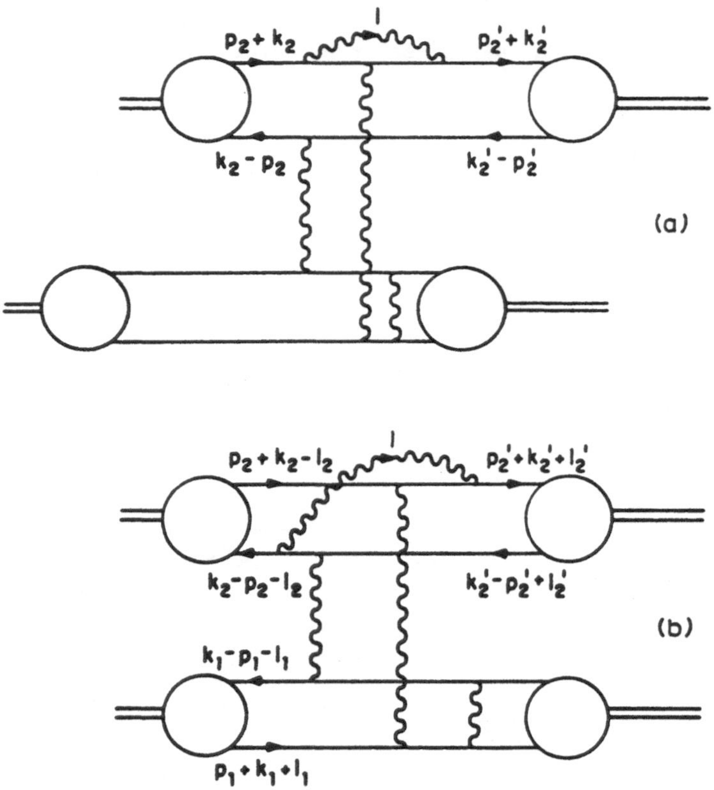

Fig. 6 : Graphs giving Sudakov corrections to the diagram in Fig. 5.

and confronting its predictions with the data. Such a comparison should be extremely helpful in guiding future attempts to put the theory on firmer ground.

Even this somewhat less ambitious program is extremely challenging from a technical point of view as soon as one considers processes more complicated than meson or baryon form factors. In order to illustrate this point all lowest order diagrams, there 204 of them, for the production of baryon pairs in two photons reactions and for Compton scattering off baryons, are presented in Fig. 7. The same counting reveals that there are roughly 480,000 diagrams in the case of proton-proton elastic scattering. It is clear that new methods must be devised in order to deal with such a large number of diagrams. These methods [13] in general are based on the choice of an explicit representation for the γ-matrices and the evaluation of the amplitude of a given process instead of directly computing the amplitude square using the Feynman trace procedure. Whenever one can neglect quark masses it is convenient to employ two component spinors in the helicity formalism and γ-matrices in the Weyl representation. Recently specialized computer programs which evaluate analytically all Born diagrams for a given reaction have been constructed. It seems clear that without such tools it will be extremely difficult to make any progress in the description of exclusive processes in QCD.

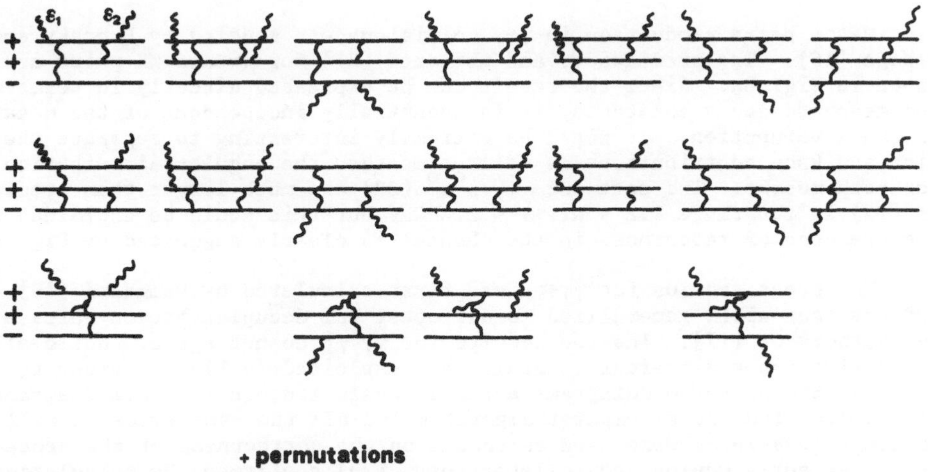

Fig. 7 : All lowest order diagrams for the production of baryon pairs in two photons reactions and for Compton scattering off baryons. All permutations of the vertices on each fermion line are understood. The + and - refer to the helicities of the quarks.

THEORY VERSUS EXPERIMENT

Various testable predictions have been produced, based on the scheme just outlined, namely proton and neutron magnetic form factors G_M^p, G_M^n [14], meson pairs production [15] and baryon pairs production in $\gamma\gamma$ collisions [16,17], Compton scattering on protons and related reactions [18].

The proton and neutron magnetic form factors turn out to be a major embarassment for the asymptotic form of the wavefunction. In fact [14] if we use a fixed coupling constant we find $G_M^p = 0$. Even if we allow α_s to run as a function of Q^2 it is impossible to reproduce magnitude and sign of both neutron and proton form factors, in the Born approximation, using the asymptotic expression [19]. On the contrary the data are reproduced quite accurately with the wavefunction proposed in [7]. This solves the problems stressed by Isgur and Llewellyn Smith [20]. Nonetheless before dismissing the asymptotic wavefunction as premature it would be desirable to know the effect of higher order corrections to the Born result.

Meson pairs production in γ-γ collisions was studied by Brodsky and Lepage [15]. The predictions and the data [21] for $\pi\pi$ and KK pairs are shown in Fig. 8a. Since the result can be expressed directly in terms of the measured decay constants, it is essentially independent of the details of the wavefunction. It would be extremely interesting to separate the pion and kaon contributions as well as measure the angular distribution for both mesons. The data for $\gamma\gamma \to \rho^o \rho^o$ [22] are much larger than predicted in [15] in the range $1.2 < W(\gamma\gamma) < 2.4$ GeV but this could be explained by the presence of resonances in the channel as clearly suggested by Fig. 8b.

The cross section for $\gamma\gamma \to p\bar{p}$ was first calculated by Damgaard [16] and was soon after generalized to all octet and decuplet baryon pairs by the authors of [17]. The two results for $\gamma\gamma \to p\bar{p}$ do not agree. While all symmetries among different diagrams were exploited in [16] in order to decrease the number of diagrams actually evaluated, in [17] all diagrams were calculated via a computer algorithm and all the symmetries as well as gauge invariance were used as checks on the correctness of the procedure. Recently Gunion and collaborators [23] confirmed the calculation of [17].

The cross section for $p\bar{p}$ pairs computed in [17] using the asymptotic wavefunction is smaller than the data [24] by a factor of roughly 50, while using the wavefunction advocated in [7] the QCD result falls within two standard deviations from the experimental result due to large error bars. Details of the predictions are presented in Fig. 9. It is to be denoted that one expects the cross section for $\gamma\gamma \to \Delta^{++}\overline{\Delta^{++}}$ to be larger than the pp cross section since photon couplings are proportional to the quark charges. TASSO [25] measured $\sigma(\gamma\gamma \to \Delta^{++}\overline{\Delta^{++}}) < 1/3\sigma(\gamma\gamma \to p\bar{p})$. In my opinion this could be an indication that mechanisms other than asymptotic QCD must explain the results at present energies and that no real conflict exists between available data and predictions. A rather modest increase in available energy should reveal if resonances are indeed reponsible for the magnitude of the TASSO results. Furthermore it would be interesting to extend the prediction obtained with the non asymptotic wavefunction to other baryon pairs.

Photon elastic scattering off baryons has been recently considered [18]. It must be emphasized that available data refer to photon energies varying from 1.7 GeV to 6 GeV. While s^6 changes by roughly three orders of magnitude $s^6 d\sigma/dt$ varies by at most a factor of three indicating that even at this relatively low energy the data scale according to the dimensional counting rules. As a consequence a comparison between our calculation and the experimental results [3a] may be justified. In Fig. 10a

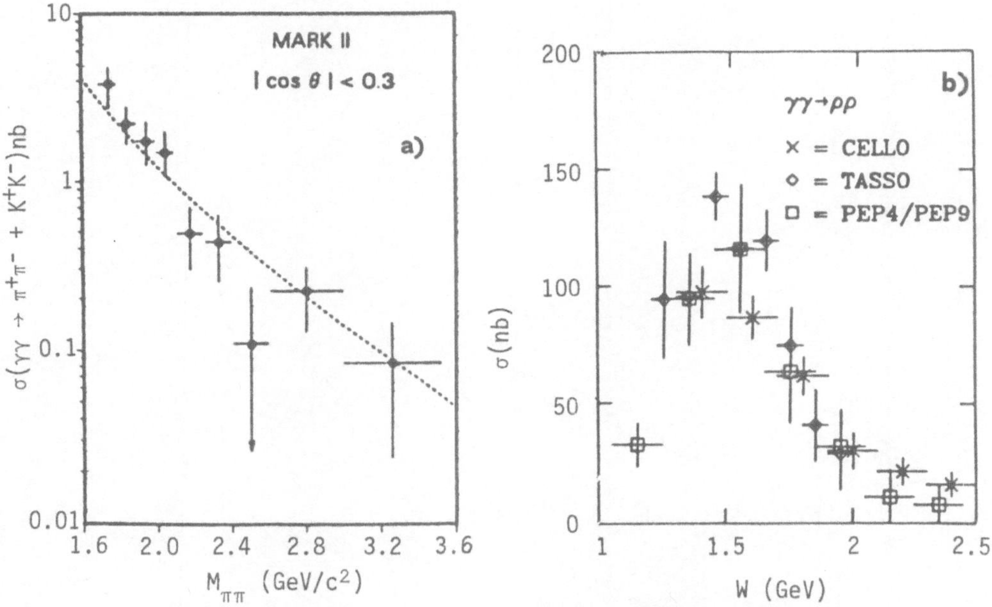

Fig. 8 : a) Cross section [21] for ππ and KK pairs in γγ collisions. The continuous line are the predictions in [15].
b) Cross section for $\gamma\gamma \to \rho^0 \rho^0$ [22].

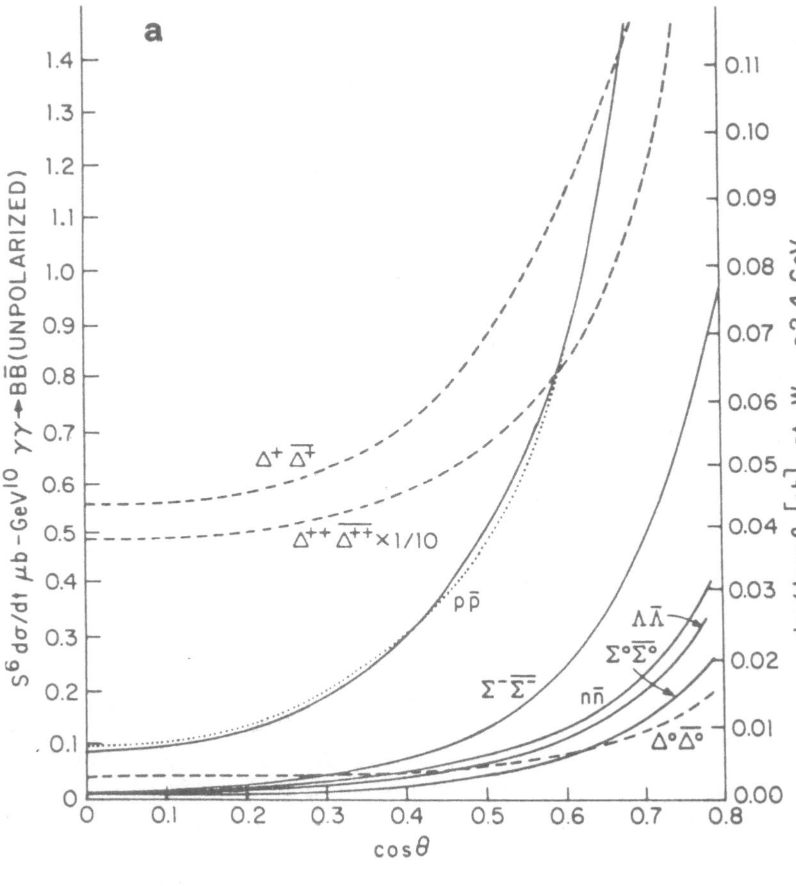

Fig. 9a : Predicted cross section for $\gamma\gamma \to p\bar{p}$ [24] using symmetric wavefunctions.

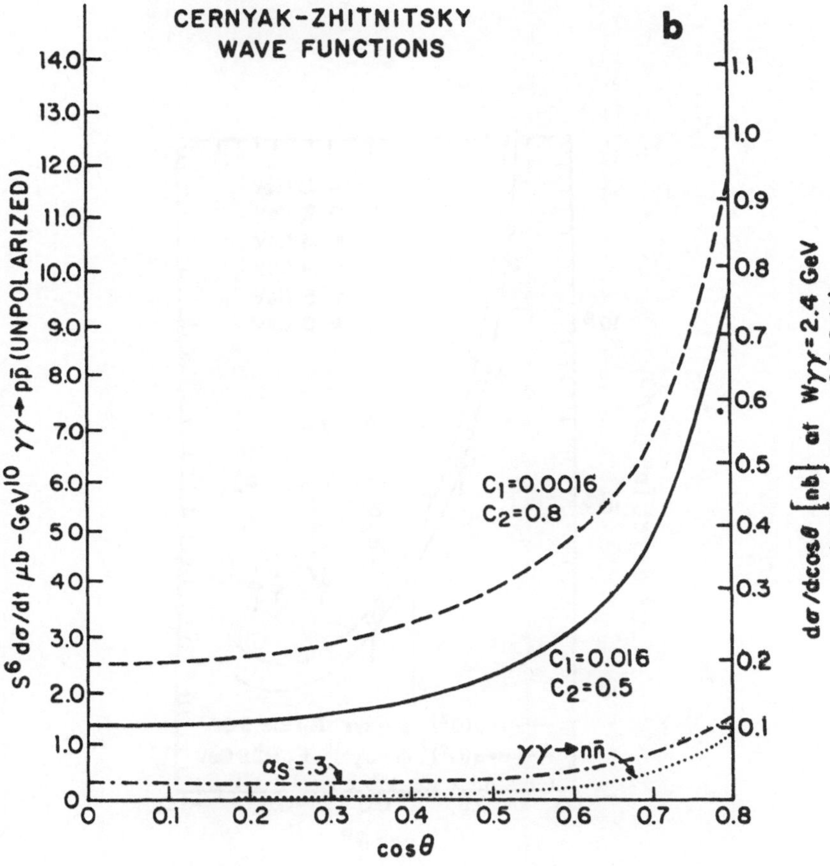

Fig. 9b : Predicted cross section for $\gamma\gamma \to p\bar{p}$ [24] using non-symmetric wave-functions.

our predictions for different choices of parameters and symmetric wavefunctions are compared with the data for unpolarized photon-proton scattering. The strong coupling α_s is parametrized as

$$\alpha_s(Q^2) = \frac{4\pi}{\beta_o \ln Q^2/SC + \beta_1/\beta_o \ln \ln \frac{Q^2}{SC}} \qquad C = \Lambda^2/S$$

for Q^2 such that $\alpha_s(Q^2) < 1$ and $\alpha_s = 1$ in all other cases.

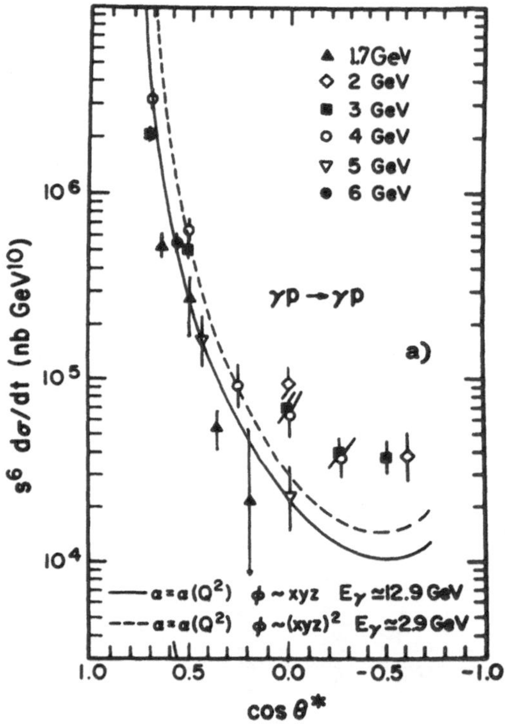

Fig. 10a : Predictions [18] versus data [3a] for γp→γp using symmetric wavefunctions.

The predictions fit the data quite well, particularly for cos θ > 0. I find no reason for concern in the fact that the agreement between the data and our results seems to worsen for cos θ < 0. While for positive values of cos θ we have data for a relatively wide range of photon energies, for negative values of cos θ the only available data refer to photon of energy between 2 and 3 GeV. Secondly, errors are larger in this range in which statistics is poorer than at larger $\cos\Theta$. In Fig. 10b the predictions obtained with the non asymptotic wavefunction [7] are compared with the data. The dotted line represents the fixed coupling result, with $\alpha_s = 0.18$, while the broken and the continuous lines represent the running coupling result with $c = 1.6\ 10^{-3}$ and $c = 6.4\ 10^{-3}$ respectively.

While the fixed coupling results are too small, the running coupling predictions seem to have a steeper slope in the forward direction than the data while the overall normalization is not too far the mark. Comparing Fig. 10a and Fig. 10b it appears that the predictions obtained with the asymptotic wavefunction, or with wavefunctions that, although not asymptotic, are symmetric under exchange of the quark momenta, fit the Compton scattering data slightly better than the predictions obtained with the non asymptotic, non symmetric one. Since the predictions are rather sensitive to the parametrization of the strong coupling constant it is impossible to determine the correct form of the baryon wavefunction on the basis of these data alone. Further work on the dependence of the predictions on the details of the coupling parametrization is in progress.

The agreement between the predictions for the completely unpolarized reaction $\gamma p \rightarrow \gamma p$ and the experimental results is certainly encouraging. However the data are plagued with rather large errors. Furthermore only data at relatively low incident energy are available for $\cos\Theta < 0$. Our results indicate that Compton scattering off neutrons is of the same order of magnitude than the scattering off protons. This prediction should be relatively easy to verify. The photoproduction reaction $\gamma p \rightarrow \gamma \Delta^+$ is predicted to have a cross section one order of magnitude smaller than the proton elastic process so it could be well within reach of experiment. The same could be true for $\gamma n \rightarrow \gamma \Delta^o$ which is approximately one order of magnitude smaller than its charged counterpart. I believe it is extremely important to add precise measurements of Compton scattering off baryons to the list of data with which theoretical suggestions can be confronted. Due to the abundant fluxes of photons which can be obtained via bremsstrahlung, Compton scattering seems particularly well suited to give extremely detailed differential cross sections which are needed to get a firmer grasp on exclusive processes in QCD. It must be noticed that since Compton scattering is the simplest exclusive process which can be measured experimentally and for which double logarithmic corrections do not identically cancel, it provides an excellent testing ground for understanding the effects of such corrections.

CONCLUSIONS

As we have seen QCD correctly predicts the scaling behaviour of exclusive large momentum transfer processes. When the comparison can be made the predictions obtained in the framework presented here agree quite well with the data, with the exceptions of $\gamma\gamma \rightarrow \rho^o \rho^o$ and $\gamma\gamma \rightarrow p\bar{p}$ where the results could be explained by resonances. This seems to suggest that higher order corrections are rather small. Of course we need a much better theoretical understanding of Sudakov corrections before being able to really control this class of phenomena. The correct form of the hadron wavefunction is still unclear and we need more precise data on a larger set of exclusive reactions in order to really settle this important issue. It is clear from what has been said that the technical difficulties involved in computing large numbers of Feynman diagram can be solved with the help of new automated calculation procedures. It still remains to be seen, if the problems connected with measuring reactions with small cross sections, that scale like $1/s^n$, can be handled for energiès large enough to exclude the presence of resonances, and with enough details to really allow stringent tests of all aspects of the theory.

Exclusive hadronic processes provide one more possibility of exploring the relationship between long and short distance physics and the properties of bound states in an asymptotically free theory like QCD. They challenge our ability of performing perturbative calculations, both because of the sheer size of the computations needed already in lowest order, and because of the necessity of resuming all higher order corrections, at least in some suitable approximation, seems unavoidable. I believe our understanding of strong interactions will not be really satisfactory until exclusive processes will be under firm control.

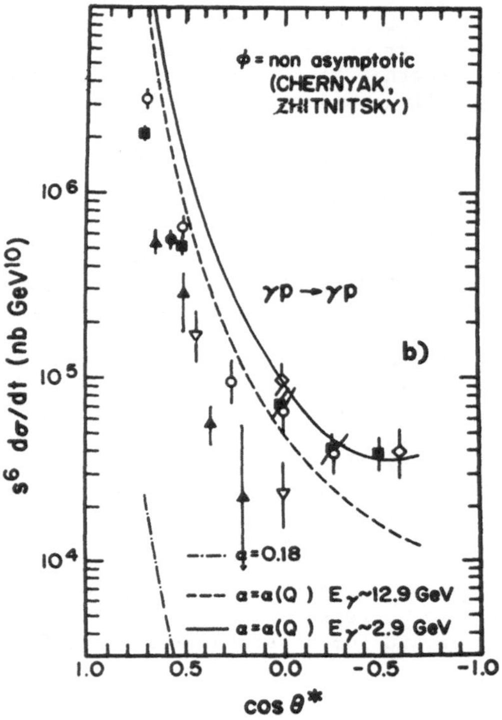

Fig. 10b : Predictions [18] versus data [3a] for $\gamma p \to \gamma p$ using non symmetric wavefunctions.

REFERENCES

[1] Many good reviews exist in the literature, see for example A.H. Mueller, Phys. Rep. 73C, (1981), 239 and references therein, C.T.C. Sachrajda in M.K. Gaillard and R. Stora eds, "Gauge Theories in High Energy Physics", North-Holland, (1983).

[2] S.J. Brodsky and G.R. Farrar, Phys. Rev. Lett., 31, (1973), 1153 and Phys. Rev., D11, (1975), 1309.
V.A. Matveev, R.M. Muradyan and A.V. Tavkheldize, Lett. Nuovo Cimento 7, (1973), 710.

[3] a) M.A. Shupe et al., Phys. Rev., D19, (1979), 1921.
J. Duda et al., Z. Phys., C17, (1983), 319.
M. Deutsch et al., Phys. Rev., D8, (1973), 3828.
b) P.V. Landshoff and J.C. Polkinghorne, Phys. Lett., 44B, (1973), 293.
c) K.A. Jenkins et al., Phys. Rev., D21, (1980), 2445.
C. Baglin et al., Nucl. Phys., B216, (1983), 1.
[4] S.J. Brodsky and B.T. Chertok, Phys. Rev. Lett., 37, (1976), 269 and Phys. Rev., D14, (1976), 3003.
[5] S.J. Brodsky and G.P. Lepage, Phys. Lett., 87B, (1979), 359 and Phys. Rev., D22, (1980), 2157.
S.J. Brodsky, G.P. Lepage, T. Huang and P.B. MacKenzie in "Particles and Fields", edited by A.Z. Capri and A.N. Kamal, Plenum, (1983), 83.
S.J. Brodsky, G.P. Lepage and T. Huang, ibidem, 143 and references therein.
[6] G.R. Farrar and D.R. Jackson, Phys. Rev. Lett., 43, (1979), 246.
A. Duncan and A.H. Mueller, Phys. Rev., D21, (1980), 1636 and Phys. Lett., 98B, (1980), 159.
[7] V.L. Chernyak and A.R. Zhitnitsky, Phys. Rep., 112C, (1984), 175.
[8] J. Weyers, these proceedings.
[9] G.R. Farrar, Phys. Rev. Lett., 53, (1984), 20.
[10] V. Sudakov, Sov. Phys., JEPT 3, (1956), 65.
[11] P.V. Landshoff, Phys. Rev., D10, (1974), 1024.
[12] S.S. Kanwal, Phys. Lett., 142B, (1984), 294.
[13] G.R. Farrar and F. Neri, Phys. Lett., 130B, (1983), 109.
P. De Causmaecker et al., Nucl. Phys., B206, (1982), 53.
[14] S.J. Brodsky and G.P. Lepage, Phys. Rev., D22, (1980), 2157.
[15] S.J. Brodsky and G.P. Lepage, Phys. Rev., D24, (1981), 1808.
[16] P.H. Damgaard, Nucl. Phys., B211, (1983), 435.
[17] G.R. Farrar, E. Maina and F. Neri, Nucl. Phys., B259, (1985), 702.
[18] E. Maina, G.R. Farrar, University of Torino preprint, IFTT-85/4, (1985).
E. Maina, Ph.D. Thesis, Rutgers U., (1984), unpublished.
[19] G.R. Farrar, Proceedings of the VIth International Workshop on Photon-Photon collisions, Lake Tahoe, CA, September 10-13, 1984 and Rutgers Univ. preprint RU-84-19.
[20] N. Isgur and C.H. Llewellyn Smith, Phys. Rev. Lett., 52, (1984), 175.
[21] PLUTO Collaboration, Ch. Berger et al., Phys. Lett., 137B, (1984), 267.
MARK II Collaboration, G. Gidal, Lawrence Berkeley Laboratory preprint, LBL-19992, (1985).
[22] TASSO Collaboration, R. Brandelik et al., Phys. Lett., 97B, (1980), 448 and M. Althoff et al., Z. Phys., C16, (1982), 13.
MARK II Collaboration, D.L. Burke et al., Phys. Lett., 100B, (1981), 153.
CELLO Collaboration, H.J. Behrend et al., Z. Phys., C21, (1984), 205.
PEP4/PEP9 Collaboration, J.G. Layter et al., Contribution to Leipzig Conf. on High Energy Phys., July 1984.
[23] J.F. Gunion, private communication to G.R. Farrar.
[24] TASSO Collaboration, M. Althoff et al., Phys. Lett., 130B, (1983), 449.
[25] TASSO Collaboration , M. Althoff et al., DESY preprint 84-015.

SOME RIGOROUS RESULTS ON POTENTIAL MODELS OF HADRONS

André Martin

CERN, Geneva
Division Thèorique
1211 Geneva 23, Switzerland

FORWORD

I shall present here two completely distinct contributions to the spectroscopy of hadrons in the framework of potential models. Both are rigorous results, which, in principle, can be applied to other fields of physics. The first one concerns the order of levels in two-body systems bound by a potential. The second one concerns "Regge" trajectories, i.e., angular excitations of two-or-three-body systems.

THEOREMS ON THE ORDER OF LEVELS IN POTENTIALS

In 1974-75, existing potential models of quarkonium all predicted a P state between the J/ψ and the ψ' and a D state above the ψ'. The question which M.A.B. Bég asked me first was to characterize the potentials for which one has such an order of levels, and H. Grosse and myself found two conditions involving the first, second and *third* derivative of the potential. One guaranteed that the lowest $\ell = 1$ would like below the first $\ell = 0$ radial excitation, while the second one guaranteed that the lowest $\ell = 2$ level would like above the first $\ell = 0$ radial excitation. This was not fully satisfactory for two reasons : (i) it looked awkward to have conditions involving the third derivative of the potential; (ii) the theorems could not be generalized to higher excitations because the method of proof made explicit use of the simple nodal structure of the radial wave functions of the low-lying levels. This meant that the theorems were insufficient, already in the case of the T system, and as we have seen in the last section, in the case of toponium.

The problem has now been solved by B. Baumgartner, H. Grosse and myself. Let us first discuss the comparison of the order of energy levels with the energy levels of hydrogen. Let us remember that in a pure Coulomb potential $V = -1/r$ we have a degeneracy of the levels : if N designates the Coulomb principal quantum number :

$$N = n \text{ (number of nodes)} + \ell \text{ (orbital ang. mom.)} + 1 \qquad (1)$$

the energy levels $E(N,\ell)$ which, for a *general* potential would depend on both N and ℓ depend only on N as whown in Fig.1.

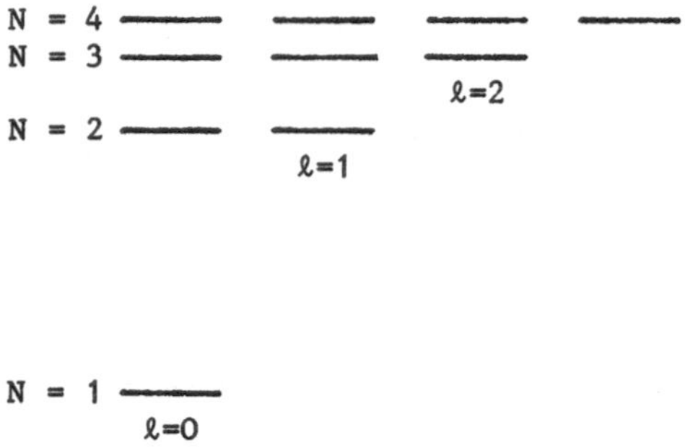

Figure 1. Each multiplet contains N states with $\ell \leq N-1$.

A characteristic of the Coulomb potential is that it is a solution of the Laplace equation $\Delta V = 0$ for $r \neq 0$. The discovery that we have made is precisely that in the case of a non-Coulomb potential the order of levels with a given N is controlled by the sign of the Laplacian of the potential. We have obtained the following theorems.

Let V be a spherically symmetric potential. Then

Theorem Ia	Theorem Ib
$E(N,\ell) \geq E(N,\ell+1)$ if $\Delta V(r) \geq 0$ for all $r > 0$	$E(N,\ell) \leq E(N,\ell+1)$ if $\Delta V(r) \leq 0$ for all r such that $dV/dr > 0$

This is illustrated by Figs. 2a and 2b.

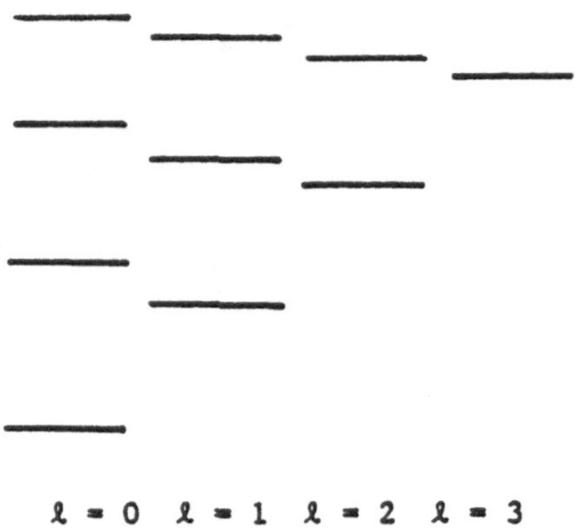

Figure 2a. $\Delta V > 0$

Figure 2b. $\Delta V < 0$

Before a complete proof of the theorem was obtained, the first indication came from a calculation in the WKB approximation (presumably good for large n) by Feldman, Fulton and Devoto[14] which gave

$$E(N,\ell) \gtrless E(N,\ell+1) \text{ if } \frac{d}{dr} r^2 \frac{dV}{dr} \gtrless 0$$

but $1/r^2\, d/dr\, r^2\, dV/dr$ is precisely the *Laplacian* of the potential ! The next indication came from the study of the limiting case

$$V = -\frac{1}{r} + \lambda v\ ,\ \lambda \to 0.$$

Here the problem seems very simple : one has to calculate

$$\int v[(U_{N,\ell})^2 - (U_{N,\ell+1})^2]\ dv, \tag{2}$$

where $U_{N,\ell}$ and $U_{N,\ell+1}$ are Coulomb wave functions, to get the sign of $E(N,\ell) - E(N,\ell+1)$. The difficulty, however, is that the integrand has many oscillations for N large. However, the Coulomb degeneracy is due to the O_4 symmetry of the problem. For this reason the wave functions with same N but different ℓ's can be obtained from one another and lowering operators of the angular momentum. One has

$$C\ U_{N,\ell} = A_\ell^-\ U_{N,\ell+1} \qquad C\ U_{N,\ell+1} = A_\ell^+\ U_{N,\ell} \tag{3}$$

with

$$A_\ell^\pm = \pm\frac{d}{dr} - \frac{\ell+1}{r} + \frac{1}{2(\ell+1)} \tag{4}$$

In (2) one can replace $U_{N,\ell}$, by the expression (3) and eliminate the derivatives of $U_{N,\ell+1}$ by successive integrations by parts. In the end, one gets exactly the *same* condition as Feldman et al.

At this point we became convinced that it was the Laplacian which controlled the order of levels, and we managed to find a proof, of which I shall give only the general ideas. The main trick was to generalize the raising and lowering operators. If you take as general form

$$A_\ell^+ = \frac{d}{dr} + g(r) \tag{5}$$

and demand that $A_\ell^+ U_{N,\ell}$ satisfies a Schrödinger equation with angular momentum $\ell+1$, you find that you have no freedom in the choice and you must take

$$g(r) = -\frac{U'_{\ell+1,\ell}}{U_{\ell+1,\ell}} \tag{6}$$

$U_{\ell+1,\ \ell}$ represents the wave function with $N = \ell+1$, i.e., the ground state wave function for angular momentum ℓ. Hence

$$\tilde{U}_{N,\ell+1} = A^+\ U_{N,\ell} = \frac{U'_{N,\ell}\ U_{\ell+1,} \quad - U'_{\ell+1,\ell}\ U_{N,\ell}}{U_{\ell+1,\ell}} \tag{7}$$

except for a normalization factor. It is possible to see that $\tilde{U}_{N,\ell+1}$ corresponds to a state of angular momentum $\ell+1$, i.e., behaves at the origin as $r^{\ell+2}$, and has n-1 nodes if $U_{N,\ell}$ has n nodes. Therefore $\tilde{U}_{N,\ell+1}$ has exactly the same "Coulomb" principal quantum number. However, there is a difference with the case of a pure Coulomb potential, which is that if U satisfies the Schrödinger equation

$$\left[- \frac{d^2}{dr^2} + \frac{\ell(\ell+1)}{r^2} + V - E \right] U_{N,\ell} (r) = 0 \tag{8}$$

$\tilde{U}_{N,\ell+1}$ satisfies a Schrödinger equation with a *different* potential

$$\left[- \frac{d^2}{dr^2} + \frac{(\ell+1)(\ell+2)}{r^2} + \tilde{V} - E \right] U_{N,\ell+1} (r) = 0 \tag{9}$$

with

$$\begin{aligned} \tilde{V} - V &= 2 \left[g'(r) - \frac{\ell+1}{r^2}\right] \\ &= 2 \left[\frac{u'^2 + (E-V)u^2}{u^2} + \frac{\ell^2-1}{r^2} \right] \\ &= 2 \left[\int_r^\infty u^2 \left(\frac{dv}{dr} - 2 \frac{\ell(\ell+1)}{r^3}\right) - \frac{\ell+1}{r^2} \right] \end{aligned} \tag{10}$$

We *thought* that we had discovered all this, but it was pointed out to us by K. Chadan that Eq.(8) had been found by Marchenko in 1955 [16]. As we said, there was no arbitrariness in the construction of the raising operator and this is why we get exactly the same.

Now we have made a decisive step forward : assume that we know that

$$\tilde{V} > V \tag{11}$$

Then from Eqs.(8) and (9), we have

$$E(N,\ell+1,\tilde{V}) = E(N,\ell,V) \tag{12}$$

But the energies are *monotonous* functions of the potential (this applies to all levels, not only the ground states). Hence

$$E(N,\ell+1,V) < E(N,\ell+1,\tilde{V}) = E(N,\ell,V) \tag{13}$$

Conversely if $\tilde{V} < V$, we get

$$E(N,\ell+1,V) > E(N,\ell,V)$$

We are left now with a relatively tough technical problem : prove that

$$g'(r) - \frac{\ell+1}{r^2} \gtrless 0 \quad \text{if} \quad \Delta V(r) \gtrless 0$$

It is impossible to give the details of the proof here. Let me say, however, that the proof is using only elementary means such as comparing the ground state wave function u in an interval 0-r or r-∞ to a Coulomb wave function and noticing that the conditions $\Delta V >< 0$ are convexity (concavity) properties, not with respect to straight lines but to hyperbolae $A + B/r$. Specifically :

i) if $\Delta V > 0$, i.e., $(d/dr)r^2 (dV/dr) > 0$, any hyperbola tangent to $r = R$ to $V(r)$ is *below* $V(r)$, or explicitly :

$$\forall R, \; V(r) > V(R) + R\frac{dV}{dR} - \frac{R^2 \frac{dV}{dR}}{r}$$

ii) if $\Delta V < 0$, it is the opposite :

$$\forall R, \; V(r) < V(R) + R\frac{dV}{dR} - \frac{R^2 \frac{dV}{dR}}{r}$$

This is illustrated in Figs.3a and 3b.

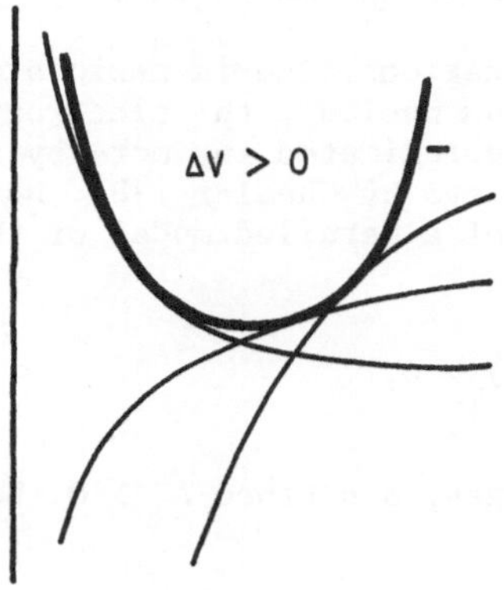

Figure 3a.

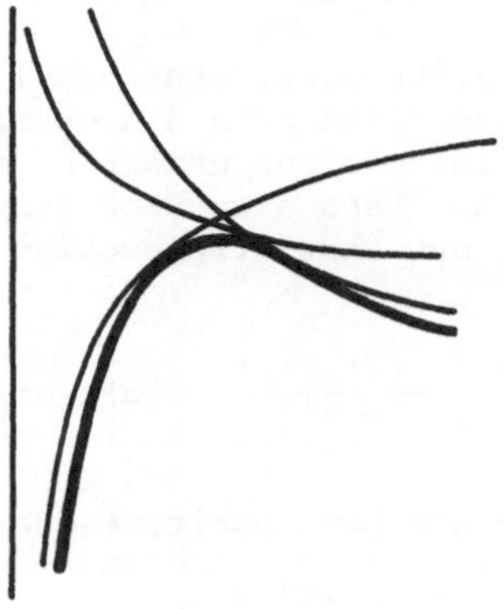

Figure 3b.

In fact, inequality

$$E(N,\ell) < E(N,\ell+1),$$

can be established under a slightly weaker condition because it suffices to have $g' < (\ell+1/r^2)$, but since

$$g'(R) = \int_R^\infty u^2 \left(\frac{dV}{dr} - \frac{2(\ell+1)\ell}{r^3}\right) dr$$

we see that if dV/dr is negative beyond a certain distance, g' is negative anf hence the inequality is satisfied.

We now turn to applications of Theorems Ia and Ib. The first application is to *quarkonium*. All quarkonium potentials have the property $\Delta V > 0$. At short distances this is clear, because asymptotic freedom is equivalent to *antiscreening*, but the fact is that in all potentials used this is true at all distances :

$$V = -\frac{a}{r} + br$$

$$V = \log r$$

$$V = A + B\, r^{0.1}$$

The Richardson [4)] and Buchmüller [6)] potentials satisfy this condition and analytical fits to lattice QCD potentials have also this property. Therefore we believe that between two successive $\ell = 0$ levels of toponium we shall find an $\ell = 1$ level, higher than the $\ell = 0$ level with the *same* number of nodes (because of the positivity of centrifugal energy), lower than the $\ell = 0$ level with one more node.

There are, however, other applications. One is mesic atoms. If a negative muon turns around a uranium nucleus, the electrostatic potential of the nucleus cannot be approximated any more by a point charge. This has been discussed long ago by Wheeler [18]. However, we want to point out that irrespective of a detailed model of the nucleus, we have

$$V = -e^2 \int \frac{\rho(x')d^3x'}{|x-x'|} \quad \text{where } \rho(x) \geq 0,$$

since the protons have positive charges, and hence $\Delta V > 0$. Hence

$$E(N,\ell) > E(N,\ell+1).$$

Another application is very old : the justification of the Mendeleev classification. Take for simplicity an alcaline atom. The outer electron sees the potential created by the nucleus ($\Delta V=0$) and the one created by the negatively charged electron cloud ($\Delta V<0$). Hence

$$E(N,\ell) < E(N,\ell+1).$$

This is visible on the spectrum of sodium (Fig.5) and explains why, for instance, the third electron of lithium is 3S instead of 3P ! Physicists above 60 generally tell me that this was obvious anyway, because electrons with large ℓ are more often far away from the nucleus. This, however, does not dispense us from giving a clean proof.

Before ending this section, let me remind you that there was another problem of ordering of levels, which was the relative position of the lowest $\ell = 2$ state and of the first $\ell = 0$ radial excitation. More generally, if $\varepsilon(n,\ell)$ designates the energy of the level with n nodes, and angular momentum ℓ, how do we compare $\varepsilon(n,\ell)$ and $\varepsilon(n-1,\ell+2)$? Taking $V = r^2 + \lambda v$, λ small, we have obtained the condition [15]

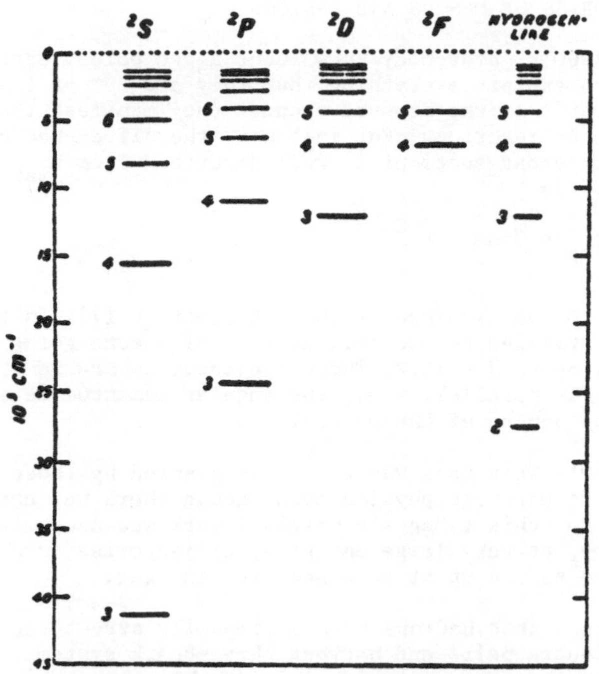

Figure 4. Energy level diagram for sodium. The S series is known to n = 14, the P to n = 59, the F to n = 5. The only doublet terms which have been resolved are the first seven of the P series.

$$\varepsilon(n,\ell) \gtrless \varepsilon(n-1,\ell+2)$$

$$\text{if } \frac{d}{dr}\;\frac{1}{r}\;\frac{dv}{dr} \gtrless 0. \tag{14}$$

Naturally if

$$v = r^2,\ \varepsilon(n,\ell) = \varepsilon(n-1,\ell+2).$$

Recently we have been able to prove (14) in full generality, without assuming v small, i.e. :

$$\varepsilon(n,\ell) \gtrless \varepsilon(n-1,\ell+2)$$

$$\text{if } \frac{d}{dr}\;\frac{1}{r}\;\frac{dV}{dr} \gtrless 0. \tag{15}$$

The proof is based on the change od variable $\rho = r^2$ which transforms, for instance, the harmonic oscillator into a Coulomb Schrödinger equation, provided the orbital angular momentum is rescaled. In fact, this field of investigation is still being explored and further results are expected.

REGGE TRAJECTORIES OF MESONS AND BARYONS

In the 1960's, everybody knew about Regge poles. Actually they were thought to explain everything. Now they are out of fashion but they still exist ! In the crossed channel they manifest themselves to describe two-body reactions. For instance, the difference between the $\bar{p}p$ and pp total cross-sections is well described from $E_{lab} = 10$ GeV to $E_{lab} = 2000$ GeV by

$$\sigma_{\bar{p}p} - \sigma_{pp} = \text{Const. } E^{-0.55}$$

corresponding to the exchange of the ρ trajectory [1]. In the direct channel it is striking to see trajectories of mesons going up to J = 6 and of baryons up to J = 17/2. These trajectories are remarkably linear and approximately parallel, i.e., the angular momentum is a linear function of the square of the particle mass.

In the 60's this fact was taken for granted by those who applied Regge's ideas to particle physics even though there was not the faintest justification for this ! Regge's original work was done with non-confining potentials, and, at very large energies, trajectories turned around in the complex J-plane to end up at some negative integer.

Now we know that hadrons have a composite structure, and that mesons are quark-antiquark pairs and baryons three-quark systems. The use of a potential interaction has met with considerable success in the description of hadrons, especially those made of heavy quarks, but also to a certain extent those made of light quarks. There is more and more support for the belief that the quark-antiquark potential is linear at a large distance, and, to the extend that the quark-antiquark system can be regarded as a relativistic string for large angular momentum, the slope of the Regge trajectory has been connected with the string tension. Here we shall do something slightly heretical and regard the potential as producing an instantaneous interaction between relativistic point-like quarks.

In the case of baryons, the most natural prescription is to take two-body forces with $V_{QQ} = 1/2\ V_{Q\bar{Q}}$. Howerver, for large separations there are good reasons to believe that the potential energy between three quarks is proportional to the length of the string of minimum length connecting the three quarks. We shall in fact consider both cases.

Several authors have proposed models in which the baryons are made of a quark-antiquark system. Then in particular the parallelism of the mesons and baryon trajectories becomes very natural. However, we still have to understand why it might be so.

Although the ground states and low-lying excited states of baryons have been well studied, the excited states with large angular momentum have only been touched upon and do not lend themselves so easily to numerical study. The main remark of the present paper is that, as for mesons - where, as we shall see, it is completely clear - the leading Regge trajectory of a baryon (i.e., the sequence of ground-state wave functions and energies with increasing angular momentum J) can be obtained in the large J limit, by minimizing the classical energy of the system for given J. Quantum effects only play a role in preventing the collapse of a subsystem caused by short-range singularities of the potential. For a linear two-body potential and for a string, one finds that the configuration minimizing the energy is a quark-diquark system. With relativistic kinematics one proves that the trajectories tend to become linear, and, unavoidably, since the colour of a diquark system in a baryon has to be a $\overline{3}$, the potential energy is the same as in a meson and the trajectories become parallel.

The two-body case : relativistic kinematics

For simplicity we take only the extreme relativistic case, i.e., the Hamiltonian is, taking c = 1,

$$H = |\vec{p}_1| + |\vec{p}_2| + V(r_{12}) \tag{16}$$

or, in the c.m. system,

$$H = 2p + V(r) \tag{17}$$

Provisionally we restrict ourselves to a purely linear potential,

$$V(r) = \lambda r \tag{18}$$

Here the minimum of the classical energy for a given angular momentum will give a lower bound for the energy of the ground state. In fact, we can give an explicit proof by generalizing an inequality of Herbst, who proves the operator inequality

$$|p| > \frac{2}{\pi}\frac{1}{|r|},$$

to be valid for any state. If we restrict ourselves to states of angular momentum J, we have the following inequality :

$$\langle J|p|J\rangle \geq 2 \left(\frac{\Gamma(\frac{J}{2}+1)}{\Gamma(\frac{J}{2}+\frac{1}{2})}\right)^2 \langle J|\frac{1}{r}|J\rangle \geq (J+\frac{1}{2}) \langle J|\frac{1}{r}|J\rangle \tag{19}$$

Therefore, the quantum energy of a system of two particles of momenta $\vec{p}$ and $-\vec{p}$ will satisfy the inequality

$$E_Q(J) \geq \text{Inf} \left\{ \frac{2J+1}{r} + V(r)\right\} \tag{20}$$

while the minimum of the classical energy is

$$E_c(J) = \text{Inf} \left\{ \frac{2J}{r} + V(r)\right\}. \tag{21}$$

So if we take the linear potential (12) we get

$$E_c(J) = 2\sqrt{2}\ \sqrt{J}\ \sqrt{\lambda} \tag{22}$$

and hence, if we believe that this gives the leading behaviour of the quantum ground-state energy for large J, then

$$J(t) \cong \frac{1}{8\lambda} t + \ldots \tag{23}$$

We can prove that Eq.(3) indeed gives the leading behaviour by bounding above the Hamiltonian (2), by using the operator inequalities

$$p < \frac{1}{2} \left(\frac{p^2}{x} + x \right) \tag{24}$$

$$r < \frac{1}{2} \left(\frac{r^2}{y} + y \right) \tag{25}$$

and get

$$2\sqrt{2}\sqrt{\lambda}\,(J + \tfrac{1}{2})^{1/2} < E(n,J) < 2\sqrt{2}\sqrt{\lambda}\,(2n + J + 3/2)^{1/2} \tag{26}$$

A more precise but less rigorous estimate is

$$t(J) \cong 4(J + 1/2) + 4\sqrt{2n} + 2\sqrt{2} + \ldots \tag{27}$$

This means that daughter trajectories are asymptotically parallel to the leading trajectory.

<u>The three-body case</u>

For the interaction between the three quarks constituting a baryon, we have taken two extreme cases :

i) A sum of two-body interactions adjusted in such a way that if two quarks are close to one another, the potential between the quark-diquark system is identical to the quark-antiquark potential,

$$V = \frac{1}{2} [V(r_{12}) + V(r_{23}) + V(r_{13})] \tag{28}$$

where V is the $q\bar{q}$ two-body potential, and in the special case (3)

$$V = \frac{\lambda}{2} (r_{12} + r_{23} + r_{13}) \tag{29}$$

There is no rigorous justification for this choice. The rule (28) holds for one-gluon exchange contributions. The other remark is that a diquark system has colour $\bar{3}$ and therefore looks like an antiquark. Finally, let us indicate that experience has shown that the application of this rule to the calculation of ground-state energies of baryons has met with remarkable success, for instance, in the calculation of the Ω^- mass by J-M. Richard.

ii) the potential energy could be proportional to the minimum length of a Y-shaped string connecting the three quarks. Again the strength is adjusted in such a way that when two quarks coincide it agrees with the quark-antiquark potential,

$$V = \lambda \, \underset{P}{\mathrm{Inf}} (r_{1P} + r_{2P} + r_{3P}) \tag{30}$$

There are good reasons for believing that Eq.(28) holds for large separations between the quarks. For instance, this is what one gets in the static Born-Oppenheimer bag model.

We shall assume that, like in the two-body case, the classical energy gives a lower bound of the ground state energy. Therefore, we wish to minimize the classical energy.

If $\vec{J}$ is the total angular momentum

$$\vec{J} = \vec{r}_1 \times \vec{p}_1 + \vec{r}_2 \times \vec{p}_2 + \vec{r}_3 \times \vec{p}_3, \quad (\vec{p}_1 + \vec{p}_2 + \vec{p}_3 = 0)$$

we notice that if we project the points 1,2 and 3 and momenta $\vec{p}_1$, $\vec{p}_2$ and $\vec{p}_3$ on a plane perpendicular to J, then J is unchanged, the kinetic energy is reduced, and for monotonous two-body potentials as well as for the string interaction (28) the potential energy is reduced. Hence we can restrict ourselves to a motion of the three particles in a plane.

We shall not give the details of the minimization in the plane, which is a very amusing geometrical exercise. One ends up with two possibilities :

1) A quark-diquark configuration (which is the only possibility for the string) with an energy

$$E = 2p + \lambda r \tag{31}$$

and $J = pr$ where p is the impulsion of the isolated quark and r the quark-diquark distance. Hence, minimizing,

$$E = 2\sqrt{2}\,\sqrt{\lambda}\,\sqrt{J} \tag{32}$$

2) An equilateral triangle, formed by the three quarks, with an energy

$$E = 3p + \frac{\lambda}{2}\, 3\sqrt{3}\ \ R \tag{33}$$

and $J = 3pR$, when R is the distance of the quarks to the centre.

Hence

$$E = 3^{3/4}\,\sqrt{2}\,\sqrt{\lambda}\ \ \sqrt{J}\ ; \tag{34}$$

now since $3^{3/4} > 2$ (because $27 > 16$), we see that the quark-diquark configuration is always favoured.

Using the same type of majorizations, as in the two-body case :

$$2p_i < A + \frac{p_i^2}{A}\quad ,\quad 2\,r_{ij} < B + \frac{r_{ij}^2}{B}\quad ,$$

it is easy to prove that the classical minimum gives the leading behaviour of the ground state energy for large J.

The conclusion is that both baryon and meson trajectories have the same slope asymptotically :

$$J(t) \sim \frac{t}{8\lambda}$$

The fact that the coefficient does not agree with what one gets from the Nambu-Goto string should not be worrying. What matters is the comparison between the baryon and meson cases.

For references we draw the reader's attention to the CERN-preprint TH. 4259/85 (1985) to appear in Zeitschrift für Physik C. However, since then I have had communications of explicit calculations in the meson case, by S. Godefrey [Phys. Rev. D31 (1985) 2375] who uses the same Hamiltonian as us, and by H. Baacke who uses a Dirac or a Klein-Gordon equation. Both find "precocious" linearity of the trajectories, as do the calculations of Schnitzer, Preparata and collaborators, Basdevant and Boukraa.

References

B. Baumgartner, H. Grosse and A. Martin, Phys. Letters 146B (1984) 363 and Nucl. Phys. B254 (1985) 528.

RECENT RESULTS OF THE UA1 COLLABORATION AT THE CERN PROTON-ANTIPROTON COLLIDER

Antoine Leveque

DPh PE CEN Saclay

INTRODUCTION

Two years after the discovery of the intermediate vector bosons W and Z at the CERN collider, large amounts of data are now available, thanks to the improvements in luminosity and the efficient operation of the CERN accelerator complex and detectors. Therefore a comprehensive description of intermediate Vector Boson production and decay has now emerged. In addition jets, which have appeared as the dominant new phenomenon at the collider, are better understood. In this presentation I will review the latest results on the UA1 collaboration concerning "classical" collider physics. The first part deals with the production and decay of intermediate vector bosons, the second part will summarize the results on hadronic jets.

The UA1 detector is a general purpose apparatus designed to measure, as systematically as possible, the secondary particles emitted in antiproton-proton collisions; we refer to reference 1 for detailed descriptions and will only briefly remind you of its features relevant to the present analysis.

i) Electron detection. Energetic electrons deposit almost all their energy in the UA1 lead scintillator electromagnetic shower calorimeter, 27 radiation lengths deep, and deposit very little energy in the hadron calorimeter beyond this. The energy resolution of the electromagnetic calorimeter is approximately $E_{rms} = 0.16 \sqrt{E}$ where energies are in GeV.

ii) Neutrino detection. The presence of neutrino emission is inferred from an apparent lack of momentum conservation in the two components transverse to the beam direction. The detection and measurement of neutrino emission by this method is made possible in the UA1 detector by the nearly complete solid-angle coverage of calorimeters down to 0.2° from the direction of the beams. The accuracy in each component of missing transverse energy is $0.4 \sqrt{E_T}$, where E_T is the scalar sum of the

UA1 collaboration, CERN, Geneva
Aachen-Amsterdam (NIKHEF) - Annecy (LAPP) - Birmingham - CERN - Harvard - Helsinki - Kiel - London (Imperial College and Queen Mary College) - Padova - Paris (Collège de France) - Riverside - Rome - Rutherford Appleton Laboratory - Saclay (CEN) - Victoria - Vienna - Wisconsin.

transverse energies deposited in all the calorimeter cells and all energies are in GeV.

iii) Muon detection. A fast muon, emerging from the interaction region, will pass in turn through the central detector, the electromagnetic and hadronic calorimeters, and will enter the muon spectrometer, consisting of two chambers of 4 planes each, 60 cm apart. One requires matching in position, angle and momentum between the track elements seen in the central detector and the muon chamber system.

iv) Momentum analysis and charged-particle determination. An accurate curvature measurement is provided by the central detector, a large-volume drift chamber surrounding the crossing point and operated in a homogeneous 0.7 T magnetic field. The central detector is also used to determine the charged-particle topology of the event and, in particular, the isolation criteria for the electron track.

v) Jet finding and reconstruction. A momentum vector is associated with each calorimeter cell. Its direction and magnitude are defined by the spatial position of the cell and the energy deposition whithin the cell. The cells with transverse energy in excess of a threshold value are grouped in clusters and their associated momentum vectors added if the distance between them in (pseudorapidity, azimuthal angle)-space ΔR is less than 1 [$\Delta R \equiv (\Delta\eta^2 + \Delta\phi^2)^{1/2}$, where ϕ is in radians]. Cells with lower transverse energy are then included in the cluster if they are within $\Delta R < 1$ with respect to the cluster axis. Jets are retained if they have a transverse momentum in excess of 5 GeV/c, and an axis within the pseudorapidity interval $|\eta| < 2.5$. A detailed study of the relationship between the four-momentum of jets reconstructed by this method and those of the underlying parton has been made using ISAJET with the fragmentation function of Field and Feynman. The results of this study suggest that the jet energy is underestimated by about 20 % and the jet momentum by about 15 %; data have been corrected accordingly. The limitations of this procedure will be further discussed in the second part of these notes.

PART I - $W^{\pm}$ AND Z PHYSICS

SELECTION OF SAMPLE

Our electron trigger required a cluster of one or two adjacent electromagnetic calorimeter cells with a transverse energy in excess of 10 GeV at an angle of more than 5° with respect to the beam direction. The muon trigger required a fast particle with a transverse momentum larger than 5 GeV/c within the acceptance of the muon spectrometer.

During the 1984 data taking period, the collider has been operated at an increased centre of mass energy of $\sqrt{S}$ = 630 GeV and the UA1 collaboration has recorded 263 nb^{-1} of data, tripling the size of previous statistics. Our total sample consists of :

172 $W^{\pm} \rightarrow e^{\pm} \nu_e$ decays
47 $W^{\pm} \rightarrow u^{\pm} \nu_{\mu}$ decays
18 $Z_o \rightarrow e^{+} e^{-}$ decays
10 $Z_o \rightarrow \mu^{+} \mu^{-}$ decays.

THE CHARGED INTERMEDIATE BOSON $W^{\pm}$

The W has been seen in its three leptonic decay modes. The

$W^{\pm} \rightarrow e^{\pm} \nu_e$ sample is observed in the detector with the largest acceptance and the fast electron is accurately measured in the electromagnetic calorimeter. The following results, essentially based on the electron sample, are in perfect agreement with those extracted from the more restricted muon sample.

THE $W \rightarrow e\nu$ SAMPLE AND ITS BACKGROUND

The electron identification requires that an isolated electron, with a transverse momentum larger that 15 GeV/c point to the electromagnetic cluster. The missing transverse energy should also exceed 15 GeV. Details of the selection procedure are given in ref [2].

We have estimated in detail the possible sources of background in our sample of 172 events.

i) Hadronic interactions Hadronic jets can fake an isolated electron signature a) if they fragment in such a way that one energetic charged pion overlaps with one or more neutral pions, or b) if they contain a genuine energetic electron arising from heavy-flavour decays which overlaps with one or more neutrals. In both these cases we would expect a two-jet topology with little or no missing transverse energy in the event. To estimate the background from these sources, we drop the last requirement imposed in the selection, namely the presence of a missing transverse energy $E_m > 15$ GeV, and examine the missing transverse energy for the enlarged sample. For those events with $E_m < 15$ GeV the E_m distribution is well described by the experimental resolution in the measurement of this quantity (Fig. 1), and furthermore almost all those events have a two-jet topology. Extrapolating the resolution curve in the region $E_m > 15$ GeV, we estimate a background in our $W^{\pm} \rightarrow e^{\pm} \nu_e$ sample of 3.4 ± 1.8 events at $\sqrt{s} = 546$ GeV and 1.9 ± 0.6 events at $\sqrt{s} = 630$ GeV.

ii) $W \rightarrow \tau\nu_{\tau}$ We have used the Monte Carlo ISAJET [3] together with a full simulation of the UA1 detector. We estimate a background of 11.8 ± 0.6 events from this source, of which 2.7 events are associated with the decay $\tau \rightarrow \pi^{\pm}(\pi^{\circ})\nu_{\tau}$ and 9.1 events are associated with the decay $\tau \rightarrow e\, \nu_e\, \nu_{\tau}$.

iii) $W \rightarrow t\bar{b}$ We have evaluated this background using the ISAJET Monte Carlo, taking $m_t = 40$ GeV/c^2. We estimate a background of 1.5 events before applying the requirement that the transverse mass of the electron-neutrino system is greater than 20 GeV/c^2. After this cut has been applied, this background becomes negligible.

iv) $t\bar{t}$ production Evaluation of background from this source is made difficult by the lack of knowledge of the $t\bar{t}$ production cross-section. A preliminary prediction from the EUROJET Monte Carlo [4] gives a production cross-section of 1.4 nb for $m_t = 40$ GeV/c^2, and a transverse momentum cut on the t-quarks of $p_T(t) > 3$ GeV/c. Passing the Monte Carlo generated events through the W selection procedure and taking 0.12 for the top quark semi-leptonic decay branching ratio $B(t \rightarrow be\nu_e)$, we find that 1 % of the generated $t\bar{t}$ events survive the W selection cuts. This leads us to an estimated background of 7 events in the $W^{\pm} \rightarrow e^{\pm}\nu_e$ sample. However, this estimate is sensitive to m_t, the $t\bar{t}$ production cross-section, and the semileptonic branching ratio for the top quark. Since these quantities are not well known we make no correction for background from $t\bar{t}$ production in the following analysis.

DETERMINATION OF THE PRODUCTION CROSS-SECTION

The integrated luminosity of the experiment is 136 nb^{-1} at $\sqrt{s}$ =

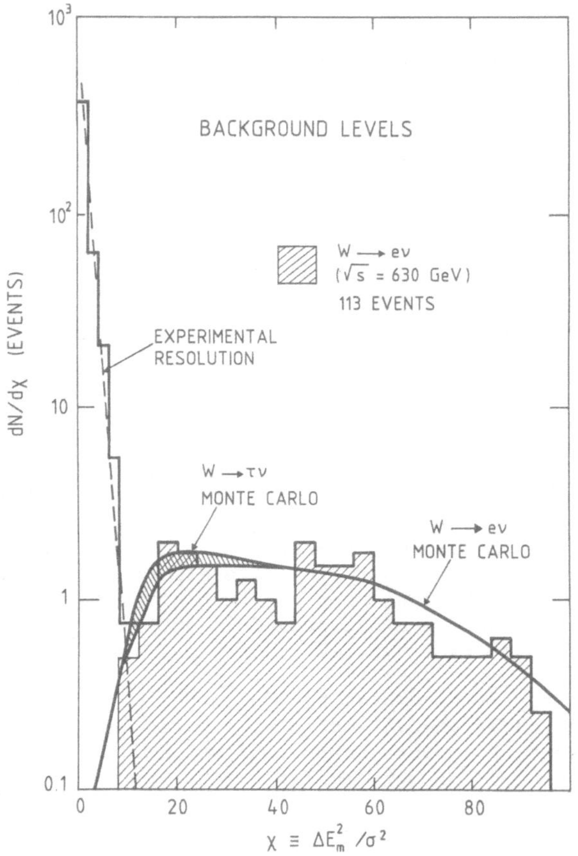

Fig. 1 : The curves are Monte Carlo predictions for electrons coming from $W^{\pm} \to e^{\pm}\nu_e$ decays. Background from two-jet fluctuations. The missing transverse-energy squared ΔE_m^2 divided by the experimental resolution on this quantity is shown for the sample of 1287 events which come from the W selection procedure (applied to the $\sqrt{s}$ = 630 GeV data) after removing the requirement that $\Delta E_m > 15$ GeV. The contribution from the 113 W events (hatched sub-histogram) agrees well with the ISAJET prediction. The contribution from the other events having $E_m < 15$ GeV (non-hatched) is well described by the experimental resolution (dashed) curve. Extrapolating this curve under the W region we expect a background of 1.9 ± 0.6 events from jet-jet fluctuations.

546 GeV and 263 nb^{-1} at $\sqrt{s}$ = 630 GeV, known to about ± 15 % uncertainty. We have used two techniques to estimate the efficiency of our selection requirements : i) the ISAJET Monte Carlo together with a full simulation of the UA1 detector, and ii) a randomized W-decay technique in which the 172 real W events are used, but where the physical electron is replaced by a generated electron coming from a random (V-A) decay of the produced W. The two methods give consistent results. We find that the efficiency of our trigger and selection requirements is 0.69 ± 0.03 for the $\sqrt{s}$ = 546 GeV data sample and 0.63 ± 0.03 for the $\sqrt{s}$ = 630 GeV sample. The cross-section is then :

$$(\sigma.B)^e_W = 0.55 \pm 0.08 \ (\pm 0.09) \text{ nb}$$

at $\sqrt{s}$ = 546 GeV, where the last error takes into account the systematics. Our result is in good agreement with the corresponding result from the UA2 experiment [5] 0.53 ± 0.10 (± 0.10) nb, and is not far from the prediction of Altarelli et al. [6] of $0.38^{+0.12}_{-0.05}$ nb, where we take the branching ratio B = 0.089.

The corresponding experimental result for the 1984 data at $\sqrt{s}$ = 630 GeV is

$$(\sigma.B)^e_W = 0.63 \pm 0.05 \ (\pm 0.09) \text{ nb.}$$

This is in agreement with the theoretical expectation [6] of $0.47^{+0.14}_{-0.08}$ nb. We note that the 15 % systematic uncertainty on these results disappears in the ratio of the two cross-sections. We obtain the result

$$\sigma^e_W(\sqrt{s} = 630 \text{ GeV})/\sigma^e_W(\sqrt{s} = 546 \text{ GeV}) = 1.15 \pm 0.17,$$

in agreement with the theoretical expectation of 1.24.

Corresponding values for $W^\pm \to \mu^\pm \nu_\mu$ are :

$$(\sigma B)^\mu_W = 0.67 \pm 0.17 \ (\pm 0.15) \text{ nb}$$

at $\sqrt{s}$ = 546 GeV and

$$(\sigma B)^\mu_W = 0.61 \pm 0.11 \ (\pm 0.12) \text{ nb}$$

at $\sqrt{s}$ = 630 GeV.

The ratio

$$(\sigma B)^\mu_W/(\sigma B)^e_W = 1.02 \pm 0.02 \ (\pm 0.09)$$

supports the assumption of electron muon universality.

W MOMENTUM DISTRIBUTIONS

The W transverse momentum is obtained by adding the electron and neutrino transverse momentum vectors. The resulting distribution is shown in Fig. 2. Its shape agrees with the expectations of the QCD improved Drell-Yan model. The solid curve superimposed to the data is calculated by Altarelli et al. [6] using the G.H.R. structure functions [7] with Λ = 0.4 GeV.

The longitudinal momentum distribution of the W boson is expected

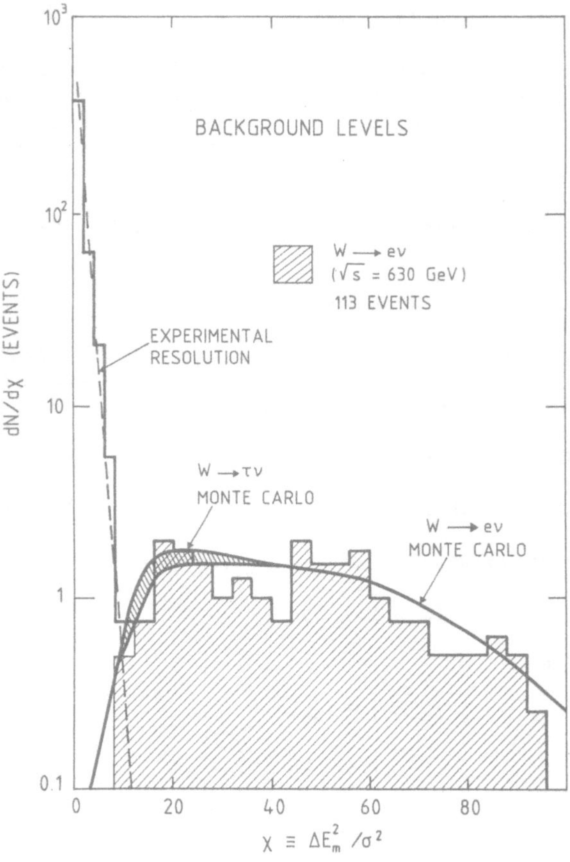

Fig. 2 : The W transverse momentum distribution. The curve shows the QCD prediction of Altarelli et al. [14]. The hatched sub-histogram shows the contribution from events in which the UA1 jet algorithm reconstructs one or more hadronic jets with transverse momentum in excess of 5 GeV/c, in the rapidity interval $|\eta| < 2.5$.

to reflect the structure functions of the incoming annihilating partons, predominantly u-quarks and d-quarks. Unfortunately we do not measure the longitudinal momentum of the W directly since we do not measure the longitudinal component of the neutrino momentum. We can overcome this difficulty by imposing the mass of the W on the electron-neutrino system. This will yield two solutions for the longitudinal component of the neutrino momentum, one corresponding to the neutrino being emitted forwards in the W rest frame, the other corresponding to the neutrino being emitted backwards. Hence we have two solutions for X_W, the Feynman X for the W. In practice, in one-third of the events one of these two solutions for X_W is trivially unphysical ($X_W > 1$), and in another one-third the ambiguity is resolved after consideration of energy and momentum conservation in the whole event. In the cases where the ambiguity in X_W is resolved one finds that the lowest X_W is nearly always selected. We shall therefore choose the lower X_W in all ambiguous cases.

Taking 84 GeV/c^2 for the mass of the W, the resulting X_W distribution is shown in Fig. 3a for the $\sqrt{s}$ = 546 GeV and $\sqrt{s}$ = 630 GeV data samples separately. There is some indication of the expected softening of the W longitudinal momentum distribution with increasing $\sqrt{s}$. The data are in reasonable agreement with the expectation resulting from the structure functions of Eichten et al. [8] with Λ = 0.2 GeV. Using the well-known relations

$$X_p X_{\bar{p}} = m_W^2/s \text{ (energy conservation)}$$

and

$$X_W = |X_p - X_{\bar{p}}| \text{ (momentum conservation)},$$

we can determine the parton distributions in the proton and antiproton sampled in W production. The results are shown in Fig. 3b. The proton- and antiproton-parton distributions are consistent with each other, and are well described by the structure functions of Eichten et al. For those events in which the charge of the electron, and hence the W, is well determined, we can identify the proton-(antiproton)-parton with a u-(d)-quark for a W^+, or a d-(u)-quark for a W^-. There are 118 events in which the charge of the electron track is determined to better than 2 standard deviations (from infinite momentum). The resulting u and d-quark distributions for these events are shown in Figs 3c and 3d, respectively. Once again there is agreement with the predictions of Eichten et al.

DETERMINATION OF THE W MASS

To measure the W mass we restrict ourselves to the subsample of $W^\pm \to e^\pm \nu_e$ decays in which both the electron and the neutrino are more than ± 15° away from the vertical direction in the transverse plane. This ensures the best accuracy in the electron and neutrino transverse energy determinations, and the subsequent mass measurement. We are left with a sample of 148 W decays for which we estimate a background of 15.6 ± 3.6 events.

The electron and neutrino transverse-energy distributions are shown in Figs 4a and 4b, respectively, for our sample of well-measured $W^\pm \to e^\pm \nu_e$ decays. The shaded parts of these distributions show the expected contributions from the various background sources. We note that i) the background events tend to have low electron and neutrino transverse energies, and ii) the background-subtracted distributions have the expected Jacobian shape. The exact shapes of the electron and neutrino transverse energy distributions depend on the underlying W mass, the experimental selection biases and transverse energy resolutions, and the underlying

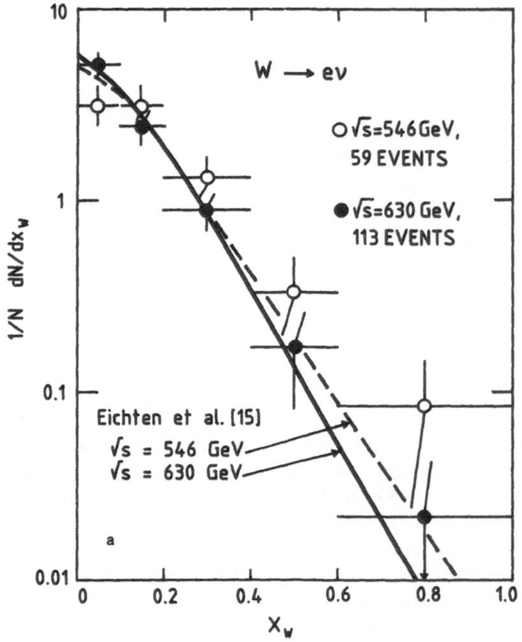

Fig. 3 : Longitudinal momentum distributions.
a) Feynman x-distribution for W's produced in $p\bar{p}$ collisions at $\sqrt{s}$ = 546 GeV (open circles) and $\sqrt{s}$ = 630 GeV (full circles). The curves show the predictions using the structure functions of Eichten et al. [15] modified in the appropriate way to take into account selection biases, experimental resolution, and the analysis bias of choosing always the low x_W solution (see text).

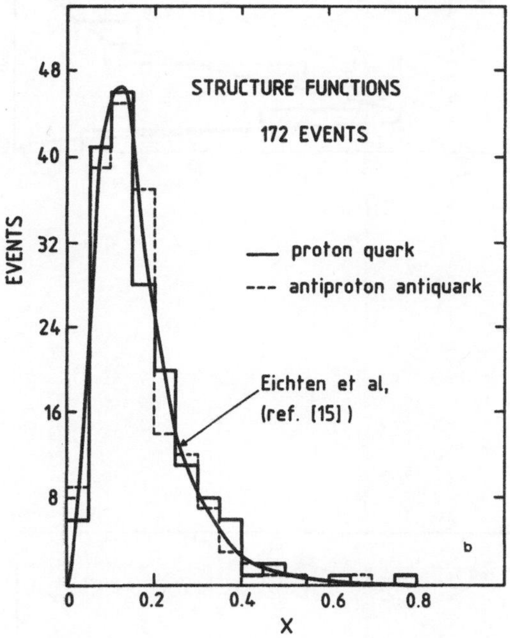

Fig. 3 : b) Feynman x-distribution for the proton and antiproton partons making the W. The curve shows the Eichten et al. prediction.

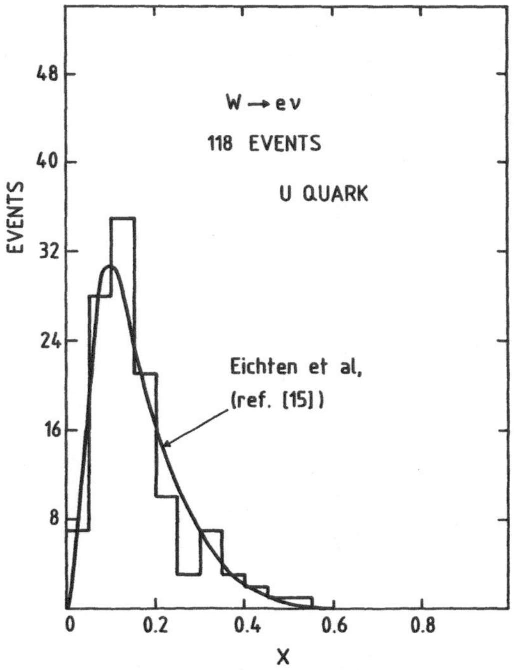

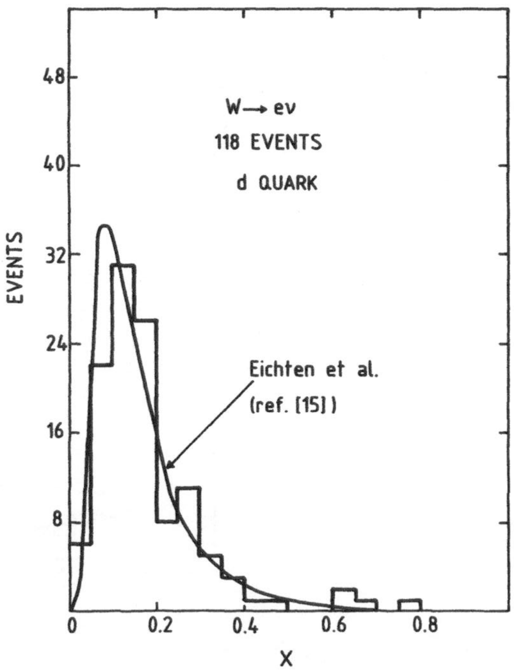

Figs 3 : c) Feynman x-distribution for u-quarks sampled in W production. The curve shows the Eichten et al. prediction.

d) Feynman x-distribution for d-quarks sampled in W production. The curve shows the Eichten et al. prediction.

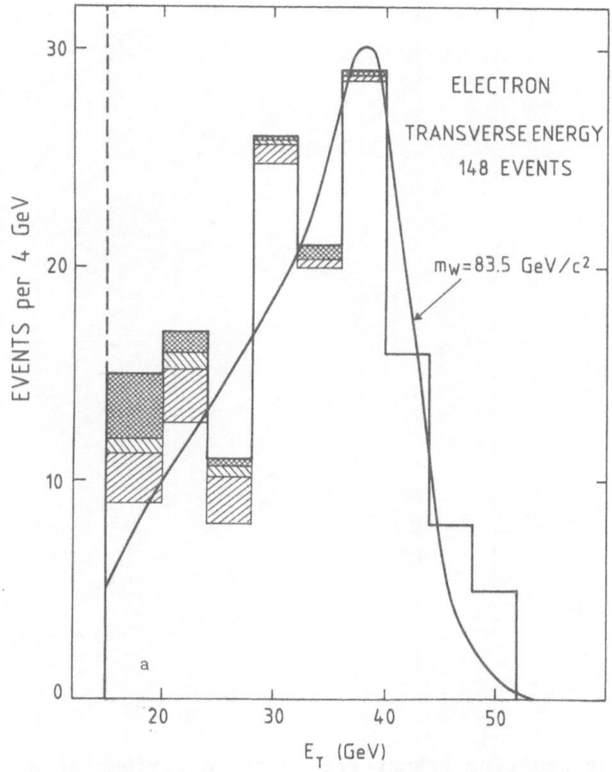

Figs 4 : The lepton transverse-energy distributions for the sample of well-measured $W^{\pm} \to e^{\pm}\nu_e$ events. The shaded parts of a) the electron transverse-energy distribution ...

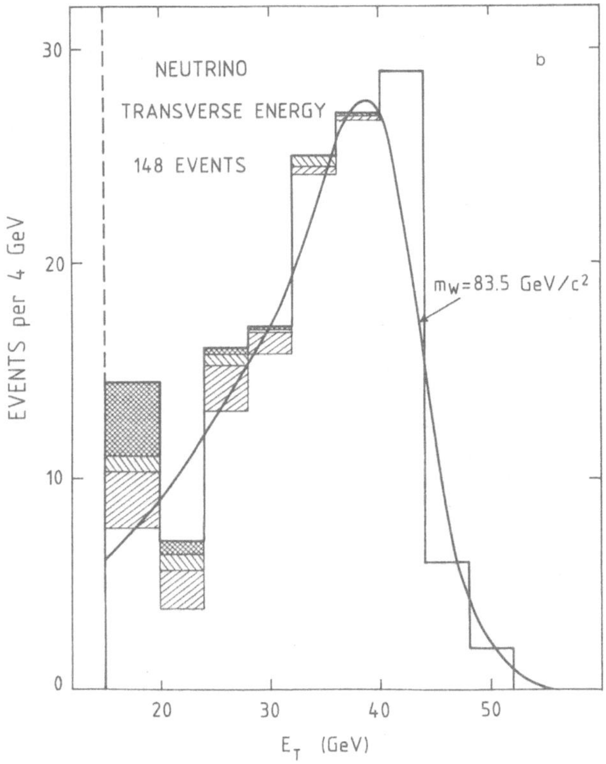

Fig. 4 : b) the neutrino transverse-energy distribution show the expected contributions from jet-jet fluctuations (cross-hatched) and $W^{\pm} \rightarrow \tau^{\pm}\nu_{\tau}$ decays with $\tau^{\pm} \rightarrow$ hadrons (top left to bottom right hatching) and $\tau^{\pm} \rightarrow e^{\pm}\nu_e\nu_{\tau}$ (top right to bottom left hatching). The curves show the predictions for the background-subtracted distributions (normalized to the data) corresponding to a W with a mass of 83.5 GeV/c^2.

W transverse-momentum distribution. To determine the W mass from our sample of 148 decays, we prefer to fit a distribution which is not sensitive to the underlying W transverse-momentum distribution. We use the electron-neutrino transverse-mass distribution,

$$m_T = [2E_T^e E_T^\nu (1 - \cos \phi)]^{1/2},$$

where E_T^e and E_T^ν are the lepton transverse energies, and ϕ is the angle between the electron and neutrino momentum vectors in the plane transverse to the beam direction. The electro-neutrino transverse-mass distribution is shown in Fig. 5a for the sample of 148 $W^\pm \rightarrow e^\pm \nu_e$ decays. The expected number of (shaded) background events decreases with increasing m_T, and is negligible for $m_T > 60$ GeV/c^2. We define the enhanced transverse-mass distribution (Fig. 5b) as the m_T distribution for those events in which both the electron and the neutrino have transverse energies in excess of 30 GeV. We are left with a background-free (< 1 event) sample of 86 $W^\pm \rightarrow e^\pm \nu_e$ decays.

Extensive Monte Carlo studies have shown that a good first approximation to the underlying W mass can be obtained by taking the point at which the upper edge of the enhanced transverse-mass distribution has fallen to half of the peak value. Thus, on the basis of Fig. 5b we see that our data correspond to a W mass of 83 GeV/c^2. A maximum-likelihood fit to the enhanced transverse-mass distribution has been performed to find most probable values for the mass m_W and decay width Γ_w of the W.

The result is

$$m_W = 83.5^{+1.1}_{-1.0} \ (\pm\ 2.8)\ \text{GeV}/c^2$$

$$\Gamma_W = 2.7^{+1.4}_{-1.5}\ \text{GeV}/c^2,$$

where the second error on the measurement of m_W comes from the systematic uncertainty in the energy scale of the experiment. The error quoted on Γ_W is the statistical error on the result. If we now include the systematic error (coming from the precision with which we know the experimental resolution function) we are only able to quote an upper limit on Γ_W :

$$\Gamma_W < 6.5\ \text{GeV}/c^2 \text{ at } 90\ \%\ \text{c.l.}$$

We note that our result for m_W is rather independent of Γ_W. The 1st. dev. and 2st. dev. contours in the (m_W, Γ_W)-plane are shown in Fig. 6. There is only a weak correlation between the two parameters.

DECAY ANGULAR DISTRIBUTIONS

The charged IVB is expected to couple to left-handed, but not right-handed, fermions. This (parity-violating) feature of the SU(2) ⊗ U(1) standard model results in a charge asymmetry in the decay angular distribution of the leptons coming from $W^\pm$ decays at the Collider. The electron (positron) from the decay of a $W^-(W^+)$ will prefer to be emitted along the incoming quark (antiquark) direction, with a decay angular distribution of the form $(1 + \cos \theta^*)^2$, where θ^* is the emission angle of the electron (positron) with respect to the proton (antiproton) direction in the W-centre-of-mass frame. To study this decay asymmetry, we restrict ourselves to the subsample of well-measured $W^\pm \rightarrow e^\pm \nu_e$ decays for which i) the sign of the charge of the electron (positron) is well determined by the measurement of the curvature of the CD track in the UA1 magnetic field, and (ii) the W centre-of-mass frame is well defined, i.e. there is no kin-

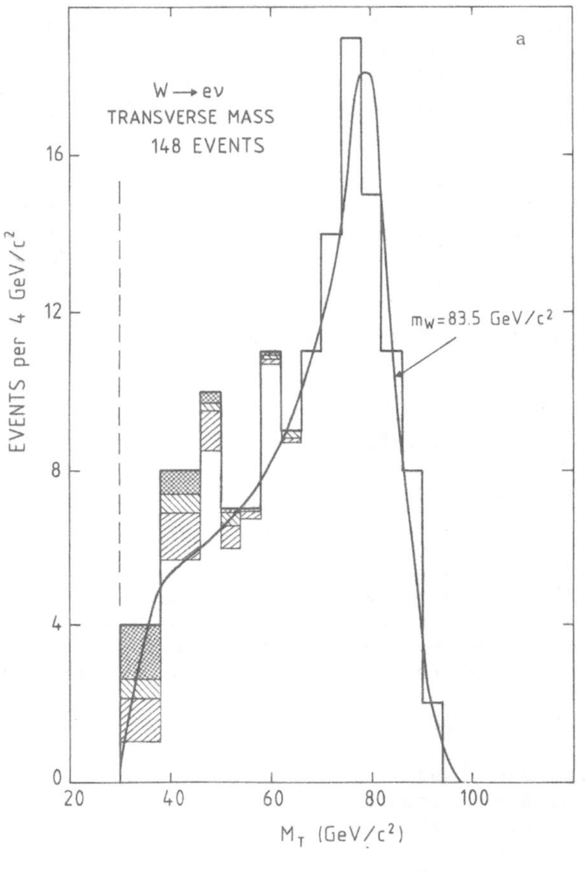

Figs 5 : The electron-neutrino transverse-mass distribution for a) the sample of 148 well-measured $W^{\pm} \rightarrow e^{\pm}v_e$ decays.

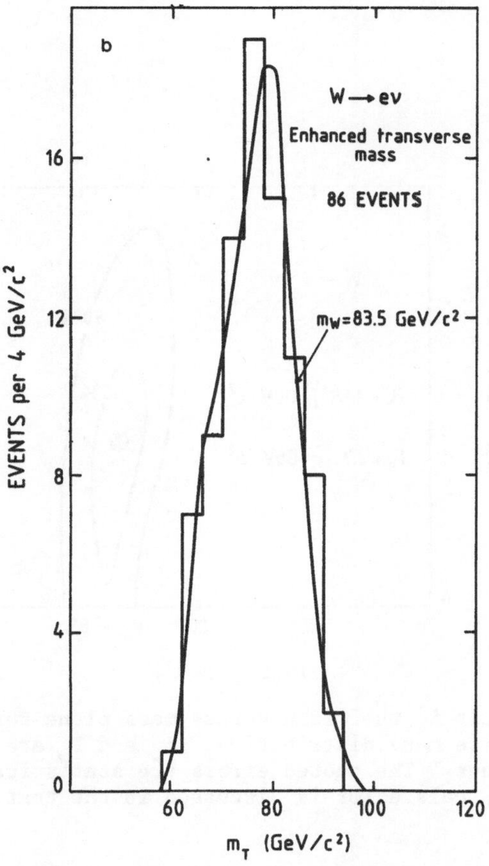

Fig. 5 : b) the subsample of these decays in which both the electron and the neutrino have a transverse energy in excess of 30 GeV (the enhanced transverse-mass distribution). The shading and curves are as for Fig. 4.

ematical ambiguity in reconstructing the longitudinal momentum of the neutrino once the W mass has been imposed on the (eν) system.

We are left with a subsample of 75 $W^{\pm} \rightarrow e^{\pm}\nu_e$ decays (26 W^+ decays and 49 W^- decays) which have no kinematical ambiguity, and for which the sign of the electron (positron) charge is determined to better than 2st.

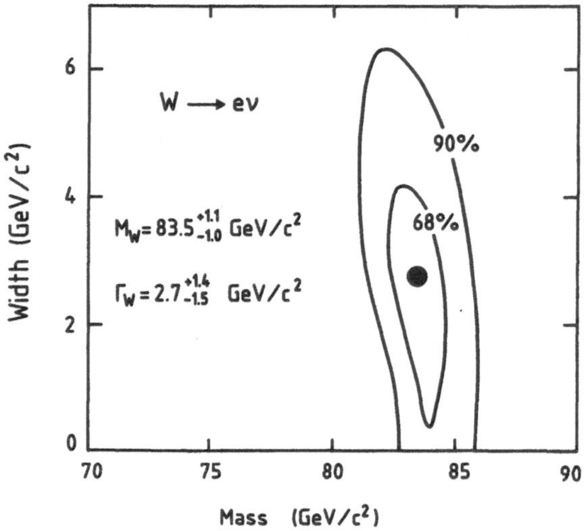

Fig. 6 : Fit result in the width versus mass plane for the enhanced transverse mass distribution. m_W and Γ_W are independent from each other. The quoted errors are statistical only. The systematic scale error is discussed in the text.

dev. (the track momentum is greater than 2st. dev. from infinity). Subtracting the expected background contribution from jet-jet fluctuations and $W^{\pm} \rightarrow \tau^{\pm}\nu_{\tau}$ decays (7.4 ± 1.2 events), the angular distribution is reasonably well described by the (V-A) expectation, which has been obtained from a Monte Carlo calculation which includes the effects of experimental resolution, geometrical acceptance, selection efficiencies, and reconstruction biases. Correcting the background-subtracted data for these effects, the resulting angular distribution is compared with the $(1 + \cos\theta^*)^2$ standard model expectation in Fig. 7. There is a striking asymmetry in this distribution which is consistent with the theoretical prediction. Defining the asymmetry parameter

$$A_W = (N_+ - N_-)/(N_+ + N_-),$$

where $N_{+(-)}$ are the number of events in the positive (negative) half of the $\cos\theta^*$ plot, we obtain

$$A_W = 0.77 \pm 0.04,$$

in agreement with the pure (V-A) value of 0.75. It has been shown [9] that for a particle of arbitrary spin J, one expects

$$\langle\cos\theta^*\rangle = \langle\lambda\rangle\langle\mu\rangle/J(J+1),$$

where $\langle\mu\rangle$ and $\langle\lambda\rangle$ are, respectively, the global helicity of the production system (ud) and of the decay system (eν). For (V-A) one then has $\langle\lambda\rangle = \langle\mu\rangle = -1$, $J = 1$, leading to a maximal value $\langle\cos\theta^*\rangle = 0.5$. For $J = 0$, one obviously expects $\langle\cos\theta^*\rangle = 0$, and for any other spin value $J > 2$, $\langle\cos\theta^*\rangle < 1/6$. Experimentally, we find $\langle\cos\theta^*\rangle = 0.43 \pm 0.07$, which supports both the $J = 1$ assignment and maximal helicity states at production and decay. Note that the choice of sign $\langle\mu\rangle = \langle\lambda\rangle = \pm 1$ cannot be separated, i.e. right- and left-handed currents, both at production and at decay, cannot be resolved without a polarization measurement.

Finally, fitting the angular distribution to the form $(1 + \alpha_W \cos\theta^*)^2$, we obtain the result

$$\alpha_W = 0.79^{+0.15}_{-0.17}.$$

THE NEUTRAL INTERMEDIATE BOSON Z°

THE Z° SAMPLE

The requirements used to select $Z° \to e^+e^-$ are :

- two isolated electromagnetic clusters.
- at least one of them satisfying the isolated electron selection criteria mentioned above the second one has $E_T > 8$ GeV and satisfies looses isolation criteria this leaves us with a sample of 39 events. The two clusters invariant mass distribution is shown in Fig. 8. We retain 18 $Z_o \to e^+e^-$ candidates for further analysis.

Using the ISAJET Monte Carlo together with a full simulation of the UA1 detector, we have evaluated the relevant background processes :

i) $Z° \to \tau^+\tau^-$, followed by $\tau^\pm \to e^\pm\nu_e\nu_\tau$ decays. We expect 0.30 ± 0.14 events in the region populated by our Z° events [$m > 80$ GeV/c^2].

ii) $Z° \to b\bar{b}$, with subsequent semileptonic decays of the $b(\bar{b})$ quark. We expect less than 8×10^{-2} events (90 % c.l.) with $m > 80$ GeV/c^2.

iii) $Z° \to t\bar{t}$, $W \to t\bar{b}$, and $p\bar{p} \to t\bar{t} + X$. These backgrounds are expected to be small (< 0.2 events) for top-quark masses in the interval 30 GeV/c^2 < m_t < 50 GeV/c^2.

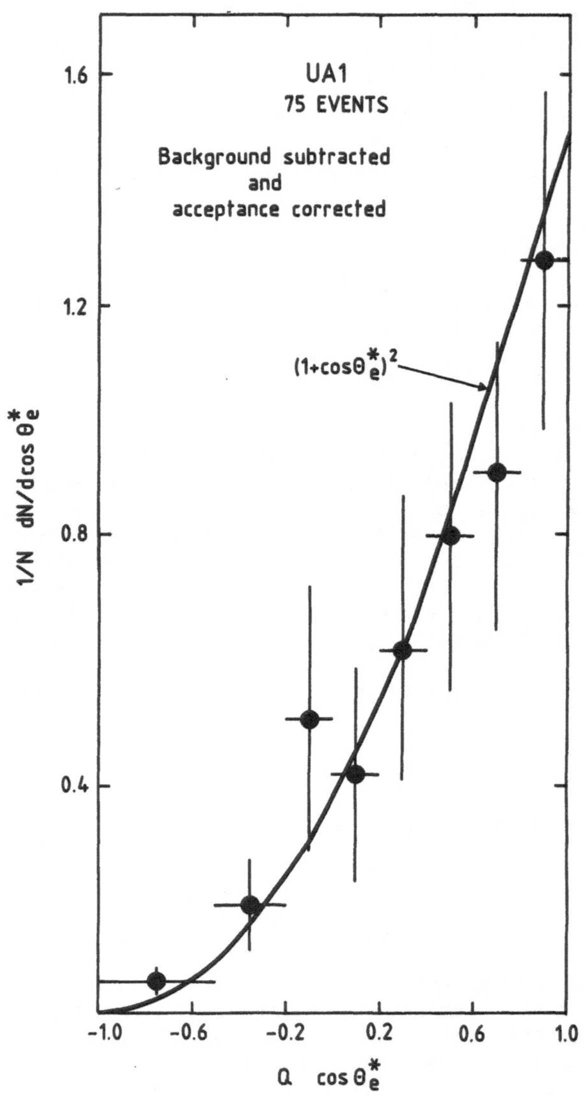

Fig. 7 : The W decay angular distribution of the emission angle θ^* of the electron (positron) with respect to the proton (antiproton) direction in the rest frame of the W. Only those events for which the lepton charge and the decay kinematics are well determined have been used. The curve shows the (V-A) expectation of $(1 + \cos\theta^*)^2$.

We conclude that we have a sample of essentially background-free $Z^\circ \to e^+e^-$ decays, the total estimated background being < 0.6 events. In calculating the Z° cross-section, we will neglect the potential (but small) contribution from t-quark related processes, leaving a total background of 0.3 events.

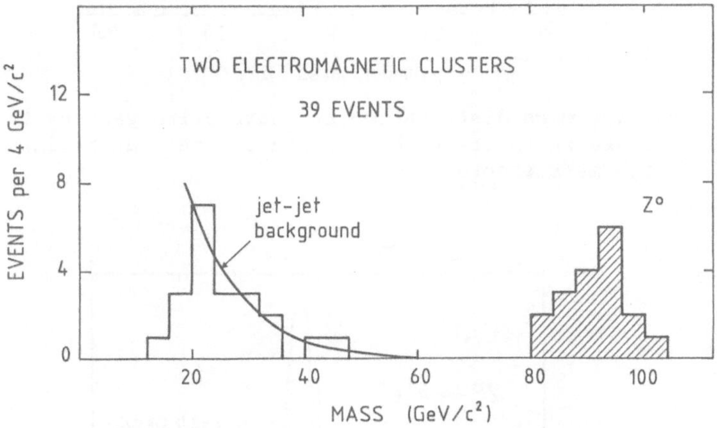

Fig. 8 : Invariant-mass distribution for two e.m. clusters passing the cuts described in the text. The curve shows the expected shape for events arising from jet-jet fluctuations. The high-mass Z° peak is clearly separated from the background.

We have selected a similar sample of 221 muons pairs of invariant mass greater than 6 GeV/c^2, requiring each individual muon to have a transverse momentum larger than 3 GeV/c. The dimuon mass distribution of Fig. 9 shows evidence for an isolated cluster of 10 high mass events.

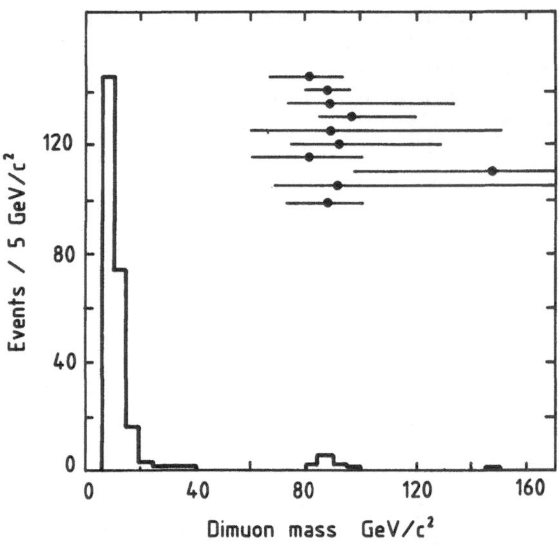

Fig. 9 : Invariant mass distribution for more pairs passing the cuts described in the text; here again one sees an isolated cluster of high mass events.

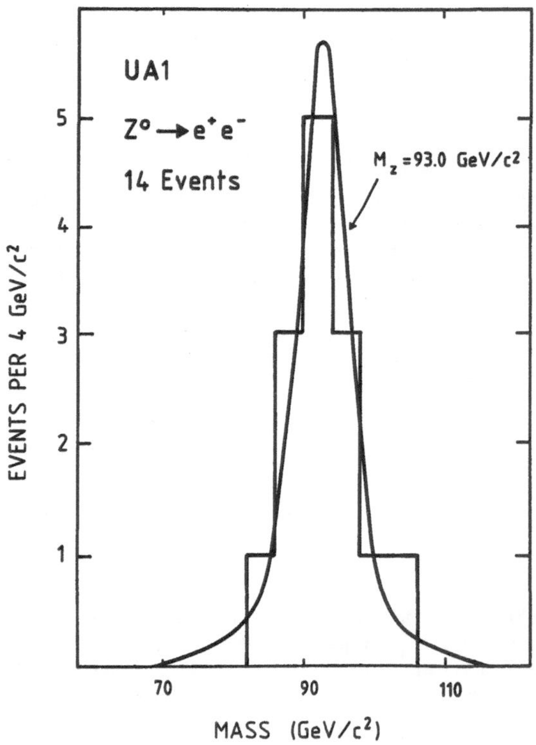

Fig. 10 : Invariant-mass distribution for two well-measured e.m. clusters with mass > 50 GeV/c^2. The curve shows the expectation for $Z^\circ \rightarrow e^+e^-$ decays (normalized to the data) with m_z = 93 GeV/c^2.

DETERMINATION OF THE Z° CROSS SECTION AND MASS

The efficiency of our selection requirements for production of Z° decaying into the electron channel is 0.7 ± 0.08 for the $\sqrt{s}$ = 546 GeV sample and 0.62 ± 0.03 for the $\sqrt{s}$ = 630 data sample. The corresponding cross sections are :

$$(\sigma B)_Z \rightarrow e^+e^- = 41 \pm 20\ (\pm 6)\ \text{nb}$$

at $\sqrt{s}$ = 546 GeV and

$$(\sigma B)_Z \rightarrow e^+e^- = 85 \pm 23\ (\pm 13)\ \text{pb}$$

at $\sqrt{s}$ = 630 GeV.

To measure the Z° mass precisely, we restrict ourselves to the sub-sample of Z° decays which pass the following cuts :

a) both electrons are more than ± 15° away from the vertical direction in the transverse plane. This removes one event.

b) an energetic charged track is associated with the second e.m. cluster. This removes two events.

In addition we remove one event where a jet enters the same calorimeter cell as one Z° decay electron track.

We are left with a sample of 14 Z° → e^+e^- decays for the Z° mass determination, with an expected background contamination of < 0.1 events.

The mass distribution for the sample of 14 well-measured high-mass e^+e^- pairs is shown in Fig. 8. The distribution has a mean value of $\langle m(e^+e^-)\rangle$ = 92.9 GeV/c^2, and an r.m.s. width of 4.8 GeV/c^2. To determine the Z° mass and decay width, a maximum likelihood fit of a Breit-Wigner distribution smeared by a Gaussian resolution function has been made to the e^+e^- mass-distribution.

The results of this fit are

$$m_z = 93.0 \pm 1.4\ (\pm 3.2)\ \text{GeV}/c^2,$$

$$\Gamma_z = 4.3^{+2.5}_{-1.6}\ \text{GeV}/c^2,$$

where the second error on the measurement of m_z comes from the systematic uncertainty in the energy scale of the experiment. The error quoted on Γ_z is the statistical error on the result. If we now include the systematic error we can, once again, only quote an upper limit on the width :

$$\Gamma_z < 8.1\ \text{GeV}/c^2 \text{ at } 90\ \%\ \text{c.l.}$$

Using the fitted values of m_z and Γ_z, the predicted shape of the Z° mass distribution (Fig. 10) gives a good description of the data. As a final check of the fitting procedure, we have generated Monte Carlo event samples with various values for the Z° mass and width. Generating 14 Z° decays in each event sample, we find that our fitting program recovers the correct mass. The decay width, however, is systematically underestimated by 10 % to 15 % in the region of interest. Our quoted value of Γ_z has been corrected for this systematic effect by using the Monte Carlo to calibrate the fitting procedure for the Z° width.

We can now compare our measurements of m_W and m_Z with the predictions of the standard model. Our result for the mass difference is

$$(m_Z - m_W) = 9.5^{+1.8}_{-1.7}\ (\pm\ 0.5)\ GeV/c^2.$$

Similar values extracted from the dimuon channel are compatible with the above

$$(\sigma B)_\mu = 51.4 \pm 17.1 \pm 8.6\ pb$$

and

$$m_Z = 88.8^{+5.5}_{-4.6}\ GeV/c^2.$$

PARAMETERS OF THE STANDARD MODEL

We now compare our measurements of m_W and m_Z in the electron decay channel with the predictions of the standard model. Our result for the mass difference is

$$(m_Z - m_W) = 9.5^{+1.8}_{-1.7}\ (\pm\ 0.5)\ GeV/c^2.$$

Defining the standard model parameters,

$$\sin^2 \hat{\theta}_W = (38.65/m_W)^2, \quad (4)$$

$$\sin^2 \theta_W = 1 - (m_W/m_Z)^2, \quad (5)$$

$$\rho = m_W^2/(m_Z \cos \hat{\theta}_W)^2, \quad (6)$$

we obtain

$$\sin^2 \hat{\theta}_W = 0.215^{+0.005}_{-0.006}\ (\pm\ 0.015),$$

$$\sin^2 \theta_W = 0.194 \pm 0.031,$$

$$\rho = 1.028 \pm 0.037\ (\pm\ 0.019).$$

In Fig. 11, we show our measurements together with the error curves reflecting the uncertainty in the energy scale (at the 68 % and 90 % confidence levels) in the ρ versus $\sin^2 \theta_W$ plane.

These results are in excellent agreement with current expectations for the standard model [11], and also with the corresponding results from the UA2 collaboration [5].

THE NUMBER OF LIGHT NEUTRINOS IN THE UNIVERSE

If the only extension of the standard model is more families and if no new quark or charged lepton is lighter than the W, the ratio Γ_Z/Γ_W will directly yield the number of those families.

Under those conditions, each additional generation opens a new $Z \rightarrow \nu\bar{\nu}$ decay mode and increases Γ_Z by 0.18 GeV while Γ_W is unaffected because of the lack of space for the corresponding decay. However direct measurements of the $Z°$ width are not at this level of accuracy and a better number is obtained from

$$R = \frac{\sigma B(W \rightarrow e\nu)}{B(Z \rightarrow e^+e^-)}$$

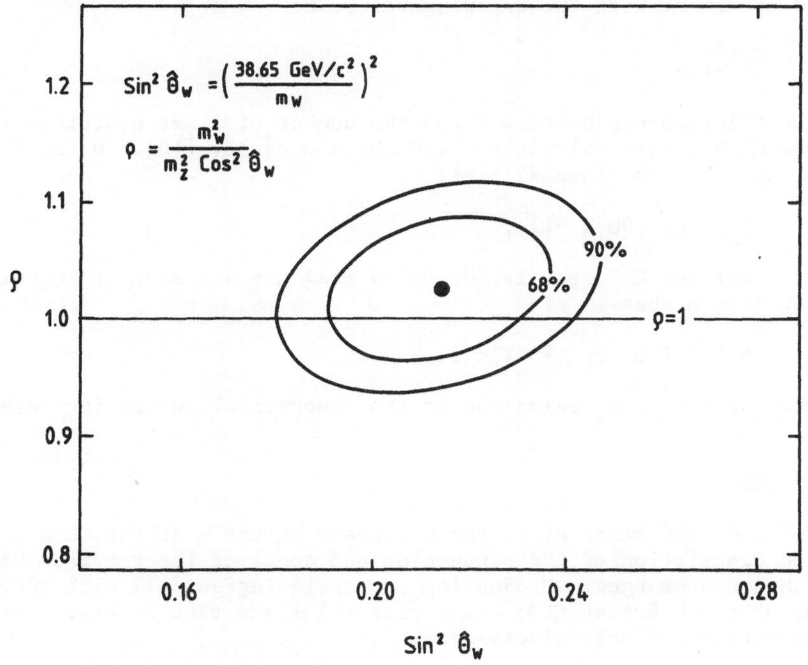

Fig. 11 : ρ versus $\sin^2 \hat{\theta}_W$ determined from the measurements of m_W and m_Z. The 68% and 90% confidence limits shown.

$$= \frac{\sigma_w}{\sigma_z} \frac{\Gamma(w \to \nu e)}{(z \to e^+e^-)}.$$

We have seen that the production of intermediate vector bosons is well described by the Drell-Yan QCD improved model; therefore we shall use thus model to evaluate σ_W/σ_Z. The theoretical error on the ratio :

$$\sigma_W/\sigma_Z = 3.3 \pm 0.2$$

is due to yet uncalculated higher order effects and the choice of QCD.

The factor $\Gamma(w \to e\nu)/\Gamma(z \to e^+e^-)$ can be calculated accurately from the standard model (10) using $M_W = 83$ GeV, $M_Z = 94$ GeV and $M_t = 40$ GeV.

The theorical expectation for R, assuming 3 families, is 9.2 ± 0.6 in good agreement with the experimental value

$$R = 9.3^{+2.9}_{-1.8}.$$

The relationship between R and the number of light neutrinos in the universe N_ν has been calculated by Deshpande et al. [10]. Using their result, and the experimental limit

$$R < 13.0 \quad \text{at} \quad 90\ \%\ \text{c.l.},$$

combining UA1 and UA2 results, de Lella [12] gives a more stringent bound on this number.

$$N_\nu < 5.4 \pm 1.0 \quad \text{at} \quad 90\ \%\ \text{c.l.}$$

the error quoted on N_ν being due to the theoretical uncertainty mentioned above.

CONCLUSIONS

Two and half years after the discovery of the W at the CERN a rather complete description of the production and decay of intermediate vector bosons has now emerged. Production properties agree well with the expectations of perturbative QCD; decay properties are also in agreement with the standard model and universality.

Progress in IVB physics clearly requires more stringent tests of the standard model : this requires both higher statistics and an improved detector. Both requirements will hopefully be met in 1987 when the antiproton collector ring ACOL will increase the collider luminosity by an order of magnitude and the UA1 upgraded calorimeter will give a factor of two improvement in resolution for jets and missing energy together with a better granularity.

PART II - JET PHYSICS

Evidence for high transverse energy jets appears as the outstanding new phenomenous at collider energies thus confirming early cosmic ray observation and predictions from QCD. The UA1 detector is well suited to the study of jets with its calorimeters covering a large solid angle. A striking feature of collider results is indeed the cleanliness of "jetty" events of which Fig. 12 shows examples. Jets are identified and measured using the clustering method mentioned above.

The percentage of events containing at least one identified jet in-

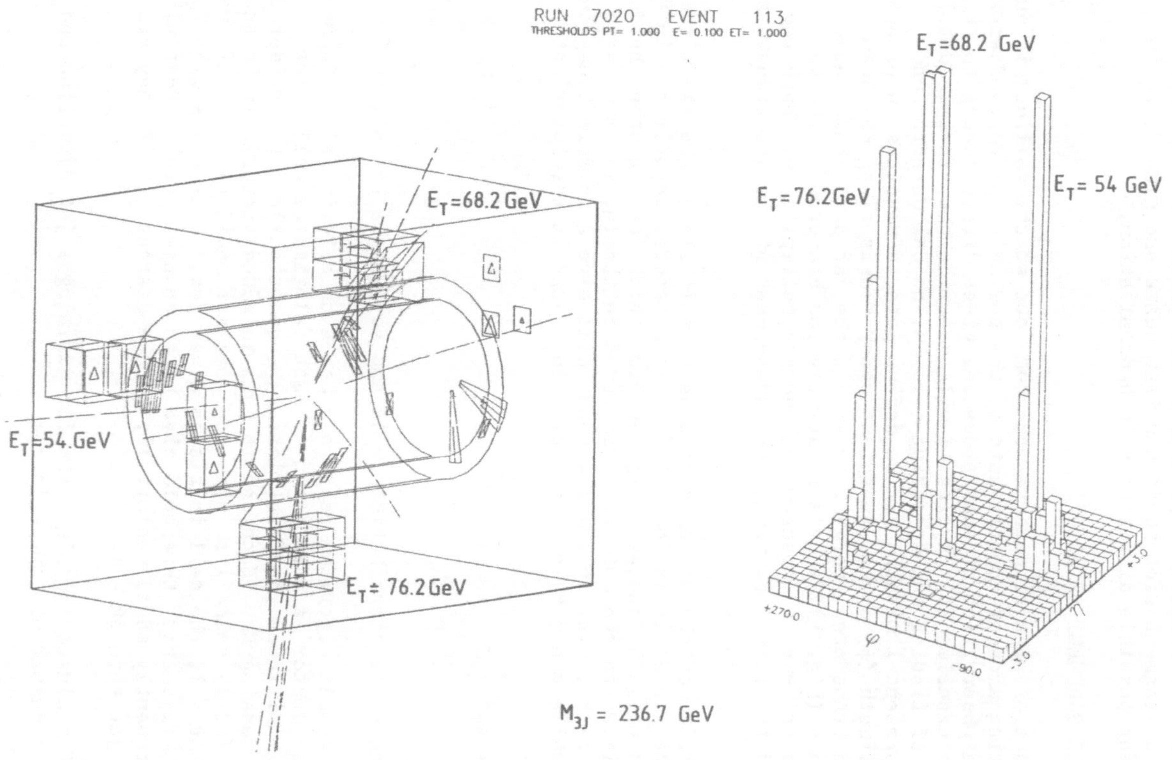

Fig. 12 : A three jet event in the UA1 detector as displayed on the interactive graphics facility ; its accompanying "Lego" plot in n.ϕ space shows the transverse energy blow as measured by calorimeters. The rectangles and sectors represent the energy depositions in the central (gondolas) and forward (bouchon) electromagnetic calorimeter elements; the cubes show the hadronic calorimeter elements recording an transverse energy in excess of 1 GeV.

creases fast with center of mass energy. The contribution of "jetty" events will be a large fraction of the non single diffractive inelastic cross section at TeV energies as shown in section 2.

The third section summarizes our results on high transverse energy jets. Inclusive cross sections and jet angular distributions in the parton-parton center of mass agree with QCD expectations; the comparison of 2 and 3 Jet final states leads to a direct determination of the strong coupling constant α_s; finally we present a very preliminary attempt to extract separate fragmentation functions from quark and gluon jets in order to look for possible differences in hadronization.

SELECTION AND MEASUREMENT OF JETS

The presence of high transverse momentum jets at the collider is due to hard scattering amongst constituants of the colliding proton and antiproton. As the observation of such processes gives direct access to the field of parton experimental physics, I will describe here in some detail the techniques of finding and measuring jets in the UA1 detector and discuss their present limitations. We define jets as clusters in pseudorapidity (η) azimuth (ϕ) space, invariant under boosts along the beam direction. The transverse energy flow around the jet axis shows indeed a sharp peak in η (Fig. 13) over a relatively low background; the same is true for the transverse momentum and charged multiplicity. Both the jet profile and the flat background are independent of the jet transverse energy.

The definition of the jet axis depends very little on the algorithm used to find the jet as it is dominated by large energy depositions in a few calorimeter cells. The jet energy determination, on the other hand, is strongly related to the cut-off parameter d in the (η, ϕ) space used in the jet algorithm. The underlying event will give a uniform background in that space while the jet is essentially contained in the invariant cone defined by

$$\Delta R = \Delta\eta^2 + \Delta\phi^2 < d$$

ϕ in radians.

Fig. 14 shows the distribution of ΔR and suggests a cut off at 1.

Our standard UA1 algorithm is purely based on calorimetric information. An energy vector is associated to each calorimeter cell. The highest E_T cell initiates the first jet; subsequent cells are considered in order of decreasing E_T. Each one in turn is added vectorially to the jet closest in (η, ϕ) space i.e. with the smallest ΔR if $\Delta R < 1$. If there is no jet with $\Delta R < 1$, the cell initiates a new jet. Only cells with E_T over a certain "initiator" threshold give birth to new "jets". Remaining cells are subsequently added vectorially to the nearest jet if they are closer to this jet than $\Delta R = 1$.

The use of a clustering algorithm for defining a jet direction and energy entails a number of corrections :

- Partons emitted within $\Delta R = 1$ of each other, from whatever source, will typically be merged into a single jet. This constitutes a part of our jet definition.

- The reconstructed jets will not contain any jet debris that has been emitted further from the jet axis than the cut parameter ΔR.

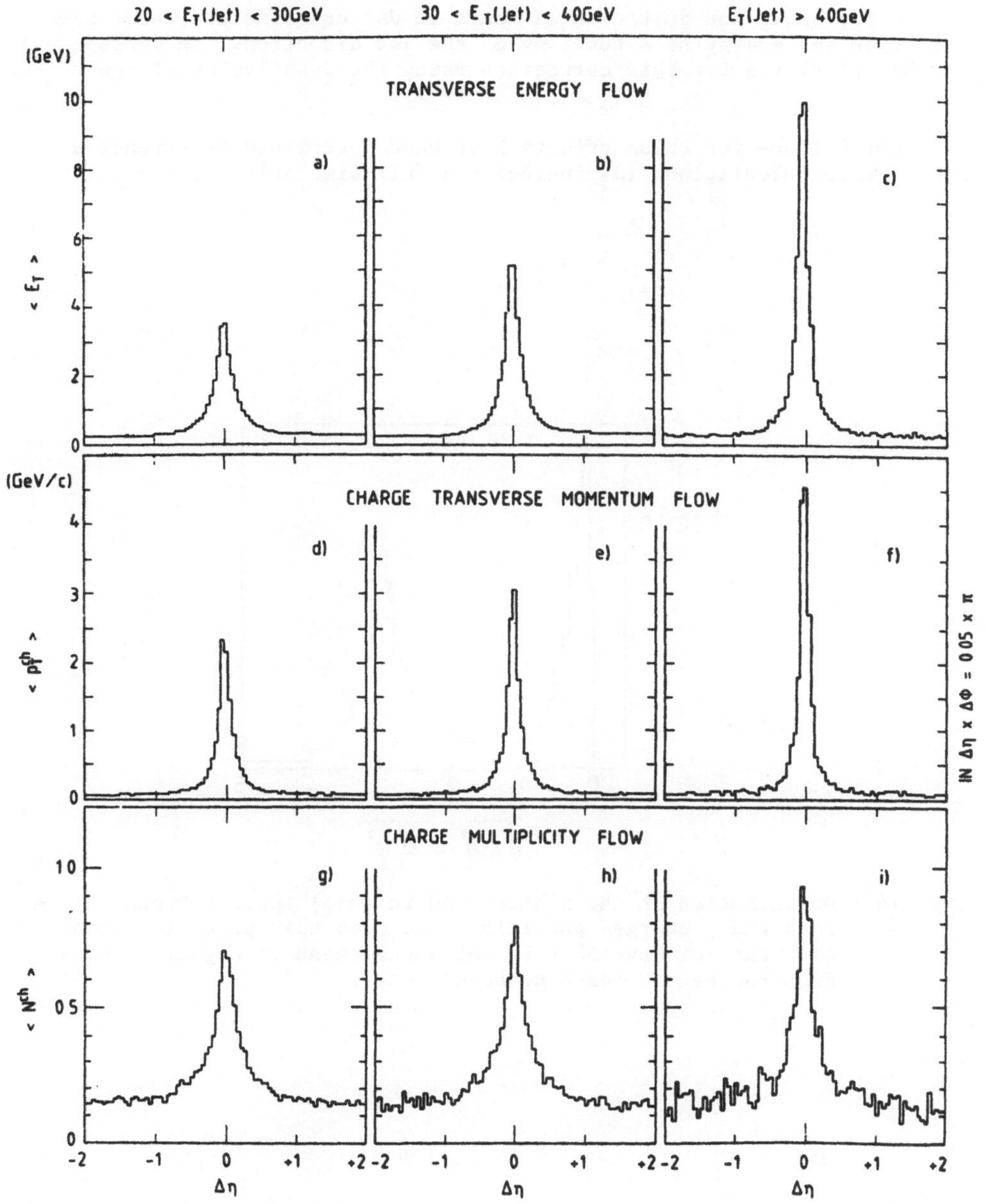

Fig. 13 : Distributions of transverse energy, transverse momentum and charged multiplicity around the jet axis. The "background" under the jet is independent of the jet transverse energy but is higher that the one of "minimum bias" events.

- Any spectator remains, i.e. parton debris not connected with the hard interaction, or any fragments from partons not sufficiently separate in ΔR, will be included in the jet by the algorithm if they happen to fall within the acceptance cone of the jet.

- The small non-instrumented zones in UA1 calorimeters cause some losses in jet energy as a function of the jet direction. An average value can be calculated for this correction using the detailed knowledge of the detector.

Corrections for these effects have been determined by extensive Monte Carlo calculations (3) including a full simulation of the detector.

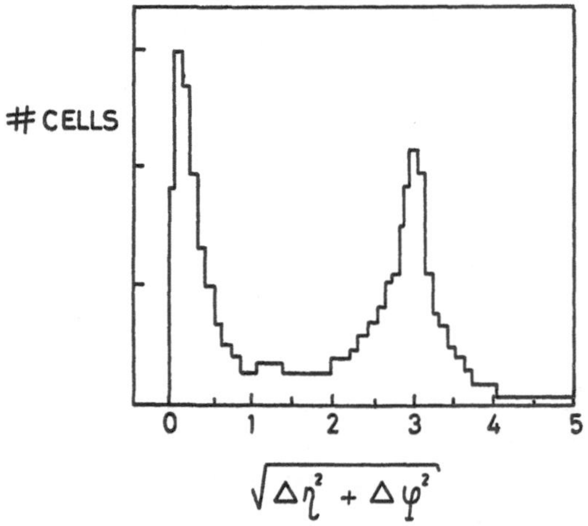

Fig. 14 : Distribution of the distance ΔR in (η.φ) space between the jet axis and a charged particle. One sees that particles associated with the jet have ΔR < 1, the enhancement at higher ΔR comes from the second (back to back) jet.

Average values and r.m.s. errors for corrections in transverse energy E_T, rapidity η, and azimuthal angle φ were evaluated as a function of E_T, and φ and data corrected accordingly. The jet reconstruction efficiency is 100 % for transverse energies in excess of 15 GeV and of the order of 20 % at 5 GeV. At very low energies, we expect our rather naive algorithm to make up jets in a random way as soon as some particle exceeds the initiator threshold. Fig. 15 shows the transverse energy flows measured in the calorimeter both for jets and high p_T single particles.

Fig. 16 shows the variations with transverse energy of the variable F

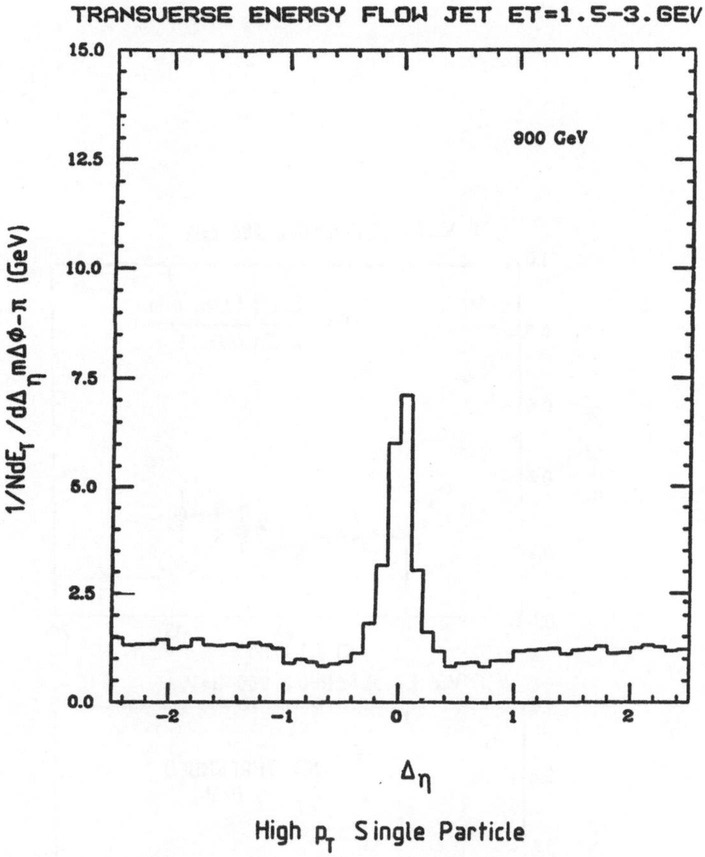

Fig. 15 : Transverse energy flow for energetic single particles and jets will a transverse energy in excess of 5 GeV.

$$F = \frac{\Sigma\ E_T(R < 0.2)}{\Sigma\ E_T(R < 1.0)}$$

measuring the lateral extension of the jet in the (η, ϕ) space. F is the order of 0.2 for a jet with a normal shape and 1 for an isolated single particle. We conclude that the UA1 algorithm, (using a jet initiator threshold of 1.5 GeV in this particular case), can be safely used down to E_T = 5 GeV.

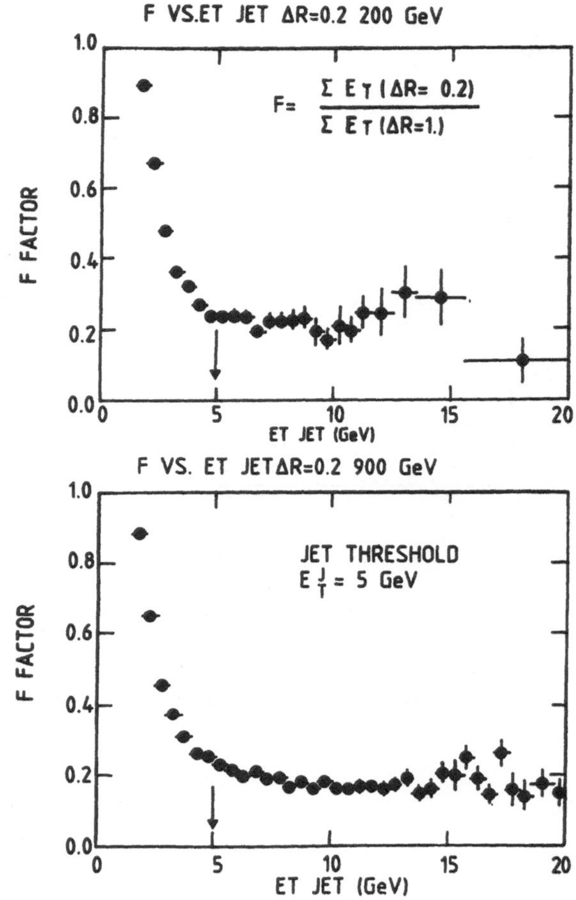

Fig. 16 : Distribution of the quantity F, defined in the text, which measures the width of the energy distribution. Fake jets are clearly generated at transverse energies under 5 GeV.

VARIATION WITH ENERGY OF THE JET PRODUCTION CROSS SECTION

In the spring of 1985, the collider has been operated in the ramping mode. The center of mass energy of the collision varying between 200 and 900 GeV. Our interaction trigger accepted all inelastic non-single diffractive events.

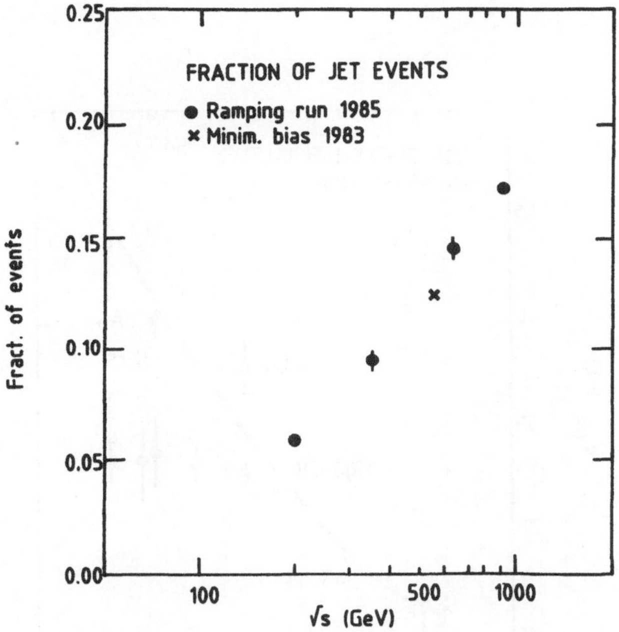

Fig. 17 : Percentage of events containing at least one detected jet over 5 GeV in the central region of the detector.

Fig. 17 shows the fraction of events containing a jet of E_T larger than 5 GeV in the central region ($-1.5 < \eta < 1.5$). The increase in the number of jet events in the ramping collider domain accounts for the major part of the rise of the inelastic non single diffractive cross section (Fig. 18). Data are consistent with a two component picture, the "jetty" and non "jetty" samples following separately a log S variation and KNO scaling (Table 1).

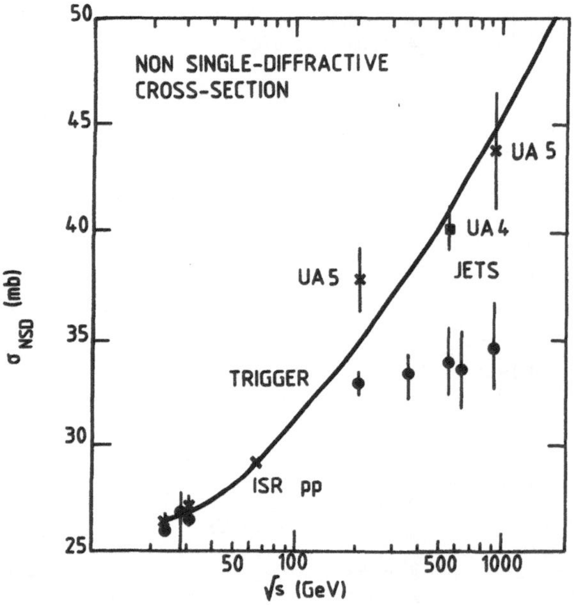

Fig. 18 : Total non-diffractive inelastic cross section (estimated to be 2/3 of the total cross section) compared to the cross section for non jetty events.

Table 1

	$<n>$		$<p_T>$	
c.m. energy	jetty*	non jetty	jetty	non jetty
200 GeV	26.5 ± 0.2	13.81 ± 0.7	0.474 ± 0.007	0.382 ± 0.005
900 GeV	32.9 ± 0.1	15.93 ± 0.07	0.516 ± 0.006	0.411 ± 0.005

* for $|\eta| < 2.5$

** a jetty event has a jet with $E_T^{JET} > 5$ GeV selected by the algorithm.

HIGH E_T JETS

Jets with a transverse energy in excess of 15 GeV are detected with full efficiency and measured with a reasonably good accuracy; the following analysis will therefore be based in those high E_T jets, selected by a localized transverse energy trigger. The data sample comes from the 1983 run with the collider operating at $\sqrt{s} = 546$ GeV.

INCLUSIVE CROSS-SECTIONS

The cross-sections for inclusive production of jets have measured in the central region $|\eta| < 1.4$. The trigger efficiency as a function of transverse jet energy has been determined using jets free of trigger bias (found in events triggered by other jets) and has been corrected for in the final results. The absolute normalization error on the luminosity, measured with a hodoscope which is part of our collision pretrigger, is estimated to be ± 15 %.

In order to compare the measured cross-sections with theoretical predictions, the effect of energy resolution has still to be corrected for in the data. A Gaussian error in the horizontal scale of a steeply falling distribution results in an average shift of cross-sections towards higher values. For a correction procedure a model is needed in order to get free of statistical fluctuations. The correction results in additional uncertainties for the low cross-section tail. We have used the QCD calculations of ref. [13] at the basis for smearing corrections. In Fig. 19, we compare the measured inclusive cross-sections, after smearing corrections, with different QCD predictions for parton transverse momenta. The curve from the COJETS Monte Carlo [14] differs in shape from the other two and appears to give a better description of the data. COJETS includes the effects of initial state bremsstrahlung which would give an increased cross-section at lower E_T as observed. Fig. 20 gives the differential cross-section integrated over jet transverse energies with our E_T threshold varying from 20 to 50 GeV. The predictions from reference [14] again describe the data in a satisfactory way.

TWO-JET CROSS-SECTION

Results on the two-jet cross-section measured in the UA1 experiment have been given in a previous paper [15]. The dominant subprocesses are

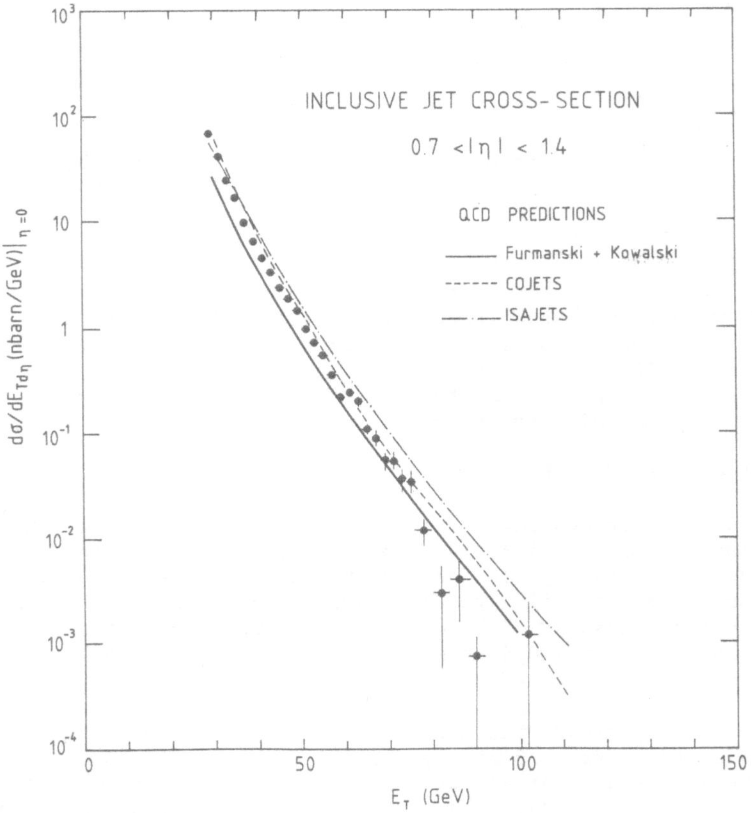

Fig. 19 : Inclusive differential jet cross-sections $d^2\sigma/(d\eta dE_T)$, after correcting for resolution smearing, and comparison with QCD calculations. Errors given are statistical only. The QCD curves have been computed using ISAJET [3], COJETS [14] and an algorithm corresponding to ref. [13] (courtesy H. Kowalski). Their normalization is independent of our data. Data and calculations for the rapidity bin 0 to 0.7.

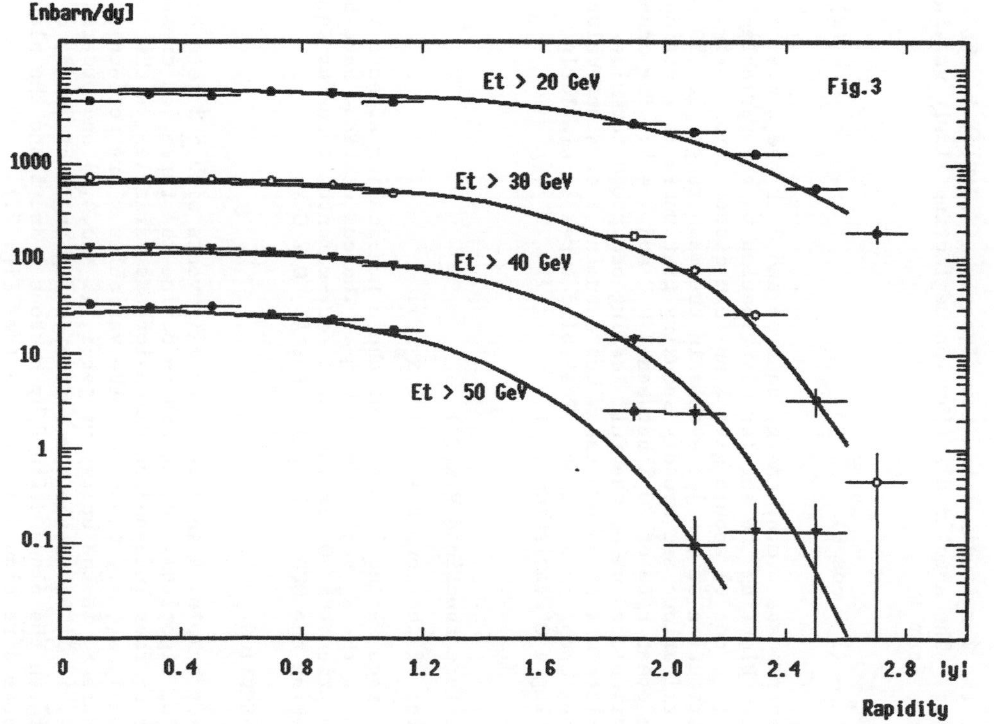

Fig. 20 : $d\sigma/d\eta$ integrating over jet transverse energy, with E_T threshold values varying from 20 to 50 GeV. The data points have been obtained from bias-free jets (i.e. not necessary for triggering), normalized in the central region ($|\eta| < 1$) to trigger jets and corrected for trigger efficiency. The regions in $|\eta|$ from 1.2 to 1.8 and > 2.8 have been left out in order to avoid jets that deposit large fractions of energy in apparatus zones having very different characteristics. The energy smearing correction was applied as explained in the text. The curves have been obtained from COJETS, and their normalization is independent of the data.

expected to be gg, .gq + $g\bar{q}$ and $q\bar{q}$ scattering. These subprocesses are predicted to have similar angular distributions and have approximately relative magnitudes in the ratio 1 : 4/9 : $(4/9)^2$ [16]. Thus in the approximation that the angular dependences of these subprocesses are identical the total 2 jet cross-section may be written in a fully factorized form :

$$\frac{d^3\sigma}{dx_1 dx_2 d\cos\theta} = \frac{F(x_1)}{x_1} \cdot \frac{F(x_2)}{x_2} \frac{d\sigma}{d\cos\theta} \tag{1}$$

where $x_1(x_2)$ are the longitudinal momentum fractions and F(x) is the structure function. The angular distribution neglecting slowly varying factors, is of the form :

$$\frac{d\sigma}{d\cos\theta} = \frac{\alpha_s{}^2}{\hat{s}} (1 - \cos\theta)^{-2} \tag{2}$$

where $\hat{s}$ is the subprocess c.m.s. energy squared and θ is the c.m.s. scattering angle. Plotting the angular distribution in the variable $\chi = (1 + \cos\theta)/(1 - \cos\theta)$ should give a uniform distribution. Fig. 21 shows the χ distribution of two jet events in the mass range 150-250 GeV. Although the distributions are broadly speaking flat there is a noticeable increase in event rate of 20 % between $\chi = 2$ and $\chi = 9$. The dashed curve has been calculated using the full leading order QCD formulae using an appropriate mix of subprocesses but assuming a constant value for α_s, the strong coupling constant. The solid curve is similarly calculated allowing a Q^2 variation of :

1) $\alpha_s(Q^2)$
2) the structure functions $F(x,Q^2)$
3) a K-factor of the form $K = (1 + 3\,\alpha_s/2)$.

Clearly the data require the inclusion of these non-scaling effects the largest of which is the α_s variation with Q^2. The calculations have been made on the assumption that Q^2 is the usual four-momentum transfer squared ($Q^2 = -\hat{t}$) and taking the QCD scale parameter $\Lambda = 0.2$ GeV.

THREE JET CROSS-SECTION

The three jet sample is selected from all events with ≥ 3 jets as defined by the jet algorithm. We choose to define the three jet cross-section in terms of four independent dimensionless variables in the c.m.s. frame, namely x_3, x_4, θ_3 and ψ. The variables x are defined $x_i = 2E_i/\Sigma E_i$ where E_i is the energy of jet-i, (i = 3,4,5) and ordered such that $x_3 > x_4 > x_5$; θ_3 is the c.m.s. scattering angle of jet 3; ψ is the angle between the plane defined by partons 4 and 5 and the plane defined by the beam direction and parton 3 (see Fig. 22).

Neglecting constant and slowly varying factors, the leading order QCD predictions for the subprocess cross-sections may be written :

$$\frac{d^4\sigma}{dx_3 dx_4 d\cos\theta_3\, d\psi} = \frac{\alpha_s^3 M^2}{\hat{s}} x_{T_3}^2 x_{T_4}^2 x_{T_5}^2 [(1-x_3)(1-x_4)(1-x_5)]^{-1}. \tag{3}$$

The exact leading order expressions for the matrix element M^2 for the processes gg → ggg, gq → gqg etc. are given in reference [17]. Note that both the two-jet and three-jet cross-sections have the same $1/\hat{s}$ dependence. The cross-sections become large for two reasons :

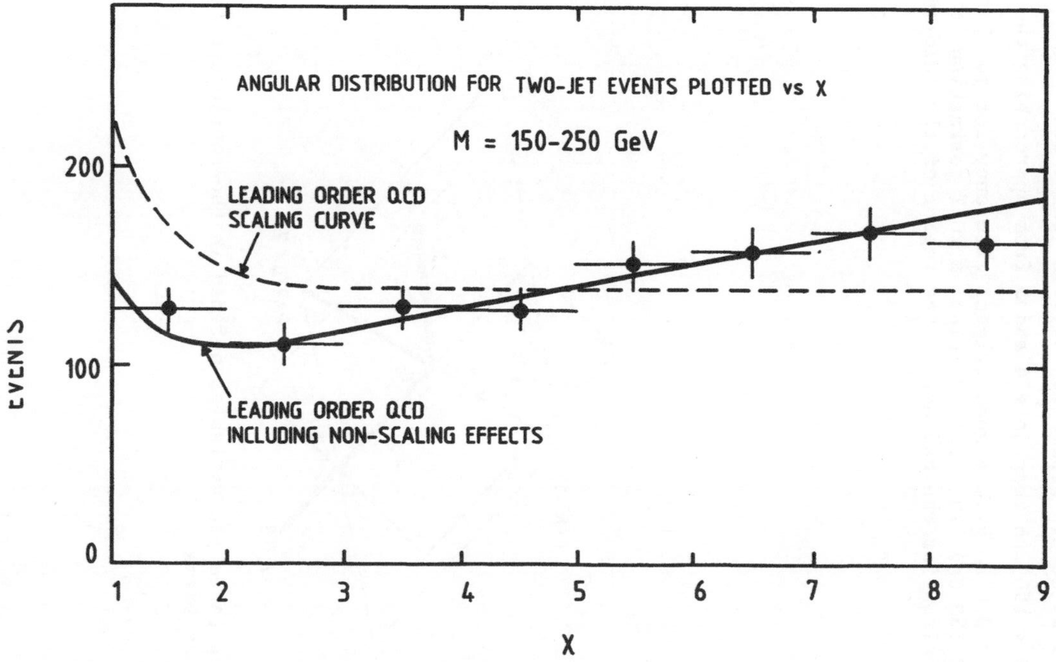

Fig. 21 : The two-jet angular distribution plotted versus $X = (1 + \cos\theta)/(1 - \cos\theta)$. The broken curves show the leading-order QCD predictions suitably averaged over the contributing subprocesses, and the solid curves include scale-breaking corrections (see text).

1) $x_{T_i} = x_i \sin\theta_i \to 0$ which is the prefered configuration for initial state bremsstrahlung.

2) x_3, $x_4 \to 1$ which is the prefered configuration for final state bremsstrahlung.

We therefore limit our analysis to a region of phase space where all jets are well resolved from the beam and from each other and where the theoretical cross-sections vary relatively slowly.

In order to ensure the above conditions the following cuts have been applied :

1) $x_3 < 0.9$ to ensure separation between jets 4 and 5.

2) $30° < \psi < 150°$ to keep jets 4 and 5 from approaching the beam.

3) $\cos\theta_3 < 0.6$ to give a good trigger efferency for jet 3.

Above a mass of 150 GeV 173 events remain. A 20 % correction is applied to the data to allow for the bad acceptance region of the calorimeter.

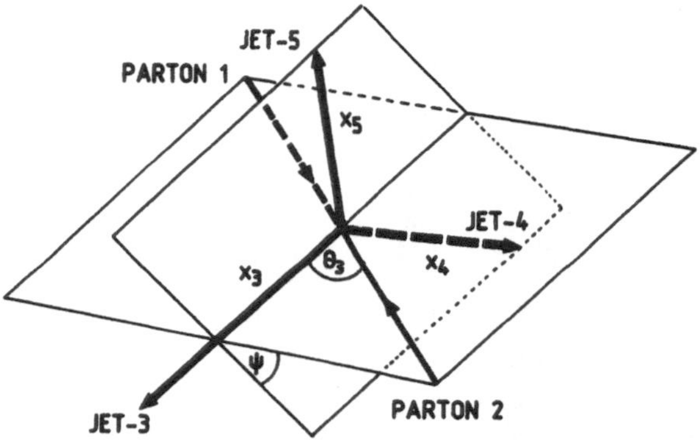

Fig. 22 : The three jet variables defined in the subprocess center of mass frame.

Fig. 23 shows the Dalitz plot of x_3 versus x_4 with the leading order QCD predictions suitably averaged over the subprocesses. Fig. 24 shows the three-jet angular distributions in $\cos\theta_3$ and ψ. The two curves are the predicted dependence of the angular variables for the dominant suprocesses which are seen to be very similar as in the two-jet case.

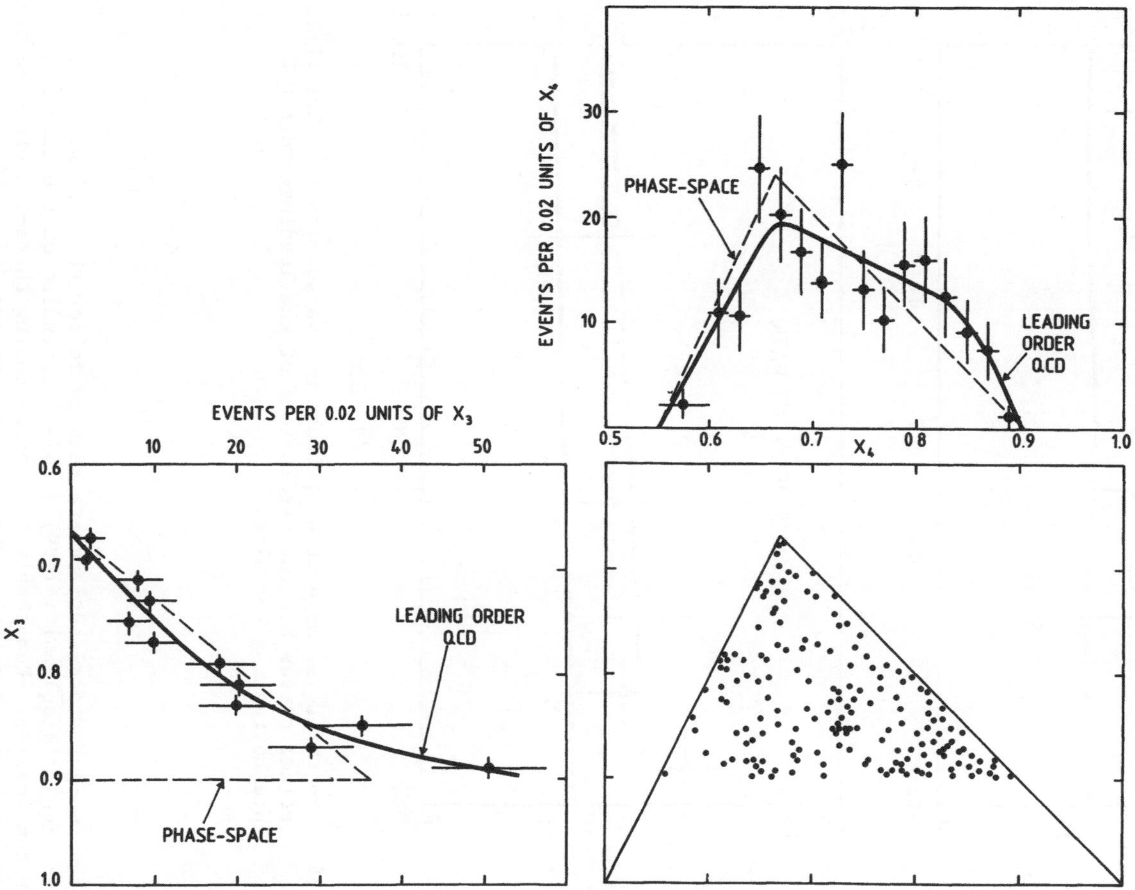

Fig. 23 : The Dalitz plot (x_3 versus x_4) for the three-jet sample. The solid curves represent the predictions of the leading-order QCD bremsstrahlung formulae [6].

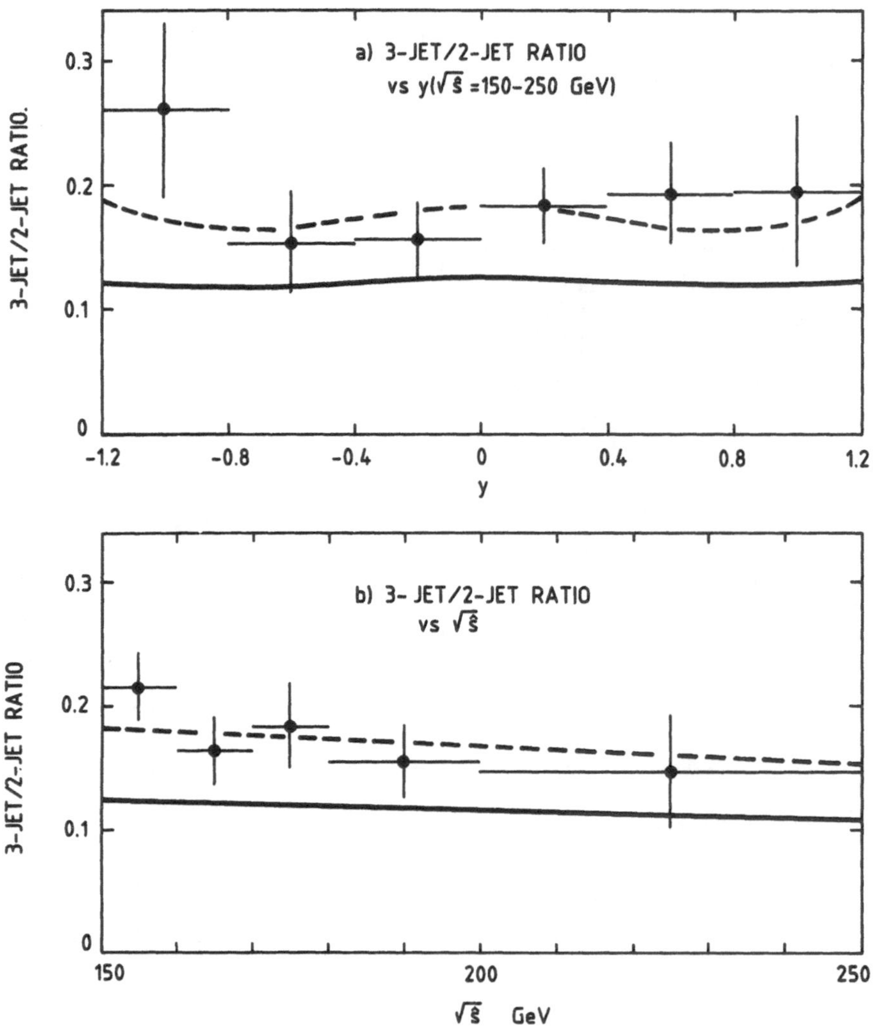

Fig. 24 : The three-jet angular distribution : versus cos θ_3. The theoretical curves are the predictions of the leading-order QCD bremsstrahlung formulae.

The two-jet cross-sections are seen to follow the well known rule $\sigma(gg) : \sigma(qg) : \sigma(q\bar{q}) \simeq 1 : (4/9) : (4/9)^2$ depending on the incoming parton combination. Remarkably the corresponding three-jet cross-sections are seen to be in essentially the same proportions (to within ± 20 %). This has the important consequence that the three-jet/two-jet ratio is predicted to be almost independent of the nature of the incoming partons. This ratio is plotted as a function of the subprocess mass ($\sqrt{s}$) in Fig. 25. The solid curve is the predicted ratio taking into account the relative contributions of each subprocess, the Q^2 dependence of α_s and assuming six effective quark flavours and Λ = 0.2 GeV. The errors are purely statistical. The estimated systematic error is ± 15 %. The broken

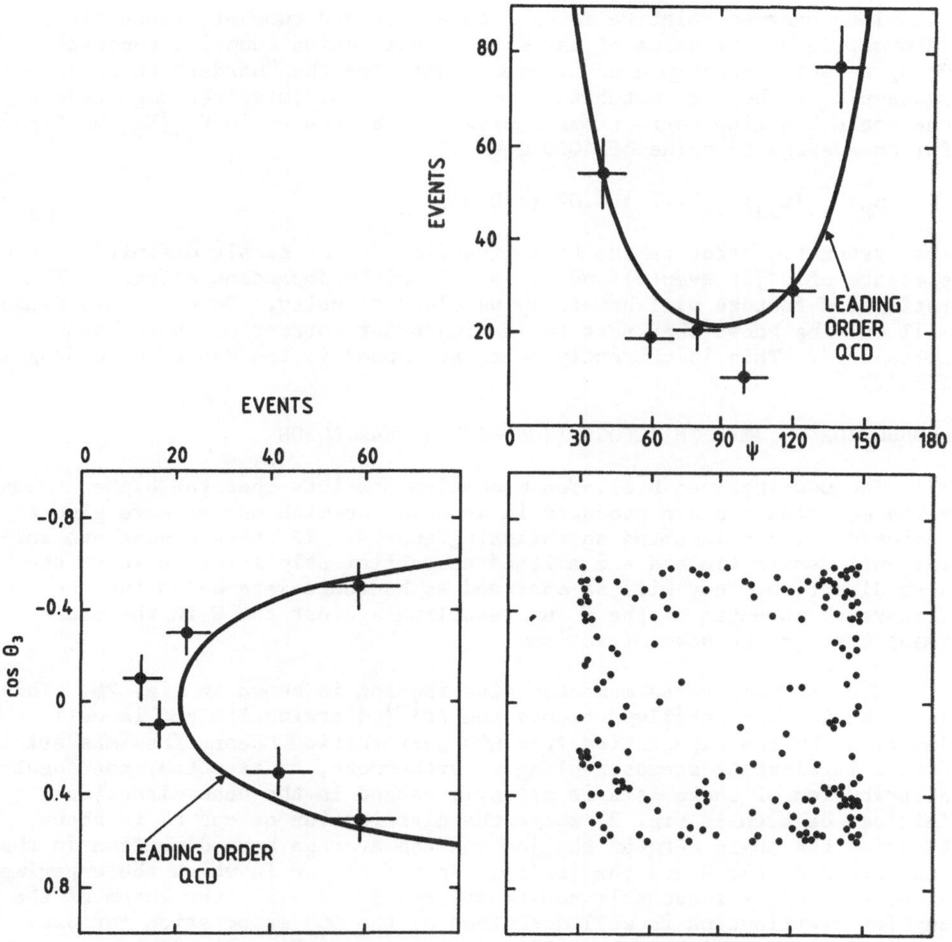

Fig. 25 : a) The three-jet/two-jet ratio plotted versus the laboratory rapidity of the three-jet (two-jet) system.

b) The three-jet/two-jet ratio plotted versus subprocess c.m.s. energy (mass). The theoretical curves have been calculated assuming six effective quark flavours and taking $\Lambda = 0.2$ GeV. The solid curves correspond to the choice of identical Q^2-scales for the three-jet and two-jet samples : $\langle Q_{2J} \rangle = 0.45m_{2J}$, $\langle Q_{3J} \rangle = 0.45m_{3J}$. The broken curves correspond to the choice of a lower Q^2-scale for the three-jet sample : $\langle Q_{3J} \rangle = 0.30m_{3J}$.

curve in Fig. 25 represents the modification of this prediction assuming a lower Q^2 scale for the 3 jet sample $<Q_{3J}> \simeq 2/3 \; <Q_{2J}>$. One might indeed argue that for comparable angular acceptance the Q_{2J} scale is naturally lower in the case where the available energy is shared among a larger numbers of partons.

The observed relative rate of three-jet and two-jet events yields information on the value of the strong interaction coupling constant α_S. Using a "soft" subsample of two-jet events and the "harder" three-jet subsample, we hope to match the Q^2-scales and minimize the magnitude of the scale breaking corrections represented by the ratio K_{3J}/K_{2J} we find, for an average Q^2 value of 4000 GeV^2 :

$$\alpha_S(K_{3J}/K_{2J}) = 0.16 \pm 0.02 \; (\pm 0.03)$$

the systematic error is due to ambiguities on the sample definition (e.g. presence of 4 jet events) and to jet algorithm dependent effects. The ratio of K factors will hopefully be close to unity. However this factor will only be known when next to leading order corrections have been calculated. This is currently being attempted in the Monte Carlo program EUROJET [4].

PRODUCTION OF JETS IN ASSOCIATION WITH A WEAK BOSON

The QCD-improved Drell-Yan mechanism predicts that the higher transverse momentum W's are produced in association with one or more gluons radiated off the incoming annihilating quarks. If these gluons are sufficiently energetic and are emitted at sufficiently large angle to the beam direction, they will be observed as hadronic jets balancing the transverse momentum of the W and recoiling against the W in the plane transverse to the beam direction.

The jet transverse momentum distribution is shown in Fig. 26. This distribution essentially reflects the $Pt^{(W)}$ distribution and is well described by the expectation from QCD perturbation theory for jets arising from initial-state bremsstrahlung. Furthermore, as expected, the angular distribution of these jets is strongly peaked in the beam directions. This can be seen in Fig. 27 where the distribution of cos θ^* is shown, θ^* being the angle between the jet and the average beam direction in the rest frame of the W and the jet(s). In the region in which the experimental acceptance is reasonably constant ($|\cos \theta^*| < 0.95$) the shape of the angular distribution is well described by the QCD expectation for bremsstrahlung jets, which is basically $(1 - |\cos \theta^*|)^{-1}$. Finally, we examine the invariant mass of the (W + 1 jet)-system for those events in which one jet has been reconstructed by the UA1 jet algorithm (Fig. 28). The shape of this mass distribution is a little broader than the expectation from ISAJET. It is, however, well described by a simple Monte Carlo in which the observed W four-vectors are randomly associated with the four vectors from our sample of jets, suggesting that the shape of the (W + 1 jet) mass plot is controlled more by the proton structure functions than by the QCD matrix element.

We conclude from this mass distribution that there is no evidence for the production of a massive-state X which subsequently decays into a W plus one jet. For a mass $m_x > 180$ GeV/c^2, with X subsequently decaying into a W and a single hadronic jet, we place an upper limit on the production cross-section for X of

$$\sigma_B(x \rightarrow W + \text{jet})/\sigma_W < 0.013$$

at 90 % confidence level.

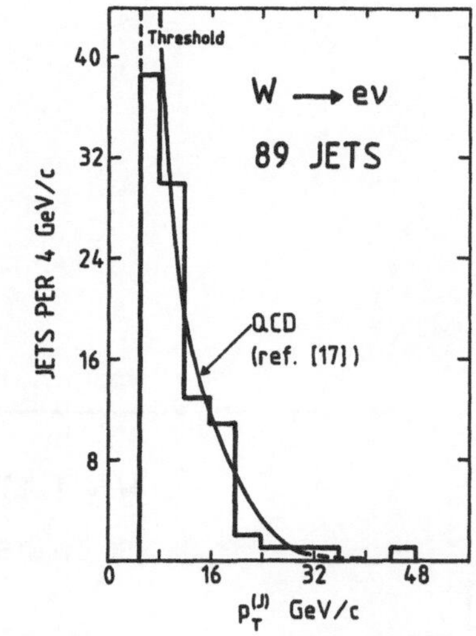

Fig. 26 : Jet transverse momentum distributions for all jets produced in association with the W. The curve shows the QCD prediction of [22] normalized to the tail of the distribution (the region in which we expect to have good reconstruction efficiency for the jets).

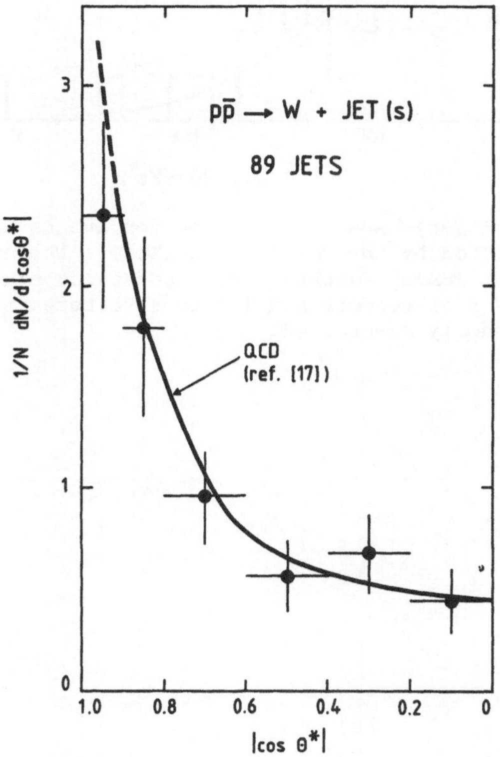

Fig. 27 : The angular distribution for jets reconstructed in W events. The distribution of cos θ* is shown, θ* being the angle between the jet and the average beam direction in the rest frame of the W and the jet(s). The curve shows the QCD prediction from [22].

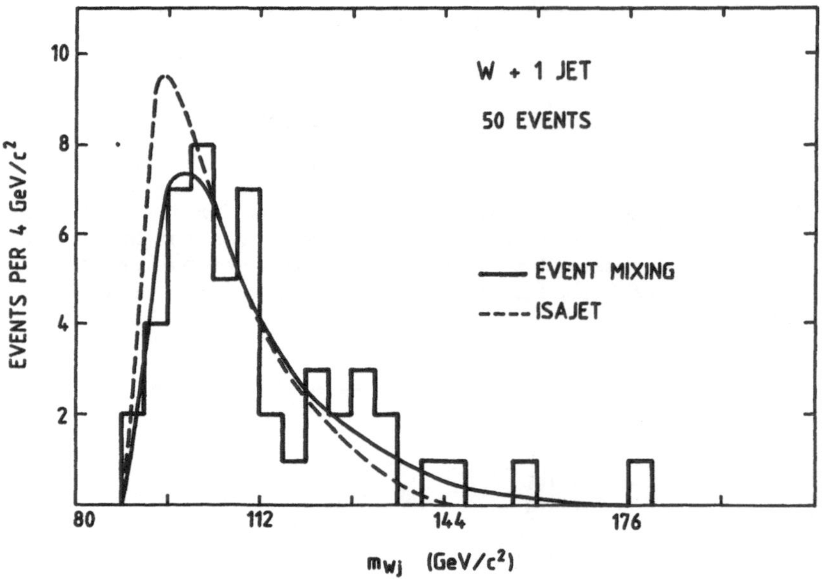

Fig. 28 : The (W + jet)-mass distribution for events in which one jet is reconstructed by the UA1 jet algorithm. The curves show the predictions of ISAJET (dashed curve) and event mixing (solid curve) in which W four-vectors and jet four-vectors from jets in W events are randomly associated.

FRAGMENTATION OF HADRONIC JETS

LONGITUDINAL FRAGMENTATION

The study of the fragmentation properties of jets is very important at collider energies, where most of the jets are gluons. A previous paper [18] from UA1 showed that there was no difference between e^+e^- jets and the Collider jets as far as fragmentation is concerned. The analysis has been redone on the 1983 data with more statistics, and is presented below.

Two-jet events have been selected with a minimum E_T of 25 GeV on the leading jet and the requirement that the two jets be back-to-back within 30° in the transverse plane. Furthermore, the axes of the jets are required to be in a good acceptance region, both for the calorimeters and the central detector, leading to a final sample of 15.000 jets. The charged tracks are associated with a jet, defined by the calorimetry and selected, if R between a track and the jet is less than 1 and if $z > 0.01$. Here z is defined as P_L/P_{jet} where P_L is the projection of the track momentum along the jet axis.

The fragmentation function is defined as

$$D(z) = \frac{1}{N_{jet}} \frac{dN_{ch}}{dz}.$$

This is shown in Fig. 29a, where the systematic errors are also seen. The dashed line is a fit to the data, of the form

$$D(z) = \frac{3.4}{z} e^{-7z}.$$

Several corrections have been applied to the raw fragmentation function. Background tracks from the rest of the event have to be subtracted, and there is a 20 % acceptance correction for track losses. The energies of the jets have been corrected for detector inefficiencies, fluctuations of calorimeter response, and the gain of energy from the rest of the event. The smearing effect due to the track momentum and jet energy errors has also been corrected for. These energy-dependent effects were underestimated in the above-mentioned UA1 publication [18] on this topic and this is the reason for the changes observed in the fragmentation function.

In Fig. 29b, the UA1 result is compared with the PETRA [19] and ISR [20] results. The fragmentation function is considerably softer at the Collider energy. This is thought to come from two effects : the Q^2 evolution of D(z) and the nature of the fragmenting parton (quarks at PETRA and the ISR, gluons and quarks at the Collider).

In order to distinguish quark jets from gluon jets, we have adopted the following method : since we have selected two-jet events, for each event we can calculate x_1, x_2, and cos θ, and compute the cross-section of each subprocess as given by :

$$\frac{d\sigma_{ab \to cd}}{dx_1\, dx_2\, d\cos\theta} = \frac{F_a(x_1,Q^2)}{x_1} \frac{F_b(x_2,Q^2)}{x_2} \frac{\pi\alpha_s^2}{2\hat{s}} |M|^2 \qquad (4)$$

we have used the parametrization of the structure functions from Eichten et al. [21], the first order matrix elements from [16] and the symmetric form for $Q^2 = 2\hat{s}\hat{t}\hat{u}/(\hat{s}^2 + \hat{t}^2 + \hat{u}^2)$. We can calculate the probability for each jet to be a gluon by summing all relevant subprocess cross-sections :

$$P(\text{jet} = \text{gluon}) = \frac{P_{\text{process}}(\text{jet} = \text{gluon})}{P_{\text{process}}(\text{jet} = \text{anything})} \qquad (6)$$

and of course P(jet = quark) = 1 - P(jet = gluon). Fig. 30 is a histogram of this probability. We observe that although gluon jets dominate, quark jets are also present. To separate gluon jets from quark jets we have chosen two separate bins (shaded areas). The fragmentation function D(z) for gluons and quarks is shown in Fig. 31. The gluon fragmentation function is the softer of the two as expected from QCD.

A simple fit to the data gives :

$$D(z)^{\text{Gluon}} = \frac{4.4}{z} e^{-8.8z}$$

$$D(z)^{\text{Quark}} = \frac{3.4}{z} e^{-6.9z}.$$

The difference between quark and gluon jets appears more clearly on Fig. 32 where the ratio of the two fragmentation functions is plotted; the solid line is the prediction of G. Ingelman. His model includes QCD evolution of the parton shower and Lund string hadronization.

Fig. 33 shows the effect of scaling violations on the gluon structure function. As expected from QCD, the longitudinal fragmentation is softer at high Q^2.

Our results suggest indeed that gluon jets have a softer fragmentation function than quark jets. Although the effect is small and hard to disantangle from scaling violations, it looks real and will probably be observed more clearly at higher energies.

CONCLUSIONS

At collider energies, jets are observed in a large fraction of events and appear to be largely responsible for the rise of the total inelastic cross-section.

- Their production properties agree well with QCD expectations. The collider really opens the way to parton physics.

- Fragmentation properties are probably different for quark and gluon jets.

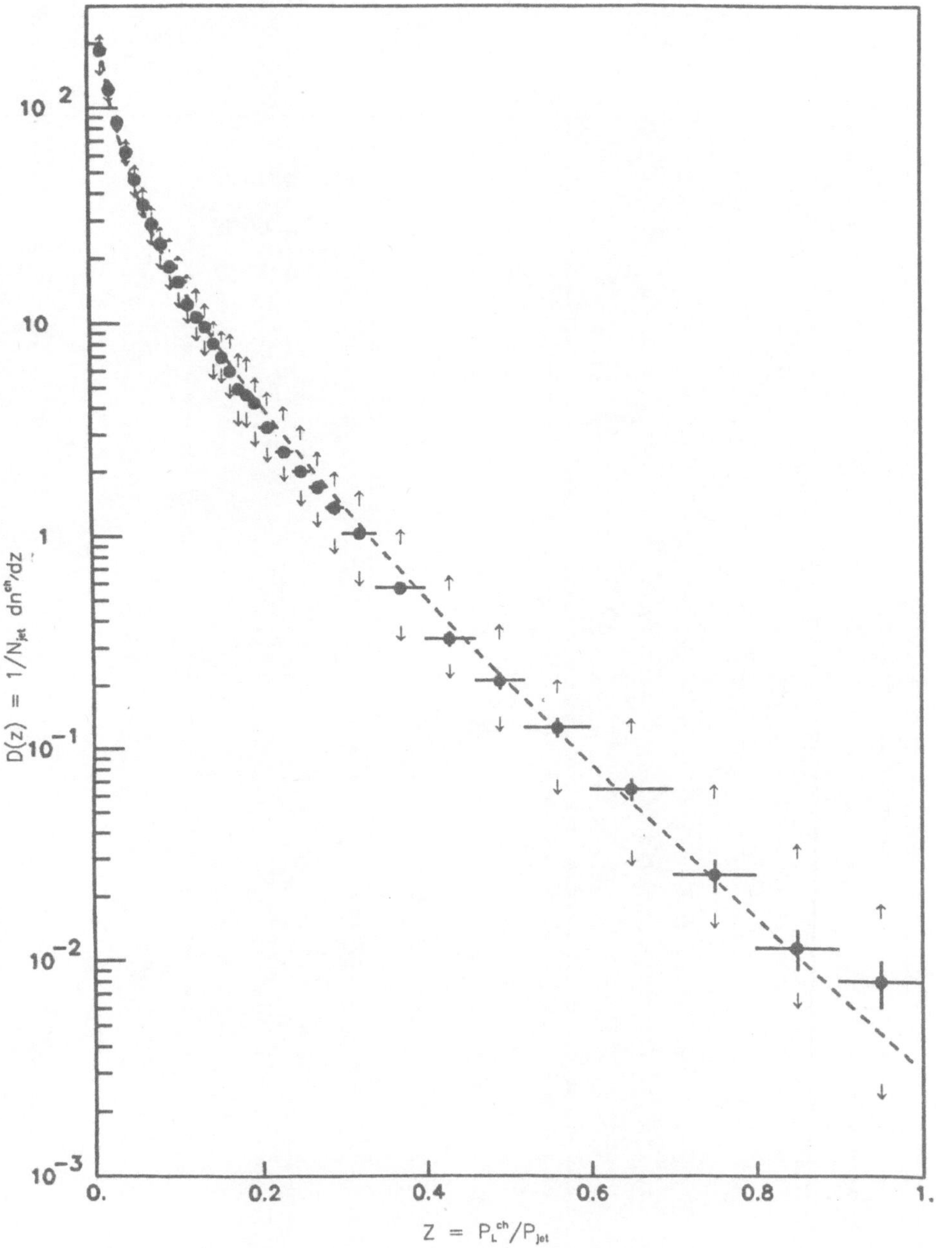

Figs 29 : a) Fragmentation function as measured by UA1. The up and down arrows are conservative estimates of systematic errors. The dashed line is a fit to the data (see text).

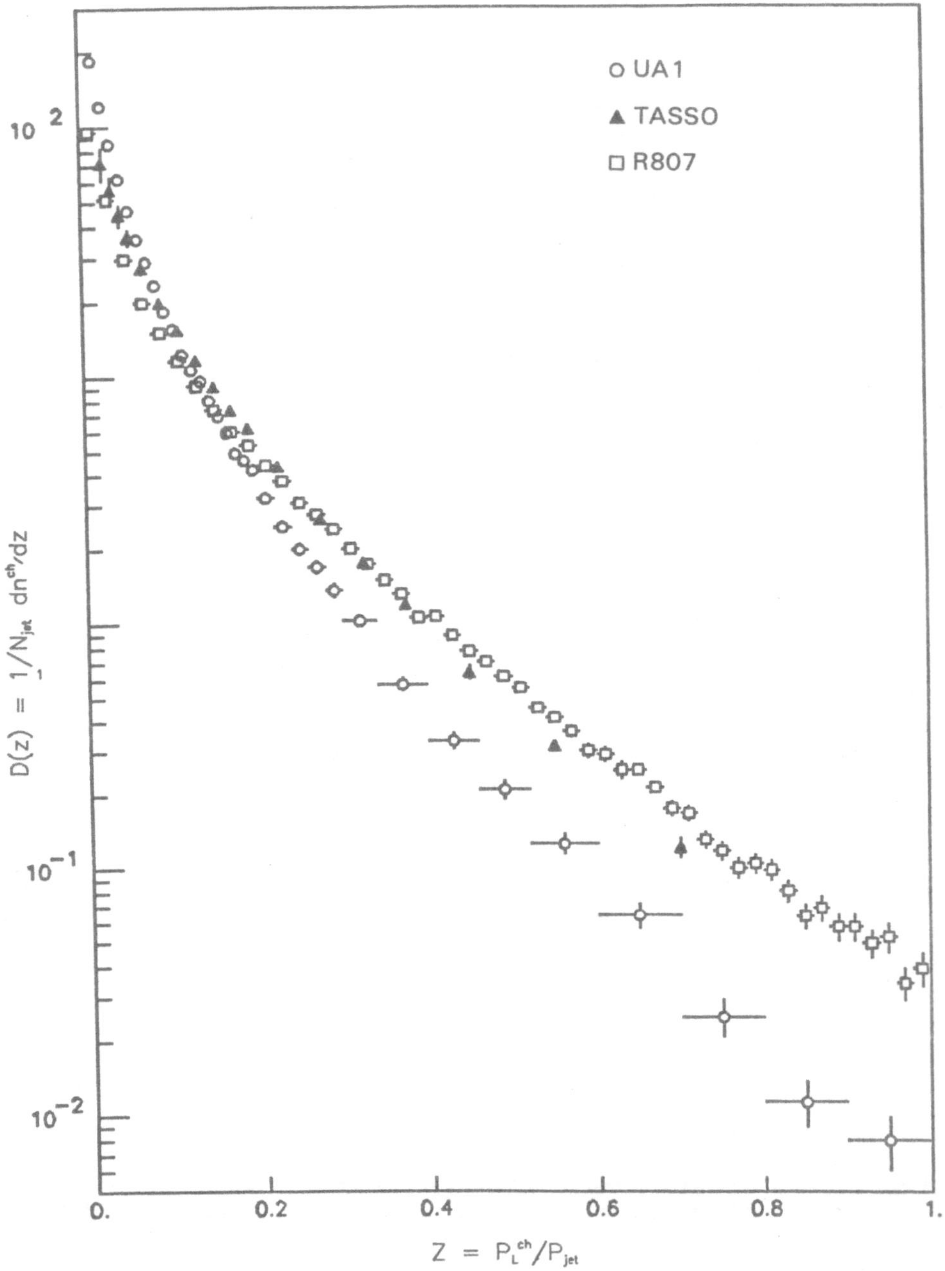

Fig. 29 : b) Comparison of fragmentation functions : UA1 open circles; PETRA [19] (TASSO at $\sqrt{s}$ = 34 GeV) full triangles; ISR [20] (R 807 at $\sqrt{s}$ = 63 GeV) open squares.

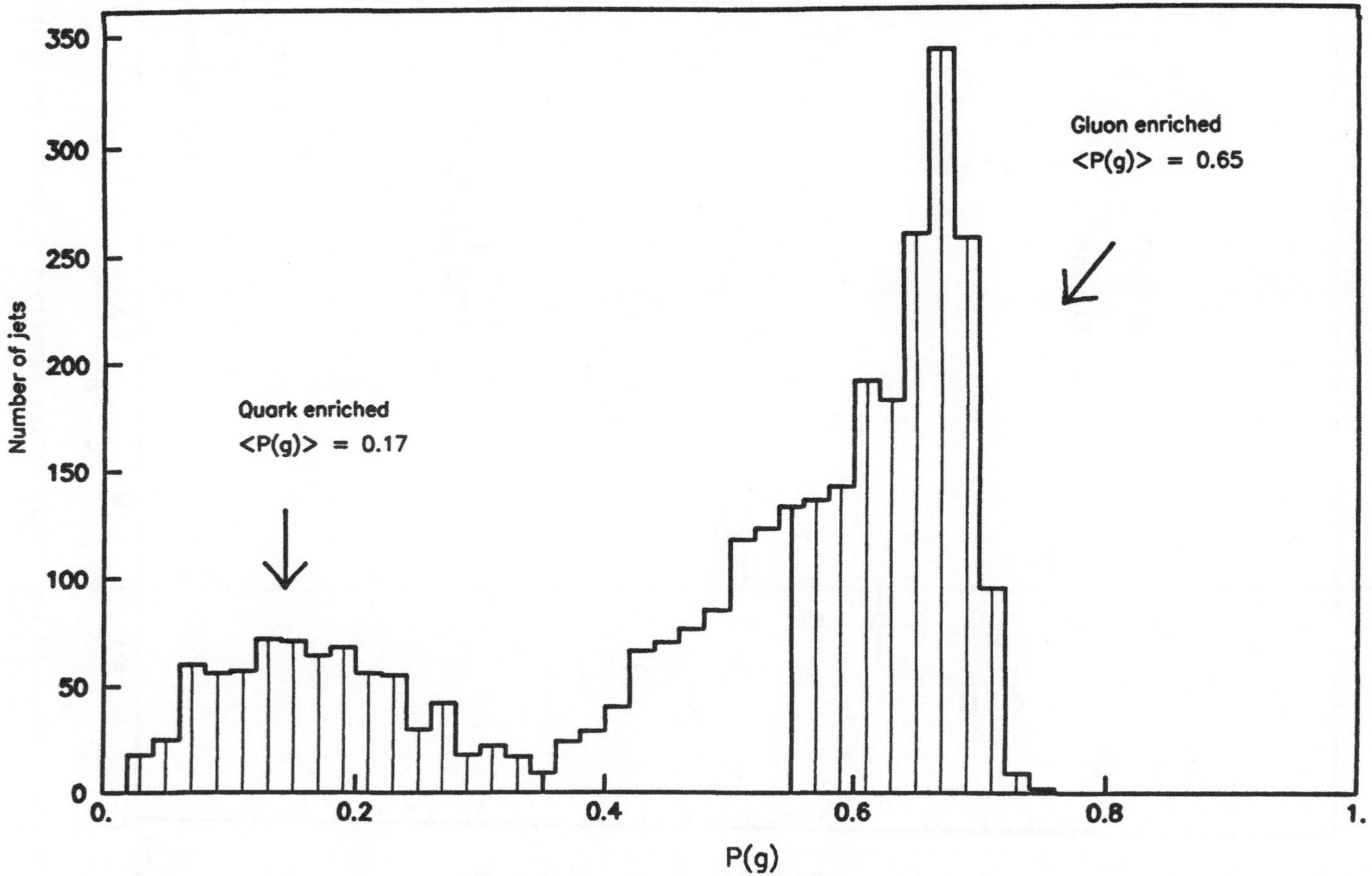

Fig. 30 : The probability distribution for individual jets to be gluons. Indicated are the two cuts [P(G) < 0.35 and P(G) > 0.55] used to define the quark and gluon enriched samples.

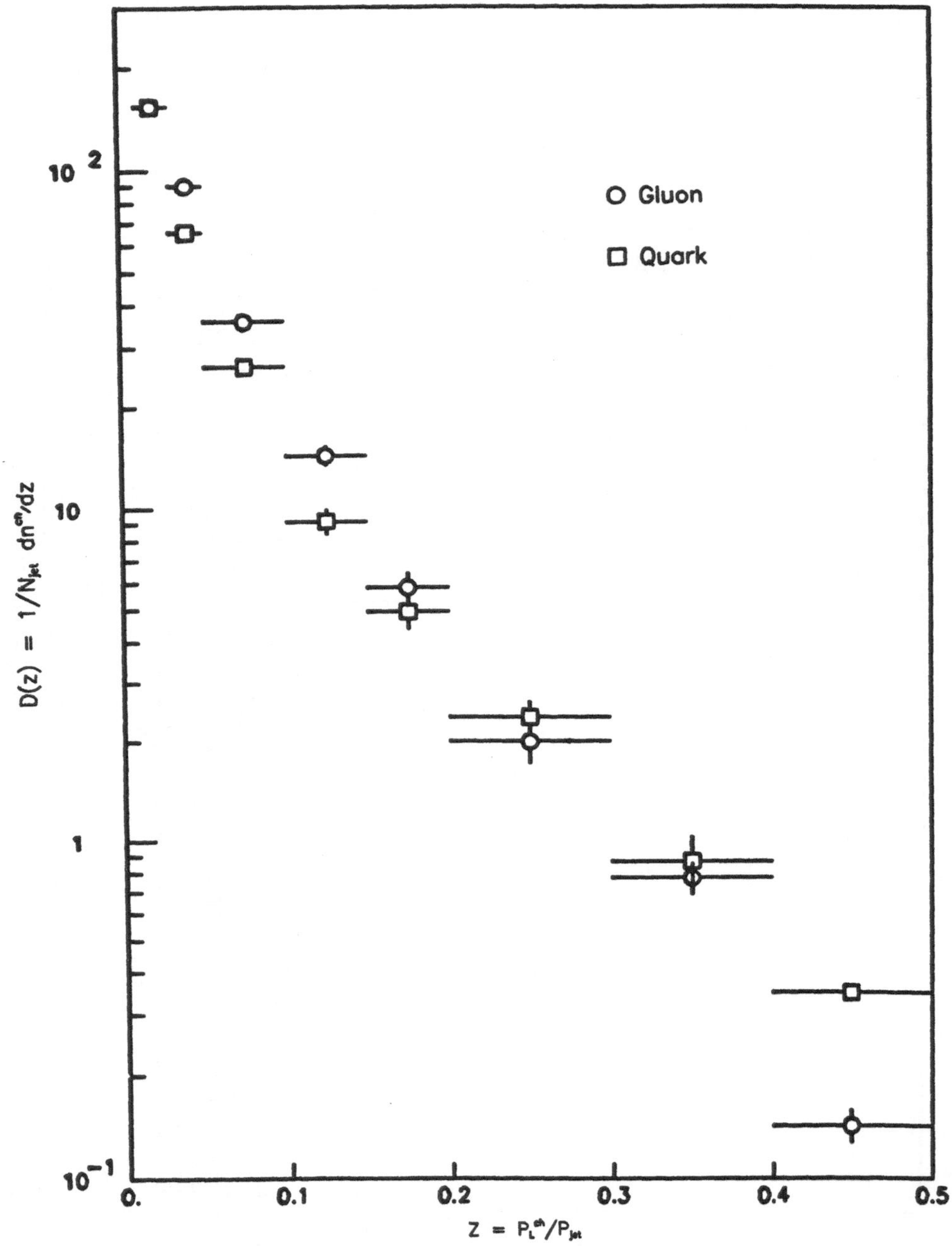

Fig. 31 : Comparison of the fragmentation functions of quarks and gluons.

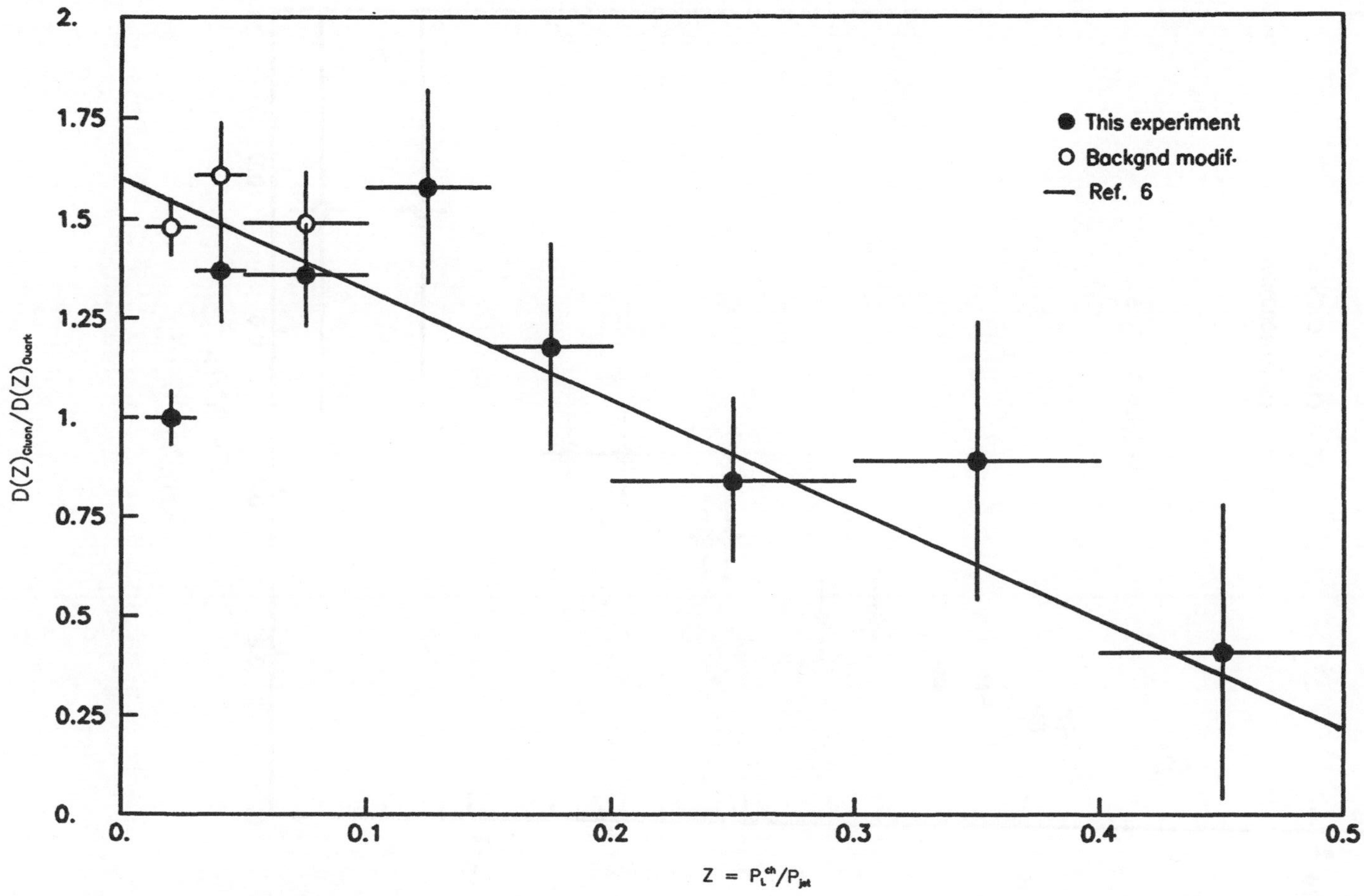

Fig. 32 : Ratio of fragmentation functions for gluons and quarks as a function of Z. "Pure" samples are defined as linear combinations of enriched samples shown on Fig. 30, using average probabilities <P(G)>.

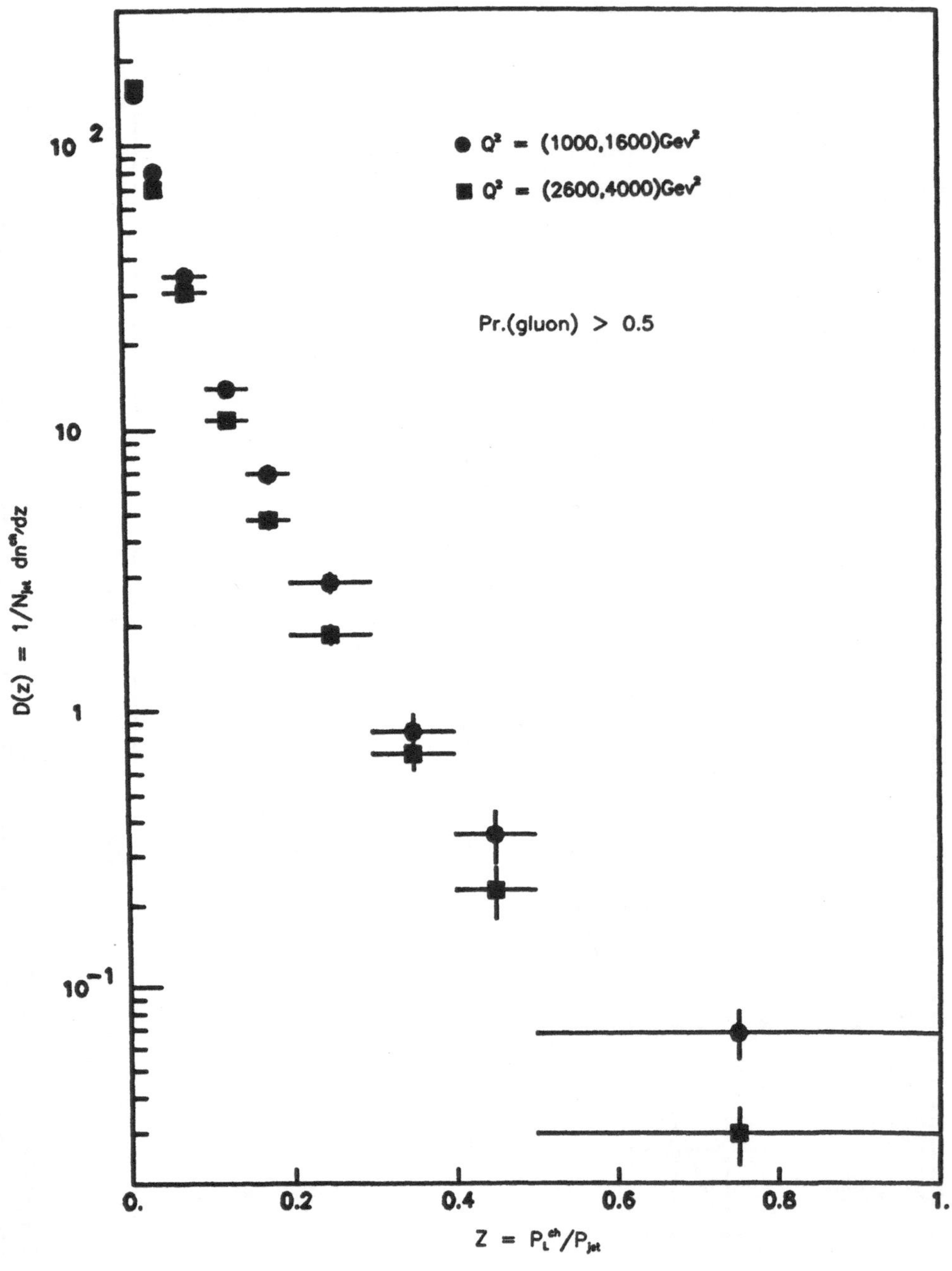

Q² = (1000,1600)Gev²
Q² = (2600,4000)Gev²
Pr.(gluon) > 0.5
D(z) = 1/N_jet dn^ch/dz
Z = P_L^ch/P_jet
10 2
10
1
10⁻¹
0.
0.2
0.4
0.6
0.8
1.

Fig 33.

REFERENCES

[1] G. Arnison et al. (UA1 Collaboration), Phys. Lett. 122B (1983) 103; 129B (1983) 273.
[2] G. Arnison et al. (UA1 Collaboration), CERN EP 85 (108) submitted to Nuovo Cimento Letters.
[3] F.E. Paige and S.D. Protopopescu, ISAJET program, BNL 29777 (1981).
[4] The EUROJET Monte Carlo program contains all first-order (α_s^2) and second-order (α_s^3) QCD processes [see B. van Eijk, UA1 Technical Note UA1/TN/84-93 (1984), unpublished, and A. Ali, E. Pietarinen and B. van Eijk, to be published]. The heavy quark is fragmented using the parametrization of C. Peterson et al., Phys. Rev. D27 (1983) 105 and the V-A decay matrix elements.
[5] M. Banner et al. (UA2 Collaboration), Phys. Lett. 122B (1983) 476. P. Bagnaia et al. (UA2 Collaboration), Phys. Lett. 129B (1983) 130; Z. Phys. C24 (1984) 1.
[6] G. Altarelli, R.K. Ellis, M. Greco and G. Martinelli, Nucl. Phys. B246 (1984) 12;
G. Altarelli, R.K. Ellis and G. Martinelli, Altarelli et al., Z. Phys. C : Particles and Fields 27 (1985) 617.
[7] M. Glück, E. Hoffmann and E. Reya, Z. Phys. C13 (1982) 119.
[8] E. Eichten et al., Rev. Mod. Phys. 56 (1984) 579.
[9] M. Jacob, Nuovo Cimento 9 (1958) 826.
[10] N.G. Deshpande et al., Phys. Rev. Lett. 54 (1985) 1757.
[11] F. Bergsma et al. (CHARM Collaboration), A precision measurement of the ratio of neutrino-induced neutral-current and charged-current total cross-sections. Contributed paper to the Int. Symp. on Lepton and Photon Interactions at High Energies, Kyoto, Japan, 1985 : preprint CERN-EP/85-113 (1985);
A. Blondel (CDHS Collaboration), Talk presented at the EPS International Europhysics Conference on High-Energy Physics, Bari, Italy, 18-24 July 1985.
W.J. Marciano and A. Stirlin, Phys. Rev. D29 (1984) 945.
[12] L. de Lella, rapporteur talk at the EPS International Europhysics conference Bari, Italy, 18-24 July 1985.
[13] W. Furmanski, H. Kowalski, Nucl. Phys. B224 (1983) 523.
[14] R. Odorico, Nucl. Phys. B244 (1984) 313. The program COJETS takes into account effects of initial state bremsstrahlung, this is not the case of ISAJET.
[15] G. Arnison et al. (UA1 Collaboration) Phys. Lett. 136B (1984) 294.
[16] B. Combridge et al., Phys. Lett. 70B (1977) 234.
[17] F. Berends et al., Phys. Lett. 103B (1981) 124.
[18] G. Arnison et al., Phys. Lett. 132B (1983) 223.
[19] K. Althoff et al. (TASSO Collaboration), DESY 83-130 (1983).
[20] T. Akesson et al. (AFS Collaboration) CERN preprint (1985).
[21] E. Eichten et al., Phys. Rev. Lett. 50 (1983) 811.
[22] S. Geer and W.J. Stirling, Phys. Lett. 152B (1985) 373.

JET PHYSICS AND STUDY OF $W^{\pm}$ AND Z° PROPERTIES IN THE UA2 EXPERIMENT AT THE CERN $\bar{p}p$ COLLIDER

R. Engelmann†

CERN and University of Heidelberg
(Presented for the UA2 Collaboration [1])

I. INTRODUCTION

This report summarizes the data presented in two lectures where more details on concepts were given. The report covers data taken by the UA2 collaboration at the CERN $\bar{p}p$ Collider during the 1982-1983 runs at $\sqrt{s}$ = 546 GeV and the 1984 run at $\sqrt{s}$ = 630 GeV with integrated luminosities of 142 nb^{-1} and 310 nb^{-1} respectively.

For the 1982-1983 data results on jet physics were reported in Ref. [2,3,4] and [5], measurements of W and Z properties were given in Ref. [6,7] and [8].

The UA2 apparatus is briefly described in Section II.

The results are presented in two parts, on jet physics in Section III and on W and Z bosons in Section IV. Some details on data taking and event analysis are given, then the measurements are discussed.

In Section III two jet dominance is recalled. Jet production is discussed in the frame of the parton model and quantum chromodynamics (QCD). The inclusive jet cross section, its s-dependence and the jet angular distribution are compared with calculations to lowest order in the strong coupling constant α_s. The parton structure function of the proton is measured. Higher order effects are discussed when presenting the two jet transverse momentum distribution and events with more than two jets. A search for structure in the two jet mass distribution is described. Some results on jet fragmentation are given.

In Section IV the production cross sections for W and Z bosons, their transverse momentum distributions and the charge asymmetry in the $W^{\pm} \rightarrow e^{\pm}\nu$ decay are compared to predictions from QCD and the standard model of the electroweak interaction. The W and Z masses are discussed in the frame of the standard model. An estimate of the width of the Z is given. The observed (quark + antiquark) structure function is compared to the prediction from neutrino collisions.

Finally in Section V conclusions are made from the UA2 results so far.

† On leave from the State University of New York at Stony Brook, N.Y.

II. APPARATUS

The UA2 apparatus has been described in detail in Ref. [9] and [10]. Here only a brief description will be given using Fig. 1.

At the centre of the UA2 apparatus is the vertex detector consisting of cylindrical proportional and drift chambers with measure particle trajectories in a region without magnetic field.

The vertex detector is surrounded by a highly segmented total -absorption calorimeter (central calorimeter). It consists of towers which point to the interaction point, subtend the intervals $\Delta\theta \times \Delta\phi = 10° \times 15°$ in polar angle and azimuth and cover the full azimuth over the range $40° < \theta < 140°$. Each tower is longitudinally segmented into a 17 radiation lenghts thick electromagnetic compartment (lead-scintillator) followed by two hadronic compartments (iron-scintillator) of ~ 2 absorption lenghts each.

The forward regions ($20° < \theta < 37.5°$ and $142.5° < \theta < 160°$) are each instrumented with twelve toroidal magnet sectors followed by drift chambers and electromagnetic calorimeters (forward calorimeter). These consist of lead-scintillator cells each covering 15° in ϕ and 3.5° in θ. Each cell is subdivided longitudinally in 2 sections 24 and 6 radiation lengths thick, the latter providing rejection against hadrons.

Both in the central and in the forward region a converter ($\simeq 1.5$ radiation lengths) followed by a proportional chamber in front of the calorimeter detects the early onset of electromagnetic showers from electrons and photons.

The energy resolution for single pions in the central calorimeter varies from $\sigma_E/E \simeq 32$ % at 1 GeV approximately proportional to $E^{-\frac{1}{2}}$ to $\simeq$ 11 % at 70 GeV [10]. In the forward region only a rough estimate of the hadronic energy from the electromagnetic calorimeter and the charged particle magnetic momentum measurement is obtained.

The stability of the calorimeters was monitored with periodic calibrations [10] over the last 4 years. The uncertainty on the absolute energy scale of the electromagnetic calorimeters is ± 1.5 % with a cell to cell variation of 2.5 % RMS. The corresponding values for the hadronic calorimeter are ± 6 % and 8 % respectively. The uncertainties on the absolute energy for the electromagnetic and hadronic calorimeter combine to a systematic error of ± 4 % on the jet energy scale.

The UA2 coverage is summarized in Fig. 2. In the central region for pseudorapidity $|\eta| < 1$ electrons (without their charge) and jets are measured, in the forward region ($1 < \eta \lesssim 2$) electrons with their charge. There only a rough estimate of the jet energy is possible. The very forward region ($\eta \gtrsim 2$) has no equipment. There is no muon detection in UA2.

Note that the coverage is full in azimuth.

III. JET PHYSICS

1. Data Taking and Jet Analysis

With the exception of Sections III.3.1 and III.3.2 all jet identification and energy measurements are done in the central calorimeter, with the tracking detector occasionally used to define the event vertex only.

Three triggers were used to select events with jets :

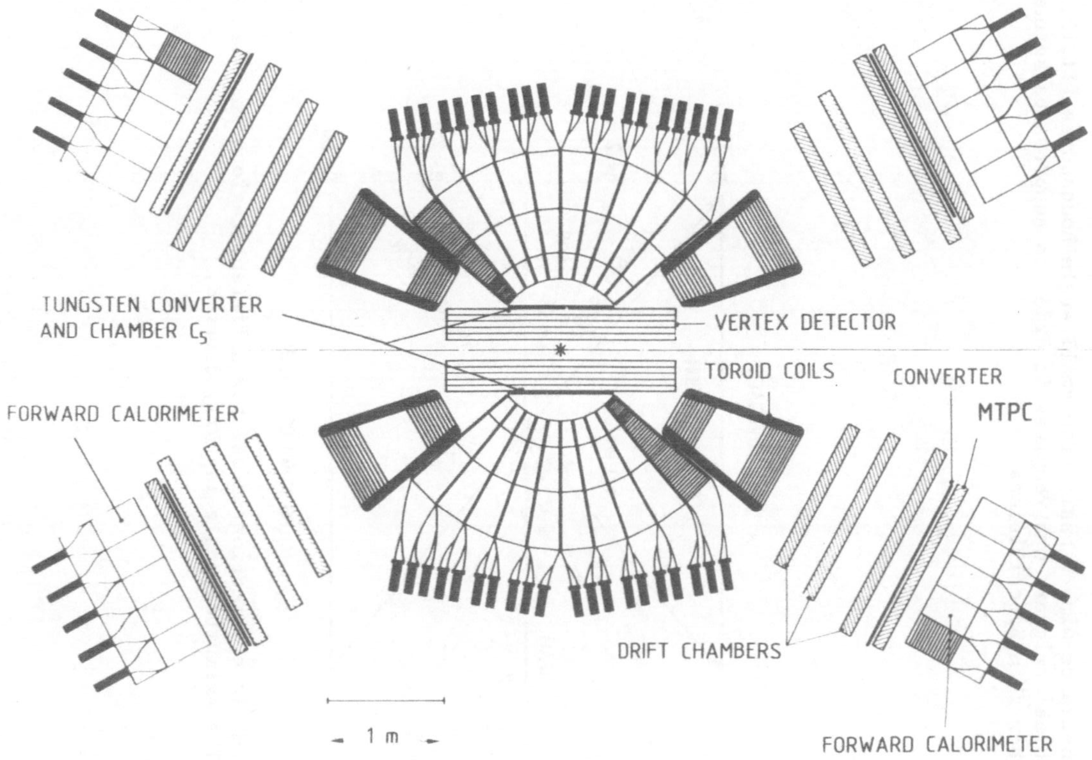

Fig. 1 : View of the UA2 detector, see text.

(1) The total transverse energy trigger required that the $\Sigma E_T = \Sigma E_i . \sin \theta_i$ exceeded a threshold (various settings, typically $\approx$ 60(30) GeV), where E_i is the energy deposited in the i-th tower, θ_i the polar angle of the tower center and the sum extends over all central calorimeter towers.
(2) The inclusive jet trigger demanded E_T in a single azimuthal wedge with a width of $\Delta\phi = 120°$ to exceed typically $\approx$ 35(20) GeV.
(3) The two jet trigger required that E_T^1 and E_T^2, deposited in any two opposite azimuthal wedges (ϕ_1 and $\phi_2 \cong \phi_1 + \pi$, $\Delta\phi_1 = \pm 30°$, $\Delta\phi_2 = \pm 60°$), both exceed typically $\approx$ 20 GeV.

A coincidence with signals from small angle hodoscopes [11,12,13] ('minimum bias' trigger) was required in order to suppress background from sources other than $\bar{p}p$ collisions.

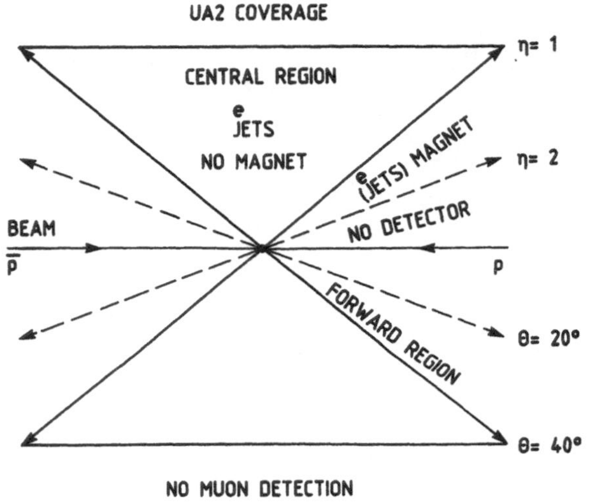

Fig. 2 : Particle detection in UA2 for different regions in polar angle. The azimuthal angle is fully covered.

The minimum bias trigger alone was used to measure the luminosity during data taking. The systematic error on the luminosity is estimated to ± 8 %.

The energy deposition pattern in the calorimeter is used to find hadron jets in the following way [2,3]. Clusters are first defined by joining cells which share a common side and have an energy > 400 MeV. The detector center is used as the event vertex to draw the cluster axis to the cluster centroid. The individual clusters are assumed to be massless [3]. Fig. 3 shows an example for two jet events. The profile of the calorimeter response in polar angle θ and azimuth ϕ demonstrates that the calorimeter cell size is well matched to the jet width.

The energy resolution for hadron jets is estimated from Monte Carlo simulations [3,13] and test beam data [10] to be 12 % at 40 GeV and 9 % at 100 GeV in agreement with a direct determination from the two-jet transverse momentum imbalance [4] (see Section III.2.5).

2. Jet Production

2.1. Two-jet dominance

Two-jet dominance of the final state in collisions with large ΣE_T will be briefly recalled.

Fig. 4 shows a ΣE_T distribution for the data as $\sqrt{s}$ = 546 GeV. The distribution falls off exponentially, then departs distinctly from this behaviour.

In order to investigate the jet character of the events the clusters are ordered according to their transverse energies ($E_T^1 > E_T^2 > E_T^3 \ldots$). In Fig. 5a) the average fractions $h_1 = E_T^1/\Sigma E_T$ and $h_2 = (E_T^1 + E_T^2)/\Sigma E_T$ are plotted versus ΣE_T. For large ΣE_T a substantial fraction is taken on the average by the two leading clusters (E_T^1, E_T^2). An event consisting of only two clusters with equal transverse energy would have $h_1 = 1/2$ and $h_2 = 1$. The fractions h_1 and h_2 rise quickly in the region where the ΣE_T distribution lifts off the steep fall off (Fig. 4), indicating a transition from soft to hard collision processes.

The two leading clusters are oriented back to back in azimuth as illustrated in Fig. 5b) which shows for ΣE_T >60 GeV and $E_T^{1,2}$ > 20 GeV the azimuthal separation between the two leafing clusters. A clear peak at $\Delta\phi_{12}$ = 180° is evident.

The clusters are well localized objects in the calorimeter illustrated in Fig. 6 where the transverse energy density for the leading cluster in events with $E_T^{1,2}$ > 15 GeV is plotted versus the azimuthal distance $\Delta\phi$ from the cluster center. More than 90 % of the energy is within $\Delta\phi$ = 30°.

The final state in collisions with large total transverse energy is dominated by the leading E_T jets. These collisions are described theoretically by the parton model and QCD : Partons (quarks and gluons) scatter and the final state partons materialize into hadronic jets. The parton density is derived from low energy neutrino interactions and the basic parton-parton scattering process is calculated by QCD. In what follows the UA2 experimental results will be discussed in the frame of this picture (referred to as 'QCD').

2.2 Inclusive jet cross section

Results for the inclusive jet cross section

$$\bar{p}p \to \text{jet} + X \qquad (1)$$

will be given for the 1983 data at $\sqrt{s}$ = 546 GeV and the 1984 data at $\sqrt{s}$ = 630 GeV taken with the UA2 apparatus.

The jets are restricted to the fiducial region $|\eta|$ < 0.85. In order to take partly account of final state gluon radiation the jet momentum

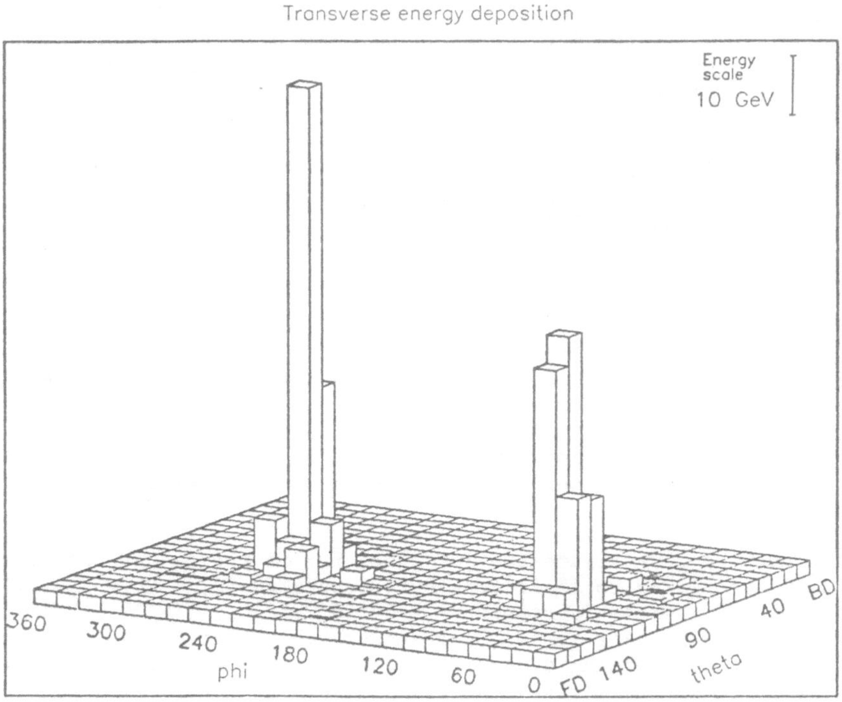

Fig. 3 : The central calorimeter energy profile in polar angle θ and azimuth ϕ for an event with two hadronic jets. The central calorimeter covers $0 < \phi < 360°$ and $40° < \theta < 140°$.

vector of the two leading jets is modified. One adds to it the momentum vectors of all clusters with $E_T > 3$ GeV within a cone around the leading jet with opening angle ω, $\cos\omega > 0.2$. The inclusion of these nearby clusters occurs in ≈ 30 % of the cases and increases the E_T distribution by ≈ 15 % roughly independent of E_T. The resulting jet acquires a mass with the inclusion of the small clusters and its transverse momentum p_T differs from E_T.

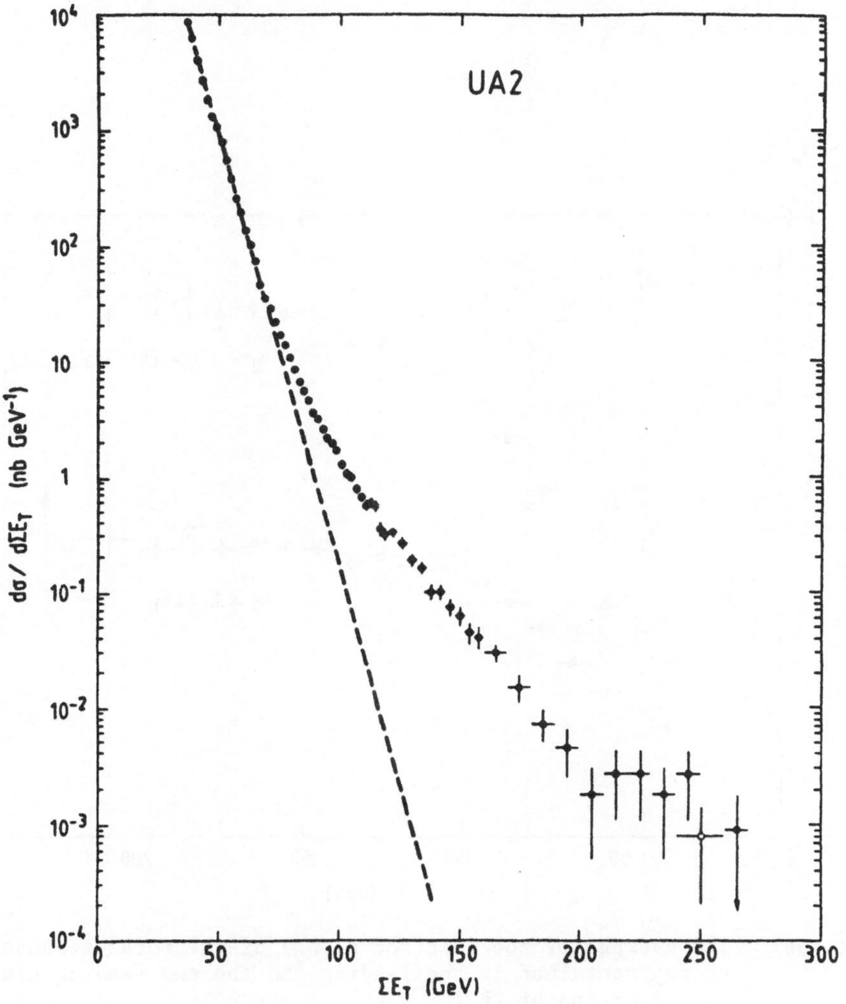

Fig. 4 : The distribution of the total transverse energy ΣE_T in the central calorimeter for $\sqrt{s}$ = 546 GeV.

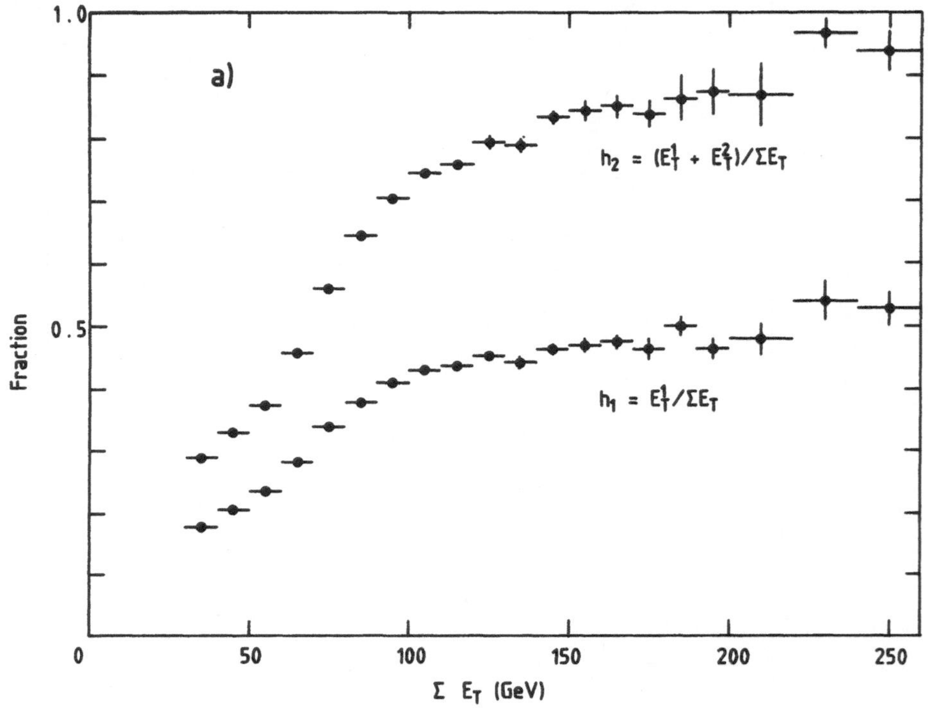

Fig. 5a) : The average of the fraction $h_1(h_2)$ of the total transverse energy contained in the leading and the two leading clusters as a function of ΣE_T.

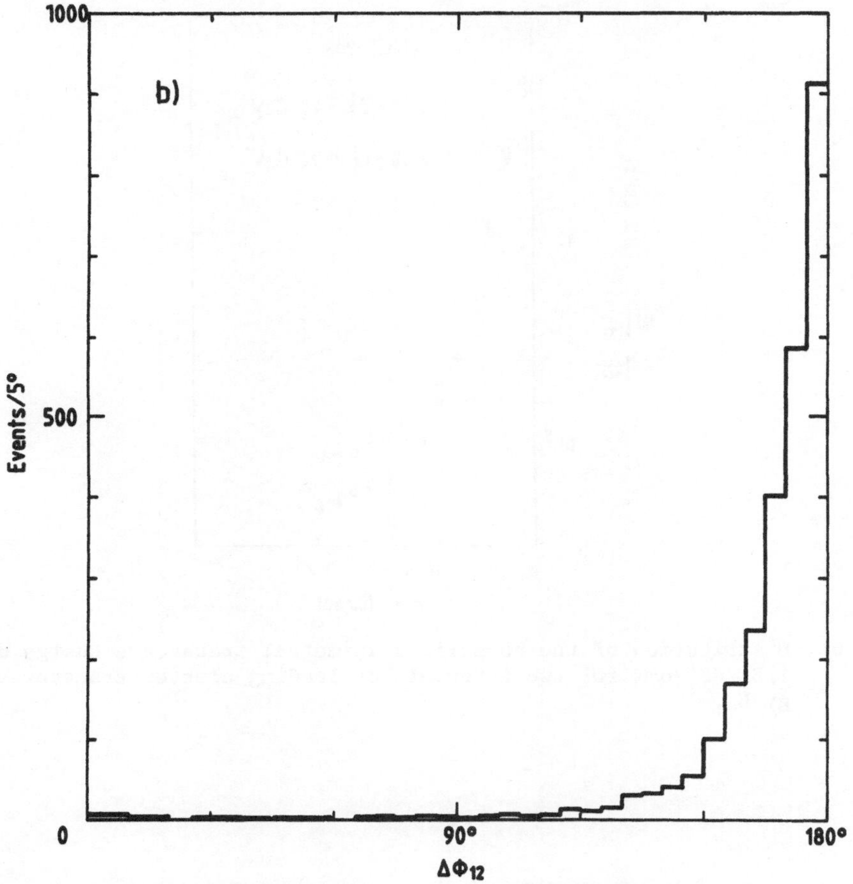

Fig. 5b) : Azimuthal separation between the two leading clusters in events with $\Sigma E_T > 60$ GeV and $E_T^{1,2} > 20$ GeV.

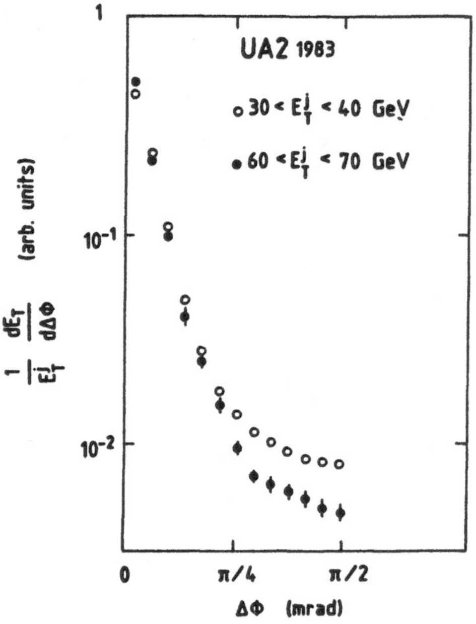

Fig. 6 : Distribution of the normalized azimuthal transverse energy density $1/E_T\ dE_T/d\Delta\phi$ for two intervals of leading cluster transverse energy E_T.

The detector acceptance is calculated with a Monte Carlo simulation [3,13] including the effects of the jet momentum definition and the smearing of the steeply falling p_T spectra with the experimental jet resolution.

An overall systematic error on the cross section is $\approx \pm 45$ % and due to effects from the acceptance, the calorimeter energy scale and the luminosity [13].

Fig. 7 gives the UA2 results [13] for the process in equation (1), $d^2\sigma/dp_T\,d\eta$ as a function of the jet transverse momentum p_T, averaged over $-0.85 < \eta < 0.85$, for the two center of mass energies, $\sqrt{s} = 546$ and 630 GeV. Also shown are results for pp collisions at the ISR at $\sqrt{s} = 63$ GeV [14]. The curves have been obtained from a leading order (in the strong coupling constant α_s) two-parton scattering QCD calculation [15] using $Q^2 = p_T^2$ of the jet and the structure functions of Ref. [16] with the scale $\Lambda = 0.2$ GeV.

The QCD calculation, an absolute calculation with the inputs denoted above, describes the cross section well over ≈ 8 decades in p_T including the dramatic rise with $\sqrt{s}$. This is quite remarkable. Actually an earlier calculation [17] before the data were taken at the $\bar{p}p$ Collider as well as at the ISR had similar success being about a factor 1.5 lower than the calculation presented here.

In the QCD calculation effects of higher order, contributions are not included which could result in a renormalization of the cross section (so called K-factor). Both the systematic uncertainty in the cross section of ± 45 % and in particular the uncertainties in the theoretical calculation [3] do not allow a determination of this K-factor to better than a factor of $\approx$ (2-3).

Since the QCD calculation in Fig. 7 follows the Collider data well it is useful to compare them with the Collider data on a finer scale looking at the s-dependence. Fig. 8 shows the cross section ratio as a function of p_T which is well described by the QCD calculation in the p_T range with sufficient statistics, including the rise with p_T which is due to a larger X-range being available at higher $\sqrt{s}$, for the production of a jet at fixed p_T.
The same consistency with the QCD calculation is found for the jet pair cross section in the reaction $\bar{p}p \to$ jet + jet + X using the two-jet mass distributions [13].

The $\sqrt{s}$ dependence has been traditionally confronted with the expectations [18] for the naive parton model with no scale breaking effects,

$$E\,\frac{d\sigma}{dp^3} = p_T^{-n}\, f(X_T) \qquad (2)$$

with $X = 2p_T/\sqrt{s}$ and $n = 4$. Fig. 9a shows to which extent this scaling is broken. In Fig. 9b, an exponent $n = 4.74$ is needed in order to bring the collider data at $\sqrt{s} = 546$ and 630 GeV together with the ISR data at $\sqrt{s} =$ 45 and 63 GeV.

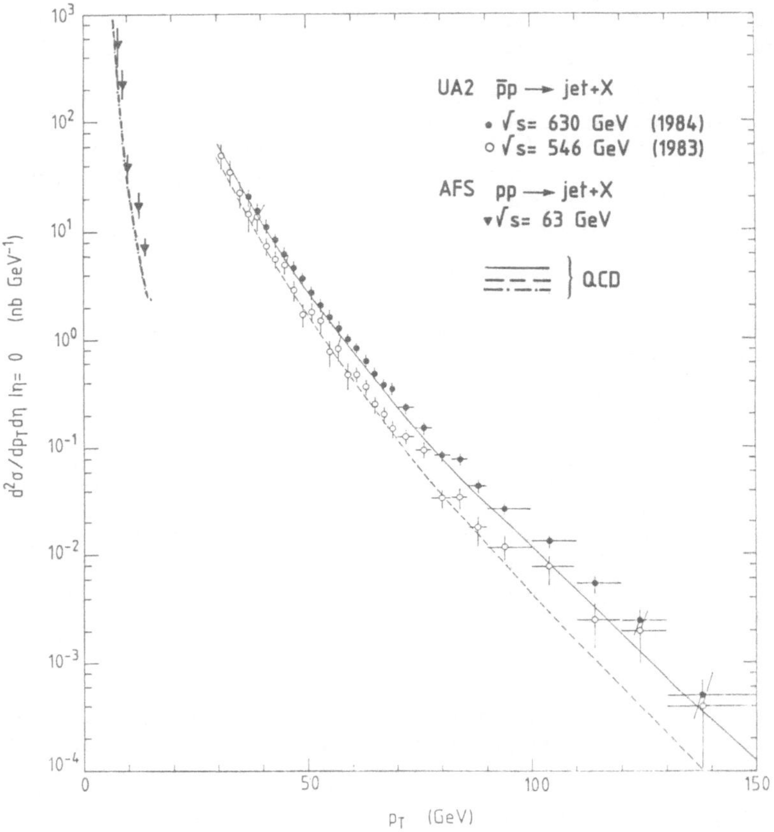

Fig. 7 : The inclusive jet cross section for $\bar{p}p$ collisions at $\sqrt{s}$ = 546 and 630 GeV and for pp collisions at $\sqrt{s}$ = 63 GeV. The curves are QCD calculations (see text).

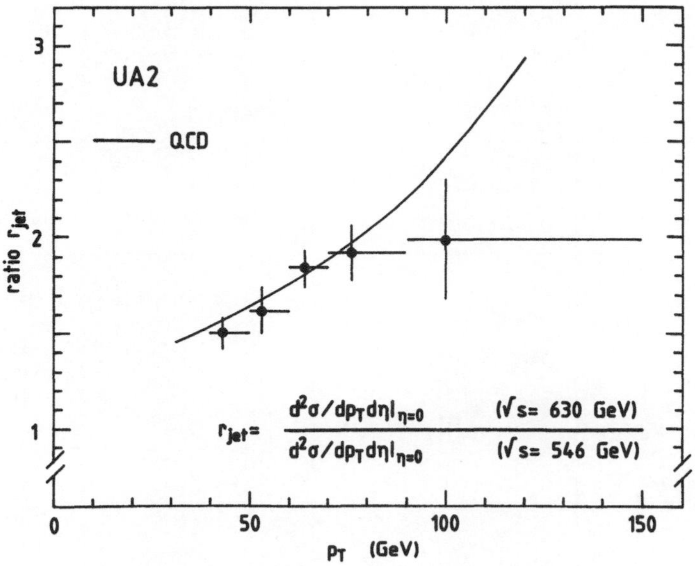

Fig. 8 : The cross section ratio for the two collider energies. For the QCD curve see the text.

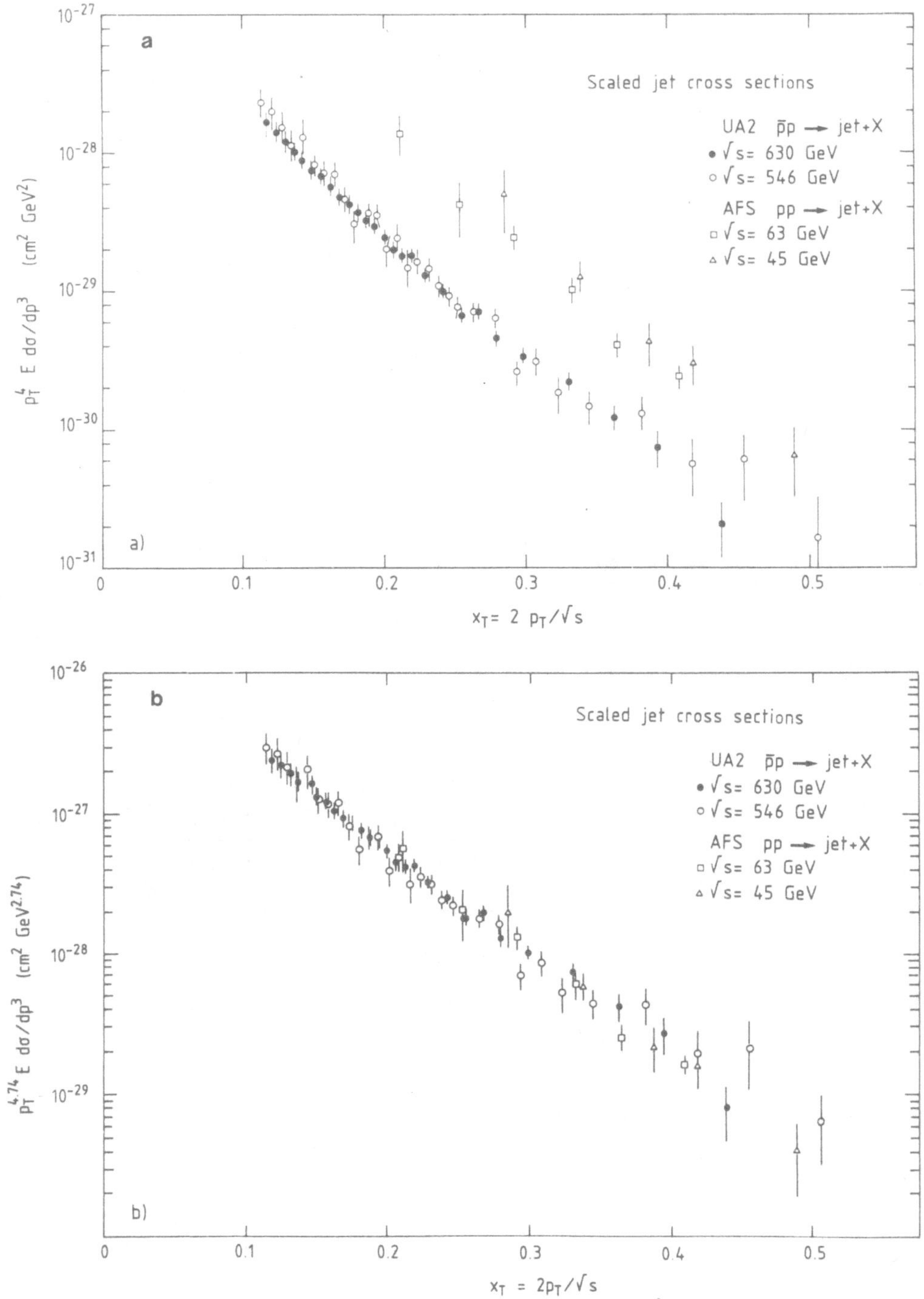

Fig. 9 : The scaled invariant cross section $p_T^{-n}.d\sigma/d\bar{p}^3$ plotted versus the scaling variable x_T for n = 4 a) and n = 4.74 b) using the UA2 data at $\sqrt{s}$ = 546 and 630 GeV for $\bar{p}p$ collisions and ISR data at $\sqrt{s}$ = 45 and 63 GeV for pp collisions.

2.3 Angular distribution

The basic parton-parton scattering process is investigated measuring the angular distribution in two-jet events.

The differential cross section is

$$\frac{d^3\sigma}{dX_1\, dX_2\, d(\cos\theta^*)} = \sum_{a,b} \frac{F_a(X_1)}{X_1} \frac{F_b(X_2)}{X_2} \sum_{c,d} \frac{d\sigma_{ab\to cd}}{d(\cos\theta^*)} \qquad (3)$$

where $F_{a,b}(X)/X$ are the structure functions describing the density of parton a,b. The sum extends over all parton-parton subprocesses ab → cd. The parton fractional momenta X_1, X_2 and the parton center of mass scattering angle θ^* are calculated from the jet p_T and the two-jet mass M_{jj} and longitudinal momentum. This is possible for events with small two-jet transverse momentum p_T^{jj} (see Section 2.5) according to the convention of Ref. [19].

Experimentally events with $p_T^{jj} < 20$ GeV/c and $M_{jj} > 45$ GeV are selected.

For the dominant subprocesses the QCD matrix elements give essentially the same $\cos\theta^*$ dependence [15] and exhibit for small θ^* the Rutherford singularity $\sin^{-4}(\theta^*/2)$ which is typical for vector gluon exchange. Fig. 10 shows the $\cos\theta^*$ distribution for the 1983 data at $\sqrt{s} = 546$ GeV. The data (normalized at $\theta = 90°$) lie in the center of a narrow band spanned by the dominant parton subprocesses (dashed curves). The average of the subprocesses using structure functions (solid curve) describes the data well. For $\cos\theta^* \geqslant 0.3$ the data follow well the $\sin^{-4}(\theta^*/2)$ behaviour. Scalar gluon exchange theories, their range indicated by the hatched area, are excluded by the data.

2.4 Parton structure function of the proton

The $\cos\theta^*$ dependence of the dominant subprocesses being very similar an average $d\sigma/d\cos\theta^*$ can be factorized in front of the sums in Equation (3).

One further replaces the sum over parton types (quarks and gluons) by one effective structure function F(X). One tests this factorization,

$$X_1 X_2 \frac{d^2\sigma}{dX_1\, dX_2} \Big/ \int \left(\frac{d\sigma}{d\cos\theta^*}\right) d\cos\theta^* = F(X_1)F(X_2), \qquad (4)$$

where $d\sigma/d\cos\theta^*$, essentially the average $\cos\theta^*$ dependence (solid curve in Fig. 10), is integrated over the acceptance. The factorization is verified by the data [4].

Fig. 11 shows the measured structure function compared to an effective structure function (dashed line) $F(X) = G(X) + 4/9\,[Q(X) + \bar{Q}(X)]$, where $G(X)$, $Q(X)$, $\bar{Q}(X)$ are the gluon, quark and antiquark structure functions with the relative strength 4/9 of the quark-gluon and gluon-gluon coupling. These are derived from neutrino scattering [20] evolved from $Q^2 = 20$ GeV to $Q^2 = 2000$ GeV^2 via QCD. One actually determines $F(X)\cdot\sqrt{K}$ with $K > 1$ (see the K-factor in Section III.2.2) which could be responsible for the overall excess of the data over the prediction from the neutrino interactions.

The dashed-dotted curve in Fig. 11 gives only the effect of the quark and antiquark contribution from the neutrino data. The clear excess of the UA2 data at low X demonstrates the dominant gluon part in this region and illustrates the fact that the jet yield at the $\bar{p}p$ Collider at low two-jet mass is dominated by gluon contribution.

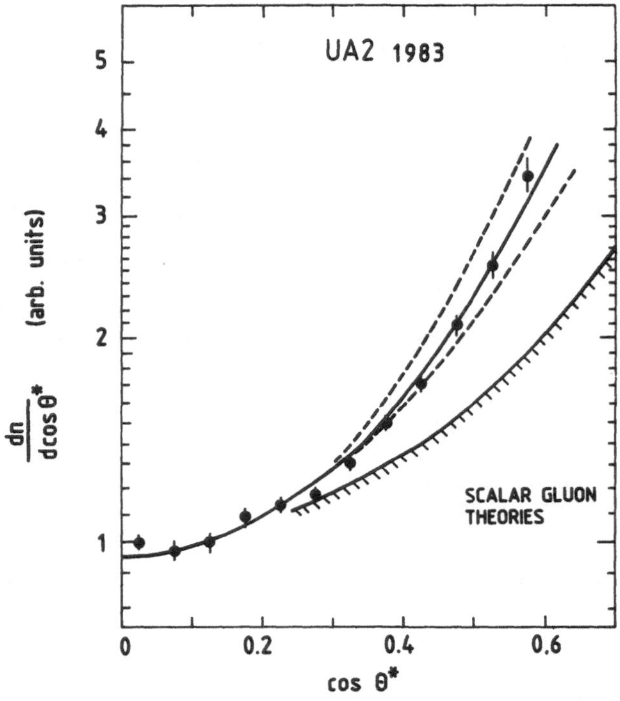

Fig. 10 : Angular distribution for two jet events for the 1983 data at $\sqrt{s}$ = 546 GeV. The dashed lines give the range for the OCD predictions for the dominant subprocesses. The solid curve is an average of the subprocesses using structure functions. Scalar gluon theories lie in the hached area. The UA2 acceptance is cos θ < 0.6.

2.5 Two-jet transverse momentum

So far the data were compared to QCD predictions to lowest order in α_s. Higher order effects via gluon radiation will be discussed in this section and in Section III.2.6.

The two-jet system acquires a transverse momentum due to a small "primordial" transverse momentum of the incident partons and, more importantly, due to initial and final state gluon radiation of the partons.

Soft gluon radiation, which makes a big contribution to the two-jet transverse momentum, is not calculable in perturbative QCD but summations

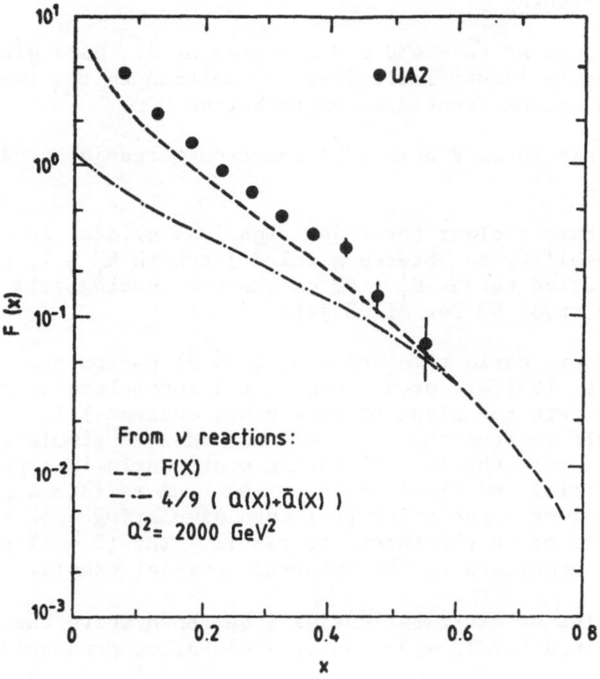

Fig. 11 : The effective structure function. The dashed line is calculated from neutrino scattering evolved via QCD to high Q^2. The dashed-dotted line includes only the quark contribution.

in the double leading log approximation are successfully done [21].

By adding the small jets to the leading jets as described in Section III.2.2 and restricting the two-jet transverse momentum p_T^{jj} to $p_T^{jj} < 20$ GeV/c the comparison to the soft gluon radiation calculation is meaningful.

Experimentally [4] one splits p_T^{jj} into two components P_ξ and P_η along the external two-jet bisector (roughly the two-jet axis) and perpendicular to it. Whereas P_ξ is strongly influenced by the jet energy resolution the effect on P_η is small. (One checks in a study of the P_ξ and P_η distributions the jet energy resolution as obtained from a Monte Carlo simulation and test beam data and finds agreement as mentioned in Section III.1.) The component P_η is used for the comparison with the QCD calculation.

Fig. 12 shows for the 1983 data at $\sqrt{s} = 546$ GeV the distribution in P_η for two-jet events with $E_T^1 + E_T^2 > 40$ GeV, azimuthal separation of the two jets $\Delta\phi > 120°$ and $E_T^{1,2} \geq 10$ GeV. The data are in agreement with the QCD prediction of Ref. [21] (dashed line and histogram after including detector effects). The dashed-dotted line gives the QCD prediction assuming the quark radiates like a gluon in clear contradiction to the data.

2.6 Multi-jet events

The 1984 data at $\sqrt{s} = 630$ GeV are presented. Hard gluon radiation results in clearly identifiable jets in addition to the two leading jets. This radiation can be treated in perturbative QCD.

The analysis checked here [22] compares three-jet production to a QCD prediction.

The data have a clear three-jet signal as evident in Fig. 13 which shows the probability to observe a third jet with $E_T^3 > E_T$ for $E_T = 10$, 15 and 20 GeV, plotted versus $E_T^1 + E_T^2$ of the two leading jets. In what follows $E_T > 15$ GeV is required for all 3 jets.

For the Monte Carlo simulation of $(2 \to 3)$ parton processes the matrix elements of Ref. [23] are used. Due to the incomplete acceptance (jets can be lost or two jets too close to each other coalesce) $(2 \to n)$ hard parton processes result in less than n jets. In order to simulate this effect for the three-jet sample the $(2 \to 3)$ parton Monte Carlo is appended with a simple model where initial and final state partons can radiate a gluon according to a bremsstrahlung probability [22] thus simulating a $(2 \to 4)$ parton process and its effect on the three-jet sample. The $(2 \to 4)$ parton Monte Carlo simulation is normalized to the observed four-jet sample.

Two examples of distributions will be shown where the Monte Carlo simulation, combined for $(2 \to 3)$ and $(2 \to 4)$ parton processes, is normalized to the data.

Fig. 14 shows the distribution in $\cos\omega_{23}$ where ω_{23} is the angle between the second (E_T^2) and third (E_T^3) jet. The Monte Carlo describes the observed shape well including the edge of the acceptance. In Fig. 15 the distribution of p_{out} is presented with p_{out} being the total energy transverse to the scattering plane, defined by the beam direction and the leading jet momentum.

$$P_{out} = \frac{1}{2} \sum_{j=2,3} \left| \vec{p}_j \cdot \frac{(\vec{p}_1 \times \vec{b})}{|\vec{p}_1 \times \vec{b}|} \right| \qquad (5)$$

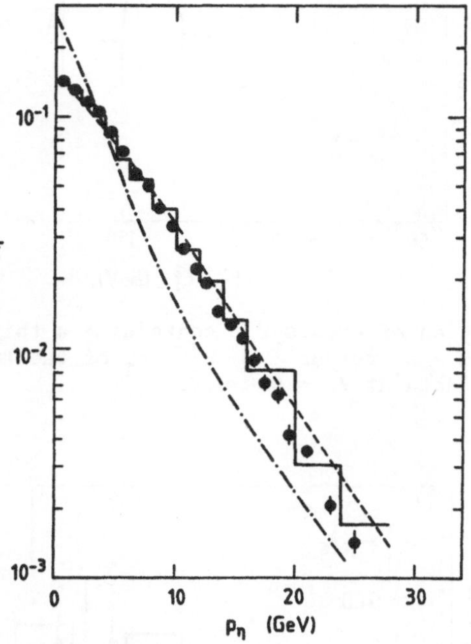

Fig. 12 : Component P_η of the two-jet transverse momentum (see text). The dashed line is a QCD prediction [21]. The histogram is the prediction with the detector effects included. The dashed-dotted line is the same QCD prediction with the assumption that gluons radiate like quarks.

where $\vec{p}_1$ is the momentum of the leading jet (E_T^1) and $\vec{b}$ a vector in the proton beam direction. For a two-jet event with no two-jet transverse momentum and ideal detector resolution $P_{out} \equiv 0$. The P_{out} distribution shape is well reproduced by the Monte Carlo simulation including the edge of the acceptance.

The QCD prediction agrees well in shape with the data.

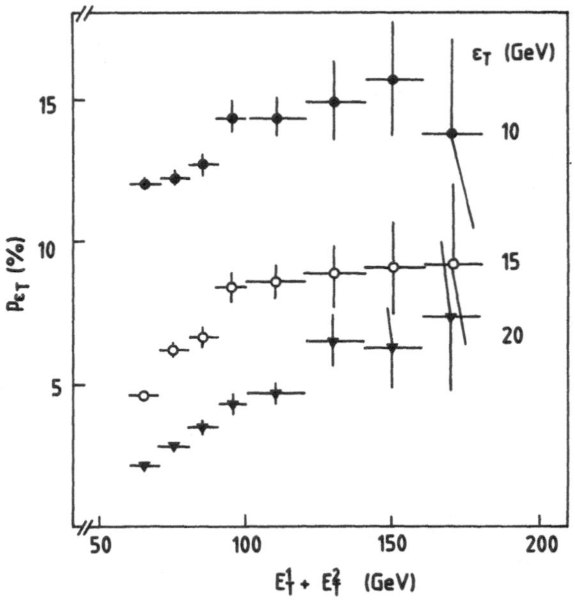

Fig. 13 : The fraction of events $P_{\varepsilon T}$ containing a third jet with transverse energy $E_T^3 > \varepsilon_T$ versus $E_T^1 + E_T^2$ of the two leading jets for the 1984 data at $\sqrt{s}$ = 630 GeV.

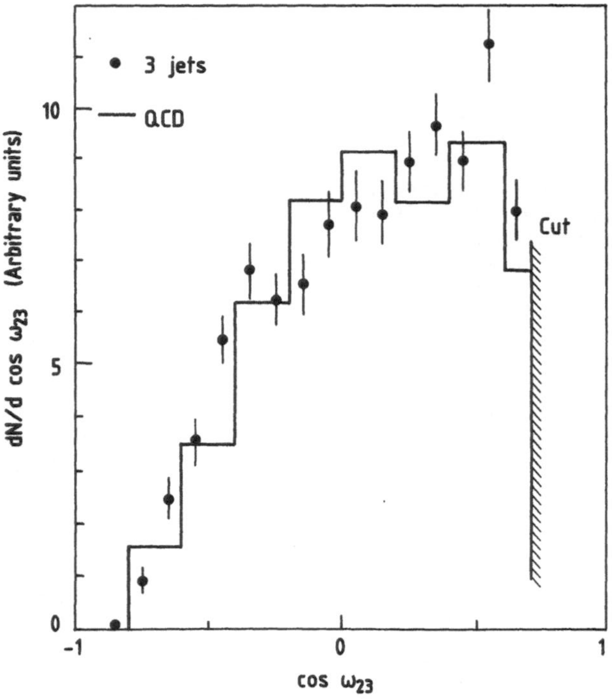

Fig. 14 : The angle ω_{23} between the second (E_T^2) and the third (E_T^3) jet for three-jet events with $E_T^{1,2,3}$ > 15 GeV. The histogram is a QCD prediction described in the text and normalized to the data.

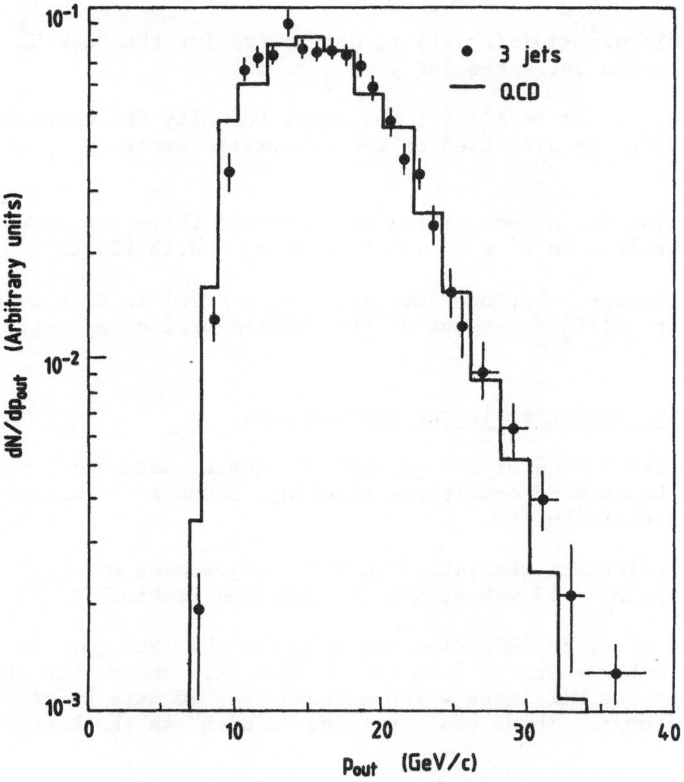

Fig. 15 : The P_{out} variable (see equation 5) for three jet events with a QCD prediction (histogram).

For a study of the absolute rate, the ratio of the observed topological cross sections $\sigma(3jet)/\sigma(2jet)$ is compared to the corresponding ratio from the QCD Monte Carlo calculation,

$$\frac{\sigma(3\ \text{jet})}{\sigma(2\ \text{jet})} = \frac{\alpha_S}{\alpha_S^{MC}} \cdot \frac{\sigma_3^{MC}}{\sigma_2^{MC}} \cdot \frac{K_3'}{K_2'} \tag{6}$$

Here σ_n^{MC} is the leading order Monte Carlo prediction for the $(2 \to n)$ parton process (Ref. [15] for n = 2 and Ref. [23] for n = 3) augmented by the next higher leading order QCD prediction for the $(2 \to (n+1))$ parton process feeding into the n-jet sample. The K-factor K_n' contains all contributions not contained in σ_n^{MC} and α_S^{MC} is the strong coupling constant used in the Monte Carlo calculation.

The ratio $\sigma(3\,jet)/\sigma(2\ jet)$ is determined for the cuts $E_T^3 > 15$, and for the angle between jet i and jet j, $\omega_{ij} > 45°$.

Assuming the ratio K_3'/K_2' to be equal to unity the resulting value for α_S is 0.2 where the estimates of the systematic error on α_S are detailed in Ref. [22].

This value for α_S describing the observed three-jet rate is close to the value obtained in $e^+ e^-$ collisions of $\alpha_S = 0.15$ [24].

Note, however, that one determines $\alpha_S \cdot K_3'/K_2'$ in this method (and not α_S alone) with K_3'/K_2' dependent on the experimental cuts quoted above !

2.7 Search for structure in the two-jet mass

The decays $W \to q\bar{q}$ or $Z \to q\bar{q}$ with the quarks materializing into hadronic jets would, background conditions allowing, cause an enhancement in the two-jet mass distribution.

The experimental resolution in the two-jet mass m is $\sigma_m = 10$ GeV at m = 80 GeV and $\sigma_m = 13$ GeV at m = 150 GeV (see Section III.1).

Fig. 16 shows the distribution in m for the 1983 data at $\sqrt{s} = 546$ GeV. There is no enhancement visible in the W(Z) mass region. In the $\pm\, 2\sigma_m$ interval around the W(Z) mass which extends over 10 data points one expects ≈ 150 W(Z) decays. These can easily be "hidden" in the big two-jet background.

A considerable improvement in jet energy resolution is needed for the detection of a $W \to q\bar{q}$ signal in the two-jet mass.

There appears a slight enhancement toward the end of the plot at m = 150 GeV, consistent in width with the resolution $\sigma_m = 13$ GeV. The combined 1983 and 1984 data, however, do not show this structure around m = 150 GeV as displayed in Fig. 17.

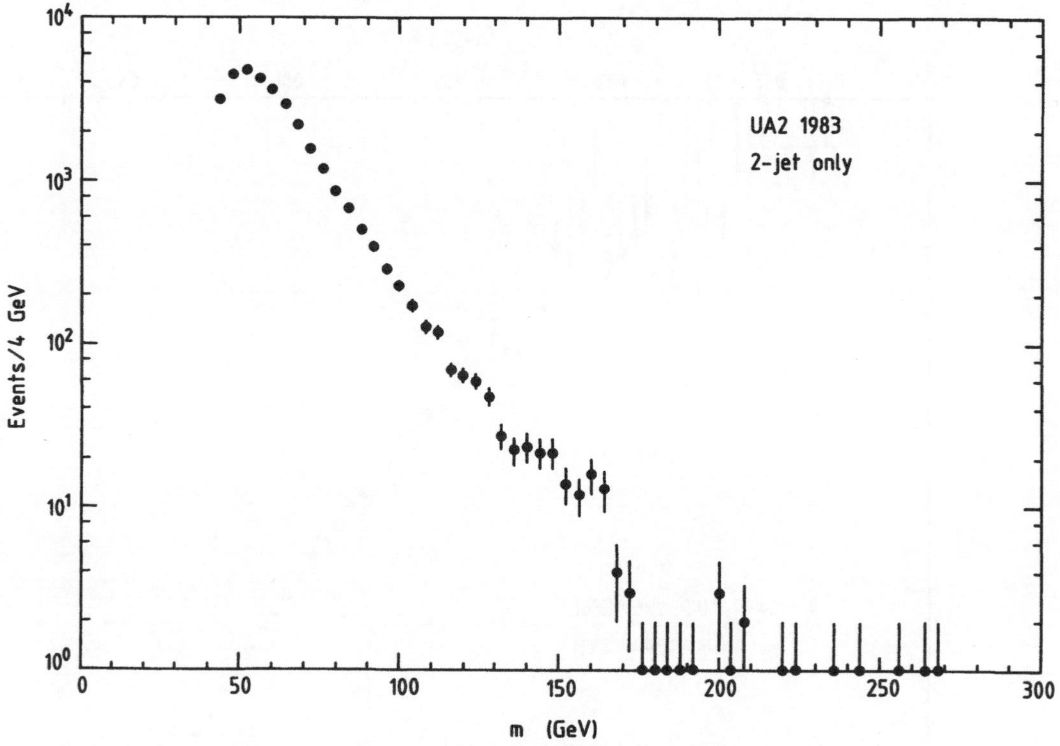

Fig. 16 : The two-jet mass distribution for the 1983 data.

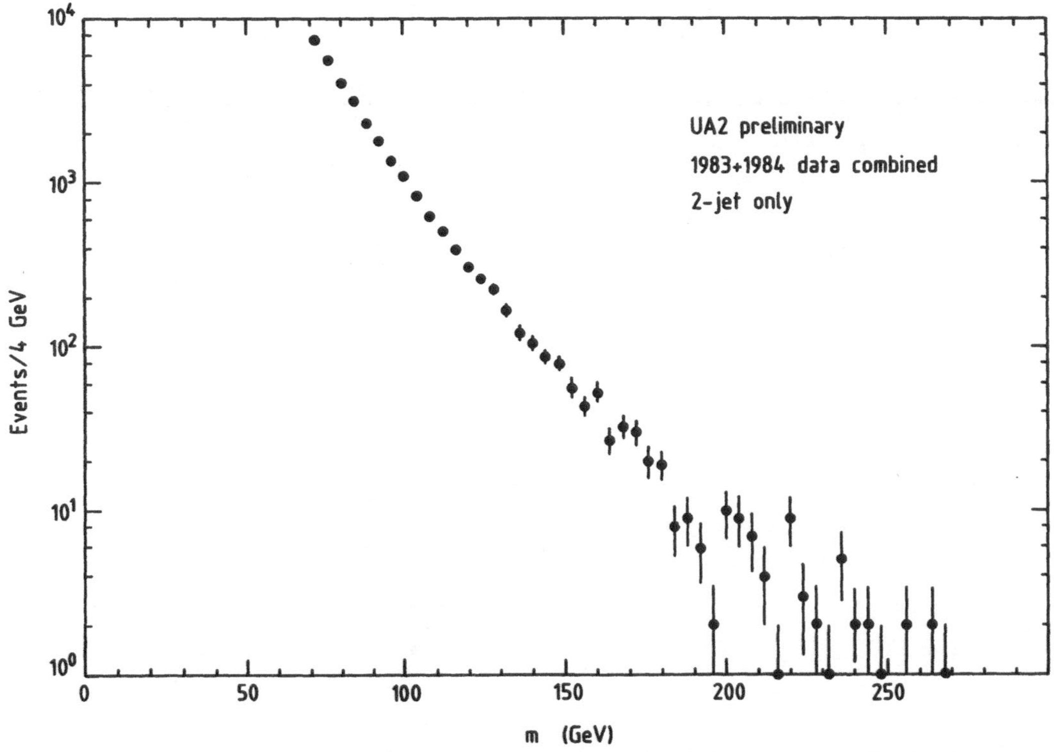

Fig. 17 : The two-jet mass distribution for the 1983 and 1984 data combined.

3. Jet Fragmentation

Two measurements are reported, one on the charged particle multiplicity in jets using the tracking detector and the calorimeter in the central region (no magnetic field), the other on jet fragmentation into leading π° using the forward electromagnetic calorimeter and tracking (with magnetic field).

3.1 Charged-particle multiplicity in jets

Fig. 18 gives a typical two-jet event viewed along the beam in the UA2 central detector. The height of the trapezoids are proportional to the transverse energy in the calorimeter for the electromagnetic (open area) and hadronic (shaded area) part. The jets are accompanied by bundles of tracks with unknown momenta (no magnetic field). The tracks outside the jet regions result from the fragmentation of the partons not participating in the hard collision, the "spectators".

The angular width distribution of such jets has been discussed in Section III.2.1 and Fig. 6.

For the measurement of the average charged-particle multiplicity, $\langle n_{ch} \rangle$, the background of the spectator particles is substracted and a jet core multiplicity is defined for the 1983 data at $\sqrt{s}$ = 546 GeV. Fig. 19a) shows the distribution $dN/d\phi$, where ϕ is the azimuthal separation between the charged particle and the axis of the leading jet (E_T^1), for two intervals of the two-jet mass m_{jj} [5]. The background distribution, normalized at $\phi = 90°$, is substracted before integrating over $dN/d\phi$. (One notes that this background level is larger than $dN/d\phi$ for ordinary soft collisions, the minimum bias events mentioned in Section III.1, as indicated by the dashed-dotted line.)

In Fig. 19b) the resulting jet core multiplicity $\langle n_{ch} \rangle$ is plotted versus the two-jet mass m_{jj} and compared to data from e^+e^- collisions [25] treated in the same way as the UA2 data. The curves are the result of a fragmentation model based on QCD [26]. The lower curve describes quark-jet fragmentation and follows the $e^+ e^-$ data well. The upper two curves give the range expected for gluon-jet fragmentation. The UA2 data, close to the higher multiplicity for gluon jets at lower m_{jj} turn at higher m_{jj} toward the lower multiplicity for quark jets. At the $\bar{p}p$ Collider energy one expects the fraction of gluon jets produced to vary from ≈ 3/4 to ≈ 1/3 as m_{jj} rises from 40 to 140 GeV, thus explaining the observed behaviour of $\langle n_{ch} \rangle$ as a function of m_{jj}.

3.2 Jet fragmentation into leading π°

The fragmentation of a high p_T jet into a leading pion is best measured with neutral pions, since their energy is well measured in the electromagnetic calorimeter and the pion p_T spectrum falls off steeply.

The UA2 forward detector with its small granularity in angle (see Section II) and its magnetic field simplifies the task of isolating the leading π° from the accompanying fragments of the jet. In order to measure an inclusive π° cross section one corrects for the effects of the isolation criteria. In the case of the measurement reported here these corrections are small, less than ≈ 40 % [27].

The π° (or γ) is identified in a fashion similar to the electron as described in Section IV.1, with the requirement that no track point to the electromagnetic shower. Fig. 20 shows the inclusive π° cross section for the 1983 data at $\sqrt{s}$ = 546 GeV (measured with the forward detector for pseu-

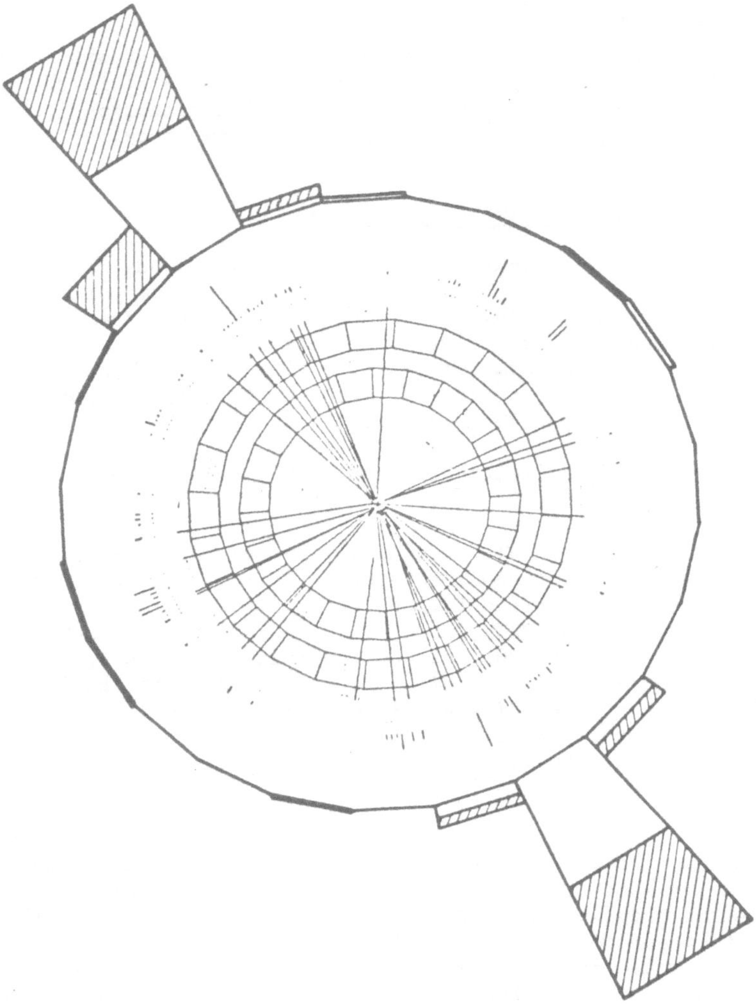

Fig. 18 : A two-jet event in the UA2 central detector viewed along the beam. The height of the trapezoids is proportional to the electromagnetic (open area) and hadronic (shaded area) calorimeter transverse energy.

dorapidity $1 < |\eta| < 1.8$) compared to the inclusive jet cross section (measured with the central detector for $|\eta| < 1$). The π°/jet ratio in the region of overlap in E_T is $\approx 10^{-3}$. The curves are QCD calculations using structure and fragmentation functions from Ref. [28]. The spectra are well described by this calculation. Note that direct photons which could be about equal to the π° rate at $P_T = 40$ GeV/C [17] are contained in the data sample.

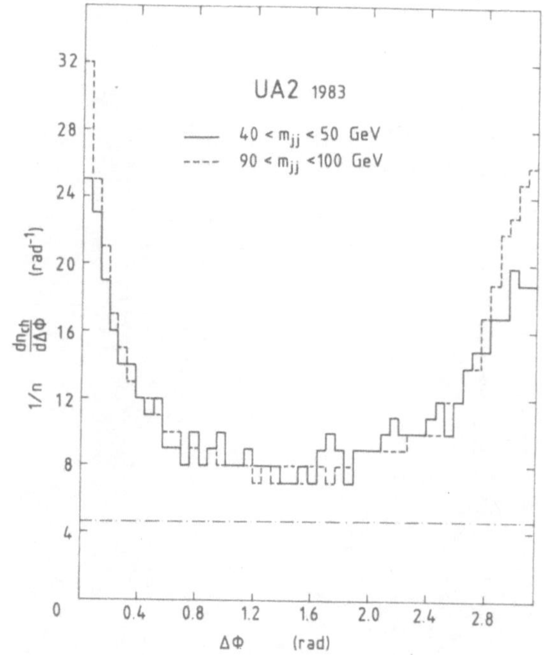

Fig. 19a) : Azimuthal separation ϕ between charged particles and the axis of the leading jet for two bins in the two-jet mass m_{jj}. The dashed-dotted curve indicates the density $dN/d\phi$ for minimum bias events.

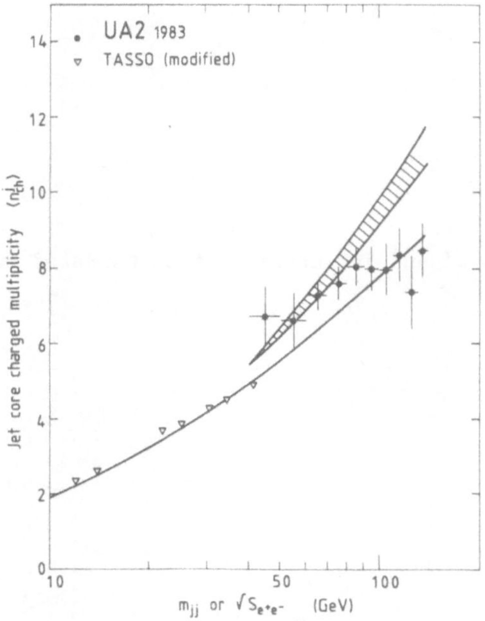

Fig. 19b) : The charged particle multiplicity in jets for the UA2 data and $e^+ e^-$ data [25]. The curves are the result of a QCD fragmentation model [26].

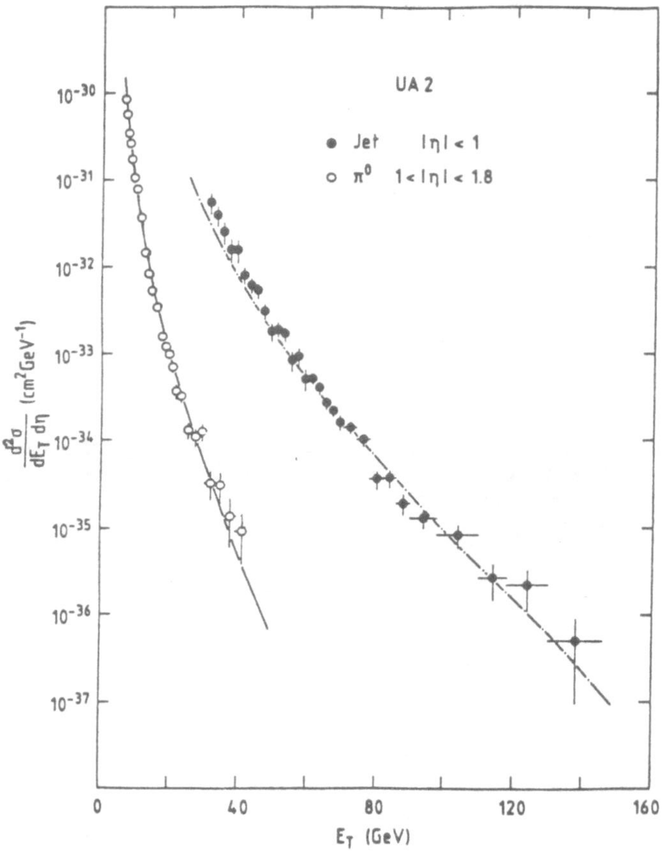

Fig. 20 : Inclusive π° and jet cross section compared to a QCD calculation [28].

IV. STUDY OF $W^{\pm}$ AND Z° PROPERTIES

1. Data Taking and Electron Analysis

The combined 1983 and 1984 data are described [29].

Two triggers were used to select events containing $W^{\pm} \rightarrow e^{\pm}\nu$ or $Z^{\circ} \rightarrow e^{+} e^{-}$ decays :

(1) The $W^{\pm}$ trigger required a cluster of transverse energy deposition $E_T > 8$ GeV in any 2 x 2 matrix of adjacent cells of the electromagnetic calorimeters.
(2) The Z° trigger required the simultaneous presence of two such clusters above a threshold of $E_T > 4.5$ GeV, separated in azimuth by at least 60°.

A coincidence with the minimum bias trigger (see Section III.1) was required.

The events considered in the subsequent analysis must fulfill a set of selection criteria as illustrated in Fig. 21 which shows a schematic representation of the signature of various particles or systems of particles in the UA2 apparatus. Shown in the figure is the transverse cross section of a quadrant of the central detector. The forward detectors present similar features with the additional measurement of the track momentum.

There are distinctive characteristics of different particles in each component of the UA2 detector :

a) in the calorimeter, the small transverse and longitudinal extension of the electromagnetic shower distinguishes electrons from particles undergoing nuclear interactions in the calorimeter material (isolated charged hadrons or hadron jets).
b) the presence of a track in the vertex chambers allows to recognize an electron from a photon (or a π°) which showers electromagnetically in the calorimeter.

The dominant two sources of background to high p_T electrons result from fragmentation of jets into a leading π° giving a narrow $e^{+} e^{-}$ pair and the geometrical overlap of an energetic π° with a soft charged hadron. The π° produces an electromagnetic shower in the calorimeter, while the charged hadron contributes a track and may fake an electron. This latter background is reduced by the high resolution preshower counter (converter and MWPC), which allows the precise localisation of the showering particle.

Therefore, the identification of high p_T electrons is based on the following main criteria :

(1) the presence of a localised cluster of energy depositions in the first compartment of the calorimeters, with at most a small energy leakage in the hadronic compartment;
(2) the presence of a reconstructed charged particle track which points to the energy cluster. The pattern of energy deposition must agree with that

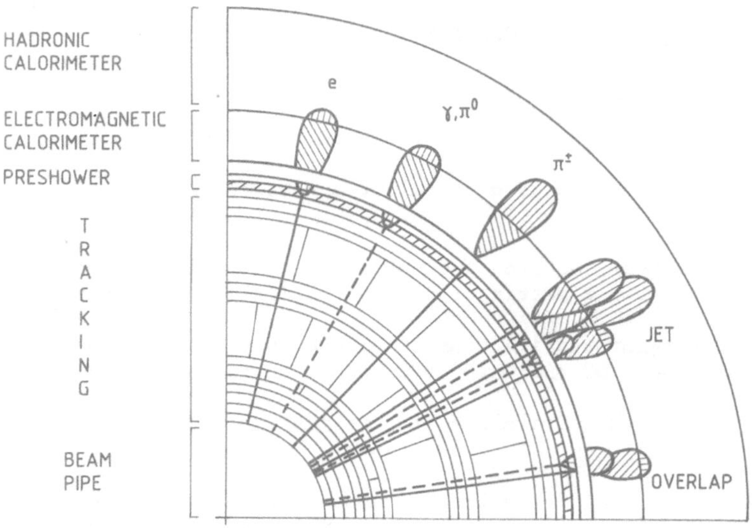

Fig. 21 : Schematic representation of particle signatures in a quadrant of the UA2 central detector.

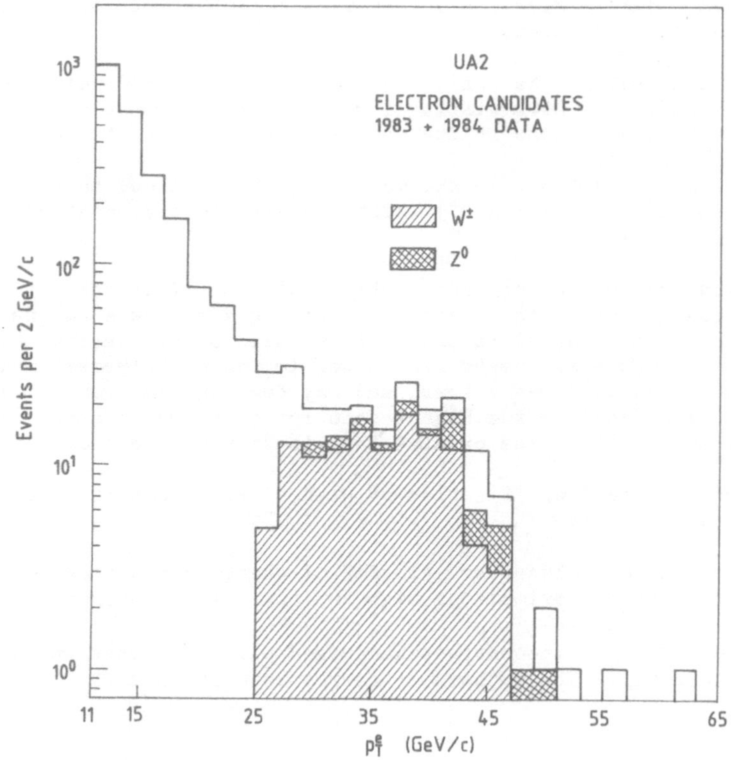

Fig. 22 : The p_T spectrum of electron candidates for the combined 1983 and 1984 data. Also shown are the electrons identified from $W \to e\nu$ (hashed area) and from $Z \to ee$ (cross hashed area) in the following analysis.

expected from an isolated electron incident along the track direction; (3) the presence of a signal in the preshower counter, the amplitude of which must be larger than that of a minimum ionising particle. The geometrical matching of the preshower hit with the projection of the track must be consistent with the space resolution of the counter itself.

These features are characteristic of a high energy electron starting a shower in the preshower counter.

In practice, because of the different instrumentation in the central and forward regions, these criteria are applied in different ways in the two detectors. In the forward region, in addition, the energy measured in the calorimeter must be compatible with the magnetically measured momentum.

(4) The number of electron pairs is reduced by requiring a hit in at least one of the two innermost chambers of the vertex detector. Furthermore, in the forward detectors the electron candidate is rejected if it is accompanied by another track of opposite sign at an azimuthal separation smaller than 30 mr.

It should be noted that most of the applied selection criteria are satisfied only by isolated electrons : the detection of high p_T electrons contained in a jet of high p_T particles is excluded by the present analysis.

Details of the electron selection are given in Ref. 8.

The corresponding overall efficiencies are $\eta = 0.74 \pm 0.04$ and $\eta = 0.79 \pm 0.03$ for the central and forward detector respectively.

The transverse momentum (p_T^e) distribution of the electron candidates passing the selection criteria for $p_T^e > 11$ GeV/c is displayed in Fig. 22. For the combined 1983 and 1984 data a shoulder in the region $p_T^e \approx 40$ GeV/c is clearly visible. Such a structure is expected from the Jacobian peak which results from the kinematics of the $W \to e\nu$ decay. Also marked are the electrons from $W \to e\nu$ (hashed area) and from $Z \to ee$ (cross hashed area) as identified in the following analysis.

2. Event Topology

The electron candidates presented in Fig. 22 contain, in addition to real electrons, fake electrons coming from misidentified high p_T hadron jets. One expects the events with real electrons from $Z \to ee$ and $W \to e\nu$ decay to contain another high p_T lepton (e or ν) and the events with fake electrons to contain another high p_T jet, both at approximately opposite azimuth.

The jet activity at opposite azimuth to the electron candidate in a 120° wedge is parameterized by the quantity

$$\rho_{opp} = \frac{-\vec{p}_T^{\,e}}{(p_T^e)^2} \cdot \Sigma\, \vec{p}_T^{\,jet} \qquad (7)$$

where the sum extends over all jets with $p_T > 3$ GeV/c.

The sample of events with an electron candidate is split into :

a) Events with $\rho_{opp} < 0.2$: this sample contains $W \to e\nu$. The ρ_{opp} cut is estimated to be (91 ± 3)% efficient, where the error reflects the uncertainty of the Monte Carlo simulation of low energy jets.

b) Events with $\rho_{opp} > 0.2$: this sample includes the $Z° \rightarrow e^+e^-$ sample. It is dominated by two-jet background with one jet misidentified as an electron and is used to estimate the background contamination to the W-sample.

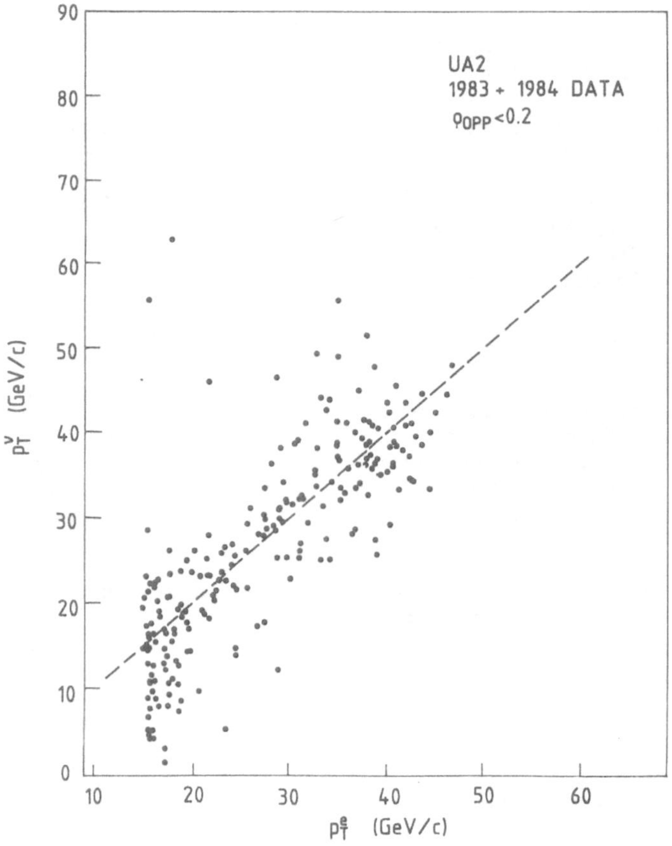

Fig. 23 : The $W \rightarrow e\nu$ sample after the topology cut $\rho_{opp} < 0.2$ in the (p_T^e, p_T^ν)-plane.

3. $W \to e\nu$ Sample

The $W \to e\nu$ events from the 1983 data have been discussed in Ref. 8.

Figure 23 shows for the 1983 and 1984 data p_T^e plotted versus the missing transverse momentum p_T^{miss} for the events with electron candidates and with $\rho_{opp} < 0.2$. The neutrino transverse momentum is calculated as

$$\vec{p}_T^{\,\nu} = \vec{p}_T^{\,miss} = -\vec{p}_T^{\,e} - \Sigma\vec{p}_T^{\,jet} - \lambda\, \vec{p}_T^{\,spect} \tag{8}$$

where $\vec{p}_T^{\,spect}$, a small fraction of $\vec{p}_T^{\,\nu}$, is the total transverse momentum of particles not belonging to jets and λ is a correction factor which takes into account the incomplete detection of the rest of the event. The value of $\lambda = 1.5 \pm 0.6$ is determined by applying the condition $\langle\vec{p}_T^{\,\nu}\rangle = 0$ to equation (8) in the sample of 16 Z^o events described in Section IV.4.

For $p_T^e > 15$ GeV/c the region populated by $W \to e\nu$ events for which one expects $p_T^\nu \approx p_T^e$ is clearly visible.

The sample in Fig. 23 is still contaminated by two-jet events with one of the two jets misidentified as an electron and the other escaping detection (either totally or partially) because of the incomplete coverage of the apparatus. In the method applied to estimate this background it is assumed that the electron candidates in the events with $\rho_{opp} > 0.2$ are misidentified jets. This method is described in Ref. 8 and results in the upper dashed curve in Fig. 24 which shows the p_T^e distribution of the $W \to e\nu$ sample for $p_T^e > 15$ GeV/c. The contributions to this spectrum from $Z \to ee$ decays with one electron undetected and from $W \to \tau\nu$, $\tau \to e\nu\bar{\nu}$ are small. The solid curve, the sum of background and the signal (W-mass of 81 GeV/c^2, see Section IV.3.1), describes the data well.

For $p_T^e > 25$ GeV/c a relatively clean sample of 119 $W \to e\nu$ events is extracted.

3.1 W mass

For each event the transverse mass variable is computed, $M_{T\to\nu}^2 = 2\, p_T^e p_T^\nu (1 - \cos\Delta\phi)$, where $\Delta\phi$ is the azimuthal separation between $\vec{p}_T^{\,e}$ and $\vec{p}_T^{\,\nu}$. Figure 25 shows the M_T-distribution for the 119 $W \to e\nu$ events with $p_T^e > 25$ GeV/c. This distribution is compared with M_T-distributions generated by Monte Carlo simulations for various values of M_W. The background contributions as shown in Fig. 24 are considered too (dotted curve). The solid curve gives the $W \to e\nu$ signal with the best fit and the dashed curve describes the sum of signal and background to be compared with the data.

The best fit value of M_W obtained is $M_W = 81.1 \pm 1.0$ GeV/c^2. The M_T-distribution depends only weakly on the W transverse momentum distribution which is influenced by the topology cut $\rho_{opp} > 0.2$. Uncertainties in p_T^ν, in the width of the W and cell to cell variations result in an additional error of $\pm$ 0.5 GeV/c^2. There is an overall systematic error of $\pm$ 1.5% on the mass scale due to the systematic uncertainty in the calorimeter calibration (Section II).

An alternate method using the p_T^e-distribution directly is used as described in Ref. 8. This method is subject to theoretical input for the longitudinal and transverse momentum distribution of the W. It yields a result consistent with the one from the M_T-distribution.

The value for M_W

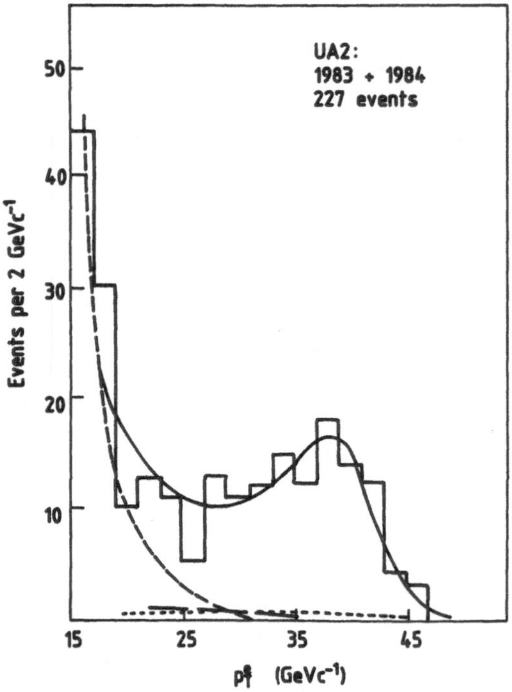

Fig. 24 :The p_T^e spectrum for the $W \to e\nu$ sample with background estimates : dashed curve for misidentified jets, dotted curve for the $Z \to ee$ decays with one electron undetected, curve with long dashes at the bottom for $W \to \tau\nu$, $\tau \to e\nu\bar{\nu}$. The solid curve describes signal (for the W mass given in Section IV.3.1) and background. There are 119 events with $p_T^e > 25$ GeV/c.

$$M_W = 81.1 \pm 1.1(\text{stat.}) \pm 1.3(\text{syst.})\ \text{GeV}/c^2 \qquad (9)$$

is quoted with the systematic error kept separately since it is expected to cancel in the difference $(M_Z - M_W)$ (see Section IV.5).

In this analysis an upper limit for the width of the W is $\Gamma_W < 7$ GeV/c^2 with 90% confidence level.

3.2 Cross-section for the W production

The cross-section σ_W^e for the inclusive process $\bar{p}\,p \to W^{\pm}$ + anything followed by the decay $W \to e\nu$ is calculated from the relation

$$\sigma_W^e = \frac{N_W^e}{L.\varepsilon.\eta} \;, \qquad (10)$$

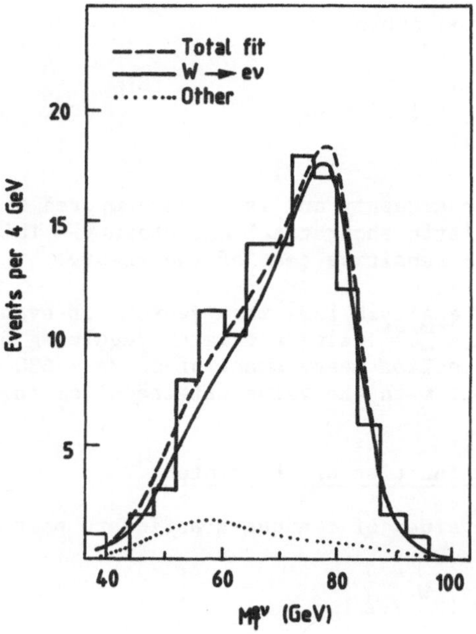

Fig. 25 : Transverse mass ($M_T^{e\nu}$) distribution of the $W \to e\nu$ events. The solid curve is the best fit for the $W \to e\nu$ signal. The background contributions (from Fig. 24) are shown by the dotted curve. The dashed curve gives the sum of signal and background for the best fit.

where N_W^e is the number of $W \to e\nu$ decays with $p_T^e > 25$ GeV/c, L is the integrated luminosity, ε is the detector acceptance including the $p_T^e = 25$ GeV/c threshold and the topology cut $\rho_{opp} < 0.2$, and η is the efficiency for electron identification.

The acceptance is estimated with a Monte Carlo simulation to ε = .61 ± .0.2.

In Table 1 the relevant quantities for the calculation of σ_W^e are listed. From these one obtains

$$\sigma(\sqrt{s} = 546 \text{ GeV}) = 494 \pm 90(\text{stat.}) \pm 46(\text{syst.})\text{pb}$$
$$\sigma(\sqrt{s} = 630 \text{ GeV}) = 526 \pm 64(\text{stat.}) \pm 49(\text{syst.})\text{pb}. \quad (11)$$

The corresponding theoretical predictions [31,32] are $\sigma_W^e = 360 \pm ^{110}_{50}$ and $460 \pm ^{140}_{80}$ pb, respectively, where the errors reflecting theoretical uncertainties are sizeable.

The cross-section ratio

$$\frac{\sigma_W^e(\sqrt{s} = 630 \text{ GeV})}{\sigma_W^e(\sqrt{s} = 546 \text{ GeV})} = 1.06 \pm 0.23$$

has no systematic uncertainty and is to be compared with the prediction [32] of 1.26 with little theoretical uncertainty. This, with improved statistics, can be a sensitive test of the theory.

An independent analysis [33] in a search for events with large missing transverse momentum $\vec{P}_T^{miss}$ using a trigger requiring $\vec{P}_T^{miss} > 30$ GeV/c results in a cross section measurement of $\sigma_W^e(\sqrt{s} = 630 \text{ GeV}) = 590 \pm 90$ (stat.) pb, in good agreement with the value obtained from the search for high P_T electrons.

3.3 Quark structure function of the proton

The Feynamn x-values of the quark-antiquark pair producing the W are calculated from

$$X_q - X_{\bar{q}} = X_W = 2P_L^W/\sqrt{s} \qquad (12)$$
$$X_q \cdot X_{\bar{q}} = M_W^2/s \quad ,$$

neglecting the transverse motion of the W.

The longitudinal momentum of the neutrino (P_L^ν) is unknown and in general one obtains two solutions from $M_W^2 = E_W^2 - (P_W)^2$ for P_L^ν and hence X_W. If there are two physical solutions for X_W the solution which gives the lower value of X_W is retained as the more probable one.

Since the electron charge is undetermined (with the exception of a small fraction of the events in the forward region, see Section IV.3.4) for 'quark' ('antiquark') one simply takes the parton coming from the proton (antiproton).

Fig. 26 shows the quark and antiquark X-distributions added up at $\sqrt{s} = 546$ and 630 GeV. The data are not corrected for acceptance. They are well described by a Monte Carlo calculation [34] for the UA2 apparatus which uses structure functions from low energy neutrino collisions evolved to collider energies supporting the hypothesis of $W^\pm$ production by $q\bar{q}$ annihilation.

3.4. Charge asymmetry

At the energies of the CERN $\bar{p}p$ collider W production is dominated by $q\bar{q}$ annihilation involving at least one Valence quark or antiquark. The V - A coupling in the production and the decay of the W produces a distinctive charge asymmetry in the decay $W \rightarrow e\nu$. The positive (negative) electron is preferentially emitted in the direction along the incoming antiproton (proton).

In the forward regions of the UA2 detector, equipped with a magnetic field, the electron momentum and the sign of the electron charge is determined. Here, one can set the threshold for P_T^e lower, $P_T^e > 20$ GeV/c, which gives 28 events in the sample for the charge asymmetry.

In Fig. 27 the electron calorimeter energy (E^{-1}) is compared with the magnetic momentum (p^{-1}) . Further the momentum is multiplied with the

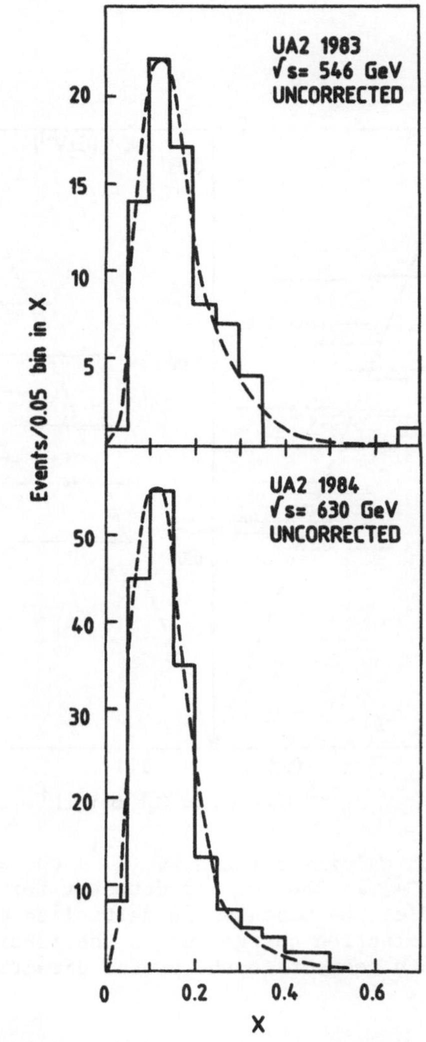

Fig. 26 : The uncorrected Feynman x distribution for quark (antiquark) in the $W \to e\nu$ sample. The curve is a Monte Carlo result using neutrino data structure functions.

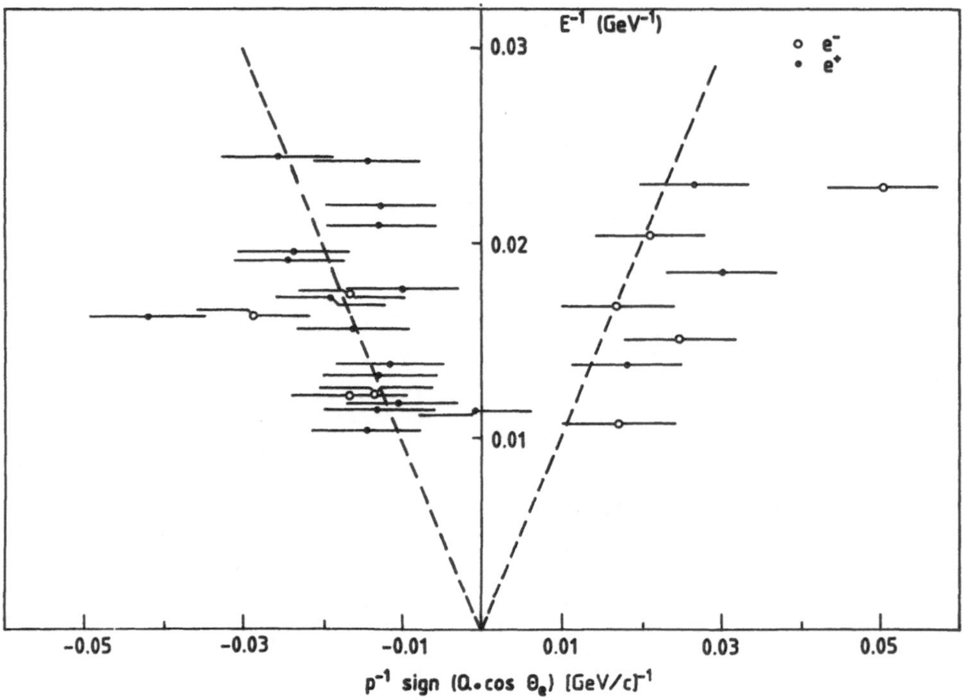

Fig. 27 : The electron calorimeter energy (E^{-1}) compared with the magnetic momentum (p^{-1}) in the forward detector for electrons with $p_T^e > 20$ GeV/c. The momentum is multiplied by sign [Q.cos θ_e] with Q the electron charge and θ_e the laboratory angle of the electron with respect to the proton direction.

Table 1

List of quantities for evaluation of the W production cross-section (see text)

	1982-83 data	1984 data
ε	0.61 ± 0.02	
η	0.75 ± 0.04	
$\sqrt{s}$ (GeV)	546	630
L(nb^{-1})	142[30]	310
Electron candidates	36	83
Background	1.7 ± 0.5	4.1 ± 1.2
$Z^\circ \to e(e)$	1.1 ± 0.3	3.3 ± 1.0
$W \to \tau \to e$	0.7 ± 0.1	1.6 ± 0.2
$W \to e\nu$	32.5 ± 5.9	74.0 ± 9.0

sign of the electron charge and the sign of cos θ_e where θ_e is the laboratory angle of the electron with respect to the proton direction. The preferred $e^\pm$ direction places the events on the left side of the plot which is clearly more populated than the right one. There are 20 events left and 8 events right which gives an asymmetry $A = (20 - 8)/28 = 0.43 \pm 0.17$. This value is in good agreement with the expected asymmetry for V - A coupling, $A = 0.53 \pm 0.06$, as obtained by a Monte Carlo calculation, also taking into account background contributions as listed in Table 1.

4. Z → ee Sample

The seven $Z^\circ \to e^+e^-$ events and one $Z^\circ \to e^+e^- \gamma$ decay of the 1983 data have discussed in Ref. [7,8].

Here the 1984 data are described. Since the efficiency of the electron selection criteria is only ≈ 0.75 (see Section IV.1) these are only required for one electron of the $Z^\circ \to ee$ candidates with the other electron passing looser selection criteria : essentially the presence of a track associated with a cluster passing the calorimeter selection criteria (Section IV.1).

Figure 28 shows in Part a) the distribution of the two cluster invariant mass, M_{ee} for events with only calorimeter cuts. An accumulation of events in the region of the Z° mass is visible. In Part b), with the full electron selection criteria required for one elctron candidate only, the Z° signal of 8 events is clearly separated from the background.

The background in the Z° signal from two jet final states is estimated from the original sample of events collected by the Z° trigger as described in Ref [7] to 0.2 events in Fig. 28b for $M_{ee} > 70$ GeV/c^2.

The $Z^\circ \to e^+e^- \gamma$ event in the 1983 data had only 11 % probability to be an internal bremsstrahlung event. No new event of this type has been found in the 1984 data and the total of 16 events yields a 22 % bremsstrahlung probability.

4.1 Z mass and width

Figure 29a shows the M_{ee} distribution for the 1983 and 1984 sample. The mass values with their error bars for the 16 events are shown in Part b).

A fit to a relativistic Breit-Wigner shape distorted by the experimental mass resolution yields for the Z° mass.

$$M_Z = 92.5 \pm 1.3 \text{ (stat.)} \pm 1.4 \text{ (syst.) GeV/c}^2. \qquad (13)$$

where the systematic error comes from the uncertainty in the energy scale of the calorimeter.

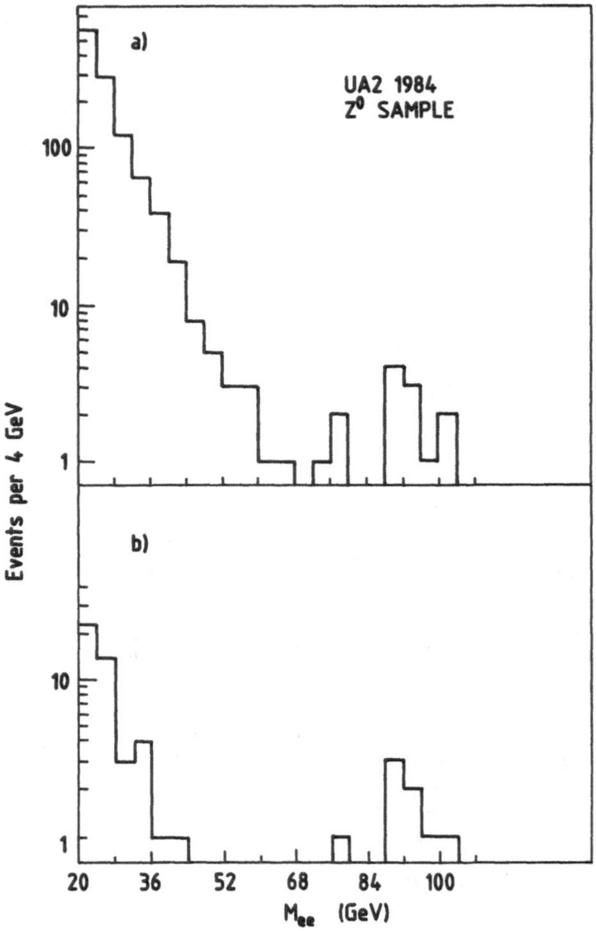

Fig. 28 : The two cluster invariant mass distributions for events which pass the calorimeter cuts only a) and b) with the full electron selection criteria required for one electron only.

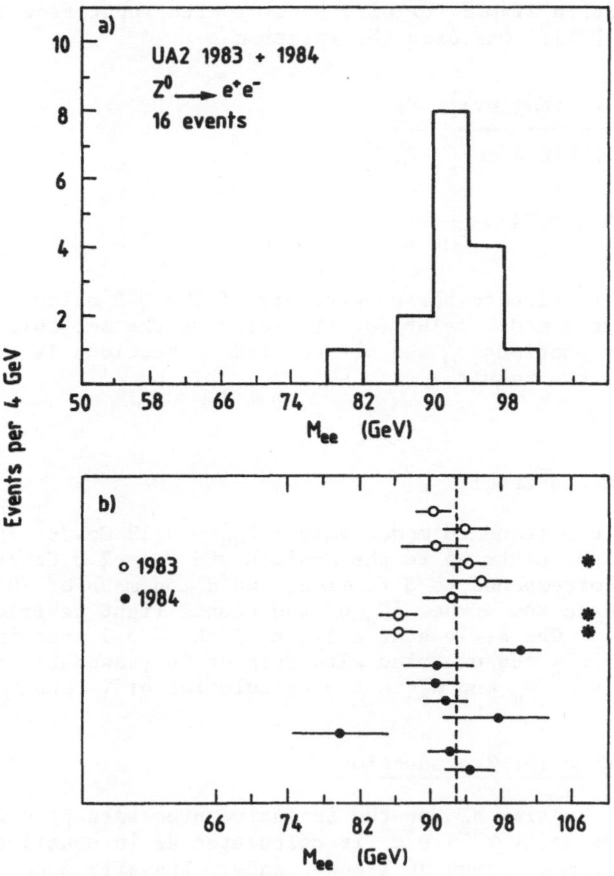

Fig. 29 : a) The two electron mass distribution.
b) The masses M_{ee} together with their experimental uncertainties. The events marked with stars from the 1983 data are excluded from the mass fit for reasons described in Ref. [7].

As shown in Fig. 29b the spread of the mass values about the mean is about equal to the average mass error. One can estimate an upper limit to the width of the Z, $\Gamma_Z \leq 3.3$ GeV/c^2 at 90 % confidence level using a method described in Ref. [8]. However, this value is very sensitive to the mass errors put in.

A more "reliable" result can be obtained from the cross section ratio σ^e_Z/σ^e_W (see Section IV.4.2 for σ^e_Z), however with input from the standard model and QCD [31]. One uses the relation

$$\frac{\sigma^e_W}{\sigma^e_Z} = \frac{\sigma_W \cdot \Gamma(W \to e\nu)}{\sigma_Z \cdot \Gamma(Z \to ee)} \cdot \frac{\Gamma_Z}{\Gamma_W} = [8.8 \pm 0.5] \cdot \frac{\Gamma_Z}{\Gamma_W} , \tag{14}$$

where the error reflects the uncertainty of the QCD calculation, together with the standard model value for the ratio of the leptonic widths [35]. With the cross sections σ^e_W and σ^e_Z reported in Sections IV.3.2 and IV.4.2 one arrives at the result

$$\Gamma_Z < 1.2\ \Gamma_W \tag{15}$$

at 90 % confidence level.

One uses the standard model values $\Gamma_{\nu\nu} = 0.18$ GeV/c^2 for the contribution of a light neutrino to the Z-width and $\Gamma_W = 2.8$ GeV/c^2 for the W-width, which corresponds to 3 fermions and a top mass of 40 GeV/c^2, in order to estimate the number ΔN_ν of additional light neutrinos in the standard model. One arrives at a limit of $\Delta N_\nu \leq 3.5$ neutrinos which is stable to within $\sim$ one neutrino with respect to reasonable variations of the top mass, $\sin^2 \Theta_W$ and α_s in the calculation of Γ_W and Γ_Z in the standard model.

4.2 Cross section for Z production

The cross section σ^e_Z for the inclusive processes $\bar{p}p \to Z° +$ anything followed by the decay $Z° \to e^+e^-$ is calculated as in equation (10). A Monte Carlo estimate gives 56 % acceptance. The efficiency of the electron selection criteria is $\eta = 0.89 \pm 0.3$. The cross sections are measured to

$$\begin{aligned} \sigma^e_Z\ (\sqrt{s} = 546\ \text{GeV}) &= 111 \pm 39\ (\text{stat.}) \pm 9\ (\text{syst.})\ \text{pb} \\ \sigma^e_Z\ (\sqrt{s} = 630\ \text{GeV}) &= 52 \pm 19\ (\text{stat.}) \pm 4\ (\text{syst.})\ \text{pb.} \end{aligned} \tag{16}$$

These are to be compared with the corresponding theoretical predictions [31, 32] $\sigma^e_Z = 42 \pm ^{13}_{6}$ pb and $51 \pm ^{16}_{10}$ pb at $\sqrt{s} = 546$ and 630 GeV respectively.

5. W and Z Mass in the Standard Model

The weak mixing angle Θ_W is in the standard model related to the W and Z masses as [36]

$$\sin^2 \Theta_W = 1 - (M_W/M_Z)^2 \tag{17}$$

from this and the measurements reported in Sections IV.3.1 and IV.4.1 one obtains

$$\sin^2 \Theta_W = .231 \pm .030 \tag{18}$$

where the systematic errors in the W and Z mass cancel. One finds good agreement with a recent compilation [37] of measurements of Θ_W (other than W and Z masses) where the world average of $\sin^2 \Theta_W = 0.220 \pm .007$ is quoted, after radiative corrections.

For a measurement of the radiative corrections the quoted errors on the W and Z masses are too big. Figure 30 shows the standard model predictions (curves b)) from Ref. [36] and the measurement of this report with one standard deviation contours (curves a)) in a $(M_Z - M_W)$ versus M_Z plot. A considerable improvement in the precision, in particular the systematic error on M_Z is needed. Also shown is relation (17) using the value of Ref. [37] for $\sin^2 \Theta_W$.

6. W and Z Production with Associated Jets

Higher ordre QCD corrections to W(Z) production increase the production cross section and give the W(Z) a sizeable transverse momentum, the hard higher order processes resulting in observable jets.

In the analysis of the W sample so far the topology cut $\rho_{opp} < 0.2$ in Section IV.2 suppresses events with jets recoiling against the W. In order to study the associated jet activity in W events the topology cut is replaced by a threshold for the neutrino transverse momentum, $p_T^\nu > 25$ GeV/c (see Figure 23). Events with $p_T^e > 15$ GeV/c are accepted. The 1983 data at $\sqrt{s} = 546$ GeV and the 1984 data at $\sqrt{s} = 630$ GeV are combined neglecting the small difference of the W and Z transverse momentum at these two energies.

Figure 31 shows the transverse momentum distribution for the W and Z events in the 1983 and 1984 data. One calculates p_T^Z directly from the two electrons and p_T^W from equation (8) with $\vec{p}_T^{\,W} = \vec{p}_T^{\,e} + \vec{p}_T^{\,\nu}$.

The QCD calculation of Ref. [32] describes the data well.

There are 31 events in the W sample with jets having $E_T^{jet} > 3$ GeV and 3 such events in the Z sample. These are marked in Fig. 31 (hatched) and populate the high p_T tail of the distribution.

The multiplicity distribution of these associated jets is shown in Fig. 32. One expects the probability of observing n jets to fall off proportional to α_s^n where α_s is the strong coupling constant. This behaviour is approximately observed in the W sample for $\alpha_s = 0.2$. For a more quantitative confrontation with QCD calculations a considerably more accurate measurement of the jet production rate at a higher jet momentum threshold is needed.

The events A, B, C (cross hatched) in Fig. 31 occurred in the 1983 data. With only 1/3 of the total 1983 + 1984 integrated luminosity available these events were rare events and not easily explained as high p_T W events [38]. With the increased integrated luminosity in 1984 only events E and F in Fig. 31 with similar features were added. Events A to F are now interpreted as W events on the high p_T^W tail. For $p_T^W > 26$ GeV/c 9 events are observed and 4 expected from the QCD calculation.

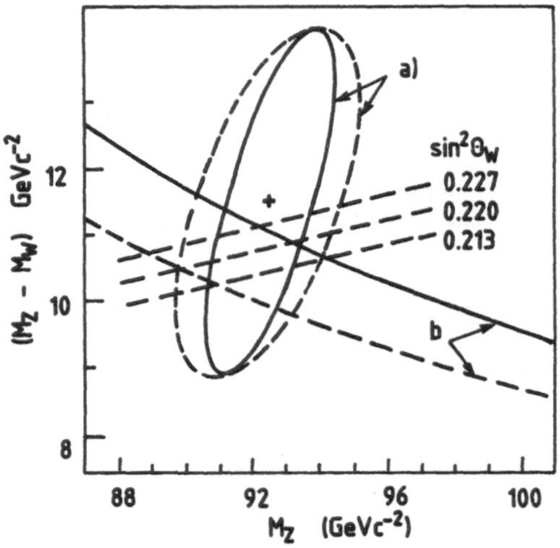

Fig. 30 : (M_Z-M_W) versus M_Z. The curves a) around the average value of the UA2 measurement give one standard deviation contours for statistical (inner curve) and total error (outer dashed curve). The curves labelled b) are the predictions of the standard model from Ref. [36], uncorrected (dashed curve) and with radiative corrections (solid curve). The dashed curves labelled $\sin^2 \Theta_W$ show relation (17) with $\sin^2 \Theta_W$ taken from Ref. [37].

V. CONCLUSIONS

Jet physics in UA2 :

The jet production cross section is in agreement with the prediction from the parton model and QCD to lowest order in α_s. The experimental uncertainty (≈ ± 45 %) and in particular the theoretical uncertainty of a factor of ≈ (2 - 3) do not allow an accurate measurement of the K-factor (the sum over all higher order contributions).

The measured parton structure function, compared to the quark structure function evolved to the $\bar{p}p$ collider energy from low energy neutrino collisions, shows a dominant gluon contribution at low x.

The predictions of higher order QCD processes are in agreement with the data in both regimes of gluon radiation, for soft and/or collinear gluons with the observed two-jet transverse momentum and for hard gluons with the observed three-jet rate.

A jet fragmentation result (charged particle multiplicity in jets) hints at the different behaviour of quark and gluon jets.

The observed π^o/jet ratio is ≈ 10^{-3}.

No structure in the two-jet mass is found.

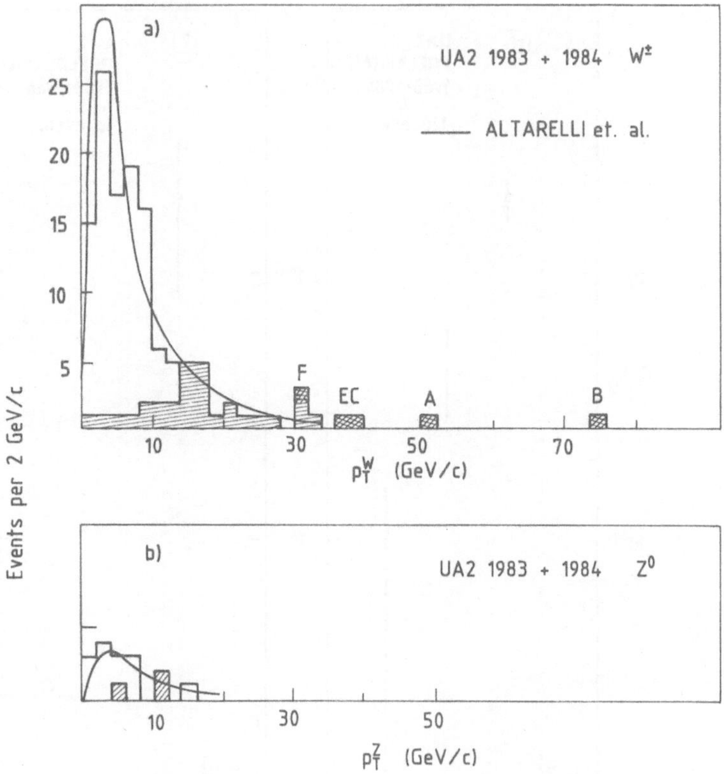

Fig. 31 : The W and Z p_T distributions. Events with jet activity (E_T^{jet} > 3 GeV/c) are hatched. The curve is a QCD calculation [32]. The cross hatched events are discussed in the text.

W± and Z° properties in UA2 :

The masses agree well with the standard model predictions. The accuracy, however, is not sufficient to measure the radiative corrections. The systematic error in the Z° mass will ultimately prevent this measurement in UA2 alone.

The cross sections and the W(Z) transverse momentum distributions are correctly given by QCD and the standard model. The observed number n of jets associated with the W± scales approximately proportionately to α_s^n as expected.

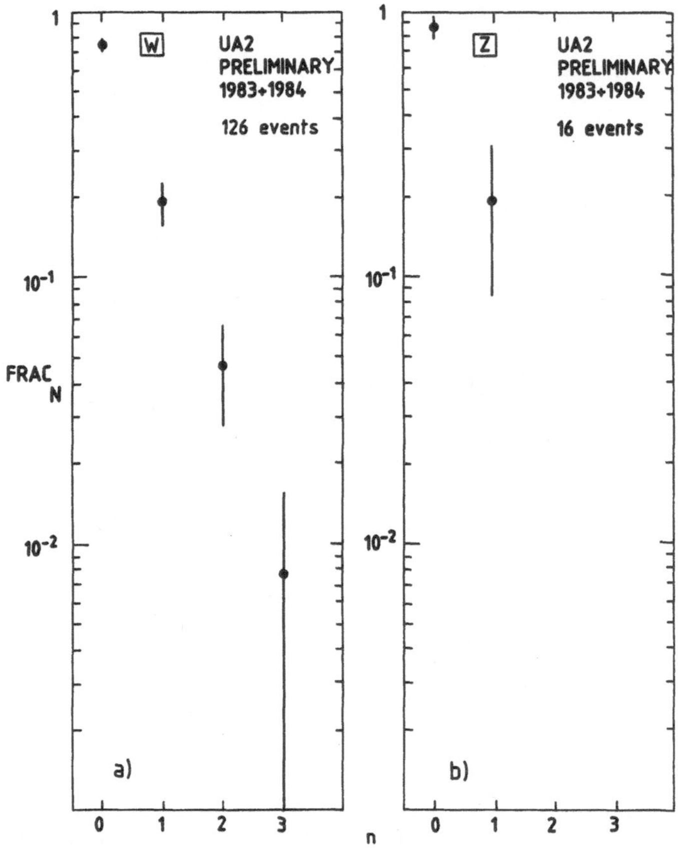

Fig. 32 : The associated jet multiplicity distribution for W and Z events. 'Frac' is the fraction of events having n jets with $p_T^{jet} > 3$ GeV/c.

The shape of the (quark + antiquark) structure function supports the hypothesis of W production via $q\bar{q}$ annihilation.

The observed charge asymmetry in the $W^{\pm} \rightarrow e^{\pm}\nu$ decay is in agreement with a (V - A) interaction in the production and the decay of the $W^{\pm}$.

A direct measurement of the width of the Z° is very sensitive to the experimental errors on the Z° mass since these have about the same value as the observed r.m.s. spread about the mean. An indirect measurement from the cross section ratio $\sigma(W \rightarrow e\nu)/\sigma(Z^{\circ} \rightarrow ee)$ yields an upper limit of 3.5 neutrinos to the number of additional light neutrinos in the standard model.

In summary the observations made with the UA2 apparatus are well described with QCD and the standard model.

The author thanks the organizers of the conference, Professors Gastman, Levy and Weyers and Marie-France Hanseler for a very pleasant stay in Cargèse.

REFERENCES

[1] UA2 Collaboration : Bern - CERN - Copenhagen (NBI) - Heidelberg - Orsay (LAL) - Pavia - Perugia - Pisa - Saclay (CEN).
[2] UA2 Collaboration : M. Banner et al., Phys. Lett. 118B (1982) 203.
[3] UA2 Collaboration : P. Bagnaia et al., Phys. Lett. 138B (1984) 430.
[4] UA2 Collaboration : P. Bagnaia et al., Phys. Lett. 144B (1984) 283.
[5] UA2 Collaboration : P. Bagnaia et al., Phys. Lett. 144B (1984) 291.
[6] UA2 Collaboration : M. Banner et al., Phys. Lett. 122B (1983) 476.
[7] UA2 Collaboration : P. Bagnaia et al., Phys. Lett. 129B (1983) 130.
[8] UA2 Collaboration : P. Bagnaia et al., Z. Phys. C Particles and Fields 24 (1984) 1.
[9] B. Mansoulié, The UA2 apparatus at the CERN $p\bar{p}$ Collider, Proceedings 3rd Moriond workshop on $p\bar{p}$ physics, Editions Frontières, 1983, p. 609;
M. Dialinas et al., The vertex detector of the UA2 experiment, LAL-RT/83-14, ORSAY, 1983;
C. Conta et al., The system of forward-backward drift chambers in the UA2 detector, Nucl. Instr. and Methods 224 (1984) 65;
K. Borer et al., Multitube proportional chambers for the localization of electromagnetic showers in the UA2 detector, Nucl. Instr. and Methods 227 (1984) 29.
[10] A. Beer et al., The central calorimeter of the UA2 experiment at the CERN $p\bar{p}$ Collider, Nucl. Instr. and Methods 224 (1984) 360.
[11] UA4, R. Battiston et al., Phys. Lett. 117B (1982) 126.
[12] UA4, M. Bozzo et al., Phys. Lett. 147B (1984) 392.
[13] UA2 Collaboration : P. Bagnaia et al., Measurement of the $\sqrt{s}$ dependence of the jet production at the CERN $\bar{p}p$ collider Ref. EP/85, CERN 1985; and to be published in Phys. Lett.
[14] T. Akesson et al., Phys. Lett. 118B (1982) 185.
[15] B.L. Combridge et al., Phys. Lett. 70B (1977) 234.
[16] E. Eichten et al., Rev. Mod. Phys. 56 (1984) 579.
[17] R. Horgan and M. Jacob, Nucl. Phys. B179 (1981) 441.
[18] See for example : J.D. Bjorken, Phys. Rev. D8 (1973) 4098;
S.J. Brodsky and G. Farrar, Phys. Rev. Lett. 31 (1973) 1153.
[19] J.C. Collins and D.E. Soper, Phys. Rev. D16 (1977) 2219.
[20] H. Abramowicz et al., Z. Phys. C12 (1982) 89; Z. Phys. C13 (1982) 199; Z. Phys. C17 (1983) 283
F. Bergsma et al., Phys. Lett. 123B (1983) 269.
[21] M. Greco, Z; Phys. C.
[22] A Publication on a detailed study of multi-jet events in UA2 is in preparation.
[23] Z. Kunszt and E. Pietarinen, Phys. Lett 123B (1983) 453.
[24] S.L. Wu, Phys. Rep. 107 (1984) 60.
[25] M. Althoff et al., Z. Phys. C22 (1984) 307.
[26] G. Marchesini and B.R. Webber, Nucl. Phys. B238 (1984) 1;
B.R. Webber, Nucl. Phys. B238 (1984) 492.
[27] M. Banner et al., Z. Phys. C (1984).
[28] R. Baier et al., Z. Phys. C2 (1979) 265.
[29] The 1983 data quoted in the chapter on W and Z properties have a small event sample from the 1982 run added in.

[30] This luminosity differs from the value used in Ref. 8 as a result of a new $\bar{p}p$ total cross section measurement [12].
[31] G. Altarelli et al., Nucl. Phys. B246 (1984) 576 and Ref. TH.3851-CERN (1984).
[32] G. Altarelli et al., Vector boson production at present and future colliders, Ref. TH.4015-CERN (1984).
[33] H. Hänni, Search for monojet and multijet events with large missing p_T in the UA2 experiment, Ref. EP/85-87, CERN (1985).
[34] S.D. Ellis, R. Kleiss and W.J. Stirling, CERN-TH. 4096/85.
[35] J. Ellis and M.K. Gaillard, Ann. Rev. Nucl. Part. Sci. 32 (1982) 443.
[36] W.J. Marciano and A. Sirlin, Phys. Rev. D29 (1984) 945.
[37] W. Marciano, Proc. of 1st Aspen Winter Physics Conf., Aspen, Colorado, 1985; also in BNL Preprint BNL 36147 (1985).
[38] UA2 Collaboration : P. Bagnaia et al., Phys. Lett. 139B (1984) 105.

PARTICLE PHYSICS WITHOUT ACCELERATORS (selected topics)

M. Spiro

Centre d'Etudes Nucléaires de Saclay
F-91191 Gif-sur-Yvette, France

ABSTRACT

Particle physics without accelerators covers a wide variety of subjects. Only the following three main topics are discussed in this report :

1) High energy cosmic ray experiments, with a particular emphasis on the observations related to Cygnus X3 and their implications.

2) Neutrino physics : Majorana or Dirac neutrinos, neutrino masses, neutrino oscillations.

3) Constancy of the coupling constants over a long period of time based on the discovery of a fossil nuclear reactor in Gabon.

It is relatively easy to define what is the study of elementary particles with accelerators. An accelerator is any device which by means of intense electric fields, accelerates charged particle beams. The process starts from a hydrogen, deuterium, or helium bottle (or heavier element). Atoms are stripped by means of an electric field, and then acceleration of either the electrons or the stripped atoms can take place. A typical experiment will be the study of the collisions between beam particles and a specific target. It is much more complicated to tell what is exactly the study of elementary particles without accelerator : all what we know about the experiments is that they do not use accelerators, but obviously this is not enough since it has to deal with elementary particle physics. So the discovery of new particles without accelerators, mass measurements, branching ratios like double beta decay belong to this field. Nobody will object to include also the study of the properties of the interactions between elementary particles : quantum number conservation, hints to ultra high energy phenomena which cannot, at least presently, be reached by accelerators.

Particle Physics without accelerators aims for fundamental discoveries. Obviously, except for cosmic ray experiments, this is a very low Q^2 physics. Nevertheless this is partly compensated by the fact that one can search for very rare processes by looking during a very long period of time at many elementary particles. In that sense this is a nice complementary way to study elementary particle physics compared to accelerator techniques.

A tentative list of the various topics is given below :

1. Hints for very high energy interactions from cosmic ray experiments.

Search for ultrahigh energy pointlike source of cosmic rays : the possible new physics involved by the recent discoveries concerning Cygnus X3.

2. Search for new particles : axions, monopoles and quarks.

3. Neutrino physics and weak interactions.

4. Experiments related to Grand Unified Theories :
   baryon number conservation (proton lifetime,
   neutron antineutron oscillations)
   electric dipole moment of the neutron
   (involves CP conservation).

5. Constancy of the coupling constants over a long period of time (fossil reactor).

It can be argued whether or not we should include quantum mechanics tests (Bell inequalities and Aspect experiments), or general relativity tests (search for gravitational waves for instance ...) and quantum electrodynamics related experiments (g-2 for the electron).

In the two lectures given at Cargèse 85, I had to choose between all these topics. Only points 1, 3 and 5 were covered and will be presented here. A more detailed report which will discuss all the above subjects will be published in Physics Reports together with M. Cribier and J. Rich [1].

This following report is not intended to be complete. Specially, some recent experimental results might be missed. The main goals are only to introduce the subjects and the motivations, to describe in a critical way the present status, and to foresee the main perspectives.

COSMIC RAY EXPERIMENTS

It is well known that most of the long lived elementary particles like π's, K's, Λ ... were discovered in the 1930-1940's in the old good time of cosmic rays. The last long lived particle which was discovered in cosmic rays was the Σ in 1953 [2]. When accelerator physics started in the 1950's it was generally believed that this was the end of the study of elementary particles with cosmic rays. Nevertheless I want to show here that most of the gross features of high energy collider experiments ($p\bar{p}$ collider specially) were already predicted some time ago by cosmic ray experiments which were being pursued despite the lack of interest from the elementary particle community. The fact that cosmic rays are the source of highest energy available is illustrated in fig. 1.1. However cosmic ray experiments have three important limitations. Firstly the energy and the type of cosmic ray particles cannot be controlled secondly the flux of primary cosmic rays falls off very rapidity with energy and thirdly different techniques (fig. 1.2 from [3]) are used to cover different energy ranges which makes it difficult to interpret energy dependent effects.

TOTAL CROSS SECTION RISE

Estimates of the p-p total cross sections up to 10^5 GeV incident laboratory energies were already given in 1972 [4]. The data which came from a compilation of proton-air showers at that time are shown in fig. 1.3. These are data prior to the evidence of rising total cross sections which came later on from ISR. The data showed already clear evidence for this effect. Moreover the fit given in this same reference for the total cross section as a function of energy was :

$$\sigma_{pp} = 38.8 + 0.4 \ln^2(s/s_o) \quad s_o = 140 \text{ GeV}^2$$

where s is the squared total center of mass energy.

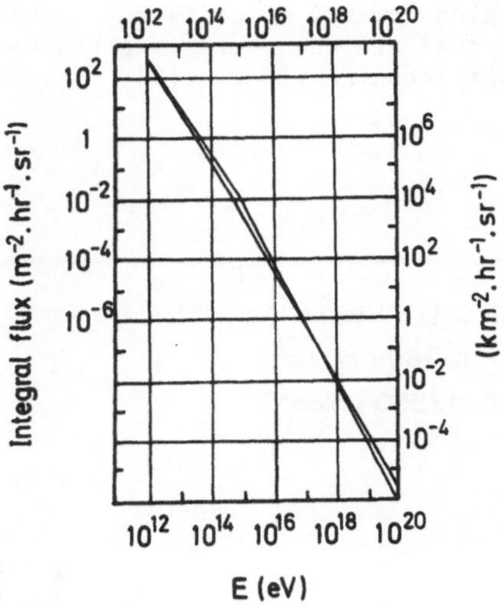

Fig. 1.1 : Integral flux of primary cosmic rays in the energy range of $10^{12} - 10^{20}$ eV. [3]

The prediction for collider energies (150 TeV laboratory energy and 540 GeV center of mass energy) were : 62.2 mb whereas last result from UA4 Collaboration is 61 mb [5].
This is certainly a very big success of very high energy cosmic ray data analysis, taking into account the complexity of the analysis which is required to obtain these figures.

New data have been collected since then. The prediction for the total p-p cross section at 40 TeV center of mass energy is around 130 mb [6].

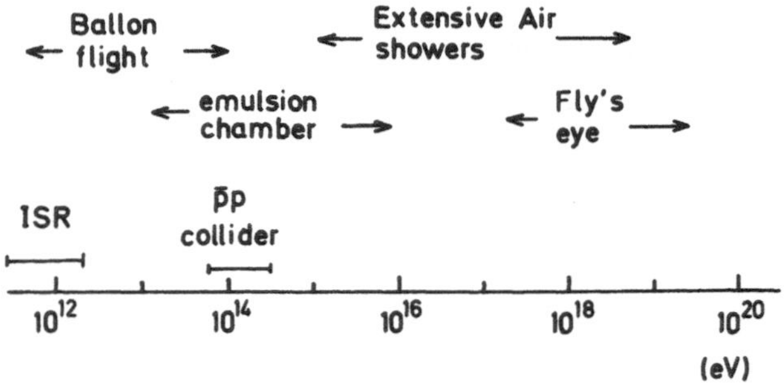

Fig. 1.2 : Energy region covered by different cosmic ray techniques - Also indicated are the energy regions by ISR and CERN SPS proton antiproton collider. [3]

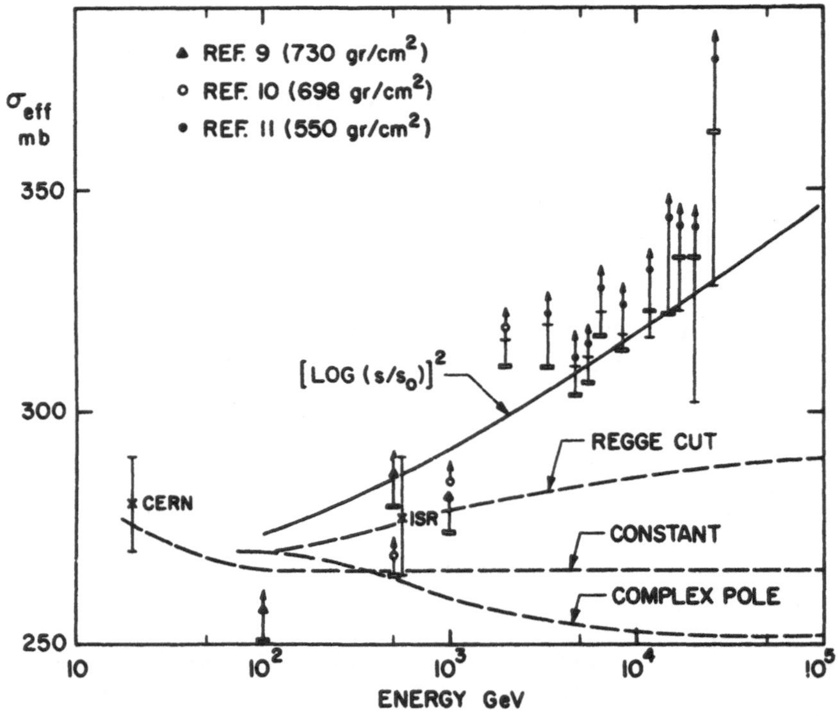

Fig. 1.3 : Effective inelastic cross sections for protons in air as a function of energy. [4]

RISE OF THE CENTRAL PLATEAU AND MULTIPLICITY DISTRIBUTION

Evidence for the rise of the central plateau comes from emulsion chamber technique cosmic ray experiments. Fig. 1.4 shows the principle of these experiments [7]. They are only sensitive to the electromagnetic part of the shower (lead emulsion sandwiches). Fig. 1.5 shows comparison between this high energy cosmic ray data (> 100 TeV) and accelerator data for the averaged charged multiplicity. The parametrisation $\langle n_c \rangle = a + b \ln s + c \ln^2 s$ is in agreement with both cosmic rays and collider data [8]. The rise and the increasing width of the central plateau is evidenced and has been confirmed at the collider (fig. 1.6a,b).

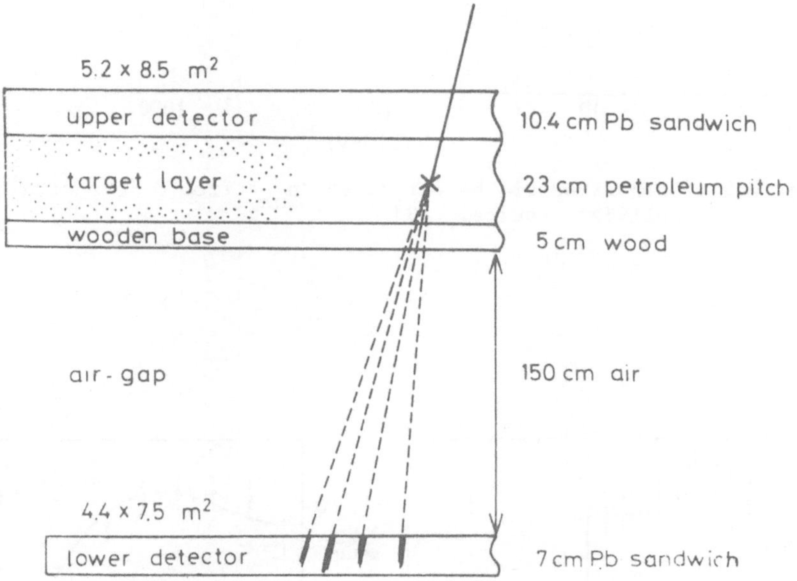

Fig. 1.4 : Chacaltaya emulsion chamber of two storey structure. [7]

Moreover very large fluctuations in multiplicity were found from event to event. Three categories of events were defined : low multiplicity ones (the so called Mirim events), medium multiplicity (Açu events) and large multiplicity events (Guaçu events). Fig. 1.7 show typical pictures of these events in terms of rapidity distribution of the produced particles [9]. We know now from collider results that this reflects the so called KNO scaling, the fact that the distribution of the number of charged particles in an event divided by the average number of charged particles at a given energy is energy independent in first approximation (fig. 1.8) [10].

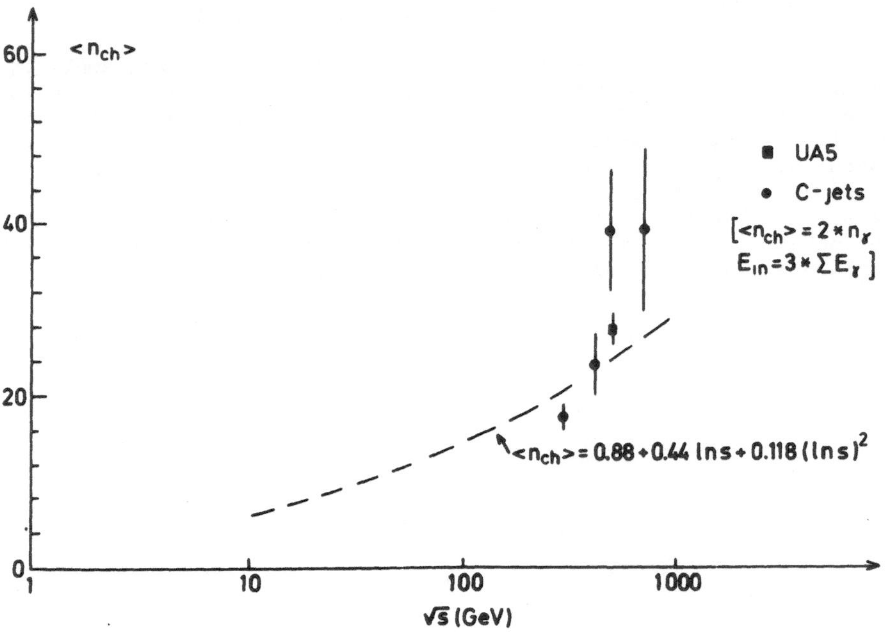

Fig. 1.5 : Comparison of the estimated $\langle n_{ch} \rangle$ from C-jets with the collider results. [3]

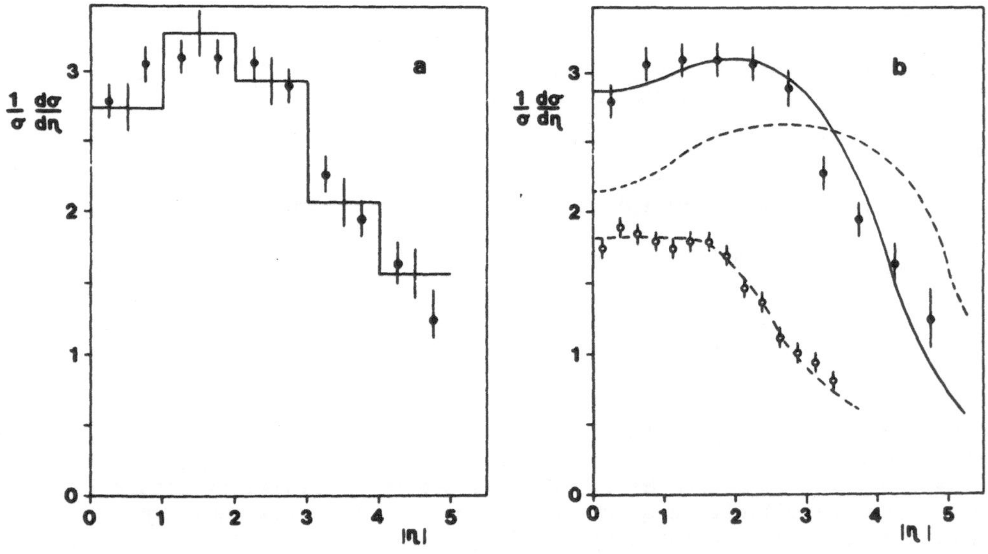

Fig. 1.6a,b : a) Pseudo rapidity distributions at $\sqrt{s}$ = 540 GeV collider energy. [8]
b) Comparison with earlier data at the ISR, $\sqrt{s}$ = 53 GeV.

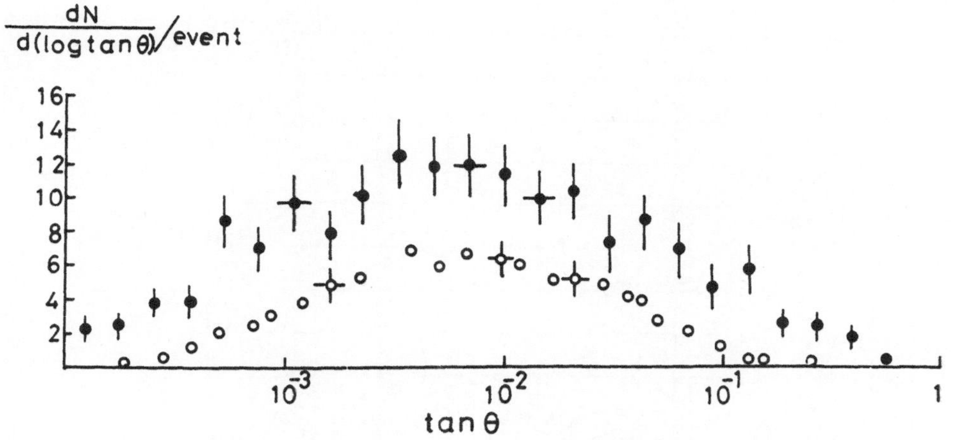

Fig. 1.6c : Pseudo rapidity distribution [ln (tan θ)] of charged particles● from Australia--Japan balloon flight, $E_0 > 7$ TeV. 0 from 400 GeV protons at FNAL. Both come from Nagoya group with chamber of similar structure. [7]

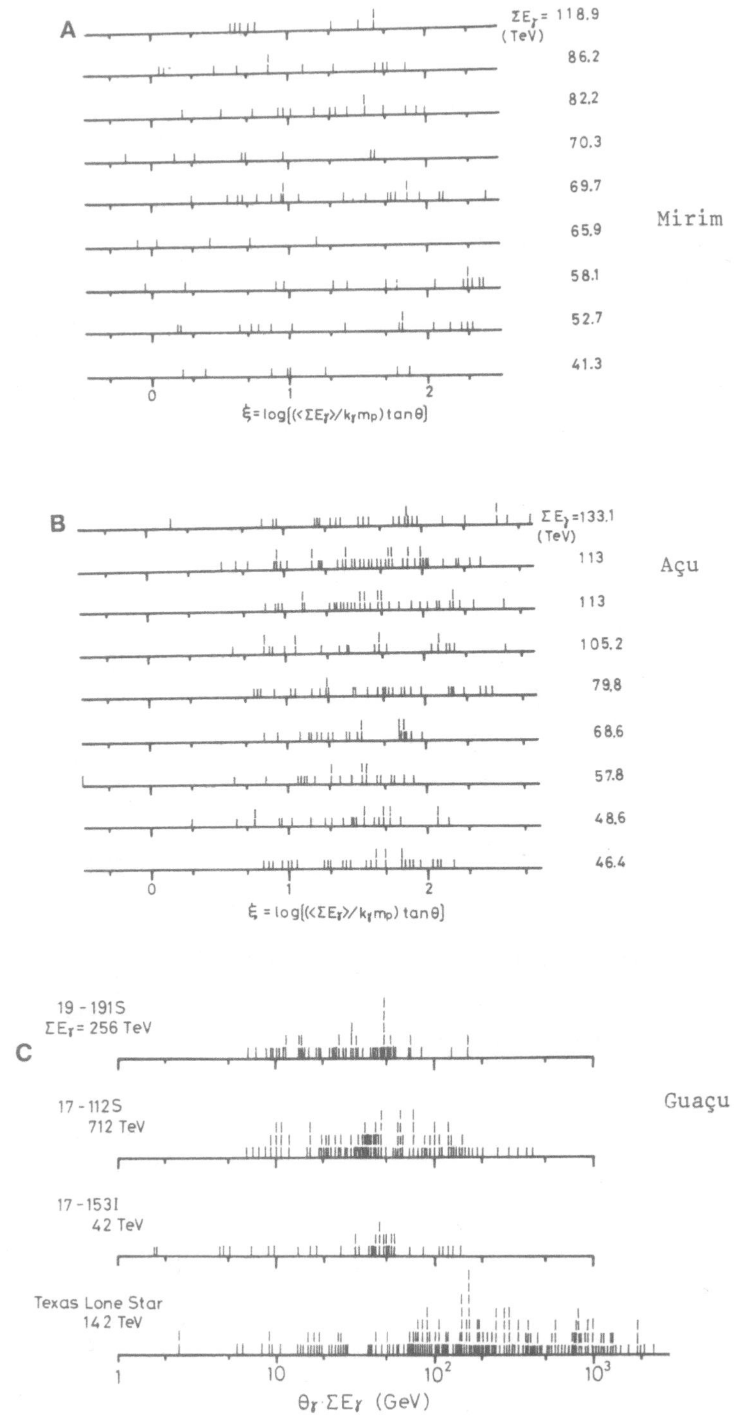

Fig. 1.7a,b,c : Pseudo rapidity display of charged tracks for 3 categories events :
A) Mirim events (low multiplicity)
B) Açu events (medium multiplicity) [7]
C) Guaçu events (high multiplicity)

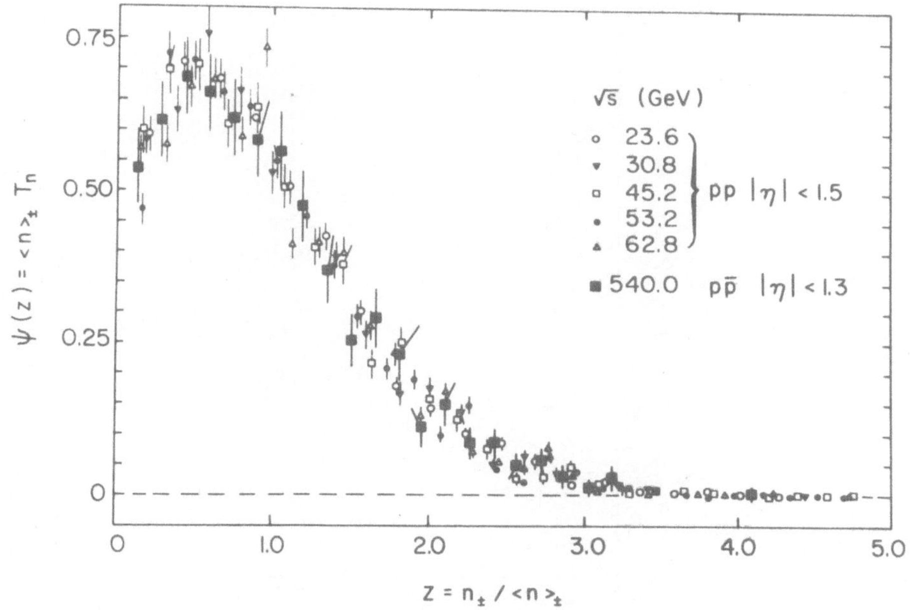

Fig. 1.8 : Distribution of the charged particle multiplicity divided by the average charged particle multiplicity for a given energy. [10]

p_t DISTRIBUTIONS

The increase of $\langle p_t \rangle$ with incident energies and the flattening of the p_t distribution at high p_t was evidenced in cosmic ray data (fig. 1.9). It has been confirmed at the collider (fig. 1.10) and the same phenomenon has been shown to be true for neutral secondaries [11,12]. It was also found for the first time that $\langle p_t \rangle$ increases when going from low multiplicity events to high multiplicity ones (fig. 1.11). This was confirmed in UA1 data in 1982 fig. 1.12 [12]. Finally double core events have been seen in cosmic ray data fig. 1.13 which should most likely be interpreted as jets as seen at the collider [13].

CENTAURO EVENTS

The Centauro events reported in cosmic rays [7] are 6 very high energy events characterized by a high multiplicity of hadrons and a multiplicity of electromagnetically showering particles which is consistent with zero (fig. 1.14). For the 2 Centauros where it has been possible to measure the emission angles of the particles, the transverse momenta have been found to be high $\langle p_t \rangle > 1$ GeV/c. The emulsion chamber shows separate signatures for electromagnetically showering particles and for other hadrons, but is only sensitive to the latter via secondary electromagnetic cascades produced by them. Hence both the total energy and transverse momenta in Centauro events depends on a knowledge of the fraction K_γ of the total energy which is converted into electromagnetic cascades. K_γ is found to be 0.2 for nucleons but might be higher for charged pions [14]. In addition there are uncertainties associated with the mean transverse momentum $\langle p_t \rangle$, since this depends on the accuracy of measuring the height of the Centauro vertex above the emulsion chamber.

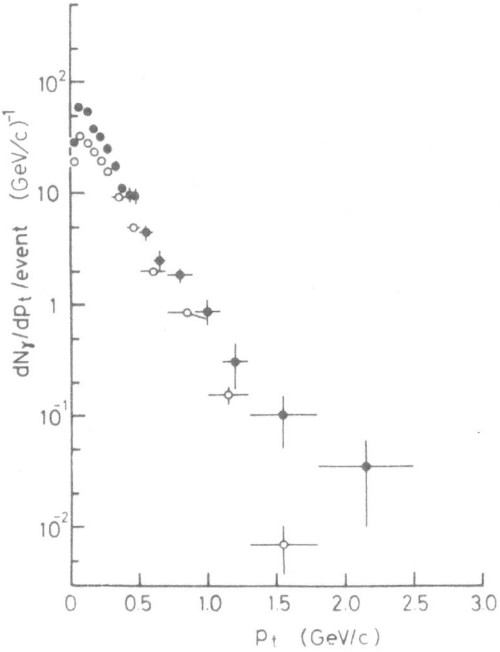

Fig. 1.9 : Comparison of P_T distributions for :
o 205 GeV proton collisions in bubble chamber [7]
● Chacaltaya data E_o > 100 TeV.

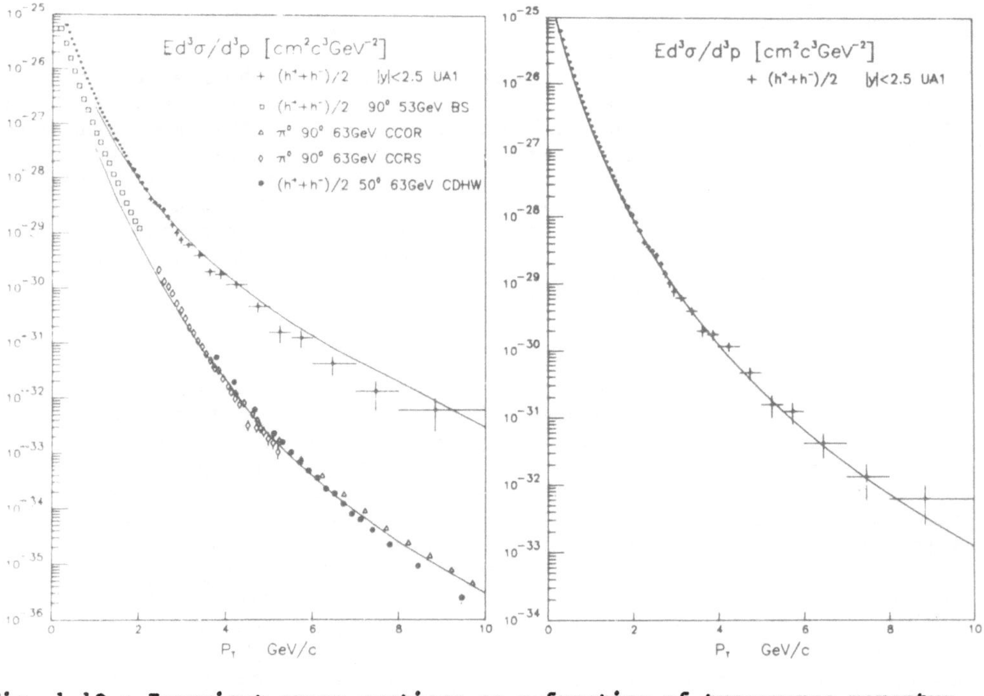

Fig. 1.10 : Invariant cross sections as a function of transverse momentum for charged particles at the proton antiproton collider and comparisons with ISR data. [12]

However it is likely that $<p_t>$ is greater than 1 GeV/c.

An analysis of the 5 Centauros found in the Mt Chacaltaya experiment by the Brasil-Japan collaboration shows that all of them have energies between $330/K_\gamma$ TeV and $370/K_\gamma$ at the vertex. This corresponds (assuming K_γ = .2) to a clustering at 1700 TeV ($\sqrt{s}$=1.8 TeV).

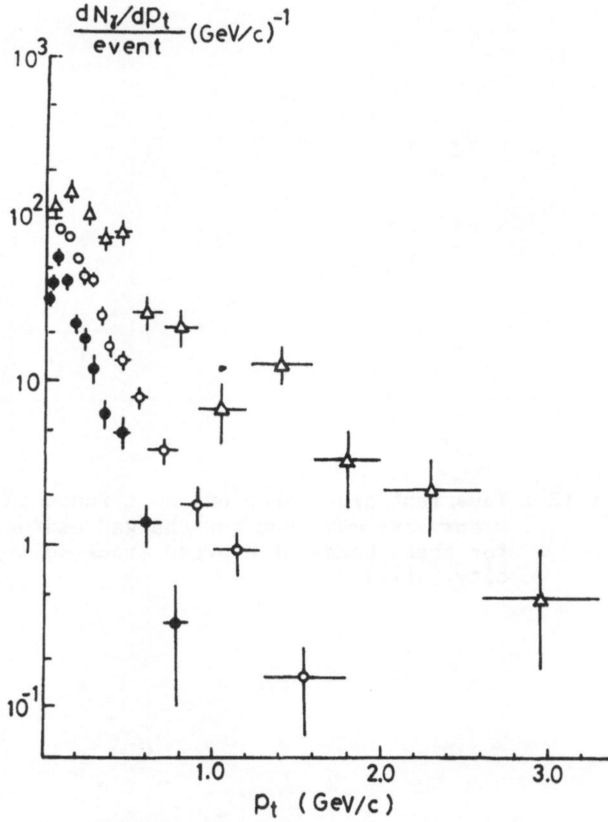

Fig. 1.11 : Distribution of transverse momentum for Mirim, Açu and Guaçu events. [7]

Non-observation of this phenomenon at the collider [15,16] together with the fact that double core events reach a frequency of 10 % near 1000 TeV ($\sqrt{s}$=1.4 TeV) in cosmic ray data suggests that a threshold may exist above the present collider energies.

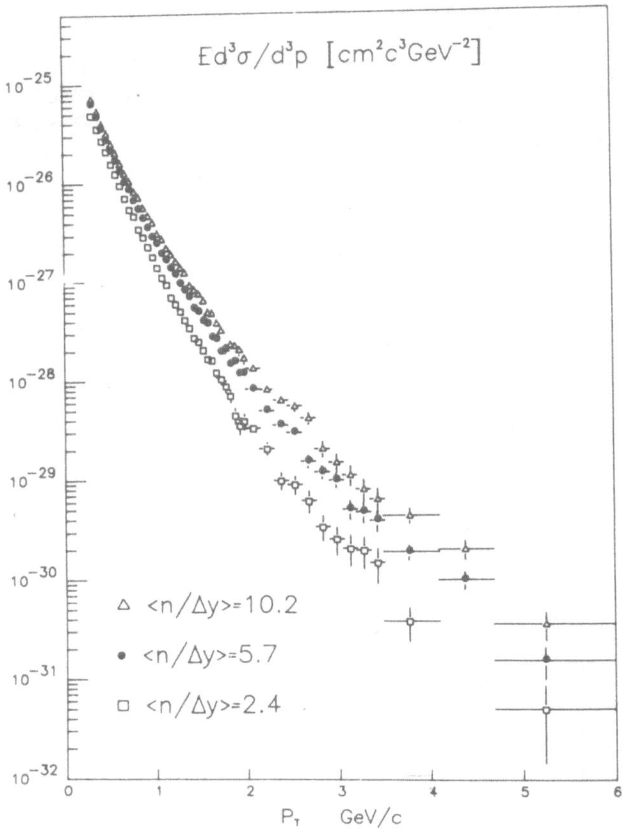

Fig. 1.12 : Invariant cross sections as a function of transverse momentum for charged hadrons for three bands of charged track multiplicity. [12]

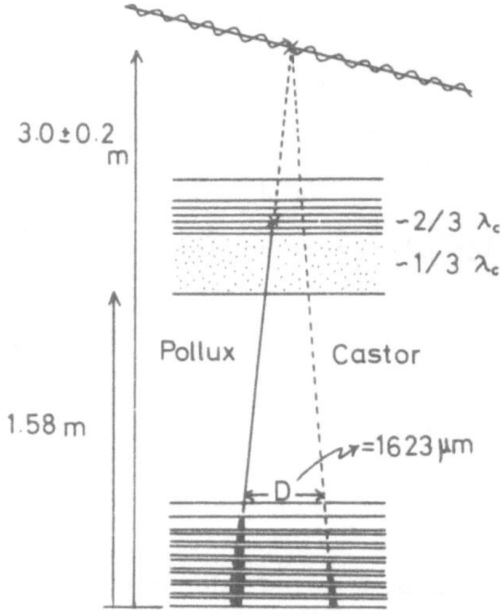

Fig. 1.13 : Schematic view of the binocular Castor-Pollux event.

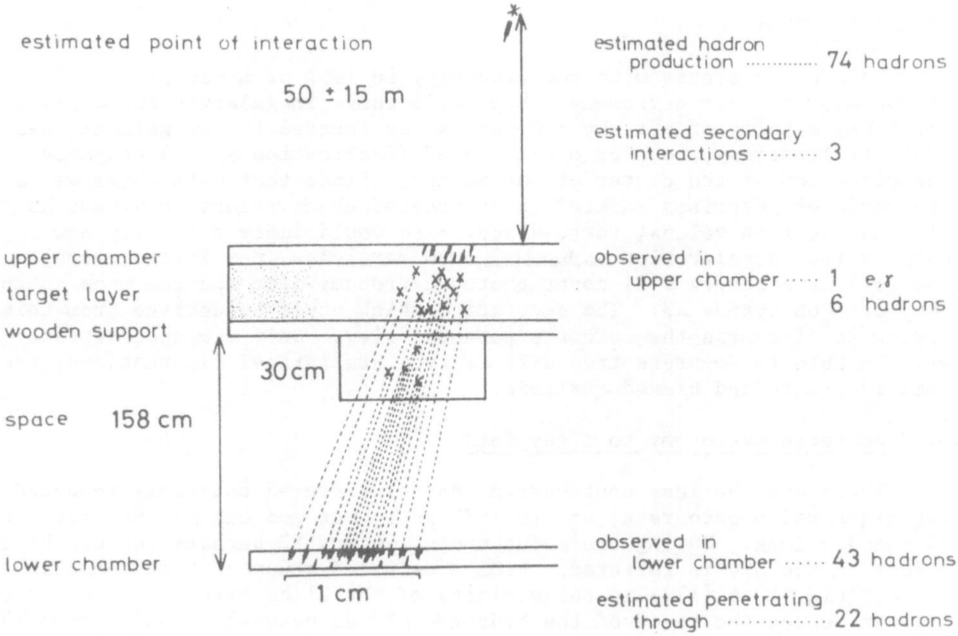

Fig. 1.14a : Illustration of Centauro I [7]

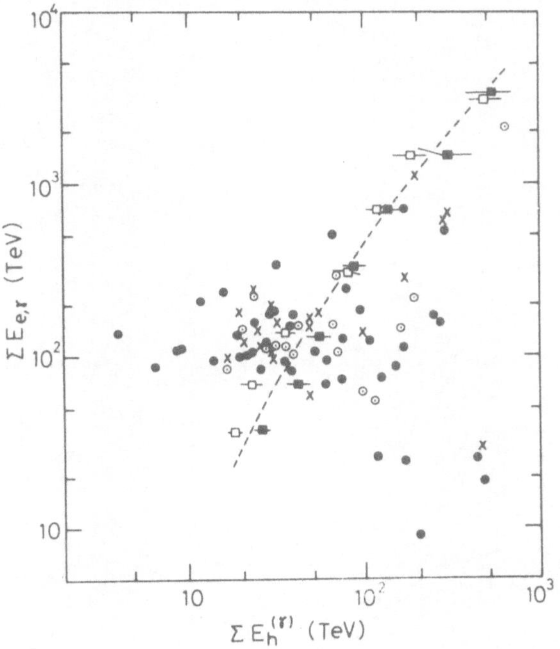

Fig. 1.14b : Diagram of hadron energy sum versus electro-magnetic energy for events with total visible energy larger than 100 TeV. [7]

CYGNUS X3 STORY

This story starts with the discovery in 1967 of a new star called Cygnus X3, in X ray astronomy. Fig. 1.15 shows in galactic coordinates the X Ray mapping of the sky : Cygnus X3 is located in the galactic plane (308° inclination), however off centered (declination = 41°) compared to the direction of the center of our galaxy. Since that date there was a crescendo of surprises related to unexpected observations on Cygnus X3. If taken at face values, these discoveries would imply a totally new physics (new acceleration mechanisms, new particles, new interactions). The following review will cover photon astronomy data and the muon astronomy data on Cygnus X3. The skepticism which could be derived from this review is of course the author's personal view. Only new observations will be able to separate true effects from statistical fluctuations, systematic errors and biased analysis.

a) From radio astronomy to X ray data

These are the less controversal data. Infrared astronomy reported two giant radio outbursts, one in 1972 september and one on 1982 september 28 one day long. During these outbursts, Cygnus X3 becomes the brightest object in the sky in infrared. From a detailed study of the absorption curve, (fig. 1.16 [17]) in the vicinity of the 21 cm wave/length emission, one can deduce the speed of the hydrogen clouds between us and Cygnus X3. Since we know the rotation curve of the hydrogen clouds around the center of the galaxy, one can deduce a minimum distance of 11 kpc from us. This places Cygnus X3 in the suburb of the galaxy fig. 1.17. It could also be extragalactic : however in that case, the luminosity would be so huge that it would be already totally impossible to conceive even at the level of radio, infrared or X ray data. At 11 kpc distance, the radio energy involved during the outbursts is about 10^{35} ergs per second. This has to be compared for instance with the well known Crab pulsar, which is the

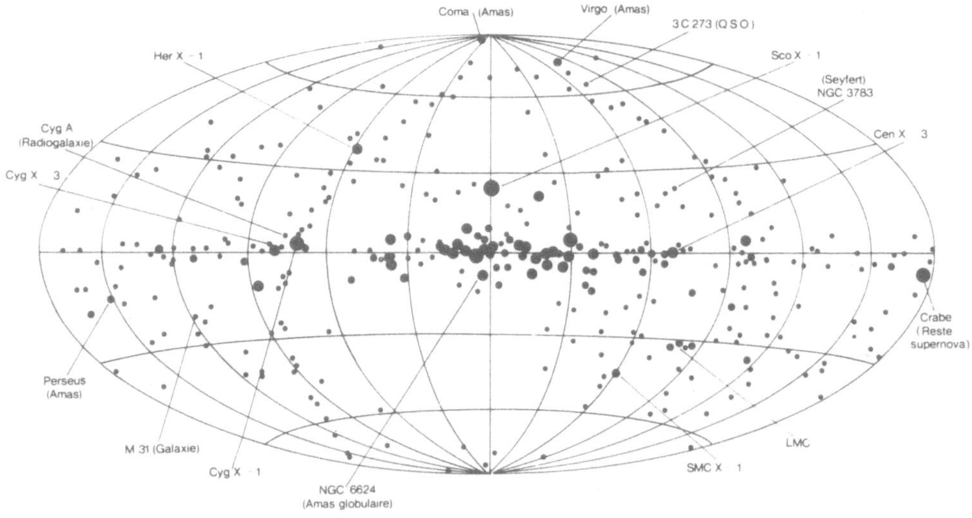

Fig. 1.15 : Position of some point like X ray sources in galactic coordinates. The size of the circles is proportional to the brightness of the sources.

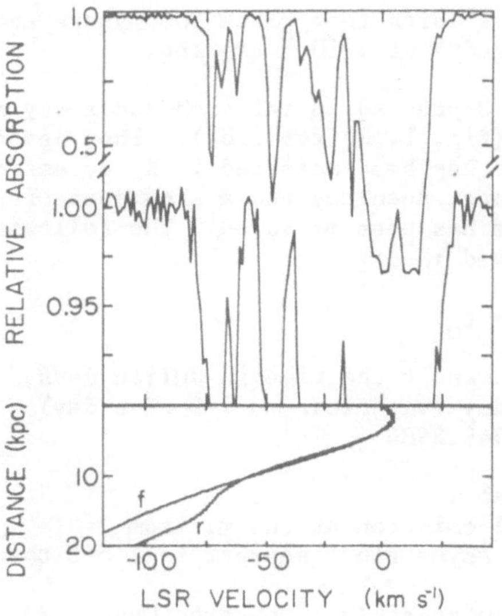

Fig. 1.16 : **Relative absorption of the radio emission of Cygnus X3 in the vicinity of 21 cm wave length. This absorption is due to the Doppler shift of the 21 cm wave length for the molecular interstellar hydrogen between us and Cygnus X3. The relation between the speed and the distance of the hydrogen clouds is also shown. [17]**

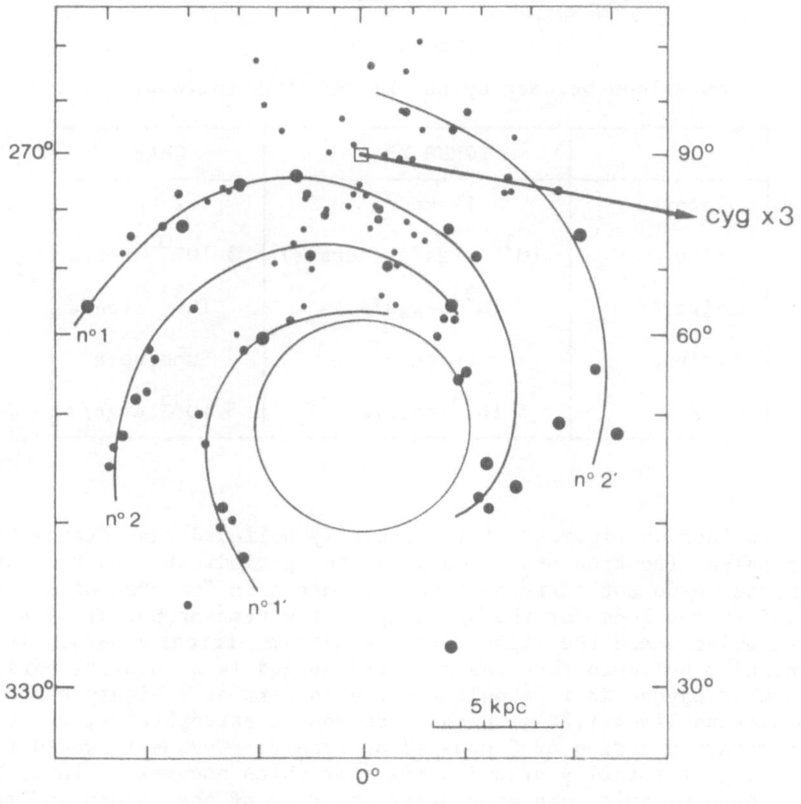

Fig. 1.17 : Position of Cygnus X3 in our galaxy.

remnant of a supernova which took place 1000 years ago, and for which the radio luminosity (steady) is 3 10^{30} erg/sec.

Observations of Cygnus X3 in infrared and X ray astronomy are in excellent agreement (fig. 1.18 from [18]). They demonstrate a periodical emission. The period has been measured in X ray astronomy (fig. 1.19 from [19]) with the best accuracy and a slow down of this 4.8 hour period as a function of time has been measured. The following formula for the period has been derived :

$$P = P_0 + P' (t - t_0)$$

where P is the period and t the time in Julian days,
with P_0 = 0.1998630 day (vey close to 1/5 of a day)
t_0 = Julian day 2440949.8986
P' = 1.18 10^{-9}
Figure 1.19 shows that :
* there is a residual emission at the minimum
* the light curve is asymetric : sharper fall off than rising.

It has never been possible to observe Cygnus X3 in visible light astronomy. This would imply the existence of a surrounding gaseous "cocoon" which absorbs the visible light.

The infrared luminosity is 10^{33} ergs/s and the X ray luminosity is 5 10^{35} ergs/s. A comparison between Cygnus X3 and the Crab pulsar is shown in the following table 1.1

Table 1.1

Comparison between Cygnus X3 and Crab luminosities.

	CYGNUS X3	CRAB
Distance	11 kps	2 kps
Radio	10^{35} ergs/s (bursts)	3 10^{30} ergs/s
Infrared	10^{33} ergs/s	10^{33} ergs/s
Visible	Unseen	Supernova
X	5 10^{33} ergs/s	5 10^{35} ergs/s

From the luminosity figures, it is generally believed that Cygnus X3 is indeed a pulsar (neutron star). However the periodicity and the shape of the light curve do not correspond to the intrinsic features of a pulsar : the period is too long for the spinning of the remnant pulsar of a collapsed object, and the light curve is not symmetrical enough. It is then generally believed that the observed period is an orbital period, and then that Cygnus X3 is a pulsar which is part of a binary system. In this picture (fig. 1.20), which is common in astrophysics, we do not know the rotation period of Cygnus X3 on itself. Cygnus X3 would then be a compact object rotating around a big star which however is invisible. The atmosphere of this companion plays the role of the cocoon and absorbs the visible light emitted by Cygnus X3.

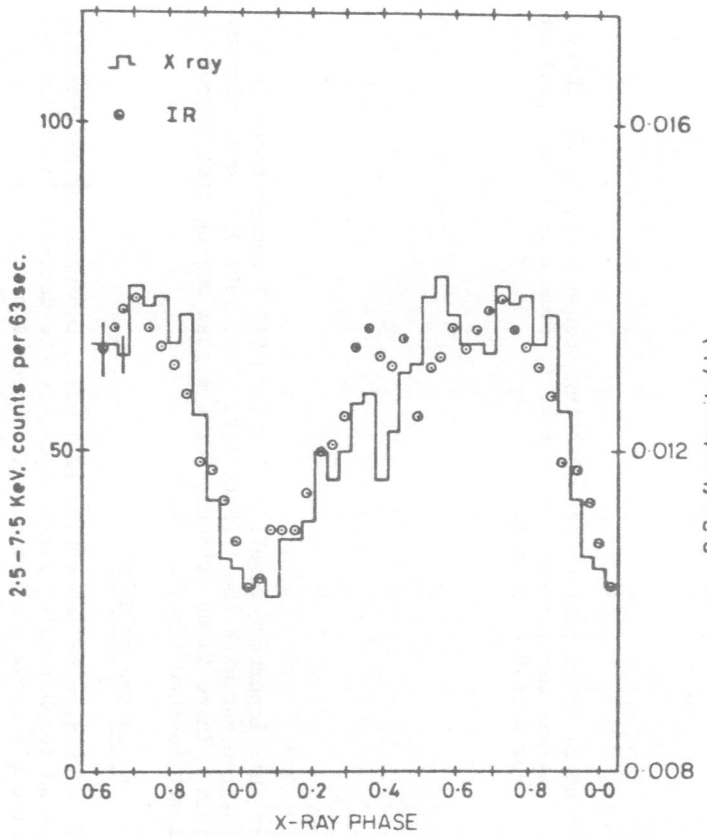

Fig. 1.18 : X Ray and Infrared periodical emission of Cygnus X3. The data are Fournier analysed and plotted modulo the period (so called phase plot). [18]

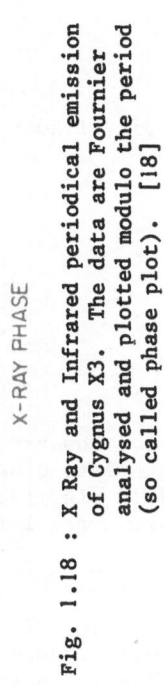

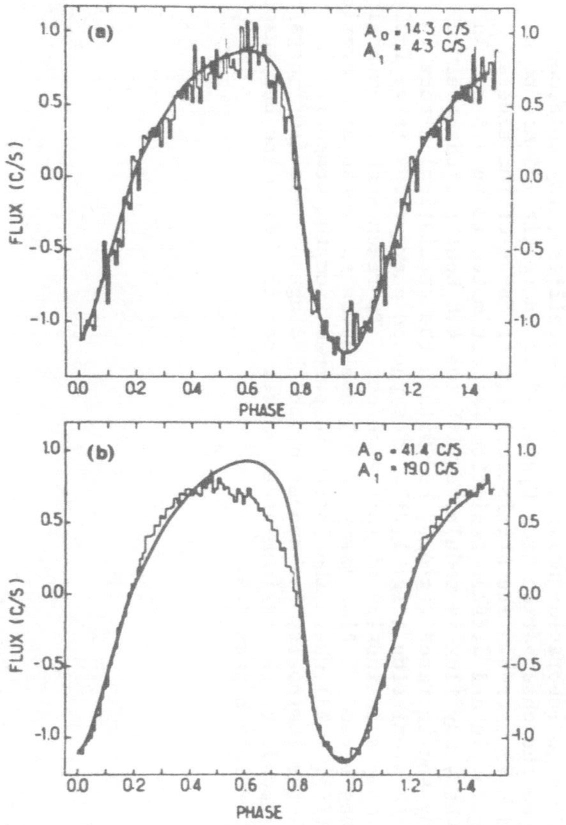

Fig. 1.19 : Most precise phase plot of Cygnus X3 from X ray data ([19] ephemerides). The slow down of the period is demonstrated in b).

b) Low energy gamma ray astronomy (MeV range)

In 1977 a collaboration using the SAS 2 satellite γ ray telescope [20] reported the observation from Cygnus X3 of energetic (35 MeV to a few hundred MeV) γ rays. The statistical significance of the excess above the galactic and diffuse radiation was estimated to be 4.5 σ. In addition, the γ ray flux is modulated with the 4.8 hour period observed in the X ray and infrared regions, and within the statistical errors in phase with this emission (fig. 1.21). The photon energy flux (fig. 1.22) would now imply a luminosity of 10^{37} ergs/sec. To reach such a luminosity during a long period of time (many years) one needs to invoke an accretion mechanism (fig. 1.23) where the energy is pumped from the companion. Furthermore the luminosity is bounded by the Eddington limit (10^{38} ergs/sec) which corresponds to the maximum speed matter can fall from the companion because of radiation pressure.

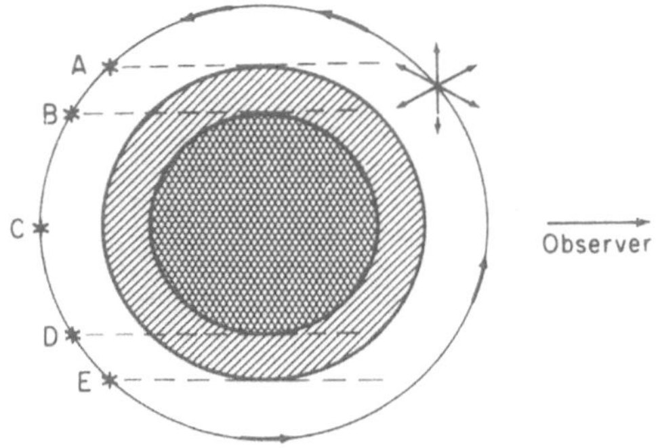

Fig. 1.20 : Model of compact object rotating around a big star. Three regimes might occur : from E to A, from B to D (eclipse) and from A to B or D to E.

However this important result coming from SAS 2 observations is in complete disagreement with the 1975 to 1982 COS-B satellite observations [21] fig. 1.24. There is no evidence for a signal coming from Cygnus X3. The upper limit is shown in fig. 1.22.

c) TeV gamma ray astronomy data

In 1978 a Crimean collaboration [22] using two pairs of parabolic mirrors coupled to photomultipliers to detect the Cerenkov light coming from extensive air showers claimed, after 6 years observation, to have evidence of a time modulated signal in the direction of Cygnus X3 and in

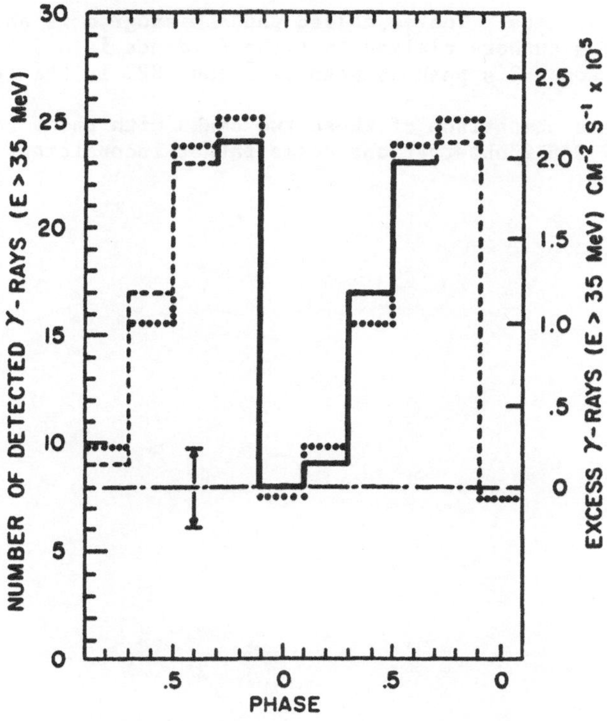

Fig. 1.21 : Evidence from SAS 2 of a time modulated signal of > 35 MeV γ rays, in phase with the period found in X ray astronomy. [20]

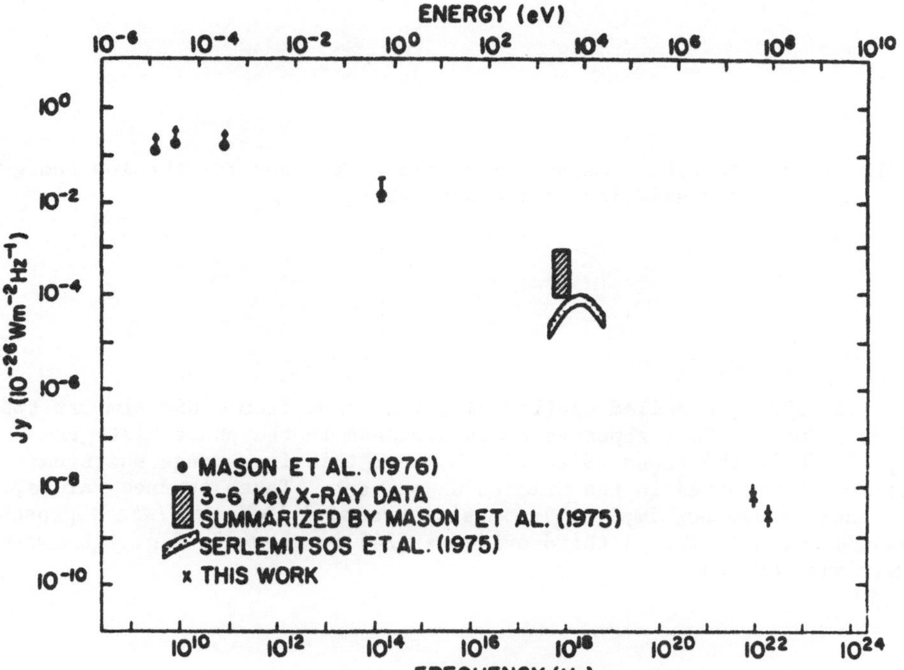

Fig. 1.22 : Cygnus X3 luminosity as a function of the light frequency.

phase with X ray data. The so called phase histogram is shown in fig. 1.25a. The authors claimed that the evidence is a 5.4 σ peak between .157 and .212 and a 3 σ peak between .768 and .823 in this plot.

However the comparison of these two peaks with the γ ray light curve for E > 35 MeV SAS 2 observations seems rather inconsistent fig. 1.21.

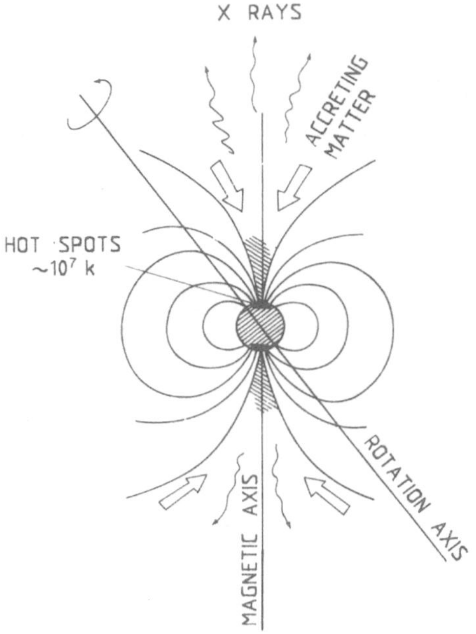

Fig. 1.23 : Accretion mechanism needed to account for the low energy γ ray emission from Cygnus X3.

In 1982 a so called confirmation [23] came from a USA similar type of experiment. They reported a 4.4 σ excess in the phase histogram fig. 1.25b in the range .5 to .7. However this is a range where no peaks were reported in the Crimean experiment. Taken at face values, these evidences would now imply a luminosity around 3 10^{37} ergs/sec approaching the Eddington limit. A third evidence [24] is shown in fig. 1.25c with a maximum near .8.

d) 1000 TeV gamma ray astronomy data

In 1983 the Kiel collaboration [25] using an experimental arrangement involving 28 scintillation counters of 1 m^2 area each at distances up to 100 m and which could investigate the electromagnetic, hadronic and also muonic nature of the core structure of extensive air showers in the primary range of 10^{15} to 10^{17} eV, reported a 4.4 σ excess in the direction of Cygnus X3 in the phase histogram fig. 1.26. The angular resolution is better than 1°. The excess is a sharp peak between 0.3 and 0.4. It would imply an extra 10^{37} erg/sec luminosity between 10^{15} and 2 10^{16} eV. This would be the first evidence for γ ray point source emitting at energies greater than 10^{15} eV. It would imply new acceleration mechanism on short distances in the vicinity of a neutron star.

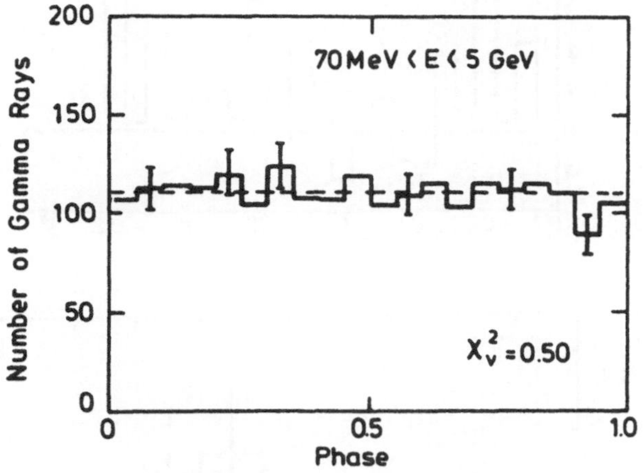

Fig. 1.24 : Negative result from COS-B observation in the search for a time modulated signal from Cygnus X3 for > 70 MeV γ rays. [21]

In october 1983 a "confirmation" came from a Leeds collaboration [26]. They use four 13.5 m^2 × 1.2 m water Cerenkov on a circle of 50 m radius. This array is sensitive to the range 10^{15} to 10^{16} eV gamma ray showers. The phase histogram is shown in fig. 1.27. A significant peak is present in the same region as the Kiel experiment. Note that the peak disappears if one does not use the ephemerides (start of the periods as a function of the date of the event) of Bonnet-Bidaud et al. [19]. In fact the Kiel collaboration used the ephemerides corresponding to fig. 1.27c where no peak emerges clearly.

e) Cygnus X3 total luminosity

A compilation of positive results on γ ray flux is shown in fig. 1.28. The spectrum ranges over more than 11 decades in energy and can be described by a power law : the number of photons, with an energy greater than E, is proportional to E^{-1}. For the local energy region E > 10^5 eV this would imply 3 10^{38} ergs/s and would violate the Eddington limit.

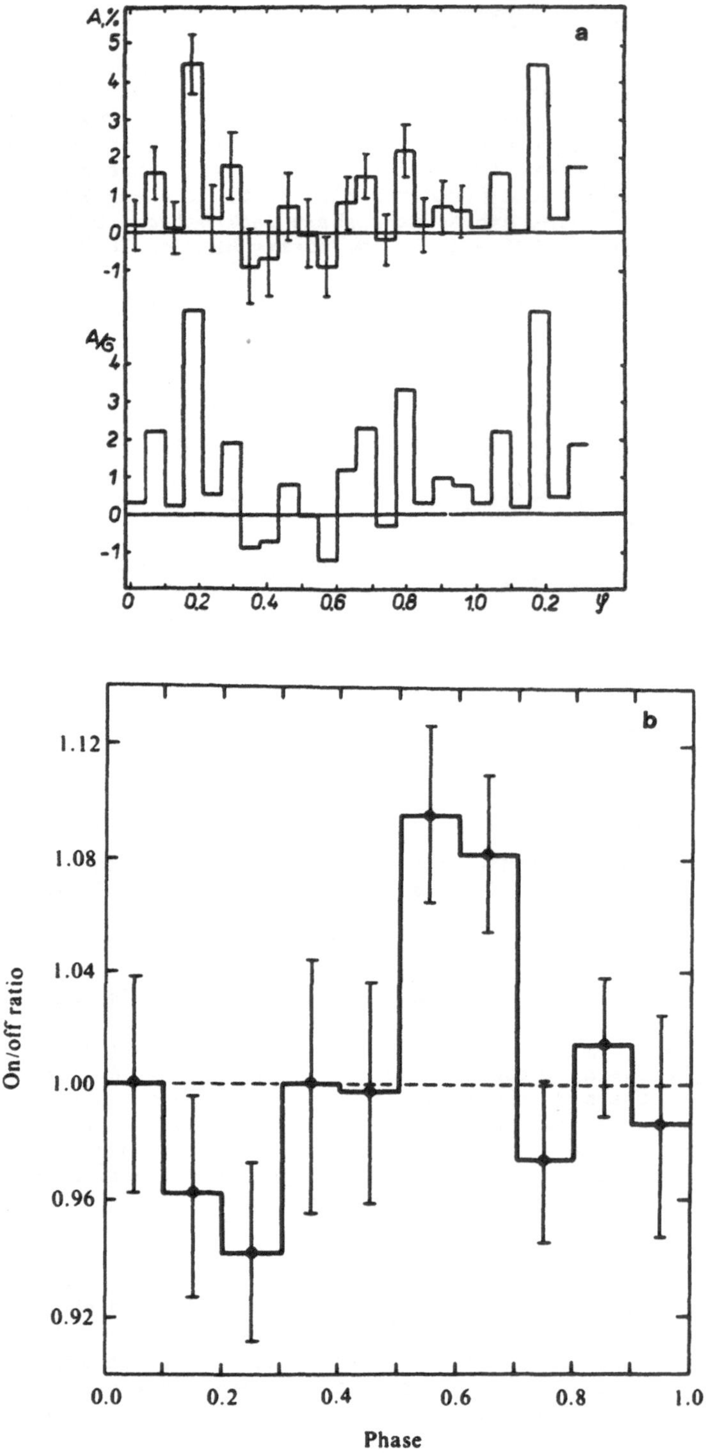

Fig. 1.25 : Evidences for modulated spikes of > 10^{12}eV γ rays from Cygnus X3 for 3 different experiments [22], [23], [24]. The agreement in the exact location of the spikes in these phase plots is not particularly good.

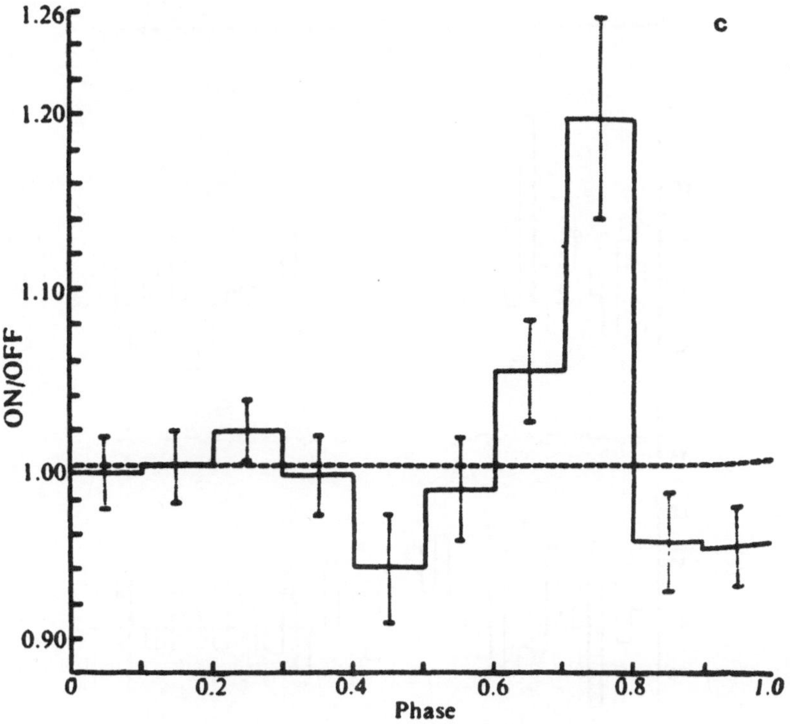

Fig. 1.25 : Continued

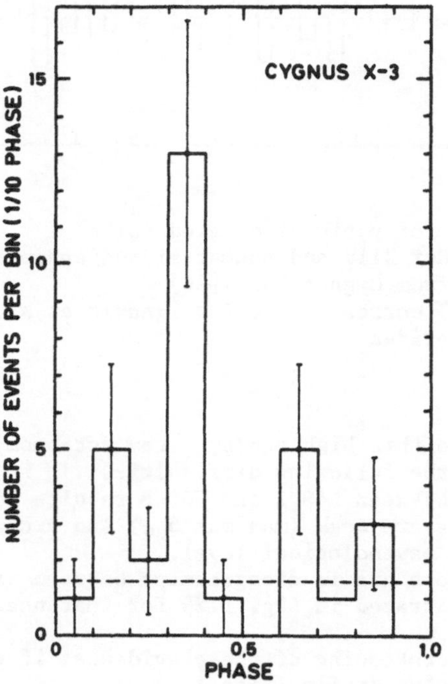

Fig. 1.26 : Evidence for a time modulated spike Parsignault et al. ephemeride) > 10^{15} eV γ rays from Cygnus X3. [25]

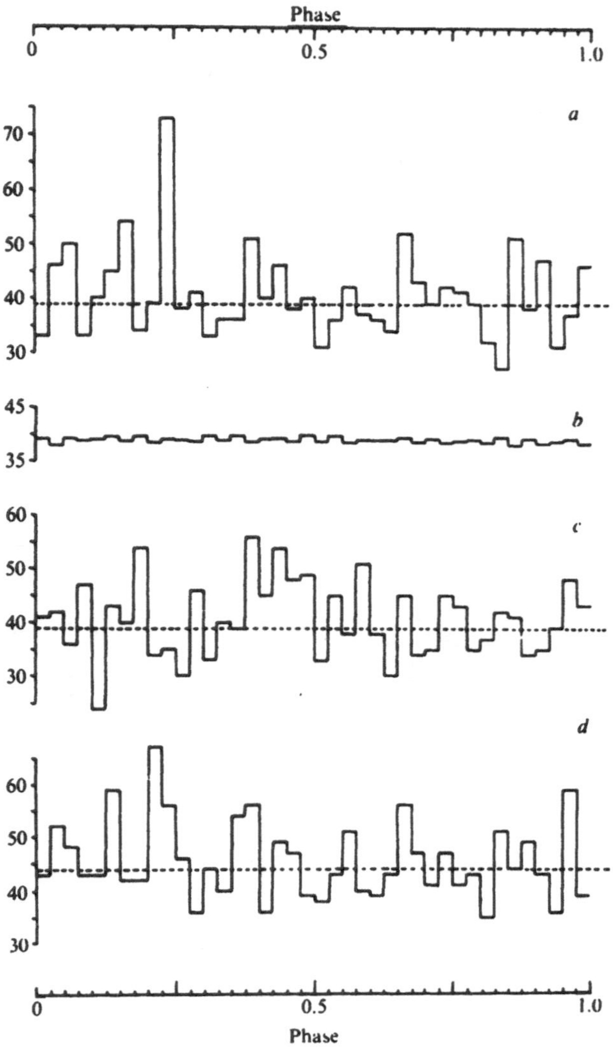

Fig. 1.27 : Evidence for a time modulated spike
a) (Van der Klis and Bonnet-Bidaud ephemerides) of > 10^{15} eV γ rays from Cygnus X3. [26]
c) and d) correspond to Parsignault et al. and Elsne et al. ephemerides.

In conclusion to this high energy γ ray astronomy, experimental evidences face with the following difficulties :

1) Discrepancy between SAS 2 and COS-B results for the MeV range.
2) All evidences are weak (maximum 5 σ) and are probably not independent on a psychological level.
3) The phase plots are in disagreement between various experiments : this is illustrated in fig. 1.25 for instance.

Theoretical understanding of these evidences if they would be confirmed hit the following difficulties :

1) New acceleration mechanisms have to be invented.
2) Eddington limit must be circumvented.

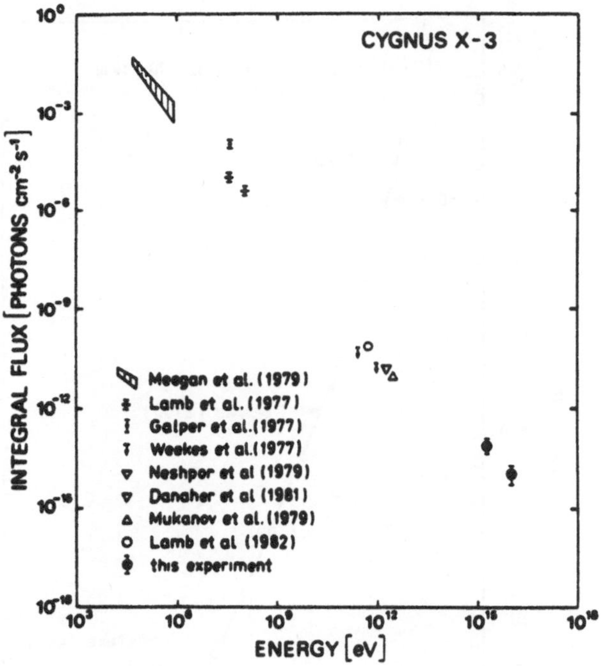

Fig. 1.28 : Plot of the integral flux of photons as a function of the energy.

f) Muon astronomy

So this is the fuzzy picture we had of Cygnus X3 until the end of february 1985. Then the focus changed to muon astronomy. Many underground proton decay detectors are presently running over the world. Fig. 1.29 shows the muon rate in these detectors. These experiments are able in principle to look for point like source giving these underground cosmic ray muons. No results are reported from the water Cerenkov IMB and Kamioka experiments due to either the on line vetoing on muons or the too poor angular resolution or both. The Kolar Gold field experiment is too small and too deep underground to accumulage statistics.

A first positive evidence for muon production by particles from Cygnus X3 came from the Soudan detector. This is a Argonne Minnesota collaboration. A schematic view of the Soudan detector is shown in fig. 1.30. The detector consists of an array of 3456 proportional tubes, each 2.8 cm in diameter arranged in 48 layers of 72 tubes each. The size of the detector is $2.9 \times 2.9 \times 1.9$ m^3. The angular resolution is 1°. A signal is observed [27] in the phase plot when selecting muons originating from a $6 \times 6°$ window which however is off centered by 3° from the nominal Cygnus X3 position : the phase plot is shown in fig. 1.31. The excess is between .65 and .9. Data were recorded between september 1981 and november 1983.

At about the same time, the observation of a time modulated muon flux in the direction of Cygnus X3 was reported by the NUSEX collaboration in the Mont Blanc tunnel. The apparatus consists of 150 t mass cube of 3.5 m side, made of 136 horizontal plates of 1 cm thickness, interleaved with planes of tubes (fig. 1.32) having 1 cm × 1 cm cross section operating in limited streamer mode. The resolution is 1°. Nevertheless the muons have in a been selected $10 \times 10°$ window this time centered on Cygnus X3. The phase distribution (fig. 1.33 from [28]) shows an excess in the .7

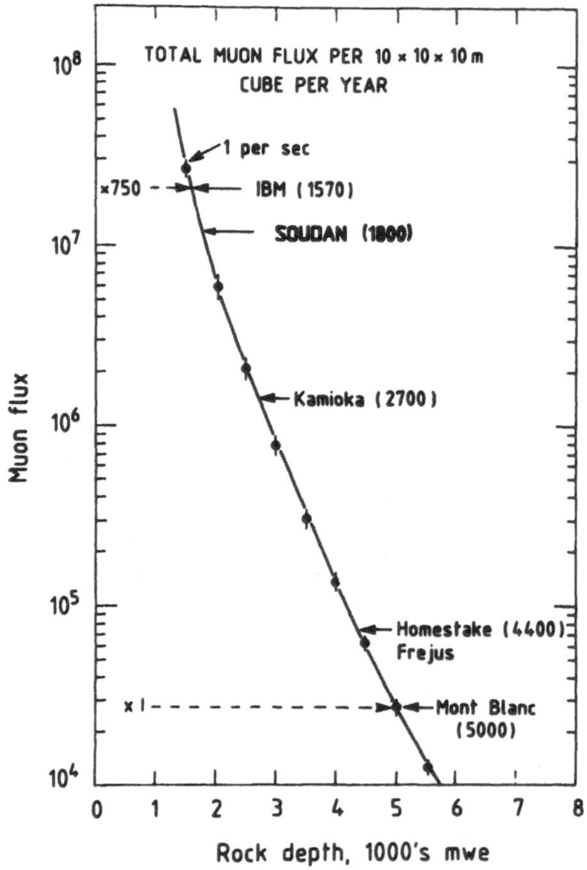

Fig. 1.29 : Muon flux in deep underground proton decay detectors.

to .8 bin where there are 32 events. However, these 32 events do not show (fig. 1.34) inside the selected window any accumulation near the exact position of Cygnus X3 but rather a depletion (the central point is not an event but the exact location of Cygnus X3). Data were recorded from 1 june 1982 to 1 february 1985.

A new proton detector is now entering the game [29]. This is the Frejus detector in the Frejus tunnel between France and Italy. It is a 912 tons detector of 12 × 6.0 m^2 horizontal surface and 6.0 m high. It is made (fig. 1.34) of 6.0 × 6.0 m^2 3 mm thick iron plates, interleaved with flash tube planes of .55 × .55 cm^2 section. Triggering is made by planes of Geiger tubes. There is a total of 1 million flash tubes (excellent granularity) and 40000 Geiger tubes. Angular resolution is better than 1°. Data taking started in june 1984. By applying the same cuts as the NUSEX experiment they obtain [29] the phase plot shown in fig. 1.35. Although the statistics is not enough to infirm the NUSEX result, they have no evidence for a peak. The only bin which sticks a little bit is not located at the same position as the NUSEX bin. They are using the same ephemerides as the NUSEX collaboration.

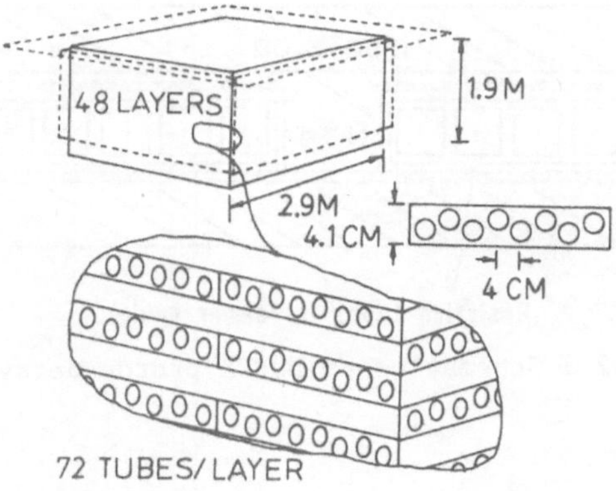

Fig. 1.30 : Schematics of Soudan I proton decay detector.

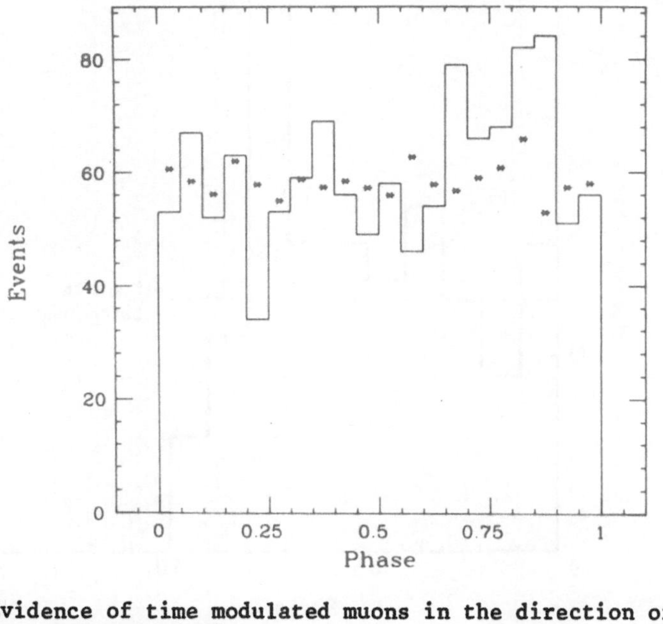

Fig. 1.31 : Evidence of time modulated muons in the direction of Cygnus X3 in Soudan I. [27]

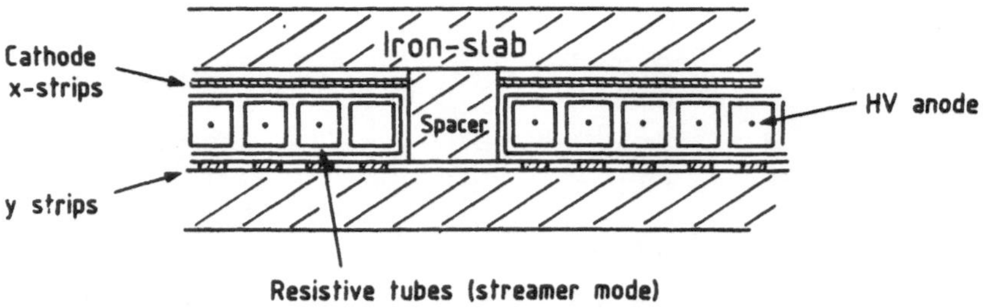

Fig. 1.32 : Schematics of NUSEX proton decay detector.

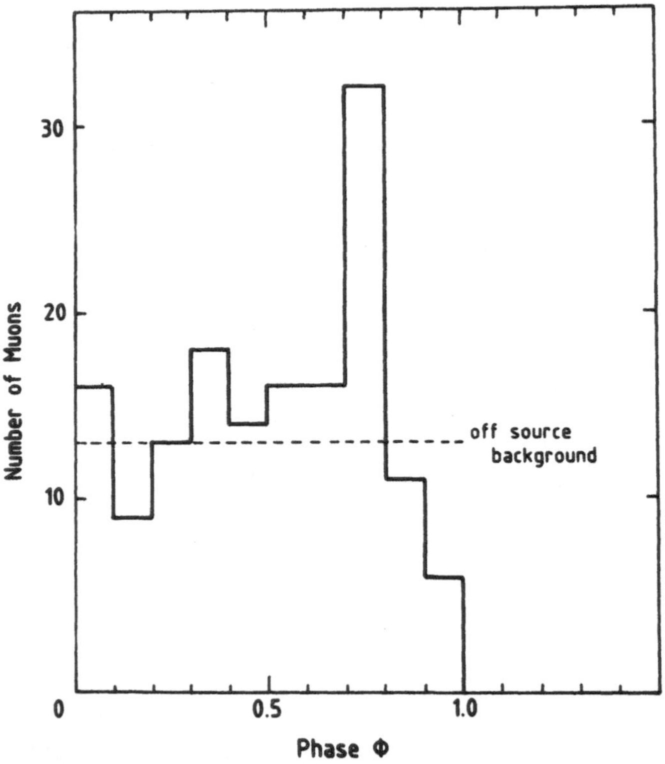

Figure 1.33 : Evidence for a time modulated spike of muons in the direction of Cygnus X3 in NUSEX. [28]

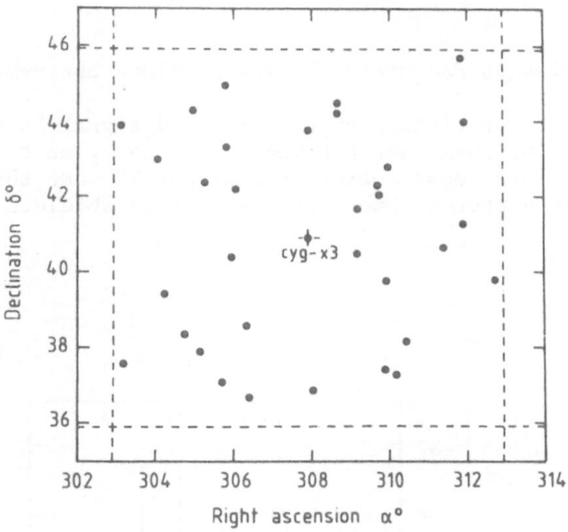

Fig. 1.34 : Angular distribution for muons in the phase interval .7 - .8 in NUSEX. [28]

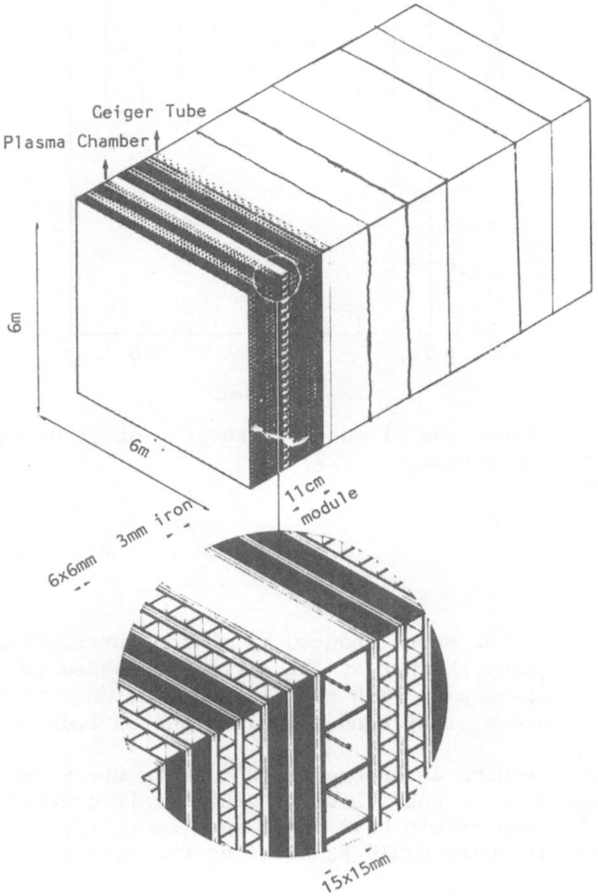

Fig. 1.35 : Schematics of the Frejus proton decay detection.

g) New physics ?

If one tries to reconcile all the previous observations, one must admit :

1) Due to the variation of the observed signal in Soudan and NUSEX with zenith angle, these are not muons induced by neutrinos interactions : there are many more muons coming from Cygnus X3 when the star is at the vertical of the detector than when the star is shielded by a big amount of earth.

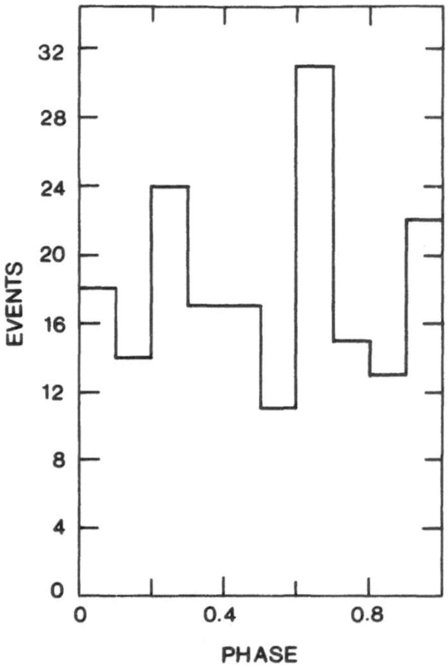

Fig. 1.36 : Phase histogram of muons in the direction of Cygnus X3 in the Frejus experiment. [29]

2) These cannot be muons induced by photon interactions because this would lead to too many photons at the surface compared to what has been reported. This agrees also with the Kiel conclusion : from the muon content of the showers, they tend to reject photon induced showers (too many muons) [30].

3) Neutrons require an energy of 10^{18} eV to reach earth from the distance of Cygnus X3 in one lifetime. This is in contradiction with the flux of all known cosmic rays above such an energy.

4) Charged particles would be swept by the galactic field.

So there must be new neutral stable particles. They must be light in order to keep the 1/2 hour phase information ($\gamma > 2\ 10^4\ \beta > 1. - 10^{-9}$).

Furthermore Cygnus X3 should be a rather fast evolving system in order to understand a copious yield in 1982-1983 seen in the Soudan and NUSEX experiment and a rather null evidence in the Frejus experiment.

Clearly this is not the end of Cygnus X3 story.

NEUTRINO PHYSICS AND WEAK INTERACTIONS

We have been accustomed to think of neutrinos and antineutrinos as distinct and massless particles. Both these 2 aspects are now questionned from the theoretical and from the experimental point of view. Grand unified theory would tend to favour neutrinos which are their own antiparticles (in relation with non leptonic number conservation) and which are massive (because there are no reasons why they should not be).

DIRAC OR MAJORANA NEUTRINOS

a) Theoretical framework

Most of the following considerations are taken from the paper of Kayser [31].

A neutrino which is its own antiparticle is referred to as a Majorana neutrino and one which is not as a Dirac neutrino.

To understand the precise physical difference between a Majorana and a Dirac neutrino, let us start with a massive neutrino with negative helicity considered at the extreme left of fig. 2.1a. CPT implies the existence of an antineutrino $\bar{\nu}_+$. If ν is massive, we can force by a Lorentz transformation the ν to go the other way while spinning in the same way as in the original frame. We have then transformed a negative helicity neutrino into a positive one. In Dirac view, this ν_+ is not the same as $\bar{\nu}_\pm$ and has its own antiparticle. The ν_- and ν_+ are expected to produce e^- or μ^- in their subsequent interactions while $\bar{\nu}_+$ and $\bar{\nu}_-$ are expected to produce e^+ and μ^+. The lepton quantum number which is defined so that an e^- or μ^- is a lepton while e^+ and μ^+ is an antilepton is conserved.

The second possibility depicted in fig. 2.1b, is that the right handed particle obtained by going to a Lorentz frame in which the momentum of ν_- is reversed is the same as the CPT image of ν_-. There are just two states with a common mass : the Majorana neutrinos. If neutrinos are massive Majorana particles, the lepton quantum number is not conserved. The apparent conservation of lepton number in high energy ν interaction is then due only to the V-A structure of the interaction. In grand unified theories, lepton number is in general violated ($p \rightarrow e\ \pi$). Thus one would expect that in such theories neutrinos will be of the Majorana type.

Why don't we know already if neutrinos are of Dirac or Majorana type ? This is due to the fact that, weak interactions being left handed ones (V-A), the ability to determine experimentally whether there are four states or two gradually disappears as the neutrino mass goes to zero. Neutrinos are produced left handed through weak interactions and the amount of right handed component in subsequent interactions is very small if the mass is very small. No realistic experiment of accelerator type can be thought of the disentangle this confusion if neutrino masses are in the ten eV range.

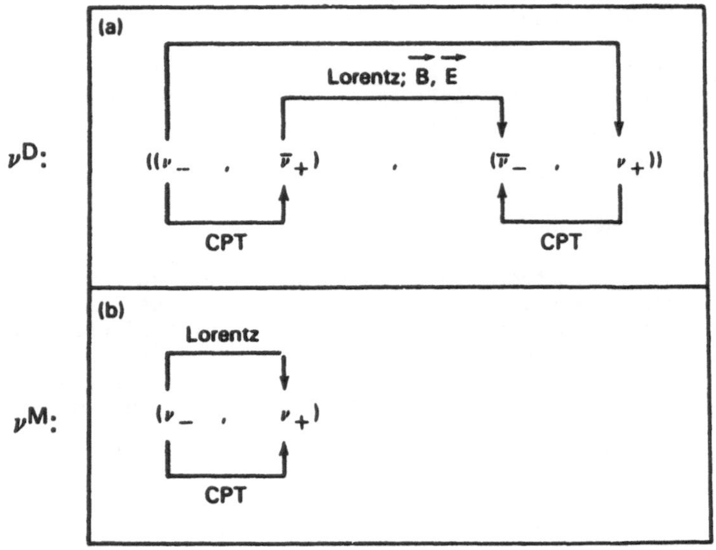

Fig. 2.1 : Illustration of the difference between a Dirac neutrino a) and a Majorana neutrino b). [31]

There is however, one special experiment which should be able to distinguish Majorana from Dirac neutrinos even if the neutrino masses are as low as 1 eV. This experiment is the search for neutrinoless double beta decay ($\beta\beta_{0\nu}$) :

$$(A,Z) \rightarrow (A,Z-2) + 2\,e^- + 0\nu.$$

The diagram is shown in fig. 2.2.

Suppose the ν_m exchanged in fig. 2.2 are Dirac particles, this ν_m must be an antineutrino at one vertex and a neutrino at the other one. Thus $\beta\beta_{0\nu}$ cannot occur if neutrinos are Dirac particles, but only if they are Majorana particles. Double β decay with the emission of 2 ν's is an allowed second order effect in the standard V-A theory.

In addition neutrinoless double beta decay cannot occur in left handed weak interactions unless the neutrino mass is non zero. This is due to the fact that at one vertex the neutrino is left handed and must be right handed at the other one. Nevertheless a neutrino with zero mass could contribute if there exists right handed currents. The hamiltonian for the double beta decay neutrinoless transition is [32]

$$\eta = -\frac{G}{\sqrt{2}} \cos\theta_c \{\bar{\psi}_e\gamma_\mu(1-\gamma_5)\psi_\nu.\bar{u}\gamma^\mu[(1-\gamma_5) + \eta_{LR}(1+\gamma_5)]d$$

$$+ \bar{\psi}_e\gamma_\mu(1+\gamma_5)\psi_{\nu'}.\bar{u}\gamma^\mu[\eta_{RR}(1+\gamma_5) + \eta_{RL}(1-\gamma_5)]d + h.c.$$

where η's are the couplings of right handed currents. This discussion is summarized in the following table 2.1.

Table 2.1

Comparison between Dirac and Majorana neutrinos.

	ν_D				ν_M			
	No right handed currents		With right handed currents		No right handed currents		With right handed currents	
	$m \neq o$	$m = o$	$m \neq o$	$m = o$	$m \neq o$	$m = o$	$m \neq o$	$m = o$
Lepton Quantum Number Conservation	Yes	Yes	Yes	Yes	No	Yes	No	No
$\beta\beta_{0\nu}$	No	No	No	No	Yes	No	Yes	Yes

b) Double beta decay experiments

1) Direct counting experiments

They are based on the search for double beta decay of ^{76}Ge in ^{76}Se. Why ^{76}Ge ?

- natural germanium contains 7.76 % of ^{76}Ge which could be potentially double beta active.

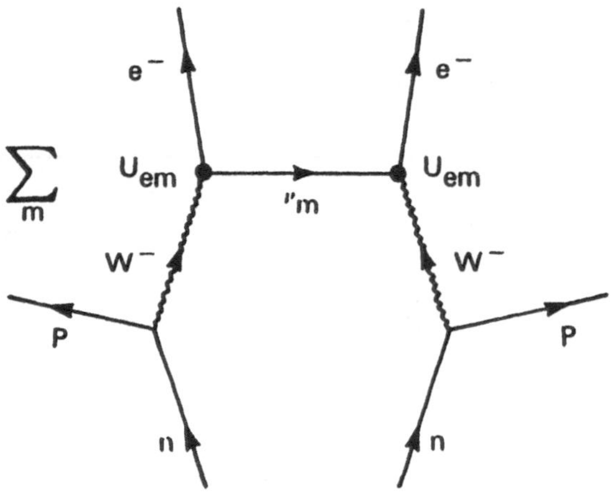

Fig. 2.2 : Double beta decay diagram. [31]

- intrinsic or Li doped Ge detectors have an excellent energy resolution (1 % at 1 MeV), can be grown in single crystals up to mass of about 1 kg or more, are reasonably free from radioactive contaminants; moreover by using several crystals one can reach few kilogram masses.

- neutrinoless decay from ground state to ground state ($0^+ \rightarrow 0^+$) would manifest itself in a sharp line at 2041 keV. Decay to the 2^+ excited state of ^{76}Se would manifest by a line at 1482 keV in coincidence with a γ ray of 559 keV. Experiments consist of a set of Ge detectors heavily shielded with a variety of materials (fig. 2.3 from [33]). Typical background counting is in the range of a few 10^{-3} count/keV/hour (fig. 2.4). A summary of present upper limits is given in table 2.2.

Table 2.2

Experimental results on double beta decay direct counting experiments using ^{76}Ge.

Collaboration	0^+ T 1/2 [Yrs]	References
Bagatelle-South Carolina	$> 1.7\ 10^{22}$	P.R.L. 50 (1983) 521
Bordeaux-Zaragoza	$> 6.5\ 10^{21}$ $(0^+ \rightarrow 2^+)$	N.C. 78A (1983) 50
Caltech	$> 1.7\ 10^{21}$	To be published
Guelph-Downsview-Kinston	$> 3.2\ 10^{22}$	To be published
Milano	$> 2\ 10^{22}$	P.R. 121B (1983) 72
Osaka	$> 2\ 10^{22}$	Int. Symp. Osaka March 84
Richland-Columbia	$> 1.7\ 10^{22}$	P.R.L. 54 (1985) 2309

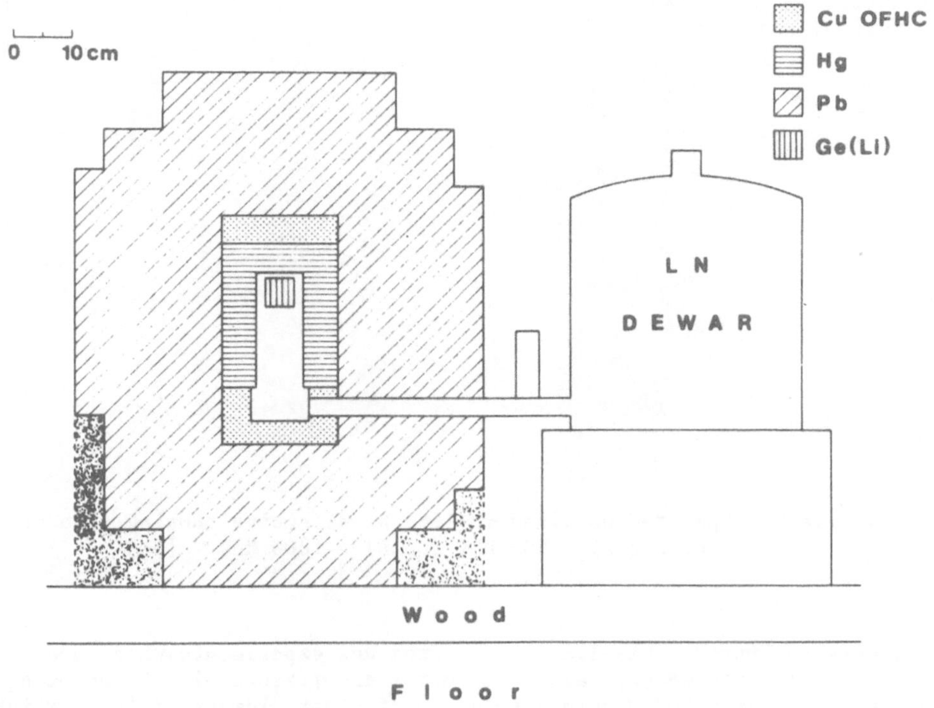

Fig. 2.3 : Experimental set-up of the Mont Blanc double beta decay experiment. [33]

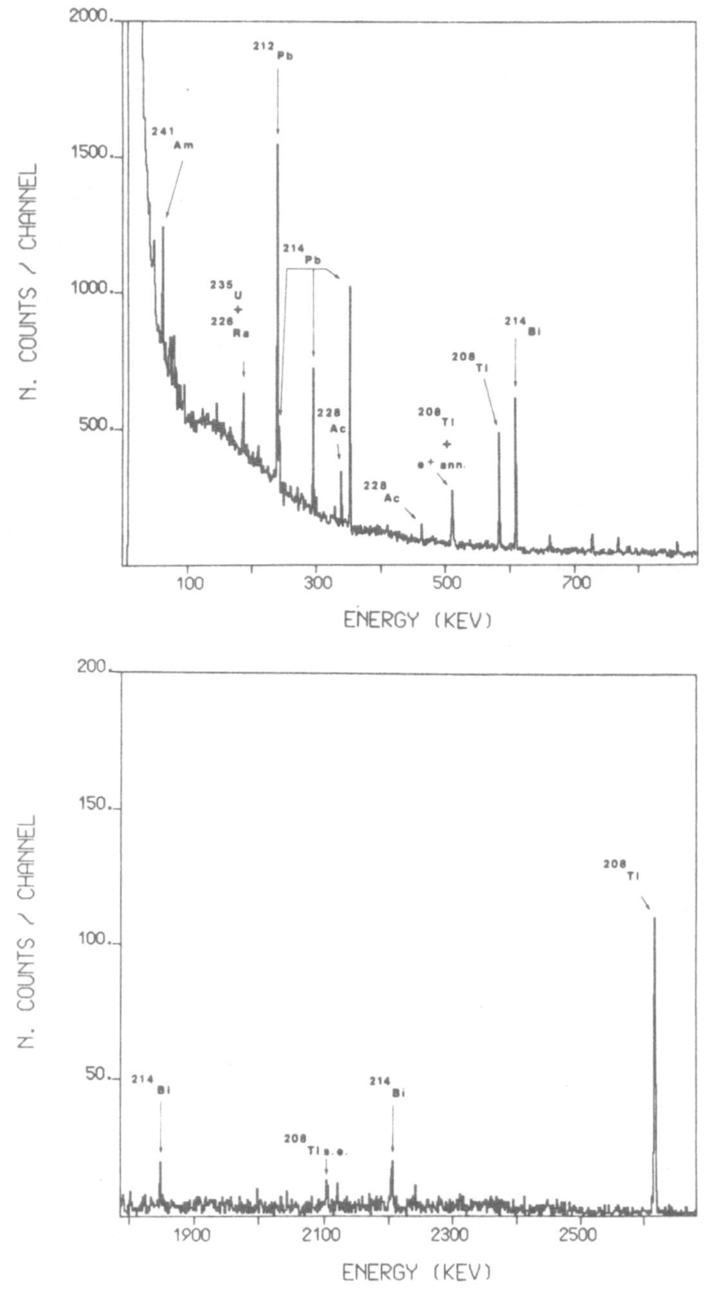

Figs 2.4a,b : Spectrum obtained in 1668 h. effective running time for energies a) < 900 KeV and b) > 1800 KeV. [33]

Prospects to improve this limit comes from new experiments which are planned using more Ge crystals. But other experiments will be devoted soon to the search for double beta decay of other elements : ^{136}Xe which can be used in TPC, or other elements using Ionisation Resonance Spectrometry or even bolometers or superconducting tunnel junctions.

2) Geochemical experiments

The direct counting experiments have the potential to distinguish between 2ν double beta decay and 0ν double beta decay, but do not have evidence for either mode. The geochemical experiments, which have not this potential, showed however an unambiguous evidence for double beta decay.

The geochemical method benefits from the accumulation of double beta decay product nuclei over geological periods (10^9 years) in ores which are rich in prospective double beta decay active parent nuclei. Well established examples are now the decays $^{130}Te \rightarrow {}^{130}Xe$ and $^{82}Se \rightarrow {}^{82}Kr$ [34]. Since 2ν and 0ν emission cannot be distinguished this gives upper limits for the 0ν mode (assuming that the entire signal comes from the 2ν mode) which are shown in table 2.3 (from [34]).

Table 2.3

Experimental results of geochemical double β decay experiments. Stated are the actual values as well as the lower limits for the 0ν-halflives after of the consideration of the 2σ-errors.

Decay, transition energy	$T^{\Sigma}_{1/2}$ [yrs]	$T^{0\nu}_{1/2}$ (2σ) [yrs]
$^{130}Te \rightarrow {}^{130}Xe$ (2.53 MeV)	$(2.55 \pm 0.2) \times 10^{21}$	$> 2.15 \times 10^{21}$
$^{128}Te \rightarrow {}^{128}Xe$ (.87 MeV)	$> 8 \times 10^{24}$	$> 8 \times 10^{24}$
$^{82}Se \rightarrow {}^{82}Kr$ (3.0 MeV)	$(1.7 \pm 0.3) \times 10^{20\,+})$	$> 1.1 \times 10^{20}$
$^{130}Te/^{128}Te$ combined	$T^{130}_{1/2}/T^{128}_{1/2} > 3030$ (2σ)	

This might suggest that 2ν rates are irrelevant to the problem but this is not so. Rather they may serve to "calibrate" the nuclear matrix elements which are difficult to calculate but which are allowed in the standard Fermi theory by the simultaneous decay of 2 neutrons. The definite upper limit which has been derived from these experiments is in the range of 6 to 9 eV for the neutrino mass (2σ) varying according to detailed assumptions. The sensitivity of the geochemical experiments is presently about the same as the direct counting germanium experiments fig. 2.5a,b.

The mass limits deduced from double beta decay experiments is in any case in contradiction with the neutrino restmass > 20 eV reported by Ljubimov et al. [35] based on the tritium single beta decay experiments. If both results stand up, this would imply that the neutrino is not a

Majorana particle and might create problems in particle symmetry schemes in GUT's.

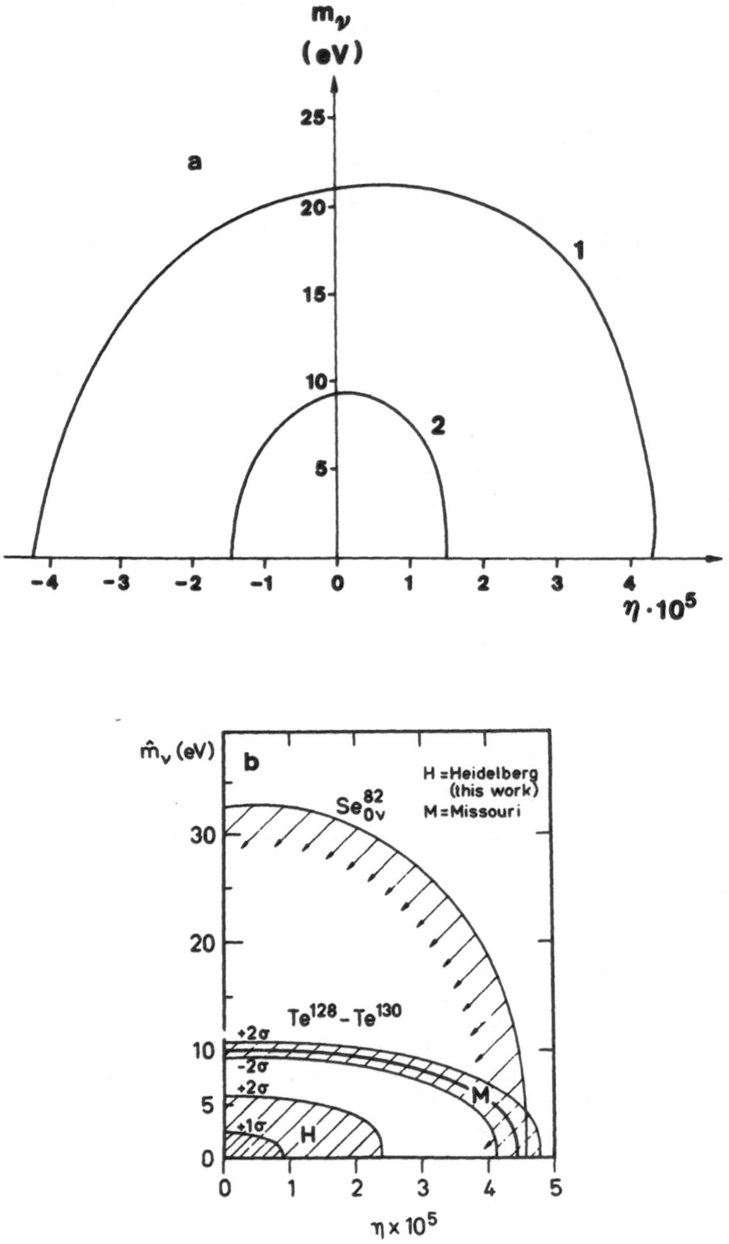

Fig. 2.5a,b : Exclusions contours in m_ν and η for a) direct counting experiments b) geochemical experiments. [33,34]

a) Magnetic spectrometers

So far the best experimental results on direct neutrino mass measurements come from β spectrum magnetic spectrometer measurements using Tritium source :

$$^{3}T \rightarrow {}^{3}He + e^{-} + \nu.$$

The shape of the β spectrum (the famous Kurie plot) for a free ^{3}T nucleus is totally determined by phase space and depends only on the end point energy value and the neutrino mass.

From the experimental point of view, the precise determination of neutrino mass from β-spectrum shape is rather difficult. For a 10 eV neutrino mass the resolving power should be 0.1 % at least since the end point value for tritium decay is around 18.5 keV. Second, because of the low counting rate in the vicinity of the end-point the radioactivity of the source must be as high as possible to increase the sensitivity in this region and improve the signal over noise ratio. Only the ITEP experiment [35] got a positive result which was reproduced in various conditions. They use a Tretyakov type of spectrometer (fig. 2.6) [36].

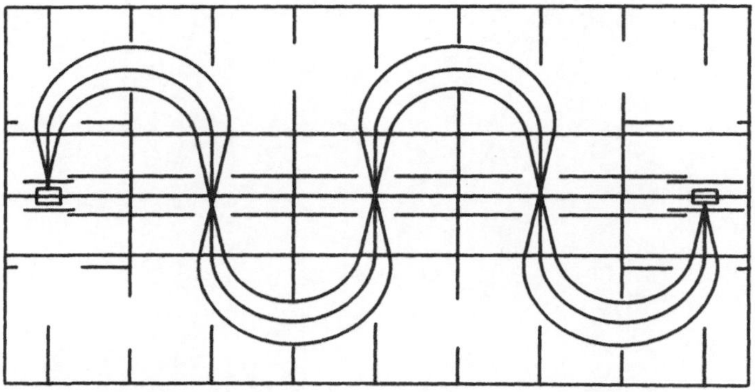

Fig. 2.6 : Tretyakov type of spectrometer. [37]

Fits to the ITEP data are shown in fig. 3.4.

Table 2.4 shows the present results of fits to the experimental data near the end point.

Table 2.4

Results of single mass fits for three different Sources (ITEP data) - E_0 is the end point - ΔE is the energy range of the data which are used in the fit.

Source	R 1	R 2	R 3	Average value
		ΔE = 1680 eV		
M_ν^2 eV^2	1364.0 ± 63	1174.1 + 81	1146. ± 140	1215 ± 130
E_0 eV	18585 ± .3	18584.1 ± .3	18485.5 ± .5	18584.2 ± 1.6
χ^2/N	317/303	523/509	471/508	
		ΔE = 330 eV		
M_ν^2 eV^2	1384. ± 175	1416. ± 156	1261. ± 283	1375 ± 140
E_0 eV	18585.2 ± 1.2	18585.5 ± .9	18584.4 ± 1.5	18585.1 ± 1.4
χ^2/N	184/165	266/318	294/316	

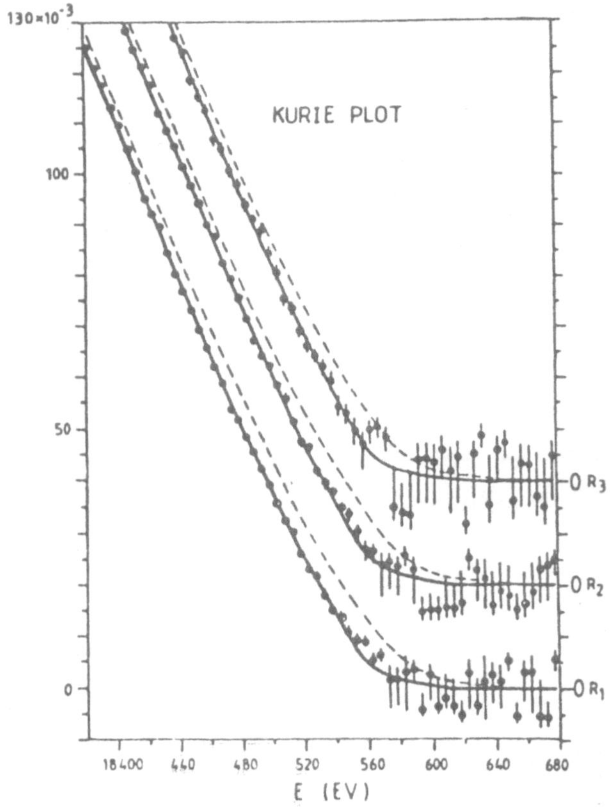

Fig. 2.7 ; Fits to the ITEP data. [35]

Two main criticisisms were made of this experiment :

1) Knowledge of the resolution function of the spectrometer

The spectrometer resolution function (SRF) is the spectrum recorded in the spectrometer when monoenergetic electrons are emitted from the source, and a bad knowledge of this function will distort the fit and so lead to wrong results. The SRF could be obtained in principle by using a conversion electron source having a sufficiently small natural width. Unfortunately, this proves to be difficult due to the occurence of accompanying shake off excitations. This is illustrated in fig. 2.8 which comes from the measurement of the Zurich β spectrometer [37] (same kind of spectrometer as Ljubymov et al.) which is presently taking data. 7290 eV K-conversion electrons from the decay of a 14403 eV state in ^{57}Fe are used. A 12 parameter function composed of 3 satellites, the no-energy loss peak, and the background all folded with the SRF give a good agreement between fit and data.

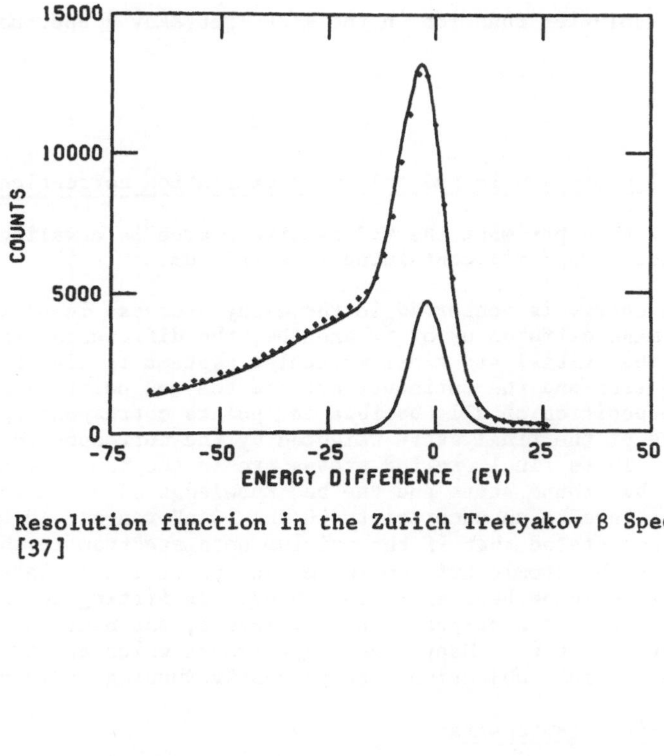

Fig. 2.8 : Resolution function in the Zurich Tretyakov β Spectrometer. [37]

One sees at least that an adequate representation of the data is obtained with a symmetric 9.6 eV resolution function. Ljubymov et al. [35] have used the resolution function shown in fig. 2.9. The spurious tail, although perhaps justified, might lead to systematic errors towards larger neutrino masses when fitting the data. These systematic errors could be comparable to the mass reported.

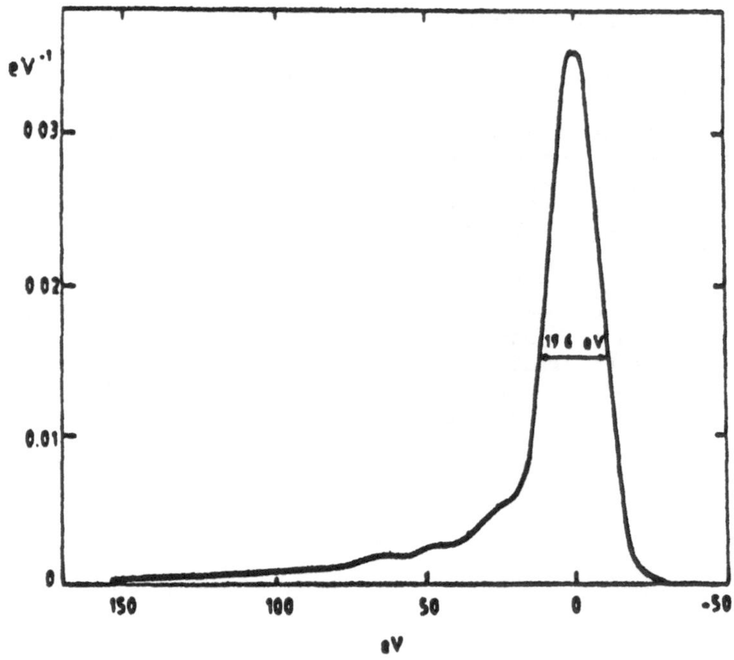

Fig. 2.9 : Resolution function in the ITEP Tretyakov β Spectrometer. [35]

2) Final state atomic and molecular excitation corrections

In the ITEP experiment the radioactive source is a valine ($C_5H_{11}NO_2$) foil ($2 \mu g/cm^2$) containing 18 % tritium.

Because energy is conserved in the decay process, in addition of the nuclear mass differences of ^3T and ^3He, the differences in the binding energies of the initial and final molecular systems is also distributed in the β electron and the antineutrino. So the end point is not unique : it is a superposition of all possible end points corresponding to each excited state of the final state weighted by the corresponding transition probability. These final excited states are in the range of a few 10 eV compared to the ground state and the bad knowledge of all transition possibilities lead to an uncertainty in the 10 eV range. In their estimation Lyubimov stated that if the tritium beta spectrum in the source corresponds to the atomic tritium (which can be well calculated), then the neutrino mass would be between 14 and 26 eV. If fitting to the data a 2 level formula, i.e. a superposition of 2 levels, the neutrino mass would be between 26 and 46 eV. Many other experiments which are trying to get rid of these two main objections are presently running or in preparation.

b) Calorimetric measurements

Neutrino mass deduced from the tritium beta spectrum recorded in a magnetic spectrometer has uncertainties caused by the molecular excitation effects of the radioactive source. Attempts to get rid of these effects led to another approach : calorimetry.

Simpson [38] measured the tritium β spectrum by implanting the tritium in a Si(Li) detector. In this experiment, because the whole energy released in each decay is measured (β energy and X rays are added because of the 6 μs charge collection time), the spectrum recorded in the crystal always correspond to ground state transition. As analyzed in [38] in great detail the initial state is expected to be atomic tritium rather strongly bound in the silicon lattice and the final state ^{3}He atom in its lowest energy level in the silicon lattice.

The resolution in this experiment is around 200 eV.

Such kind of experiments are still in progress and there may be some hope to reach the few eV resolution by using bolometers [39] or superconducting tunnel junctions [40].

The present mass limit obtained by Simpson is $m_\nu < 65$ eV.

In the same experiment Simpson found evidence [41] for a 10 keV mass neutrino by observing a distorsion at the beginning of the Kurie plot. This distorsion if not a spurious effect would be accounted for by a mixture of 97 % low mass neutrino (< 65 eV) and 3 % 17 keV mass neutrino. This result is not presently confirmed by the Princeton experiment [42] which studies the decay of ^{35}S and do not find any evidence for this neutrino in their data.

NEUTRINO OSCILLATIONS

a) Reactor experiments

The hypothesis of neutrino oscillation between their different flavours (ν_e, ν_μ, ν_τ) is now well known. It is based on the idea that neutrinos, eigenstates of the weak interaction, are not eigenstates of the mass but linear combinations of ν_1, ν_2 and ν_3 which are the eigenstates of the mass :

$$|\nu_l\rangle = \Sigma\, U_{li}|\nu_i\rangle \qquad l = e,\mu,\tau \qquad i = 1,2,3.$$

The fraction of flavour m neutrinos observed at a distance L from the flavour l neutrinos source can be written :

$$P(\nu_l \rightarrow \nu_m) = \Sigma\, U_{li}^2.U_{mi}^2 + \Sigma\, U_{li}.U_{mi}.U_{lj}.U_{mj}\, \cos(2\pi L/l_{ij})$$

where $l_{ij}(m) = 4\pi p_\nu/|m_i^2-m_j^2| = 2.48\ p(\mathrm{MeV}/c)/|\Delta m^2|$

is the oscillation length.

If we consider a mixture of two flavours only the U matrix is a rotation matrix (angle θ) and we have :

$$P(\nu_e \rightarrow \nu_e) = 1 - \sin^2 2\theta.\sin^2(1.27\ \Delta m^2 L/p_\nu).$$

A number of dedicated experiments have been performed both at reactors and at accelerators setting limits on δm^2, $\sin^2 2\theta$ for ν_e and ν_μ oscillations.

Best sensitive experiments near reactors are the ones near power reactor plants Bugey (France) [43] and Gosgen (Switzerland) [44] which allow neutrino detectors to be placed at two positions a few tens of meters apart and compare with high statistics the neutrino flux coming from the reactor (few MeV range) in the 2 positions. This probes small neutrino mixings for a mass difference δm^2 in the range .1 to 4 eV^2.

Nuclear reactors are pure sources of $\bar{\nu}_e$ which are emitted isotropically and originate from the beta decay of the fission products. The experimental area has to be well shielded from the reactor core and associated activities. This would favor pushing the detector further and further away from the reactor (Gosgen option). However high statistics is needed and also a good ratio signal over noise which favors keeping the detector near the reactor (Bugey option).

The characteristics of the two main experiments are shown in table 2.5.

Table 2.5

Comparison between Bugey and Gosgen neutrino oscillation experiments.

	Bugey	Gosgen
Detector Position	14 m, 18 m	38 m, 46 m
Nb. of events	63000	11000
Neutron identif.	^{3}He chambers	^{3}He chambers
Target	321 1 liquid Scintillator	377 1 liquid Scintillator

I describe in the following the Bugey experiment. The principle of detection in both experiments is the same (fig. 2.10).

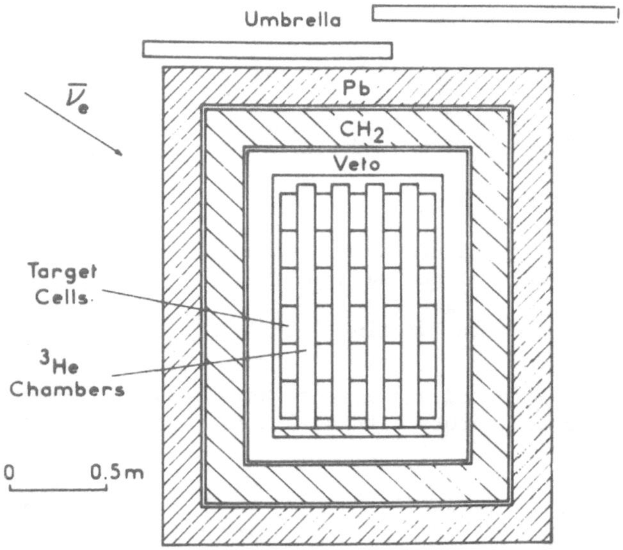

Fig. 2.10 : Principle of $\bar{\nu}_e$ detector in experiments near a reactor.

Anti-neutrinos are detected using the reaction $\bar{\nu}_e + p \rightarrow e^+ + n$ and their energy is given by $E_\nu = E_{e^+} + 1.8$ MeV (+ neutron recoil corrections). The detector is made of 321 l of liquid scintillator segmented in 5 planes of 6 cells. The target acts as positron calorimeter and neutron moderator. The neutrons, once thermalized, are captured in the neighbouring ^3He chambers.

Fig. 2.11 shows the positron energy spectrum measured at position 1 and 2 with reactor on and reactor off. The expected spectrum from Monte Carlo studies is shown as shaded bands, delimited by the point to point error.

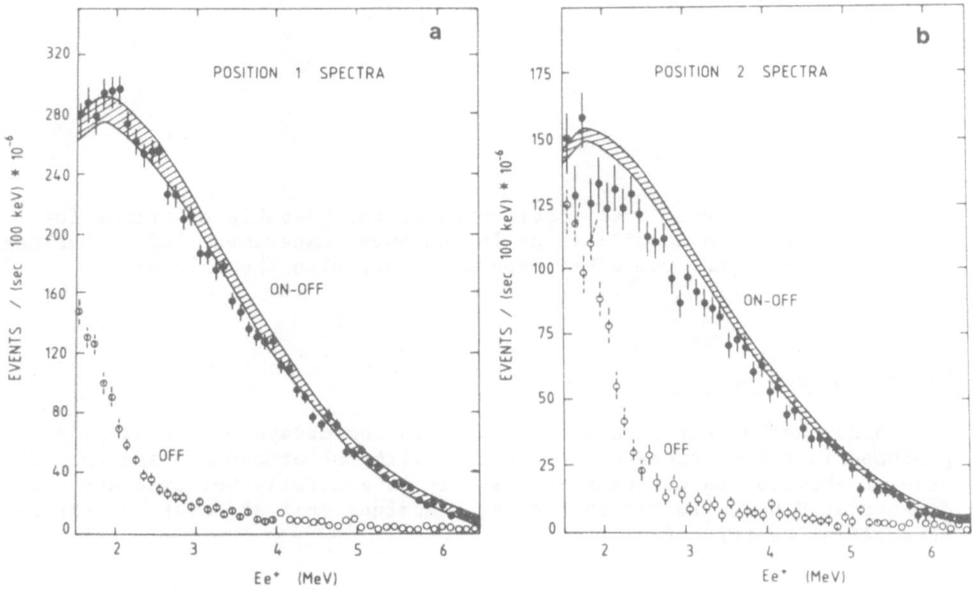

Fig. 2.11 : Positron energy spectra measured at position 1 and 2 with reactor ON or OFF. [43]

The Bugey experiment showed a positive evidence of $\bar{\nu}_e$ disappearance [43] in the second position compared to the first one while no such effect has been seen in the Gosgen experiment. Fig. 2.12 shows the combined probability contour in the ($\sin^2 2\theta$, δm^2) plane for the Bugey experiment compared to exclusion contours from other experiments. Accelerator data are sensitive to smaller values of $\sin^2 2\theta$ but to higher values for δm^2. We may note that part of the solution region around $\delta m^2 = 0.2$ eV^2 is not excluded by the Gosgen experiment, the only experiment using two positions and sensitive to this range.

Both experiments are improving their sensitivity.

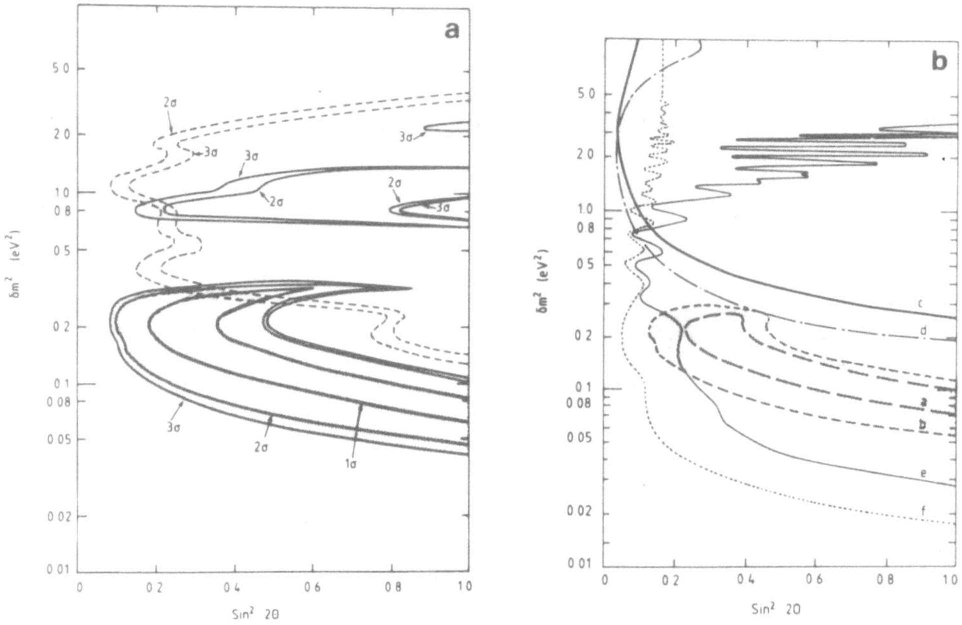

Fig. 2.12 : The shaded areas correspond to the possible solutions for neutrino oscillations in the Bugey experiment [43]. Excluded regions from other experiment are also shown in b).

b) Atmospheric neutrinos

Atmospheric neutrinos are produced in the decays of π's and μ's produced in interactions of cosmic rays with the atmosphere and the subsequent showers and desintegrations. It is generally believed and now demonstrated below 10 GeV that these neutrinos dominate over the extra-terrestrial neutrino flux.

Table 2.6
Neutrino event rates observed in proton decay detectors.

Experiment	KGF	Kamioka	IMB	NUSEX	FREJUS
Latitude	2°	27°	50°	46°	45°
N°. of (neutrino) events	~ 17 (a)	29	112	10	21
Kiloton yrs	0.22	0.27	1.20	0.113	.3
Rate/kton yr (corrected for efficiency)	77 ± 19	123 ± 23	125 ±12	118 ±37 (b)	100 ± 20
Prediction	85	99	132	132	125

(a) Total vertex rate.
(b) Assumes 75 % containment efficiency.

Atmospheric neutrinos are a source of background in proton decay detectors because they can induce contained events in these detectors which can ultimately fake a proton decay. So a great attention has been put to look for these events in all proton decay detectors. Table 2.6 shows that there is a good agreement between the expected (calculations) number of neutrino induced interactions per kiloton of detector and per year with the indeed observed number of events [45]. One notes a slight change with latitude which is well understood in terms of geomagnetic effects (cutoff). The shape of the energy distribution agrees well between the various experiments and agrees with expectations fig. 2.13. The water Cerenkov IMB experiment which is the biggest fiducial mass proton decay detector (3300 ton fiducial) has accumulated 401 contained events. The detector is based on the detection of Cerenkov light produced in water by 2048 photomultipliers fig. 2.14. The pattern of the hit photomultipliers plus the timing permits the reconstruction of the direction fig. 2.14b. ν_μ interactions produce less light (for charged currents) than ν_e interactions. Furthermore muons produce a delayed decay signal which can be identified for 55 % of the muons.

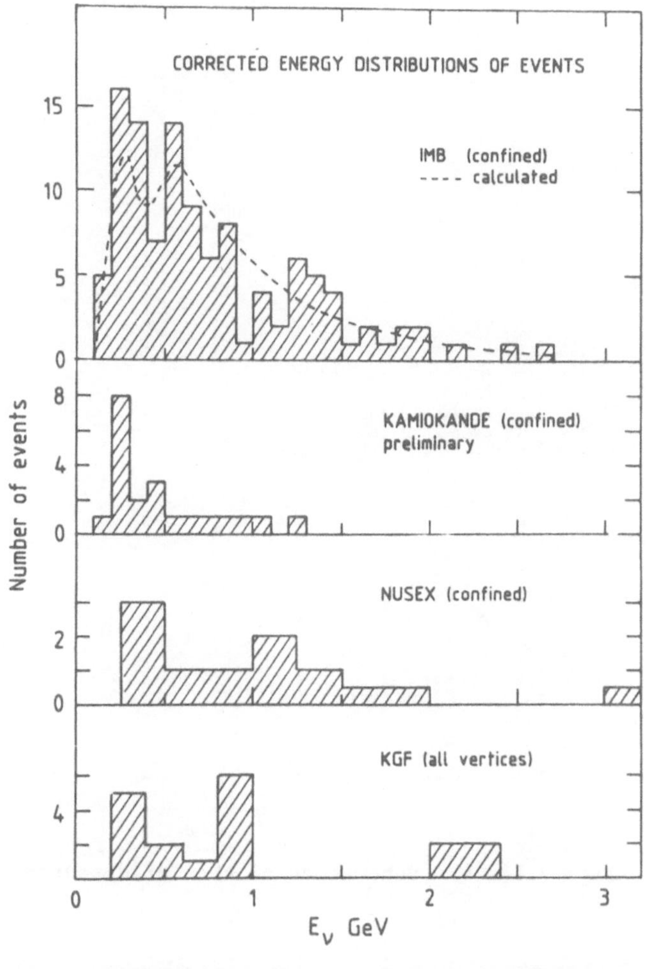

Fig. 2.13 : Energy distribution of atmospheric neutrinos detected in proton decay experiments. [45]

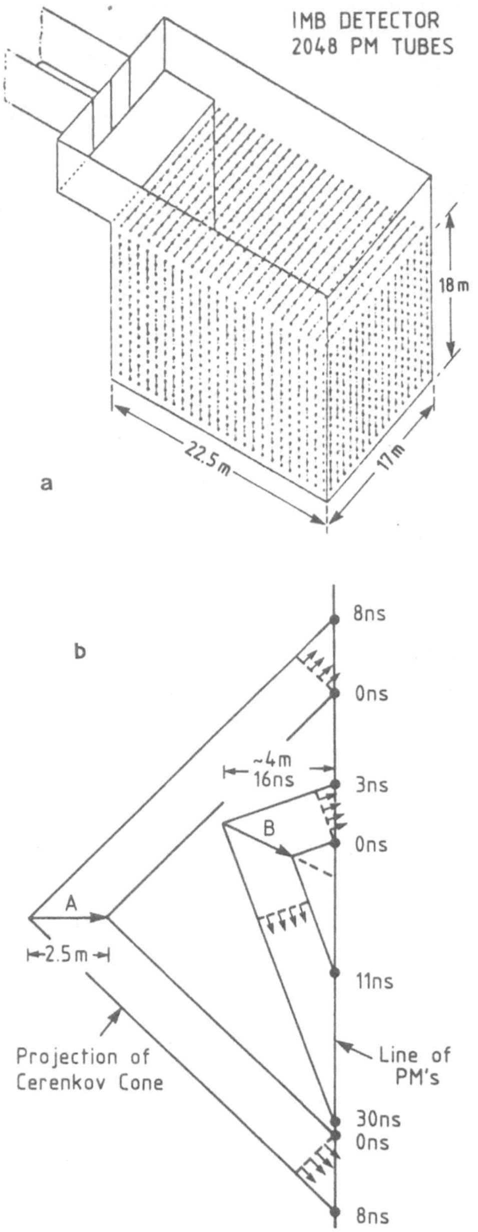

Figs 2.14a,b : Schematics of the IMB detector.

The distance a neutrino has traveled is a function of its zenith angle. If neutrino oscillations occur we might expect differences (for δm^2 in the range of a few 10^{-5} eV^2) between the neutrinos in the upward going 1/5 of the solid angle which travel a mean distance of 10^7 m and those in the downward going 1/5 of the solid angle that have only traveled about 10^4 m. By comparing these two classes of events one can look for oscillations.

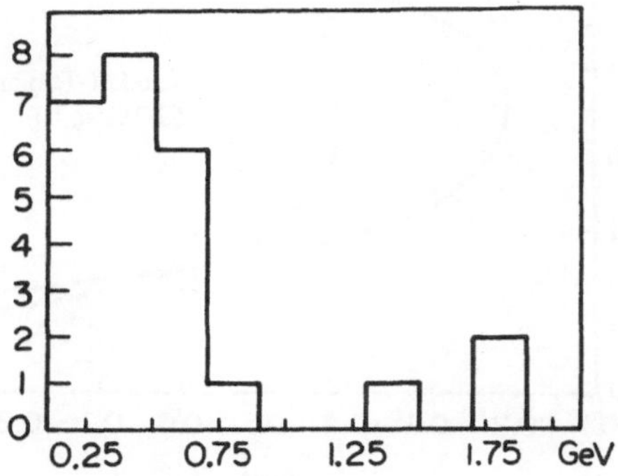

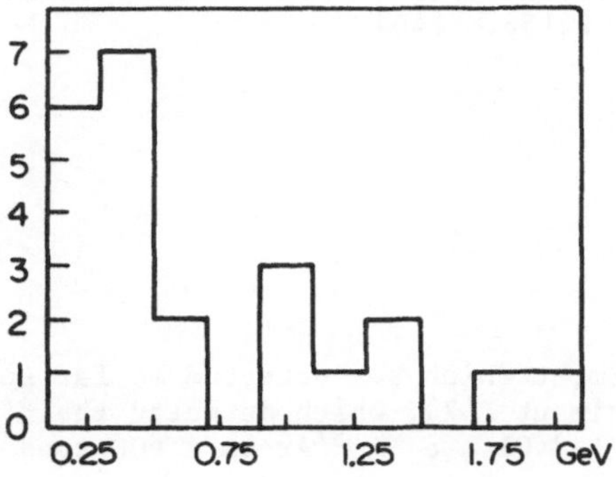

Fig. 2.15 : Visible energy of a) upward going neutrino interactions b) downward going neutrino interactions. [46]

The visible energy distributions of the 25 events (after fiducial cuts) in the upward 1/5 of solid angle and the 25 events (it happens to be the same number) in the downward solid angle are shown [46] in fig. 2.15. These distributions are very similar. By using the experimentally checked relation $E_\nu = E_{vis} \times 0.758 + 410.$ (MeV) and by the fact that 40 ± 15 % of both the upward and downward sample have a muon decay signature the authors have put a flux independent limit for an exclusion contour in $\sin^2 2\theta$ Δm^2 plot shown in fig. 2.16. Also shown is the exclusion contour one can obtain by comparing the observed energy distribution with an explicit Monte Carlo calculation using calculated flux.

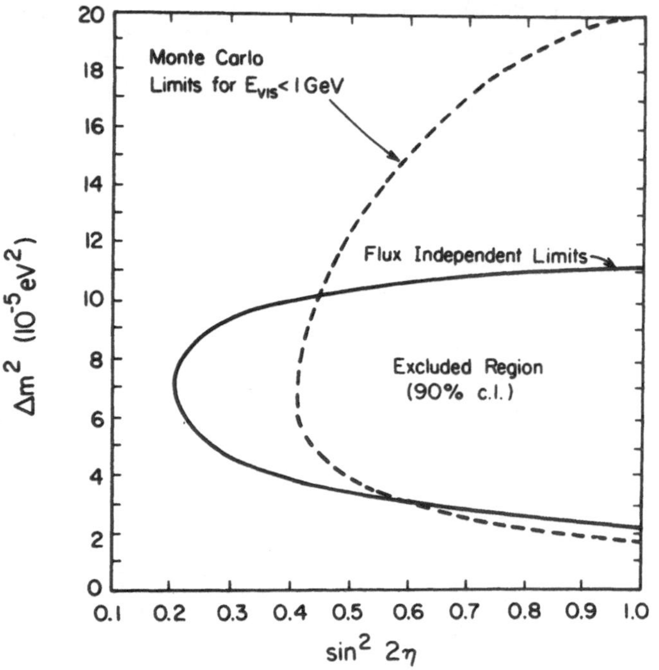

Fig. 2.16 : Exclusion contour from the comparison of fig. 2.15a and fig. 2.15b. [46]

c) Solar neutrinos

The only experiment which has detected so far solar neutrinos is the Brookhaven experiment [47], which detected the ^{37}Ar atoms created through reaction $\nu_e + {}^{37}Cl \rightarrow e^- + {}^{37}Ar$ in a 600 tons tank filled with C_2Cl_4.

The schematic arrangement of the Brookhaven solar neutrino detector is shown in fig. 2.17 [48]. It is installed in the Homestake gold mine at a depth of 4400 m water equivalent (fig. 2.18). The tank is exposed for periods of 35 to 100 days with 0.1 cm^3 inactive ^{36}Ar or ^{38}Ar carrier. Argon is removed by helium purge and trapped in charcoal. The argon is purified by gas chromatography and by getters. ^{37}Ar decay is measured by a proportional counter. The measurement relies on the K capture of ^{37}Ar which induces a 3keV fast energy deposit from Auger electrons. Fig. 2.19 shows as a function of the energy a quantity which is inversely proportional to the pulse rise time. The box correspond to the expected ^{37}Ar pulses. The background in the counter is due either to electronic noise or Compton events in the edges of the counter. A mass spectrometer allows the analysis of the ^{36}Ar or ^{38}Ar yield to derive the extraction efficiency.

However, other types of background which are impossible to get rid of and which amount to about 20 % of the signal are due to the production of ^{37}Ar atoms by other sources than neutrinos. These sources are listed in table 2.7.

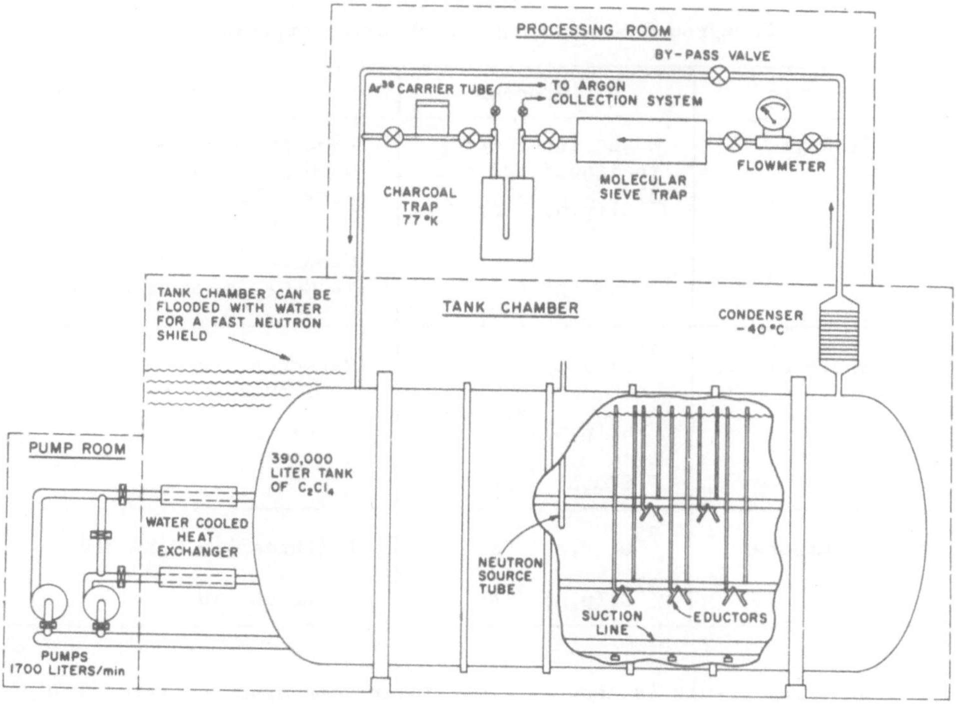

Fig. 2.17 : Schematic arrangement of the Brookhaven solar neutrino detector. [48]

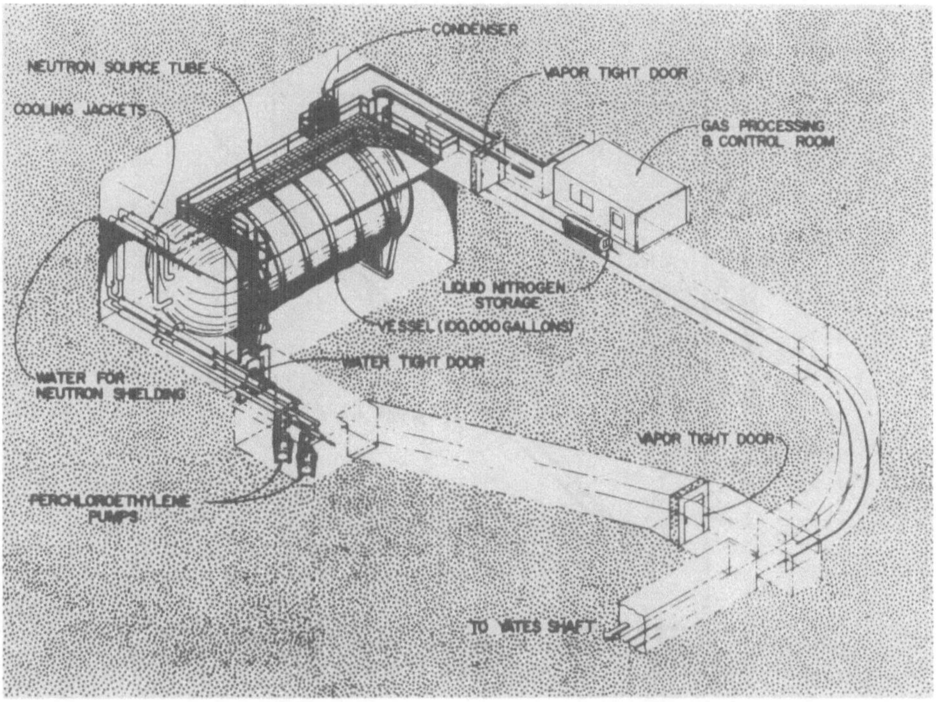

Fig. 2.18 : Homestake laboratory.

Table 2.7

Background effects in the chlorine experiments.

Source	Process	Control
Cosmic Rays	Muons interacting in liquid and rock $^{37}Cl(p,n)\ ^{37}Ar$	4850 fr underground 4400 m.w.e. Flux = 3.4 μ's/m^2 day $\bar{E}_\mu$ = 300 GeV Measure ^{37}Ar production in C_2Cl_4 at shallow depths
Internals Alphas	$^{35}Cl(\alpha,p)\ ^{38}Ar$ $^{37}Cl(p,n)\ ^{37}Ar$ $^{34}S(\alpha,n)\ ^{37}Ar$	Measures α-emission in liquid and walls Total α rate < 10^8 day^{-1}
Fast Neutrons	$^{35}Cl(n,p)\ ^{35}S$ $^{37}Cl(p,n)\ ^{37}Ar$	E (thresh) = 0.99 MeV Water shield

The final result is about 3 times lower than expected (fig. 2.20) :

2.1 ± 0.3 SNU instead of 6.6 ± 0.7 SNU.

1 SNU corresponds to the formation of one atom ^{37}Ar per 10^{36} ^{37}Cl atoms and per second.

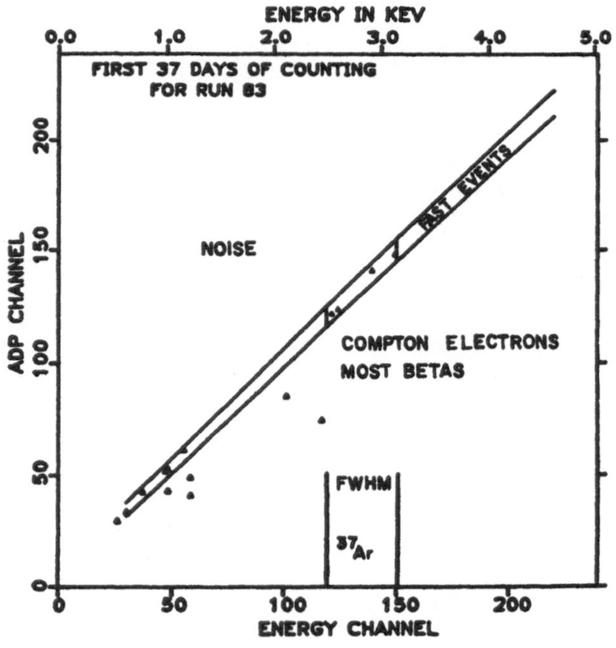

Fig. 2.19 : Identification (energy and rise time) of ^{37}Ar decays. [47]

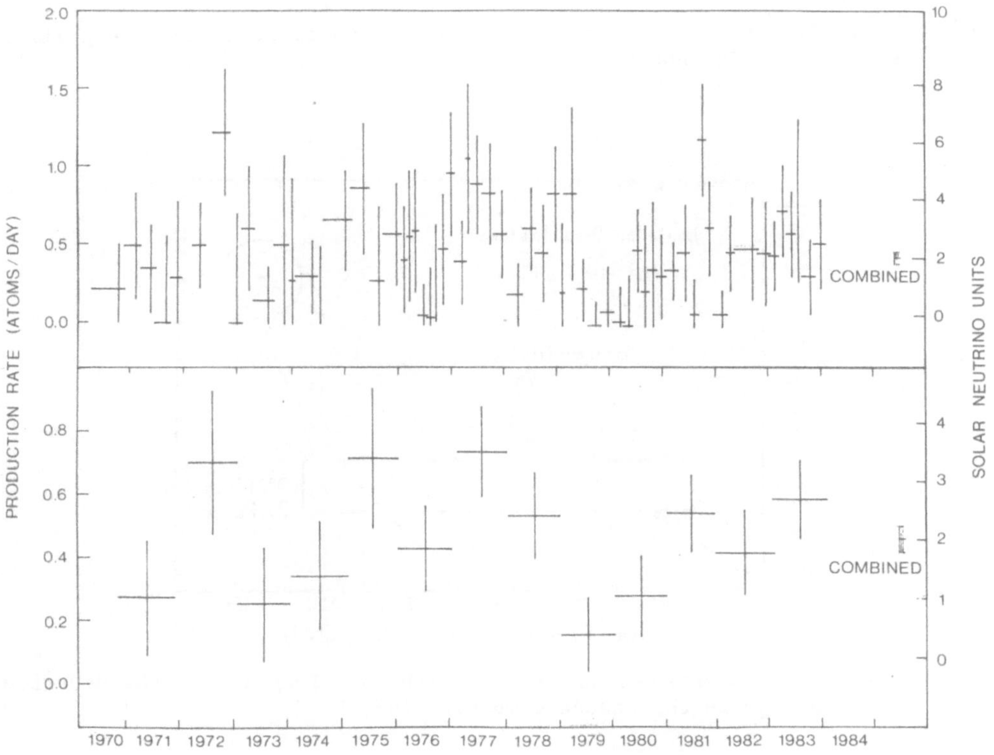

Fig. 2.20 : Production rate of ^{37}Ar since 15 years. [47]

6.6 SNU are expected within the so called Sun Standard Model [49] which accounts for all other observations on the sun.

While the threshold of the reaction is 814 keV, most of the sensitivity in the chlorine experiment is for ν above 5 MeV due to a strong contribution of a 4 MeV ^{37}Ar excited state. This corresponds to a sensitivity to the high energy tail of the neutrino energy spectrum, (fig. 2.21), which is very "temperature" dependent (T^{20}). By temperature, we mean the temperature in the center of the sun. Attempts to modify the sun standard model, i.e., to decrease the temperature in the center of the sun, seem all very ad hoc although cannot be rejected (convection, turbulent diffusion [50]).

However, although it is possible to "play" a little bit with the high energy tail of the neutrino spectrum, it is almost impossible to act on the total expected integrated number of neutrinos emitted from the sun. This is due to the fact that the luminosity of the sun is directly related to the number of $4p \rightarrow {}^4He$ fusion reactions which take place in the sun. This is so because the sun is not able (yet) to burn its helium or heavier elements. The $4p \rightarrow {}^4He$ provides a 27 MeV mass defect and 2 ν which are unavoidable in any cycle of reactions. The total number of neutrinos is then $L/27\ MeV/4\pi d^2 = 6.6\ 10^{10}$ neutrinos/cm^2/sec on earth.

A Gallium experiment [51] is now planned in Europe, in the Gran Sasso laboratory. It is based on the counting of ^{71}Ge atoms produced in the reaction $\nu_e + {}^{71}Ga \rightarrow e^- + {}^{71}Ge$ which has a threshold of 233 keV and is almost sensitive to the whole spectrum of solar neutrinos. A one megacurie monochromatic (746 keV) neutrino source will be used to compare directly the solar neutrino flux to a well monitored source flux

(^{51}Cr source). The prediction of the standard solar model for the gallium experiment is 115 ùa. SNU.

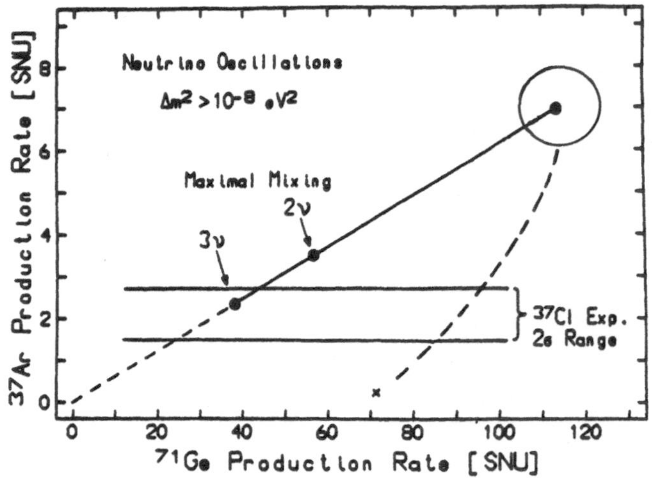

Fig. 2.21 : Solar neutrino flux at the earth as a function of the neutrino energy in the standard model. [49]

Fig. 2.22 shows the ability this experiment will have to disentangle the solar neutrino puzzle coming from the chlorine experiment :

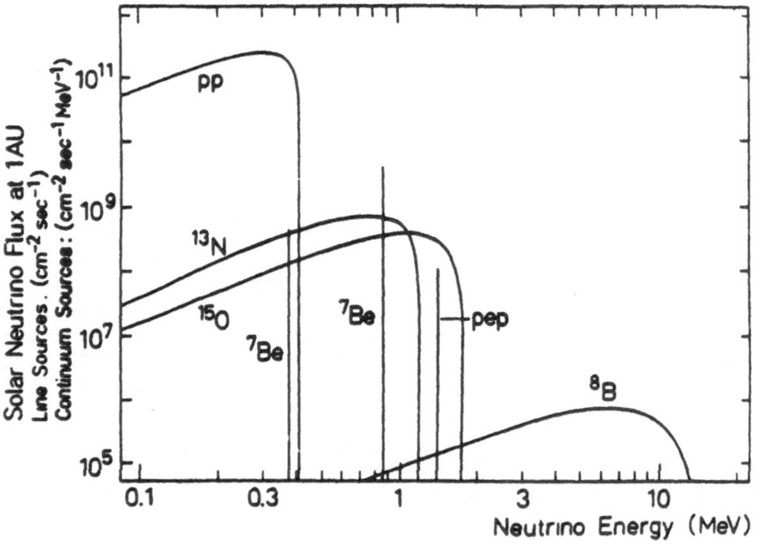

Fig. 2.22 : Ability of a Gallium experiment to disentangle the solar neutrino puzzle. [51]

* If the result of the gallium experiment is between 70 and 100 SNU, this means that indeed the sun standard model has to be modified in order to decrease the central temperature in the sun (broken line).

* If on the contrary, the result is below 50 SNU, the only way to understand it is presently neutrino oscillations (solid line) induced by a $\Delta m^2 > 10^{-8}$ eV^2 and a near maximal mixing angle.

d) Conclusions

Although some experiments have reported positive results on neutrino masses and neutrino oscillations, these results are not yet convincing. We do not know yet if neutrinos are Majorana or Dirac particles, if they are massless or not, if they oscillate between the various flavours or not. In all these fields, the best sensitive experiments are to be done without accelerators. Many experiments are presently proposed or even running. This neutrino physics is the good quality garantied physics for elementary particle study without accelerator.

CONSTANCY OF THE COUPLING CONSTANTS OVER A LONG PERIOD OF TIME

The temporal invariance of the coupling constants is one of the basic postulates of physics, more precisely it is the framework in which operates cosmology.

The best evidence of temporal invariance of the laws of physics [52] comes from the discovery in 1972 of a fossil reactor at Oklo, (Gabon) following an extraordinary combination of chance, technical care and scientific ingenuity by a team of the French Atomic Energy Commission (CEA).

Bouzigue et al. [53] were controlling on a routine basis the input of Uranium at the French uranium gas-diffusion plant at Pierrelatte. All Uranium isotopes are radioactive and we find in nature all isotopes whose lifetimes are longer than or at least comparable to the time since their formation. The solar system is approximately 4.6 10^9 old and hence only the two longest lived uranium isotopes have survived in measurable abundances. ^{238}U has a lifetime of 5 10^9 years while that of ^{235}U is only 7 10^8 years. The relative abundance of ^{235}U to ^{238}U simply reflects this difference in decay rates. Table 3.1 shows the percentage of ^{235}U as we can track it back in time. Note that about 2 10^9 years ago this percentage was of the same order as the percentage needed in pressurized water reactor.

Table 3.1

Percentage of ^{235}U in natural maximum in past.

Age in 10^6 years	0	700	1400	2100	2800
$\%^{235}U$	0.72	1.3	2.3	4.0	7.0

Today the percentage of ^{235}U is 0.7 ± .1. Suddenly in january 1972 natural Uranium standards were found to be heavily depleted ^{235}U : 0.3 %. This discrepency although small was taken seriously and was further investigated. The resulting "detective" investigation was then run backward in time through the complex chain of processes involved in the preparation of uranium in several centers of CEA and up to the original sources of Uranium used to prepare the suspect standards, some ores of the Oklo deposit in Gabon. Search for rare earth fission products (Nd) showed that they were abundant in the Oklo ores. Therefore in september 1972, Neuilly et al. [54] felt sufficiently confident to announce that the Oklo deposit had been the site of a natural chain reaction in the distant past.

An analysis of the neodymium isotopic abundances from the Oklo samples confirms this statement. The isotopic abundances are compatible with $^{235}U^f$ yields (uranium fission products) as it can be seen from table 4. This fixes the $^{235}U^f$ total rate and then the date the reactions went critical : $1.8 \pm .07\ 10^9$ years with an integrated flux of 1 to $2\ 10^{21}$ neutrons cm^{-2}. How long did it last ? This comes from ^{239}Pu (24000 years) → ^{99}Ru (213000 years) and subsequent $^{99}Ru \rightarrow {}^{100}Ru$ (stable) reaction. Best favoured values for the duration period of the fossil reactor is 600000 years. The last line in table 1 shows that the Okloneodymium abundances are excellently reproduced by a small plutonium fission admixture. This relative abundance of plutonium 239 and uranium 235 is very sensitive to the neutron flux in the reactor core and suggests a typical flux of $10^9\ 10^{10}$ n $cm^{-2}\ s^{-1}$ which is very low compared with the situation of man made reactors where fluxes of $10^{12}\ 10^{13}$ n $cm^{-2}\ s^{-1}$ are typical. This is again consistent with an integrated flux of $1 2\ 10^{21}$ neutrons cm^{-2} at the Oklo site.

We can thus compare the workings of a nuclear reactor $2\ 10^9$ years ago with those of present days.

For this we shall use the samarium isotopes 147 and 149. The relative natural abundances of these isotopes is :

$(\text{samarium } 149/\text{samarium } 147)_{nat} = .92$ while the average ratio coming from uranium fission fragments is :

$$(\text{samarium } 149/\text{samarium } 147)_{fission} = .5.$$

We would expect something intermediate for the Oklo site while it is found :

$$(\text{samarium } 149/\text{samarium } 147)_{Oklo} = .1.$$

This distorsion is due to the fact that samarium 149 has a neutron resonance at .098 eV of width .063 eV. This resonance corresponds to a thermal neutron capture cross section (fig. 4.1) which is almost 3 orders of magnitude bigger than neutron thermal capture cross section for samarium 147.

We can in fact deduce the thermal neutron capture cross section in samarium 149 as it was $2\ 10^9$ years ago, from the depletion of samarium 149 and the integrated thermal neutron flux 10^{21} cm^{-2}. The result is within 10 % of its present day value. Since the temperature range is in the 500-1000K, kT is .043-.086 eV. A change of less than 10 % in the thermal cross section implies a shift of less than .01 eV in the position of the neutron capture resonance in samarium 149 over the past $2\ 10^9$ years.

A shift of less than .01 eV in the position of a neutron resonance in a potential well 50 MeV deep over $2\ 10^9$ years represents a variation of less than one part in 10^{19} per year [55].

The 50 MeV deep neutron potential comes approximately 95 % from strong, 5 % from electromagnetic and 10^{-5} % from weak interactions. So

α_s, α_{em}, α_w, the coupling constants from strong, electromagnetic and weak interactions should have varied by less than one part in 10^{19}, one part in 5 10^{17} and one part in 10^{12} per year respectively in the past 2 10^9 years.

ACKNOWLEDGEMENTS

I would like to thank particularly Drs. J. Bouchez, G. Chardin, J. Paul, M. Cribier, J. Rich and D. Vignaud for many fruitful and helpful frequent discussions.

REFERENCES

[1] Particle Physics without Accelerators, M. Cribier, M. Spiro, J. Rich, to be published in Physics Reports.
[2] A. Bonnetti et al., Nuovo Cimento 10 (1953) 345.
[3] N. Yamdagni, Event Structure in collider and cosmics ray experiments, Physics in Collision (Stockholm, June 82).
[4] G.B. Yodh et al., Phys. Rev. Lett. 28 (1972) 1005.
[5] M. Bozzo et al., Phys. Lett. 147B (1984) 392.
[6] G.B. Yodh, Workshop on Elastic and Diffractive Scattering at the Collider and Beyond, Blois (June 1985).
[7] C.M.G. Lattes et al., Phys. Rep. 65 (1980) 151.
[8] K. Alpgard et al., Phys. Lett. 107B (1981) 315.
[9] K. Alpgard et al., Phys. Lett. 107B (1981) 310.
[10] G. Arnison et al., Phys. Lett. 107B (1981) 320.
[11] M. Banner et al., Phys. Lett. 115B (1982) 59.
[12] G. Arnison et al., Phys. Lett. 118B (1982) 167.
[13] M. Banner et al., Phys. Lett. 118B (1982) 203.
[14] S.I. Nikolovsky et al., Proc. 17th Intern. Cosmic Ray Conf. (Plovdiv, 1977);
V.I. Yakovlev et al., Proc. 17th Intern. Cosmic Ray Conf. (Plovdiv, 1977).
[15] K. Alpgard et al., Phys. Lett. 115B (1982) 71.
[16] G. Arnison et al., Phys. Lett. 122B (1983) 189.
[17] J.M. Dickey, Astrophysical Journal 273 (1983) 71.
[18] K.O. Mason et al., Astrophysical Journal 207 (1976) 78.
[19] M. Van der Klis and J.M. Bonnet-Bidaud, Astronomy and Astrophysics 95 (1981) L5.
[20] R.C. Lamb et al., Astrophysical J. 212 (1977) L63.
[21] COS-B Coll., paper presented at the 19th Int. Cosmic Ray Conf., La Jolla (USA), August 1985.
[22] Yu.I. Nespor et al., Astrophysics and Space Science 61 (1979) 349.
[23] R.C. Lamb et al., Nature 296 (1982) 543.
[24] S. Danaher et al., Nature 289 (1981) 568.
[25] M. Samorski et al., Astrophysical Journal 268 (1983) L17.
[26] J. Lloyd-Evans et al., Nature 305 (1983) 784.
[27] M.L. Marshak et al., Phys. Rev. Lett. 54 (1985) 2079.
[28] G. Battistoni et al., Phys. Lett. 155B (1985) 465.
[29] Aachen-Orsay-Palaiseau-Saclay-Wuppertal Collaboration. Paper presented at the 19th Int. Cosmic Ray Conf., La Jolla (USA) (August 1985);
G. Chardin, private communication.
[30] M. Samorski et al., Proc. 18th Int. Cosmic Ray Conf., Bangalore, India.
[31] B. Kayser, Comments on Nucl. Phys. 14 (1985) 69.
[32] H. Nishiura, Kyoto University, RIFP-453 (1981);
M. Doi et al., Prog. Theor. Phys. 69 (1983) 602.
[33] E. Bellotti et al., Phys. Lett. 121B (1983) 72.
[34] T. Kirsten, Proc. 11th Int. Conf. on Neutrino Phys. and Astrophys., Dortmund (1984) p. 145.

[35] V. Lubimov et al., Proc. 22th Int. Conf. on High Energy Phys. Leipzig (1984), p. 259.
S. Boris, Phys. Lett. 159B (1985) 217.
[36] E.F. Tretyakov et al., Izv. Akad. Nauk SSSR Ser. Fiz. 40 (1976) 20.
[37] J.W. Petersen, Moriond Workshop on Massive Neutrino in Astrophys. and Particle Physics (1984), p. 261.
[38] J.J. Simpson, Phys. Rev. D23 (1981) 649.
[39] CERN-COURRIER 25 (1985) 182.
[40] D. Twerenbold et al., SIN, PR-84-07.
[41] J.J. Simpson, Phys. Rev. Lett. 54 (1985) 1891.
[42] T. Altzitzoglou, Phys. Rev. Lett. 55 (1985) 799.
[43] J.F. Cavaignac et al., Phys. Lett. 148B (1984) 387.
[44] K. Gabathuler et al., Phys. Lett. 138B (1984) 449.
[45] D.H. Perkins, Ann. Rev. of Nucl. Science 34 (1984) 1.
[46] J.M. Secco et al., Phys. Rev. Lett. 54 (1985) 2299.
[47] J.K. Rowley et al., Solar Neutrino and Neut. Astronomy Conference, Homestake (1984) 1.
[48] R. Davis et al., Phys. Rev. Lett. 20 (1968) 1205.
[49] J.N. Bahcall et al., Review of Modern Physics 54 (1982) 767;
J.N. Bahcall et al., Astrophysical J. 292 (1985) L79.
[50] E. Schatzman, A. Maeder, Astron. Astrophys. 96 (1981) 1.
[51] Members of the Collaboration : T. Kirsten (spokesman), W. Hampel, G. Eymann, G. Heusser, J. Kiko, E. Pernicka, B. Povh, M. Schneller, K. Schneider, H. Volk (Heidelberg) - K. Ebert, E. Henrich, R. Schlotz (Karlsruhe) - R.L. Mossbauer (Munchen) - I. Dostrovsky (Rehovot) - M. Cribier, G. Dupont, B. Pichard, J. Rich, M. Spiro, D. Vignaud (Saclay) - G. Berthomieu, E. Schatzman (Nice) - E. Fiorini, E. Bellotti, O. Cremonesci, C. Liguori, S. Ragazzi, L. Zanotti (Milano)- L. Paoluzi, S. D'Angelo, R. Bernabei, R. Santonico (Roma).
T. Kirsten, "The gallium solar neutrino experiment", in Proc. of the Resonance Ionization Spectroscopy meeting, Knoxville, Tennessee, April 1984;
W. Hampel, "The gallium solar neutrino detector", in Solar Neutrinos and neutrino astronomy, Homestake (1984), AIP Conf. Proc. n° 126, p. 162;
D. Vignaud, "The gallium solar neutrino experiment GALLEX", paper presented at the 5th Moriond Astrophysics meeting, Les Arcs (1985).
[52] J.M. Irvine, Contemp. Physics 24 (1983) 427.
[53] See M. Maurette, Ann. Rev. Nucl. Sc. 26 (1976) 319.
[54] M. Neuilly et al., C. R. Acad. Sci. Paris 275 (1972) 1847.
[55] F.J. Dyson, Rev. Mod. Phys. 51 (1979) 447.

BEYOND THE STANDARD MODEL

G. Altarelli

Dipartimento di Fisica, Università "La Sapienza", Roma, Italy

I.N.F.N. - Sezione di Roma, Italy

A LARGE SCALE MAP OF PARTICLE PHYSICS

The standard model of particle interactions is a complete and relatively simple theoretical framework which describes all the observed fundemental forces. It consists of quantum chromodynamics (QCD) [1] and of the electro-weak theory of Glashow, Salam and Weinberg [2]. The former is the theory of coloured quarks and gluons, which underlies the observed phenomena of strong interactions, the latter leads to a unified description of electromagnetism and of weak interactions. The inclusion of the classical Einstein theory of gravity completes the set of established basic knowledge. The standard model is in agreement with essentially all of the experimental information which is very rich by now. The recent discovery [3] of the charged and neutral intermediate vector bosons of weak interactions at the expected masses has closed a really important chapter of particle physics. Never before the prediction of new particles was so neat and quantitatively precise.

Yet the experimental proof of the standard model is not completed. For example, the hints of experimental evidence for the top quark at a mass ~40 GeV have not yet been firmly established. The Higgs sector of the theory has not been tested at all. Beyond the realm of pure QED, even remaining within the electro-weak sector, the level of quantitative precision in testing the standard model does not exceed 5 % or so.

Furthermore, the standard model does not look as the ultimate theory. To a closer inspection a large class of fundamental questions emerges and one finds that a host of crucial problems are left open by the standard model. First of all, a large number of coupling constants and masses is left unspecified. Furthermore, the pattern of observed mass ratios and quantum numbers is rich of nontrivial unexplained features. The masses of charged fermions span several orders of magnitude. The iteration of three identical generations of quark and lepton flavours is impressive and without explanation. The quantization of electric charge, that is the integer ratios among quark and lepton charges, is not implied by any principle within the standard model, although it leads to the cancellation of chiral anomalies.

Some other basic features are merely described or accommodated in the theory, but their profound origin is mysterious. Among these properties we may mention the violation of P and C, and also of CP, (for example, the problem of the smallness of the CP violation by the strong interactions or "θ - problem"), the conservation of the baryon number B and of the three lepton numbers L_e, L_μ, L_τ, the masslessness of the neutrinos. The conservation of B and of L_e, L_μ, L_τ, is not protected by a gauge principle as opposed to the conservation of the electric charge which is guaranteed by the massless photon. These conservation laws might be violated by some new interactions not yet discovered but predicted in all grand unified theories as necessary ingredients for the unity of all forces. The separate conservation of L_e, L_μ and L_τ is even more delicate than that of their sum and is presumably directly invalidated if neutrino masses are nonvanishing and different.

From another point of view, a very important fact is that quantum gravity is not described by the standard model. Actually, until very recently the formulation of a consistent theory of quantum gravity appeared very much out of reach. Clearly, in principle, microscopic physics cannot do for ever without quantum gravity. By the indetermination principle ($\Delta r \gtrsim \frac{\hbar}{Mc}$) more and more energy density is needed, in order to probe matter at smaller and smaller distances, until finally space-time is curved by a non negligible amount. The scale where quantum gravity effects become important is specified by the value of G_N, the Newton constant, which has the physical dimensions of M^{-2}. That is the relation

$$G_N = \frac{\hbar c}{M_{pl}^2} \tag{1}$$

defines M_{pl}, the Planck mass, being of order

$$M_{pl} \simeq 10^{19} \text{ GeV}. \tag{2}$$

It is increasingly clear that the problem of quantum gravity can no more be ignored or evaded by particle physics. For example, grand unified theories (GUT) [4] are a natural extrapolation of the standard theory of particle interactions and directly lead to consider an energy domain remarkably close to the Planck scale. The basis for GUTS is that, on one hand, there is no unification of the fundamental forces in the standard model, because a separate gauge group and coupling is introduced for each interaction. On the other hand, the structural unity implied by the common, restrictive, property of gauge invariance strongly suggests the possibility that all the observed interactions actually stem from a unified theory at some more fundamental level.

In GUTS at sufficiently large energies a simple gauge group

$$G \supset SU(3) \otimes SU(2) \otimes U(1) \quad (G=SU(5), SO(10), E6, SU(3) \otimes SU(3) \otimes SU(3) \ldots)$$

describes all particle interactions. The effective couplings depend on the energy scale. Thus the observed difference of couplings is washed out when the energy scale is increased. As we shall see in sect. 4 if no new physics arises below M_{GUT} ("the desert assumption") then the logarithmic running of the three couplings can be predicted. From these calculations and the observed couplings at low energies one immediately derives an estimate of M_{GUT} :

$$M_{GUT} \simeq (10^{14} \div 10^{16}) \text{ GeV}. \tag{3}$$

Also, since quarks and leptons are in the same GUT multiplets, B and L violating interactions are in general induced by exchange of superheavy gauge bosons with mass $\sim M_{GUT}$. The transition $q + q \to \bar{q} + \bar{l}$ is in fact allowed by colour and electric charge conservation. The resulting lifetime of the proton is clearly of the order :

$$\tau_p \simeq \frac{M_{GUT}^4}{M_p^5} \frac{(\text{factors})}{\alpha_{GUT}^2} \tag{4}$$

where M_p is the proton mass and α_{GUT} is a suitably defined GUT gauge coupling. From the experimental constraints on τ_p, which are much stronger now than a few years ago, (for example, $\tau_p/B(e^+\pi^o) > 2.10^{32}$ years) it follows that :

$$M_{GUT} \gtrsim (10^{15} \div 10^{16}) \text{ GeV}. \tag{5}$$

Note that this result is largely independent of the "desert" assumption.

We see that M_{GUT} and M_{pl} are indeed very close. Thus the domain of particle physics naturally extends up to the edge of quantum gravity. Actually, the question is immediately posed of whether unification without the inclusion of quantum gravity is at all plausible. In any case it is unlikely that quantum gravity can be simply added without profound effects on particle interactions at least at nearby scales. Thus quantum gravity can no more be ignored by particle physicists.

We have seen that M_{GUT} and M_{pl} set the horizon of particle physics. The question is : is it plausible that the standard model remains unchanged up to an energy scale of order M_{GUT} or M_{pl} as contemplated by the "desert" scenario ? The answer to this question is negative because of the well known "hierarchy problem".

To discuss this point we recall that in the standard model the fermion and vector-boson masses are all specified in terms of the vacuum expectation value of the Higgs field V. The value of V is determined by the curvature scale of the Higgs potential :

$$V(\varphi) = -\frac{1}{2}\mu^2\varphi^+\varphi + \frac{\lambda}{4}(\varphi^+\varphi)^2, \tag{6}$$

according to

$$V = \frac{\mu}{\sqrt{\lambda}}. \tag{7}$$

The observed values of the masses require for V (and, therefore, roughly for μ as well) that $V \sim 100$ GeV.

If $\Lambda \sim (10^{15} - 10^{19})$ GeV, then one faces the problem of justifying the presence of two so largely different mass scales in a single theory (the so-called hierarchy problem). In general, if Λ is very large in comparison with μ, then, even if we set by hand a small value for μ at the tree level, the radiative corrections would increase μ up to nearly the order of Λ. This problem is particularly acute in theories with scalars, as in the standard model, because the degree of divergence of mass corrections is quadratic, while the same divergences are only logarithmic for spin 1/2 fermions.

One general way out would be that the limit $\mu \to 0$ corresponds to an increase of the symmetry of the theory. In fact, the observed value of μ^2 can be decomposed as

$$\mu^2 = \mu_o^2 + \delta\mu^2, \tag{8}$$

where μ_o^2 is the tree level value and $\delta\mu^2$ arises from the loop quantum corrections. If no new symmetry is induced when $\mu \to 0$ and no other nonrenormalization theorem is operative, then a small value for μ^2/Λ^2 can only arise by an unbelievably precise cancellation between μ_o^2/Λ^2 and $\delta\mu^2/\Lambda^2$. If, however, $\mu = 0$ leads to an additional symmetry, then $\delta\mu^2$ must be proportional to μ_o^2, because for $\mu_o = 0$ both the tree diagrams and the loop corrections must respect the symmetry. Then, if one starts from a small value of μ_o^2/Λ^2, the radiative corrections preserve the smallness of μ^2/Λ^2.

For fermions chiral symmetry is added when $\mu \to 0$, because the axial currents are also conserved in this limit, as their divergence is proportional to the fermion mass. Chiral symmetry and the logarithmic degree of divergence of fermion masses considerably alleviate the hierarchy problem in theories with no fundamental scalars.

In the standard model no additional symmetry is gained for $\mu = 0$. This is also seen from the explicit formula for μ^2 at the one-loop level, which shows that μ^2 is not proportional to μ_o^2 :

$$\mu^2/\Lambda^2 = \mu_o^2/\Lambda^2 + \frac{1}{128\pi^2}\frac{d}{d\eta^2} \sum_J (2J + 1)(-1)^{2J} M_J^2(V) + \ldots, \tag{9}$$

where terms which vanish with $\Lambda \to \infty$ are indicated by the dots. The sum over J includes both particles and antiparticles (counted separately) of spin J and mass M, (expressed as a function of V). Note the opposite sign of the contributions from bosons and fermions. Thus one is forced to the conclusion that in the standard model the natural value for μ^2/Λ^2 is of order one or so. Therefore, as Λ can be interpreted as the energy scale where some essentially new physical ingredient becomes important, one is led to expect that the validity of the present framework cannot be extended beyond $\Lambda \sim (1 - 10)$ TeV.

One sees that the problem of explaining the Fermi scale is much connected with the Higgs mechanism and the consequent presence of scalars, which makes the problem of testing the Higgs sector particularly crucial.

The main ideas to solve the hierarchy problem are compositeness and/or supersymmetry (SUSY). For example, if the Higgs bosons are composite of fermions, as in technicolour theories [5], there are only fundamental fermions in the theory and the hierarchy problem is eased, as it follows from the previous discussion. A more radical step is to construct theories where the weak bosons $W^{\pm}$ and Z^o are composite [6]. The SU(2) $\otimes$ U(1) gauge symmetry is only an appearence in this case. When the energy is increased one would not find the scalar Higgs, but instead the W/Z^o would be resolved into their constituents and the theoretical framework would change completely.

The only way out of the hierarchy problem that maintains fundamental scalar fields is provided by SUSY, which we now introduce.

SUSY [7] relates bosons and fermions, so that in a multiplet which forms one representation of supersymmetry there is an equal number of bosonic and fermionic degrees of freedom. This implies that SUSY generators are spin - 1/2 charges, Q_α. SUSY leads to an extension of the Poincaré algebra. Besides the obvious algebraic relations between Q_α and the Poincaré generators, which specify the spinorial transformations of Q_α under Lorentz transformations and its invariance under translations, the essentially new relation is the anticommutator

$$\{Q_\alpha, \overline{Q}_\beta\} = -2(\gamma_\mu)_{\alpha\beta}\ P^\mu, \qquad (10)$$

where P^μ is the energy-momentum four-vector, which generates spacetime translations.

If all fundamental symmetries are gauge symmetries, then also SUSY is presumably a local symmetry. This immediately leads to the realm of gravity. In fact, the product of two local SUSY transformations is a translation with space-time-dependent parameters, as follows from eq.(10). But a translation with space-time-dependent parameters is a general co-ordinate transformation. As ordinary gravity can be seen to arise from gauging the Poincaré group, similarly gauging the Poincaré algebra enlarged by SUSY generators leads to an extended version of gravity, called supergravity [8].

The SUSY generators can be extended to form a set of N mutually anticommuting spinorial charges Q_α^A (A = 1,2,...,N), and one speaks of N-extended SUSY or N-extended supergravity in the local case (the bound $N \leq 8$ follows from restricting to helicities not larger than 2).

Theorists like SUSY because of several reasons. SUSY is the maximum symmetry compatible with a nontrivial S-matrix in a local relativistic field theory. The powerful constraints among couplings and masses implied by SUSY drastically reduce the degree of singularity of field theory as deduced from power counting. In some cases a finite field theory is even obtained. Thus N = 4 extend SUSY Yang-Mills theories are finite in 4 dimensions. More in general powerful nonrenormalization theorems are deduced for SUSY theories. This property may solve some naturalness problems of the standard model. In particular, the hierarchy problem is solved in SUSY theories because the mass divergences of bosons and fermions become the same, and the quadratic divergences of scalars are reduced to logarithmic divergences as for spin - 1/2 fermions. Thus in SUSY theories one automatically obtains

$$\frac{d^2}{d\eta^2} \sum_J (2J + 1)(-1)^{2J}\ M_J^2(V) = 0, \qquad (11)$$

due to a cancellation between bosons and fermions. Recalling eq.(9), we see that this mechanism can explain the relative smallness of observed particle masses as generated by the Higgs vacuum expectation value. This remains true even in the presence of SUSY breaking, provided that the related effects on observed masses are controlled down to a size of order M_W or so.

In the domain of physics at M_{pl} and beyond the impact of supersymmetry has generated a host of very fertile ideas culminated with the present versions of superstring theories. A short summary of these ideas is given in the following.

Although not free of problems, supergravity is definitely less singular than conventional quantum gravity. The study of the complexity of N = 8 extended supergravity [9], the most comprehensive theory of this sort, has revealed new unexpected concealed symmetries. In particular, it turns out that a SU_8 local symmetry is implied by the theory (the explicit O(8) gauge symmetry is too small to contain $SU(3) \otimes SU(2) \otimes U(1)$). There have been unsuccessful attempts to interpret the hidden SU_8 as the gauge symmetry at the preon level, with the corresponding gauge bosons generated dynamically as composites of the same preons. Extended supergravity theories have also been considered in d > 4 space-time dimensions, with the aim of recovering the observed 4-dimensional universe by spontaneous compactification of space-time, along the lines originally proposed by Kaluza-Klein in a simpler context.

But higher dimensions and extended supergravity are not sufficient by themselves to produce a viable theory unifying all particle interactions including gravity. A new physical idea is needed. This is provided by the string theory [10] approach to gravity.

Strings were originally introduced in the context of dual models of hadrons [11]. In this case, the string connects a $q\bar{q}$ pair in a meson. The string tension is inversely proportional to the universal Regge trajectory slope, i.e. $T \sim 1\ GeV^2$. String theory is a relativistic field theory of extended objects. There are an infinite number of normal modes in the string, which in the quantum theory correspond to an infinite sequence of excited states. The mass gap separating the ground state from the excited states is $\Delta^m = O(\sqrt{T})$. At small energies (with respect to $\sqrt{T}$) the excited modes of the string decouple and one is left with an effective pointlike field theory. One can construct open string theories such that, in the limit $T \to \infty$, among the particles in the ground state there are massless spin 1 Yang-Mills bosons (i.e. with the right gauge couplings for a gauge group G). In the case of a closed string a massless spin 2 particle is found among other states. It was observed by Scherk, Schwarz and Yoneda in 1973-74 that the couplings of this spin 2 particle are exactly proportional to the gravitational couplings of the graviton. Thus it was suggested that the strings could be applied to describe a theory of gravity. In this context $T \sim (M_{pl})^2$. What has been recently discovered is that the combination of supersymmetry, higher dimensions, strings and definite gauge groups can lead to possibly finite theories which unify all known forces including gravity. It has been known for a long time that string theories are only consistent for given numbers of space-time dimensions : d = 2,10,26 for different types of string theories : e.g. d = 26 for the bosonic string theory, d = 10 for the Neveu-Schwarz-Ramond model. The latter is suitable for describing both bosonic and fermionic degrees of freedom. Implementing SUSY, Green and Schwarz constructed superstring theories in d = 10. These theories admit chiral fermions. The step that has recently brought these theories at the center of everybody's attention, was made by Green and Schwarz in 1984 [12]. They proved that there is a unique dimension of the Yang-Mills gauge group, N = 496, such that the chiral and gravitational anomalies cancel each other. The resulting theory is one-loop finite and may be all-loop finite. SUSY is crucial for finiteness. Two types of superstring theories survive. They are based on the gauge groups SO(32) or $E8 \otimes E8$ (the heterotic string) [14].

The dimensional reduction from 10 down to 4 space-time dimensions is induced by spontaneous compactification (a mechanism in some sense analogous to spontaneous symmetry breaking) [13]. As a result 6 out of 10 dimensions are curled up with a radius of order $M_{pl} \sim 10^{-33}$ cm :

$$M_{10} \to M_4 \oplus M_6. \tag{12}$$

In the process of compactification there is a subtle relation between the curvature tensor $R^{mn}_{\mu\nu}$ and the Yang-Mills curl $F^{nm}_{\mu\nu}$. If a number of phenomenological requirements are imposed, e.g. that a N = 1 SUSY survives in 4 dimensions, then the holonomy group (displayed by the spin connection ω related to the gravitational field, hence to $R^{mn}_{\mu\nu}$) is not the real group U(6) but the complex group SU(3). Compact 6-dimensional spaces with SU(3) honolomy have been studied by Calabi and Yau.
As a consequence of the above mentioned relation between $R^{mn}_{\mu\nu}$ and $F^{mn}_{\mu\nu}$, the holonomy group SU(3) is a subgroup of the original Yang-Mills symmetry.

In the $E_8 \otimes E_8$ superstring theory, upon compactification

$$E_8 \otimes E_8 \to E_8 \otimes E_6 \otimes SU(3). \tag{13}$$

The unbroken E_8 corresponds to a hidden sector which only interacts with the ordinary world via gravitational interactions (the consequences for cosmology are still to be studied). E_6 is the effective (supersymmetric) grand unification group.

The program is to obtain all low energy properties from the topology of M_6. For example, the number of families is determined by the number of handles and holes of M_6, the breaking of E_6 is due to topological effects and even the observed mass ratios should be determined by the topology of M_6. No realistic realization of this program exists at the moment. Thus, at present, the superstring scenario is not yet a definite theory, but the sketch of a giant architecture.

Many problems remain open. Some of them are quite fundamental. For example : prove finiteness, find the right M_6 for phenomenology, explain why $d = 4$ in the observed world, clarify the connection between strings and Yang-Mills theories etc.

Anyway, in trying to go beyond the standard theory now there are two opposite routes : from above and from below. From above one tries to construct a truly unified theory of all interactions. At present, the steps are as follows. At energies $E \gg M_{pl}$ the correct description is in terms of extended strings in $d = 10$ dimensions. At $E \sim M_{pl}$ after decoupling of the higher string excitations an effective pointlike supergravity plus Yang-Mills theory is obtained. Also spontaneous compactification leads to $d = 4$, $N = 1$ SUSY and a definite grand unified theory, presumably E_6. Finally, if SUSY is to be relevant at low energy as well then SUSY breaking by gravity should take place at scales $M_s \sim (M_W M_{pl})^{1/2}$, as in the models which were already constructed before the string explosion. The effective theory at low energies is described by the sum of a globally supersymmetric extension of the standard model Lagrangian plus soft SUSY breaking terms.

The opposite route is to start from the standard theory, study its limitations and try to extrapolate at higher energies. This has been done a lot in the past few years. Now one can try to merge the two approaches. In the rest of these lectures I shall reconsider the theoretical work on possible extensions of the standard theory by keeping in mind the indications that arise from the superstring speculations.

FAMILY REPLICAS ARE NOT FOR EVER

There are three sources of information that suggest that the proliferation of fermion families is close to an end. Each argument is not completely tight. But the collection of the three is quite restrictive. The three arguments are given in the following :

a) The ratio of weak neutral couplings to charged weak couplings defines the well known parameter ρ. In the Born approximation, if only Higgs doublets are present, then

$$\rho = M_W^2/M_Z^2 \cos^2\theta_W = 1. \tag{14}$$

Radiative corrections induced by largely split weak isospin doublets produce deviations from 1. These corrections arise from one loop vacuum polarization diagrams in the neutral vector boson sector. They depend quadratically on the mass splittings whithin isospin doublets. For example, for $M_t \gg M_b$, the top quark contribution is :

$$\delta\rho = \frac{3G_F}{8\sqrt{2}\pi^2} \cdot M_t^2 \simeq 1.25\ 10^{-2}\ (\frac{M_t}{200\text{GeV}})^2 . \qquad (15)$$

Experimentally $\delta\rho < 0.015$ (2σ), so that one obtains $M_t \lesssim 300$ GeV, and similarly for possible higher generations. The limit of this argument is that it only forbids widely split weak isospin multiplets.
b) Primordial light element abundances in the universe depend on the expansion time scale at $t \sim 3$ minutes. The parameters needed are n_b/n_γ and n_ν, where n_b, n_γ and n_ν are the baryon, photon and neutrino numbers. Since n_b/n_γ is known ($\frac{n_b}{n_\gamma} = (4 \pm 2)\ 10^{-10}$) one derives n_ν from, say, helium abundance. This is the number of relativistic neutrinos at nucleo-synthesis, i.e. neutrinos with $m_\nu \lesssim 100$ KeV. The result which is obtained is

$$2 < n_\nu < 5. \qquad (16)$$

The argument does not forbid heavy ($m_\nu \gtrsim 1\text{-}2$ GeV) neutrinos.
c) A third constraint is provided by the observation of W/Z^o production at the CERN $p\bar{p}$ Collider. One starts from the identity

$$R = \frac{(\sigma B)\ W \to e\nu}{(\sigma B)\ Z \to ee} = \frac{\sigma_W}{\sigma_Z}\ \frac{\Gamma(W \to e\nu)}{\Gamma(Z \to ee)}\ \frac{\Gamma_Z}{\Gamma_W}. \qquad (17)$$

The ratio of Γ_W/Γ_Z can be predicted from QCD with more precision than the individual cross sections (the estimated error on the ratio is about 7 %). The partial widths $\Gamma(W \to e\nu)$ and $(Z^o \to ee)$ also are computed from the standard electroweak theory. Then, by assuming that no heavy charged leptons contribute to the W total width, one obtains Γ_Z from the experimental value of R. The combined UA1 and UA2 results lead to :

$$\delta n_\nu \lesssim 2.4 \pm 1. \qquad (18)$$

While heavy neutrinos are included up to $m_\nu \lesssim M_Z/2$ this argument clearly cannot exclude the possibility of neutrals not coupled to the Z^o.

NEW GAUGE INTERACTIONS. LEFT-RIGHT SYMMETRY

We proceed in our discussion of possible developments beyond the standard model by pointing out that new gauge interactions could exist and manifest themselves at larger energies. The gauge group could be enlarged by addition of a new factor G, leading to a basic group of the form $SU_3 \otimes SU_2 \otimes U_1 \otimes G$. In this section we shall discuss the most attractive possibility of this type, namely left-right symmetric models [15]. In the next section the approach toward unification will be considered.

The motivation for left-right symmetric models is the attempt of understanding P and C violations in weak interactions. In these theories the electroweak Lagrangian is P and C conserving before spontaneous symmetry breaking and the observed violations of P and C are attributed to a noninvariance of the vacuum.

The electroweak group $SU_{2L} \otimes U_1$, is replaced by $SU_{2L} \otimes SU_{2R} \otimes U_1$, with a discrete symmetry under left-right interchange, so that $g_L = g_R$, where $g_{L,R}$ are the $SU_{2L,R}$ gauge couplings. The right-handed quarks and leptons which are singlets under SU_{2L} now become doublets under SU_{2R}. The complete assignments of quarks and leptons are thus given by

$$q_L = (\frac{1}{2}, 0, \frac{1}{3}), \quad q_R = (0, \frac{1}{2}, \frac{1}{3}),$$
$$\ell_L = (\frac{1}{2}, 0, -1), \quad \ell_R = (0, \frac{1}{2}, -1). \tag{19}$$

The electric charge becomes

$$Q = I_{3L} + I_{3R} + \frac{B - L}{2} \tag{20}$$

where B and L are the baryon and lepton numbers, respectively. Note that the U_1 generator now acquires an elegant physical meaning, being proportional to B - L.

The minimal set of Higgs scalars required to break $SU_{2L} \otimes SU_{2R} \otimes U_1$ down to U_{1Q} is given by

$$\Delta_L(1, 0, 2) + \Delta_R(0, 1, 2) \tag{21}$$

and

$$\varphi(\frac{1}{2}, \frac{1}{2}, 0). \tag{22}$$

Under left-right symmetry $\Delta_L \leftrightarrow \Delta_R$ and $\varphi \leftrightarrow \varphi^+$. It is interesting that for a range of values of the relevant parameters one obtains parity-violating minima for a left-right symmetric potential. These correspond to the vacuum expectation values

$$\langle\Delta_{L,R}\rangle = \begin{pmatrix} 0 & 0 \\ v_{L,R} & 0 \end{pmatrix}, \quad \langle\varphi\rangle = \begin{pmatrix} k & 0 \\ 0 & k' \end{pmatrix} \tag{23}$$

with

$$v_L = \gamma \frac{k^2}{v_R}, \tag{24}$$

where γ depends on the Higgs self-couplings. The constraint eq.(24), which automatically arises from the minimization procedure, ensures that v_L vanishes when $v_R \to \infty$.

The charged-W mass eigenstates are mixtures of W_L and W_R

$$W_1 = \cos z\, W_L + \sin z\, W_R, \quad W_2 = -\sin z\, W_L + \cos z\, W_R \tag{25}$$

with the mass values and the mixing angle given by

$$\mathrm{tg}\, z \simeq \frac{kk'}{k^2 + k'^2 + 8v_R^2}, \quad m^2_{W_{1,2}} = \frac{g^2}{2}(k^2 + k'^2 + 2v^2_{L,R}). \tag{26}$$

In order to agree with observations, one needs $m_{W_1} \simeq m_W$, $m_{W_2} \gg m_{W_1}$ and $z \ll 1$. These relations imply $v_R \gg (k^2 + k'^2)^{1/2} \gg v_L$.

Once this is true, the neutral-current sector also does not deviate much from the standard situation. There are two Z^os, the light one (always lighter than in the standard model) approximately satisfying $m^2_{W_1}/m^2_{Z_1} \cos^2\theta_w \sim 1$ with departures of order $v_L^2/(k^2 + k'^2)$ and the heavy one with mass $m_{Z_2} \sim g_{L,R}\, v_R$. The observed neutral-current phenomenology is reproduced within the accuracy of the data provided that

$$m_{Z_2} \gtrsim 4m_{Z_1}, \tag{27}$$

which implies

$$m_{W_2} \gtrsim 220 \text{ GeV}. \tag{28}$$

These bounds also ensure agreement within errors with the measured values of m_W and m_{Zo}. A precision of order 1 % in m_W and m_{Zo} is, in fact, necessary, in order to detect the deviations from the standard values if the above bounds are satisfied.

The bounds on m_W and z that can be obtained from changed-current data depend on the mass of the right-handed neutrino. If ν_R is allowed by phase space in μ decay, then

$$m_{W_2} \gtrsim 380 \text{ GeV} \tag{29}$$

and

$$z \lesssim 0.05. \tag{30}$$

However, the ν_R mass could be large, as seems theoretically more resonable, and in this case one goes back to the limit of eq.(28), while z must be small anyway ($z \lesssim 0.095$).

Stringent bounds on m_{W_2} have recently been obtained from non-leptonic amplitudes, which, however, necessarily include some model dependence. In particular, the $K_L - K_S$ mass difference leads to an important bound with the assumptions that

a) the box diagram approximation is reasonable,
b) the K -M mixing matrices for left and right quarks are the same or one the complex conjugate of the other.

Assumption b) is true in the simplest and most interesting versions of left-right symmetric models. Barring unplausible cancellations with top-quark and Higgs exchanges, one then obtains

$$m_{W_2} \gtrsim 1.6 \text{ TeV}. \tag{31}$$

In conclusion left-right symmetric models are interesting and cannot be ruled out even with m_{W2} and m_{Z2} as low as a few hundred GeV, although there are solid indications that the right-handed gauge bosons, if they exist, are to be searched above the TeV region.

Before closing this section we make a comment on neutrino masses in left-right symmetric theories. The possible neutrino mass terms are of the form :

$$\left.\begin{array}{l}\bar{\nu}_L \cdot \nu_R \\ \bar{\nu}_R \cdot \nu_L\end{array}\right\} \quad \text{(Dirac)} \tag{32}$$

or

$$\left.\begin{array}{l}\nu_L^T \cdot \nu_L \\ \\ \nu_R^T \cdot \nu_R\end{array}\right. \qquad \text{(Majorana)}. \tag{33}$$

The Dirac mass terms are the usual ones, while the Majorana mass terms violate lepton number. With the standard assignments of weak isospin and only Higgs doublets, the $\nu_L^T\,\nu_L$ mass term is absent and the mass matrix is given by (in the L, R basis) :

$$M_\nu = \begin{pmatrix} 0 & m \\ m & M \end{pmatrix} \tag{34}$$

where m and M are the Dirac and Majorana masses respectively. If $M \gg m$ the approximate eigenvalues are :

$$\lambda_1 \simeq \frac{m^2}{M} \tag{35}$$

$$\lambda_2 \simeq M. \tag{36}$$

Now, in left-right theories, due to eq.(20), the breakings of B-L and of I_{3R} are related when the electric charge and I_{3L} are conserved :

$$\Delta I_{3R} = \frac{1}{2}\,\Delta(B-L). \tag{37}$$

Thus the scale of the Majorana mass M is related in these theories to that of the breaking of the right handed group : $M \sim M_{W_R/Z_R}$. Note that from eq.(35) the mass of the light ν_e would be :

$$m_{\nu_e} \simeq \frac{m_e^2}{M} \simeq 1\ \text{eV}\left(\frac{1\ \text{TeV}}{M}\right).$$

GRAND UNIFICATION AFTER THE DESERT

In this section we discuss the present status of grand unification. Let us first consider GUTS constructed under the assumption of no new physics up to M_{GUT}. Later we shall describe SUSY GUTS which arise from a particularly attractive way of populating the desert.

As a preliminary observation we remark that the group G commutes with the Lorentz group, and consequently all states in a given multiplet of G must have the same helicity. It is, therefore, useful to describe fermions in terms of a pair of Weyl two-component spinors, one for the left-handed fermion and one for the left-handed charge-conjugated fermion (instead of the right-handed fermion). In this sense we shall replace, say, e_R^- with e_L^+ and so on, and the label L can simply be omitted.

The content of one generation of fermions in terms of $SU_3 \times SU_2 \times U_1$ is

$$\begin{pmatrix} u \\ d \end{pmatrix} \qquad \begin{pmatrix} \nu \\ e- \end{pmatrix} \qquad \bar{u} \qquad \bar{d} \qquad e^+, \tag{38}$$

$$(3,2)_{2/3,-1/3} \qquad (1,2)_{0,-1} \qquad (\bar{3},1)_{-2/3} \qquad (\bar{3},1)_{1/3} \qquad (1,1)_1$$

where the charge Q has also been indicated. This makes fifteen Weyl spinors. Note that $\bar{\nu}$ has not been included (massless neutrinos) but it could be added. Then the rank of G must be not smaller than 4 and G must presumably admit complex representations (that is non-self-conjugate representations, because the spectrum of fermions listed above is not self-conjugate). Now the only simple groups that admit complex representations are SU_n (n > 2), SO_{4n+2} and E_6. Products of identical simple groups connected by a discrete symmetry (in order to have a single gauge coupling) are also possible. At present the most interesting candidates are

$$G = SU(5),\ SO(10),\ E6,\ SU(3)^3 \tag{39}$$

Of these SU(5) has rank 4 (the other possibility of rank 4, $SU(3)^2$ [2] is immediately seen not to work : for example, $\sin^2 \theta_W(M_{GUT}) = 3/4$ etc.), SO(10) has rank 5, E6 and $SU(3)^3$ have rank 6. We do not consider here the possibility of including horizontal symmetries in G. Recently interesting models of this sort have been considered. For example those based on SO(18) [16]. These models can be compatible with a string theory based on SO(32), while there is apparently no space for them in E8 × E8. Models based on SO(32) have difficulties in maintaining chiral fermions upon compactification.

In SU(5) each generation of fermions is accomodated in a sum of two representations :

$$\text{one generation} = \bar{5} + 10. \tag{40}$$

The content of the 1st generation $\bar{5}$ is $(\bar{d}_1, \bar{d}_2, \bar{d}_3, \nu, e^-)$ where the three coloured $\bar{d}_i$ sit together with the electron doublet. The electron charge is three times the d charge because there are three colours. This is the GUT explanation of charge quantization.

The superstrong breaking from SU_5 down to $SU_3 \times SU_2 \times U_1$ is most simply obtained by a 24 multiplet of Higgs bosons. The standard breaking down to $SU_3 \times U_{1,Q}$ is induced by a different set of Higgs, which for economy must also provide the fermion masses. The only representations that can contribute to fermion masses are those in $\bar{5} \times 10 = 5 + 45$. Both are needed for a general mass matrix. But an interesting possibility is to only include a 5. This implies the same mass for d and e and

$$\frac{m_d}{m_e} = \frac{m_s}{m_\mu} = \frac{m_b}{m_\tau} \tag{41}$$

in the symmetry limit. As we shall see each ratio is modified going down to low energy.

Note that in the Higgs pentaplet 3 components are superheavy and 2 are at the Fermi scale. This is unnatural (the 3-2 problem) and is an aspect of the hierarchy problem.

SO(10) admits the following interesting maximal subalgebras

$$\begin{aligned} SO(10) &\supset SU(5) \otimes U(1) \qquad (16 = \bar{5} + 10 + 1) \\ &\supset SU(4) \otimes SU(2) \otimes SU(2) \qquad (16 = (4,2,1) + (\bar{4},1,2)). \end{aligned} \tag{42}$$

The 16 is the representation of each fermion generation. We have also indicated the content of the 16 in the two subalgebras. By adding the right handed neutrino a whole generation is accomodated in a single irreducible representation, which is nice. The vanishing of ν masses is no longer guaranteed which is not necessarily bad. It depends on the symmetry breaking whether or not a given subalgebra is relevant for physics.

The routes through SU(4) ⊗ SU(2) ⊗ SU(2) or SU(3) ⊗ SU(2) ⊗ SU(2) ⊗ U(1) are compatible with left-right symmetry. The first SU(4) factor corresponds to the idea of lepton number as a fourth colour. The symmetry breaking could occur through a 45 (the adjoint representation of SO(10)) while the fermion masses can arise from $(16\times16)_{symm} = 126 + 10$.

E_6 is suggested by superstrings. It admits the maximal subalgebras :

$$E6 \supset SO(10) \otimes U(1) \qquad (27 = 16 + 10 + 1)$$
$$\supset SU(3)^3 \qquad (27 = (3,\overline{3},1) + (\overline{3},1,3) + (1,3,\overline{3})) \tag{43}$$

where $SU(3)^3$ = SU(3) colour ⊗ SU(3) left ⊗ SU(3) right can itself be a unifying group. A new interesting model based on $SU(3)^3$ has been recently discussed [17]. As the smallest representation of E6 is the 27 and one 27 is needed for each generation, many unknown fermions are predicted.

In $SU(3)^3$ language the quarks are right handed singlets with charges 2/3, -1/3, -1/3 :

$$(3,\overline{3},1) : \begin{pmatrix} u_1 & d_1 & B_1 \\ u_2 & d_2 & B_2 \\ u_3 & d_3 & B_3 \end{pmatrix} \tag{44}$$

the antiquarks are left handed singlets

$$(\overline{3},1,3) : \begin{pmatrix} \overline{u}_1 & \overline{u}_2 & \overline{u}_3 \\ \overline{d}_1 & \overline{d}_2 & \overline{d}_3 \\ \overline{B}_1 & \overline{B}_2 & \overline{B}_3 \end{pmatrix} \tag{45}$$

and the leptons are colour singlets

$$(1,3,\overline{3}) : \begin{pmatrix} E^o & E^- & e^- \\ E^+ & \overline{E}^o & \nu \\ e^+ & N & \overline{N} \end{pmatrix}. \tag{46}$$

In E6 realistic symmetry breakings can be realized by Higgs systems transforming as 78 + 2(27) or $27 \otimes 27 = \overline{27} + 351_s + 351_a$ [18]. If $SU(3)^3$ is considered in itself, them 2(27) are sufficient [17]. Note that in E6 and $SU(3)^3$ the Higgs can be chosen in a way which is compatible with fermion-antifermion bound states. The unwanted fermions can be made superheavy (in $SU(3)^3$) some neutral leptons are left at relatively low masses). String theories have their own way of breaking E6 by topological effects, which is similar to Higgs in the adjoint representation.

We now consider the constraints imposed by the unifying group G on the couplings in the symmetric limit and at present energies [19]. In fact, as already mentioned, those we measure are effective scale-dependent couplings (and masses) in the renormalization group sense. Otherwise, a unique gauge coupling could not be reconciled with the observed strength of the different interactions. The symmetric limit applies at an energy scale larger than the symmetry-breaking superheavy masses of gauge bosons and Higgs. For the general case of a unifying group G, let us normalize the generators T_i such that the charge generator is given by

$$Q = T_3 + C_1\, T_o, \tag{47}$$

where T_3 corresponds to the weak isospin and T_o is proportional to the weak hypercharge. These generators are normalized to $\mathrm{Tr}(T_l T_m) = N\,\delta_{lm}$. Consequently the colour generators T_c are related to the Gell-Mann colour matrices by

$$C_3 T_c = \frac{\lambda_c}{2}. \tag{47}$$

In the symmetric limit only one coupling constant g_G exists, coupled to the generators T_i, i.e. the product $g_G T_i$ specifies the couplings. When G is broken into $SU_3 \times SU_2 \times U_1$, three couplings can be defined through $g_3 T_c$, $g_2 T_{weak}$, $g_1 T_o$. When the scale of masses is increased up to the symmetric region, $g_{1,2,3}$ all approach g_G. The physical couplings g_s, g_W and g'_W are related to $g_{3,2,1}$ by

$$g_s C_3 = g_3, \qquad g_W C_2 = g_2, \qquad g'_W C_1 = g_1, \tag{48}$$

with $C_2 = 1$, or equivalently in terms of squared couplings ($\alpha = g^2/4\pi$)

$$\alpha_s C_3^2 = \alpha_3, \qquad \alpha_W = \alpha_2, \qquad \alpha'_W C_1^2 = \alpha_1. \tag{49}$$

Recalling the relations among $\sin\theta_W$, e^2, g_W, g'_W, we also have

$$\alpha_2 = \frac{\alpha}{\sin^2\theta_W}, \qquad \alpha_1 = \frac{C_1^2\alpha}{\cos^2\theta_W}, \tag{50}$$

where α is the electromagnetic fine-structure constant. The constants C_1 and C_3 only depend on the group G and on the embeddings of $SU_3 \times SU_2 \times U_1$ in G and not on the representation. They are given by

$$\frac{\mathrm{Tr}\,T_3^2}{\mathrm{Tr}\,Q^2} = \frac{1}{1 + C_1^2}, \qquad \frac{\mathrm{Tr}\,T_3^2}{\mathrm{Tr}(\lambda_c/2)^2} = \frac{1}{C_3^2}. \tag{51}$$

For all interesting groups SU(5), SO(10), E6, $SU(3)^3$ one obtains

$$C_2^2 = C_3^2 = 1, \qquad C_1^2 = 5/3. \tag{52}$$

In the symmetric limit α_1 and α_2 are equal. Thus we see from eqs.(50), (52) that

$$(\sin^2\theta_W)_{symm} = 1/(1 + C_1^2) = 3/8 \tag{53}$$

and

$$\alpha_G = C_3^2(\alpha_s)_{symm} = (1 + C_1^2)(\alpha)_{symm} = \frac{8}{3}(\alpha)_{symm}. \tag{54}$$

If we assume that :

(a) only one step of symmetry breaking leads directly from G down to $SU(3) \otimes SU(2) \otimes U(1)$. This implies, in the region 10^2 GeV $< M < M_{GUT}$

$$\frac{1}{\alpha_i(M)} = \frac{1}{\alpha_i(\mu)} + \frac{b_i}{2\pi C_i^2}\ln\frac{M}{\mu} \qquad (i = 1,2,3). \tag{55}$$

(b) the only light particles are the standard ones, with n_G fermion generations and H Higgs doublets. This fixes the coefficients b_i/C_i^2 :

$$\frac{b_3}{C_3^2} = -11 + \frac{4}{3} n_G$$

$$\frac{b_2}{C_2^2} = -\frac{22}{3} + \frac{4}{3} n_G + \frac{1}{6} H \tag{56}$$

$$\frac{b_1}{C_1^2} = \frac{4}{3} n_G + \frac{1}{10} H.$$

For $M = M_{GUT}$ the three $\alpha_i(M_{GUT})$ approach the same limit, and one obtains from eqs.(55,56) :

$$\sin^2 \theta_W(\mu) = \frac{1}{6} \frac{1+\frac{H}{22}}{1+\frac{H}{66}} + \frac{5}{9} \frac{1-\frac{H}{110}}{1+\frac{H}{66}} \frac{\alpha(\mu)}{\alpha_s(\mu)} \tag{57}$$

$$\ln \frac{M_{GUT}}{\mu} = \frac{2\pi}{11(1+\frac{H}{66})} \left[\frac{1}{2\alpha(\mu)} - \frac{4}{3\alpha_s(\mu)}\right]. \tag{58}$$

With assumptions (a) and (b) these results are valid for all the interesting groups we have mentioned. However, contrary to SU(5), all other unifying groups of rank > 5 allow the possibility of intermediate symmetry stages with an additional scale of mass.

If one takes $\mu \sim 100\ \text{GeV}^2$, $\alpha_s(\mu) \sim 0.10 - 0.13$, $\alpha(\mu) \simeq 1/128$, $H = 1$ one obtains

$$\sin^2 \theta_W(\mu) \simeq 0.204 - 0.214$$
$$M_{GUT} \simeq (2.4 \, . \, 10^{14} \div 1.4 \, . \, 10^{15})\text{GeV}. \tag{59}$$

Also, for three generations, the symmetric value of the coupling, defined in eq.(54) is given by

$$\alpha_G \simeq 0.024. \tag{60}$$

More careful calculations, including two loop effects, thresholds etc. do essentially confirm the above estimates. The experimental value of $\sin^2 \theta_W \simeq 0.22 \pm 0.01$ (from ν-nucleon scattering and UA1, UA2 data) is consistent with the "desert" GUT prediction.

In a unified theory both quarks and leptons are in one multiplet of G. Then in general some of the gauge bosons turn quarks into leptons. There are exceptions to this statement. For example in $SU(3)^3$ the gauge bosons cannot do it. Higgs exchange also may induce baryon and lepton number nonconservation. Note that the unbroken symmetry $SU_{3,colour} \times U_{1,Q}$ allows the transition $q + q \to \bar{q} + \bar{l}$ because in colour $3 \times 3 \supset \bar{3}$. In $SU_3 \times U_{1,Q}$ the following transitions are allowed :

$$u + u \to \bar{d} + e^+$$
$$u + d \to \bar{u} + e^+, \bar{d} + \bar{\nu}. \tag{61}$$

The proton lifetime is given by [29]

$$\tau_p \simeq \frac{(\text{factors})}{\alpha_G^2(M)} \frac{M_{GUT}^4}{m_p^5}. \tag{62}$$

Neglecting the factors in front, I find from eqs.(59,60,62) :

$$\tau_p \simeq (1.5\ 10^{29} \div 2.10^{32})y. \tag{63}$$

For models, like SU(5), where the most likely proton decay channel should be $p \to e^+\pi^o$, the prediction eq.(63) is alarming because of the experimental limit

$$\tau_p/B(p \to e^+\pi^o) > 2\ .\ 10^{32}y \tag{64}$$

and the fact that all possible refinements of our crude estimate bring the predicted lifetime of the proton further down.

We may add that the mass ratios in eq.(41), predicted by light Higgs transforming as a 5 (and no 45) of SU(5), are renormalized according to

$$\frac{m_b(\mu)}{m_\tau(\mu)} = \left(\frac{\alpha_s(\mu)}{\alpha_G(M)}\right)^{12/(33-2f)} \left(\frac{\alpha_1(\mu)}{\alpha_G(\mu)}\right)^{3/2f} \frac{m_b(M)}{m_\tau(M)}. \tag{65}$$

With $m_b(M) = m_\tau(M)$ this leads to a value of about 3. This is good for b, disputable for s and certainly bad for d. The failure could be blamed on the too small value of the s and d masses, or taken as a further indication that a 5 is not sufficient.

The possible problem of SU(5) with proton decay are not present for higher rank unifying groups, if more then a single stage of symmetry breaking is allowed. For example, many different chains of symmetry breaking steps are possible in SO(10) along the left-right symmetric route. Parity (i.e. $L \leftrightarrow R$ discrete symmetry) and $SU(2)_R$ can be broken at the same scale or separately. In the first case one can have, in obvious notation :

$$\begin{aligned} SO(10) &\to 224P \to 2213P \to 2113 \to 213 \\ &\to 224P \to 214 \quad \to 2113 \to 213 \end{aligned} \tag{66}$$

etc. In the second case, one can have :

$$\begin{aligned} SO(10) &\to 224P \to 224 \to 2213 \to 2113 \to 213 \\ &\to 224P \to 224 \to 2113 \to 213. \end{aligned} \tag{67}$$

etc.

As a result of recent studies [20] on these different routes of symmetry breaking (assuming the minimal Higgs system required in each case) it is found that confortable consistency with experiments on τ_p,

$\sin^2\theta_W$ etc. is no problem. However it is very difficult to obtain a relatively light charged right handed $W_R (M_{W_R} \sim (10^9 - 10^{10})\mathrm{GeV})$. On the contrary a neutral Z_R could be allowed in the range $M_{Z_R} \sim (0.5 \div 10)\mathrm{TeV}$.

In conclusion the great virtues of grand unification remain aesthetical appeal, an elegant explanation of charge quantization, the possibility of explaining the observed value of $\sin^2\theta_W$ and the set-up of a coherent framework for studying new interactions as, for example, B- and L-violating transitions. The shortcomings, in part connected with the "desert" assumption, are essentially the hierarchy problem but also the proliferation of free parameters and many of the problems of the standard model which are left unsolved after grand unification.

MODEL INDEPENDENT ANALYSIS OF B AND L VIOLATION

At low energy ($M \ll M_{GUT}$) one can write down an effective lagrangian, sum of the standard model lagrangian and a number of possibly non renormalizable interactions arising as residuals of heavy particle exchanges :

$$\mathcal{L} = \mathcal{L}_{S.M.} + \sum_i C_i \, M_{GUT}^{4-di} \, O_{di} \tag{68}$$

where C_i are dimensionless coefficients, O_{di} are local operators of dimension di. Obviously baryon and lepton number violation can only arise from the latter terms. In the logic of the desert the relevant operators with $\Delta B \neq 0$ must necessarily involve quark fields (the only ones with $B \neq 0$). Then the requirement that the relevant O_{di}'s are colour singlets leads to products of at least three quark fields. An even number of fermions are needed for making Lorentz scalar operators. It follows that the only $\Delta B \neq 0$ operators have $d \geq 6$ if the desert is true. The dominant contributions arise from the minimum dimension, because higher dimension operators are suppressed by powers of $(M_{GUT})^{-1}$. One obtains :

$$\begin{array}{lll} d = 6 & qqql & \Delta B = \Delta L = 1 \\ d = 7 & Dqqql^c & \Delta B = -\Delta L = -1 \end{array} \tag{69}$$

etc., where D indicates a bosonic field of the presence of a derivative. We see that operators with d = 6 conserve B-L.
Thus B-L violation can only occur through d = 7 operators. This means that B-L violation is invisible unless an intermediate scale $M_x \simeq 10^{10} - 10^{11}$ GeV is also present in the symmetry breaking. Also it happens that B-L is an exact global symmetry in SU(5). Thus B-L violation can only occur in SO(10) (at the intermediate scale where $SU(2)_R$ is broken, because of eq.(20)) or in E6.

SUSY - GUTS

One great advantage of SUSY GUTS is that the hierarchy problem is directly solved, at least in its technical aspect : after fine tuning supersymmetry guarantees the stability of the result as far as quantum corrections are concerned. However the physical problem of explaining the origin of the huge mass ratios involved is not solved. But once the internal consistency of the "low energy" theory is guaranteed one can leave that problem for the "high energy" theory to solve.

For all GUTS one simply extends each multiplet into the simplest corresponding SUSY multiplet. For example, in SU(5) each $\overline{5} + 10$ becomes a chiral SUSY multiplet describing quarks, leptons, s-quarks and s-leptons.

Assuming that all SUSY partners of the standard particles have masses not larger than a few TeV's, as demanded for an explanation of the Fermi scale, then the calculation of the running of the three gauge couplings up to M_{GUT} can be immediately repeated. For example, the value of b_3/C_3^2 given in eq.(56) is modified by the presence of gluinos which transform -11 into -9, and of 2 kinds of s-quarks for each flavor of quarks which modify $+\frac{4}{3} n_G$ into $2n_G$. Similarly for the other couplings. One then obtains :

$$\frac{b_3}{C_3^2} = -9 + 2n_G$$

$$\frac{b_2}{C_2^2} = -6 + 2n_G + H/2 \tag{70}$$

$$\frac{b_1}{C_1^2} = 2n_G + \frac{3}{10} H$$

where C^2 are given in eqs.(52), n_G and H are the number of fermion generations and of Higgs scalar doublets (in SUSY theories H is at least 2). If no intermediate mass scales are present, then we now find :

$$\sin^2 \theta_W(\mu) = \frac{1}{6} \frac{1+\frac{H}{6}}{1+\frac{H}{16}} + \frac{5}{9} \frac{1-\frac{H}{30}}{1+\frac{H}{18}} \frac{\alpha(\mu)}{\alpha_s(\mu)}. \tag{71}$$

$$\ln \frac{M_{GUT}}{\mu} = \frac{2\pi}{9(1+\frac{H}{18})} \left[\frac{1}{2\alpha(\mu)} - \frac{4}{3} \frac{1}{\alpha_s(\mu)}\right]. \tag{72}$$

For lack of better information we still keep $\mu \sim 100$ GeV, the same values for α and α_s as before and H = 2. We then obtain :

$$\begin{aligned} \sin^2 \theta_W &= 0.228 - 0.236 \\ M_{GUT} &= (6.7\ 10^{15} - 4.6\ 10^{16})\text{GeV}. \end{aligned} \tag{73}$$

Furthermore, for $n_G = 3$:

$$\alpha_G = 0.040 - 0.042. \tag{74}$$

We see that the value of $\sin^2 \theta_W$ is increased but is still in an acceptable range. M_{GUT} is also increased, which makes the contribution to proton decay of superheavy gauge bosons unobservable by present (and probably also future) experiments.

But proton decay through Higgs and Higgs-ino exchange could still be observed. In the language of the previous section, new operators with d = 5 come into play. These are obtained by replacing two fermions with two scalars in the operator with d = 6 :

$$qqql \rightarrow \tilde{q}\tilde{q}ql \quad \text{or} \quad \tilde{q}qq\tilde{l} \tag{75}$$

where by $\tilde{q}$, $\tilde{l}$ s-quarks and s-leptons are denoted. The external s-quarks and/or s-leptons must be riconverted into ordinary fermions by light gaugino exchange. The expected rate cannot be predicted. However the tendence of Higgs and Higgsinos to couple stronger to heavier particles and the structure of (75) imply that the expected dominant channel would be $p \to \bar{\nu}K^+$.

PHENOMENOLOGICAL SUPERSYMMETRY

In this section we describe the phenomenology of the most promising class of broken SUSY theories, where the boson-fermion symmetry is broken by gravitational effects [21]. In particular we discuss the present limits on SUSY particles.

In general, in the presence of spontaneous symmetry breaking of SUSY, the size of the mass splittings inside a SUSY multiplet is ruled by a relation of the form :

$$\Delta m^2 \simeq g_G M_S^2 \tag{76}$$

where g_G is the coupling, between a fermion and its bosonic partner, of the goldstino, the fermionic analogue of a Goldstone boson, and M_S is the scale of mass of SUSY breaking. One wants Δm^2 to be of order M_W for naturalness, as discussed in connection with the hierarchy problem. Then, either g_G is very small, or M_S is itself of order 1 TeV or so. Attempts at implementing SUSY breaking at such nearby mass scales have met with phenomenological and theoretical problems. Instead, it has been realized that one can start from N = 1 supergravity with matter superfield and a spectrum divided into a heavy sector of particles of mass of the order of M_{pl} and a light sector of ordinary particles. In the heavy sector $g_G \sim O(1)$, but the heavy sector can only communicate with the light sector by gravitational, hence very weak, interactions, so that in the light sector SUSY breaking is comparatively much smaller and $g_G \sim O(M_W/M_{pl})$. Therefore

$$\Delta m^2 \simeq O(\frac{M_W}{M_{pl}})M_S^2. \tag{77}$$

If $\Delta m^2 \sim O(M_W)$ is imposed, one obtains

$$M_S^2 \simeq O(M_W M_{pl}) \simeq [(10^{11} - 10^{12})\mathrm{GeV}]^2. \tag{78}$$

Thus, in these theories M_S is very large, being the geometric mean of M_W and M_{pl}, but, due to the smallness of g_G in the light sector, the large size of M_S is influential at accessible energies.

After switching off gravitational interactions, by taking $M_{pl} \to \infty$, one obtains a low-energy effective Lagrangian of the form

$$\mathcal{L}\ \mathrm{eff} = \mathcal{L}\ \mathrm{global\ SUSY} + V\ \mathrm{soft}. \tag{79}$$

Here, $\mathcal{L}$ global SUSY is the globally SUSY symmetric Lagrangian of ordinary particles and interactions. V soft introduces what is technically named a soft breaking of SUSY, namely a symmetry breaking with interactions of dimension d < 4. It has the following form :

$$V(\varphi)_{soft} = m_{3/2}^2\varphi^+\varphi + m_{3/2}\, f(\varphi) - \frac{1}{2}\sum_i m_{1/2}^i\, \lambda^i\lambda^i \tag{80}$$

where φ symbolizes the s-particle fields, $f(\varphi)$ is a dimension 3 polynomial operator, $m_{3/2}$ is the spin 3/2 gravitino mass, λ^i (i = 1,2,3) are the spin

1/2 gauginos (λ_3 describes the gluinos, λ_1, λ_2 the electroweak gauginos) with masses $m_{1/2}$. The gravitino itself is only gravitationally coupled to matter, so that it is essentially unobservable and its mass arises from a super Higgs mechanism (with disappearance of the goldstino from the physical spectrum), and is in general given by

$$m_{3/2} \sim o(\frac{M_S^2}{M_{pl}}) \tag{81}$$

which, in the present case, as follows from eq.(78), implies

$$m_{3/2} \sim o(M_W). \tag{82}$$

Gaugino masses are in principle independent of $m_{3/2}$. The gaugino Majorana mass terms do in fact violate R symmetry. Since R symmetry is in general broken when SUSY is violated one might well expect that $m_{1/2} \sim o(m_{3/2})$. On the other hand there could be a memory of this symmetry so that $m_{1/2} \ll m_{3/2}$ or even $m_{1/2} = 0$.

The s-particle spectrum is largely unconstrained. Purely as an indication we report on the results following from some tentative assumptions. If at scales $M \sim M_{pl}$ the effective masses of s-quarks and s-leptons are all degenerate :

$$m_{\tilde{q}} \simeq m_{\tilde{e}} \simeq m_{3/2} \tag{83}$$

(as suggested by the fact that gravity is flavour and colour blind) then the observed masses at $M \sim M_W$ can be evaluated by ren. group techniques. If electro-weak effects and fermion masses are neglected one finds [22] :

$$m_{\tilde{q}}^2 \simeq m_{\tilde{l}}^2 + 0.8\, m_{\tilde{g}}^2. \tag{84}$$

The numerical value 0.8 refers to 3 generations (for 4 generations it becomes 1.6). Similarly, if at $M \simeq M_{pl}$, $m_{\tilde{\gamma}} = m_{\tilde{g}} \neq 0$, then at $M \sim M_W$:

$$m_{\tilde{\gamma}} \simeq \frac{8}{3}\frac{\alpha}{\alpha_s} m_{\tilde{g}} \simeq (\frac{1}{5} \div \frac{1}{6}) m_{\tilde{g}}. \tag{85}$$

Two physical $\tilde{W}^+$ (and two $\tilde{Z}$) exist which are mixtures of charged gauginos and Higgs-inos. At least in the case of not too heavy $\tilde{\gamma}$'s and $\tilde{g}$'s, one $\tilde{W}^+$ and one $\tilde{Z}$ should be lighter than the W and the Z^o respectively.

It is interesting to recall that in SUSY theories there must necessarily be two distinct Higgs doublets that give masses to up and down fermions respectively. As a consequence, one is led to predict the existence of charged scalar Higgs bosons and of additional neutral ones.

Experimental bounds on SUSY particle masses are obtained from direct production channels and from virtual effects. Charged s-particles must be heavier than about 20 GeV from their non observation in e^+e^- annihilation. The bounds are weaker for charge $-1/3$ quarks, and can be much better for s-electrons, depending on the $\tilde{\gamma}$ mass. In addition to pair production : $e^+e^- \rightarrow \tilde{e}^+\tilde{e}^-$, the production of a single $\tilde{e}$ in association with a $\tilde{\gamma}$, via t channel photon exchange (fig. 1) can have more phase space if $M_{\tilde{\gamma}} < M_{\tilde{e}}$. Pair production of $\tilde{\gamma}$'s via e exchange (fig. 2) is also a very convenient channel :

$$e^+e^- \to \tilde{\gamma}\,\tilde{\gamma} \tag{86}$$

$$e^+e^- \to \tilde{\gamma}\,\tilde{\gamma}\,\gamma. \tag{87}$$

The first channel, with no photons, can only be detected if $\tilde{\gamma}$'s are unstable and decay in the apparatus according to $\tilde{\gamma} \to \gamma + X$. If photinos are instead sufficiently stable to escape detection, then only the second process is observable.

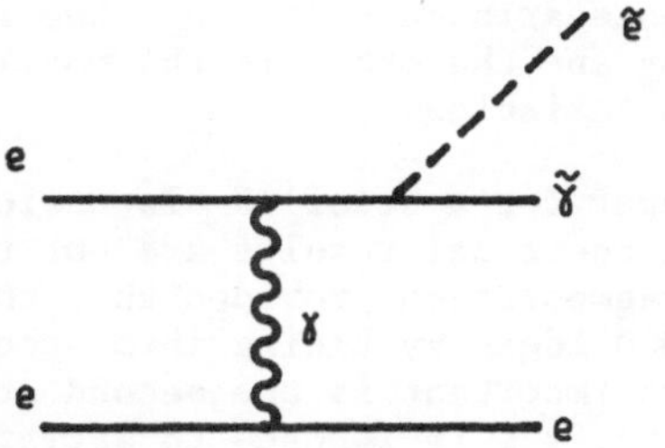

Fig. 1

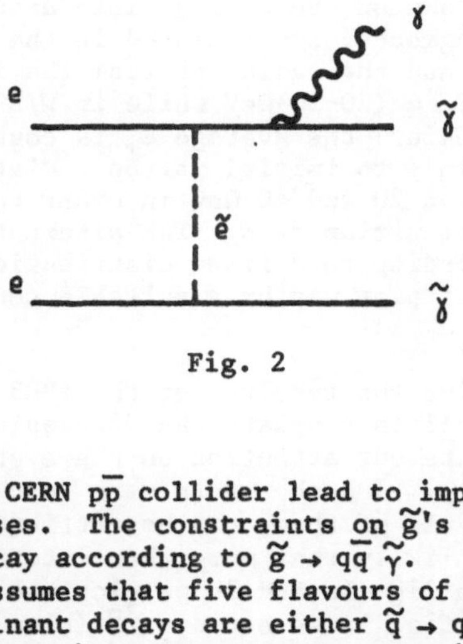

Fig. 2

The experiments at the CERN $p\bar{p}$ collider lead to important bounds on gluino and s-quark masses. The constraints on $\tilde{g}$'s are obtained under the assumption that $\tilde{g}$'s decay according to $\tilde{g} \to q\bar{q}\,\tilde{\gamma}$.
As for s-quarks, one assumes that five flavours of s-quarks are nearly degenerate and the dominant decays are either $\tilde{q} \to q\tilde{g}$ or $\tilde{q} \to \tilde{\gamma}q$ depending on whether $m_{\tilde{g}} \gtrless m_{\tilde{q}}$. The main processes that contribute to missing transverse energy events as studied especially by UA1 are :
a) $p\bar{p} \to \tilde{g}\tilde{g}\,X$, b) $p\bar{p} \to \tilde{q}\tilde{q}\,X$, c) $p\bar{p} \to \tilde{g}\tilde{q}\,X$. For s-quark masses the conclusion is that :

$$m_{\tilde{q}} > (50 - 60)\ \text{GeV}. \tag{88}$$

For gluinos, while it is clear that the existing data exclude $\tilde{g}$'s with masses in some intermediate range, it is much more delicate to decide whether or not light $\tilde{g}$'s are also to be discarded. It is thus appropriate to briefly discuss the problem of whether or not light gluinos, with decay $\tilde{g} \to q\bar{q}\tilde{\gamma}$, are excluded by the available data from the CERN $p\bar{p}$ collider. We recall that beam dump experiments impose $\tilde{g}$ masses larger then a few GeV's if gluinos mostly decay according to $\tilde{g} \to q\bar{q}\tilde{\gamma}$.

In view of the importance of this issue we have carried through in Rome [23] a complete analysis of the number and the characteristics of missing energy events from $\tilde{g}+\tilde{g}$ production (as functions of $m_{\tilde{g}}$) as they would be observed by the UA1 experiment.

There are two particularly delicate steps which are crucial for the final results of the Montecarlo calculation. One is the fragmentation of $\tilde{g}$'s before they decay and the other is the modeling of the experimental cuts in terms of parton variables.

We refer to our paper for a detailed discussion of these aspects. Here we only state that the final results are not terribly sensitive to the details of the $\tilde{g}$ fragmentation provided that this effect is described according to standard QCD ideas by taking into account our experience with heavy quarks. More important is the second step. The missing p_T ($\not{p}_T$) carried by the two $\tilde{\gamma}$'s is imposed to satisfy $\not{p}_T$ > max (15 GeV, 4σ). In reality $\sigma = 0.7\sqrt{(E_T)_{exp}}$ (σ and E_T in GeV) where $(E_T)_{exp}$ is the total transverse energy of the event. In our case we take :

$$(E_T)_{Th} = \sum_{q,\bar{q}} (E_T)_{q,\bar{q}} + \sum_{fr} (E_T)_{fr} + C \qquad (89)$$

where the contributions of the final quarks from the decays and the transverse energy of gluino fragments are added to a constant term C. This last term is needed to take on the average into account the transverse energy of the proton fragments, not included in the partonic subprocess but present in reality, and the radiation from the initial parton legs. In minimum bias events $E_T \simeq$ (20-30)GeV while in W/Z^o production (a process with large momentum transfer) the average E_T is considerably larger : $E_T \cong$ (35-40)GeV, due mainly to initial parton radiation. We thus take C in eq.(89) to vary between 20 and 40 GeV in order to study the dependence of the results on the definition of σ. The alternative procedure of letting C fluctuate according to a fixed distribution with given $\langle E_T \rangle$ can be reproduced in a simpler way by a suitable constant C in an appropriate interval.

We now first describe the results for the 1983 run, on which our paper was based. We shall then update the discussion by including the 1984 data. We concentrate our attention on the events with $\not{p}_T^2 < 10^3$ GeV^2. In the 1983 run (113nb^{-1}), 15 such events were reported, 10 monojets and 5 dijets. Most, if not all of them, can be attributed to a conventional origin, so that it is unlikely that more than a total of 10 events can be allowed for gluinos. In fig. 3a and 3b we plot the results for two extreme situations. In fig. 3a we set $d = \frac{36}{23}$ (slow evolution of $D_{\tilde{g}}(z,Q^2)$) and C = 20 GeV. In fig. 3b we have $d = \frac{36}{17}$ (fast evolution) and C = 40 GeV. We expect the real situation to lie somewhere in between. The curves shown in each figure refer to the total number of events with $\not{p}_T^2 < 1000$ GeV^2 in the experimental acceptance, the number of those which pass the trigger (the most relevant curve for setting limits) and the number of monojets, for an integrated luminosity of 100nb^{-1}. We see that the upper edge of the band of event numbers that we find is close to the results obtained in ref. 24 and would almost exclude gluinos with $m_{\tilde{g}} \gtrsim 5$ GeV. However, at the other edge of the band, gluinos with $m_{\tilde{g}} \lesssim$ (10÷15)GeV are certainly allowed. On the side of relatively large gluino masses, we

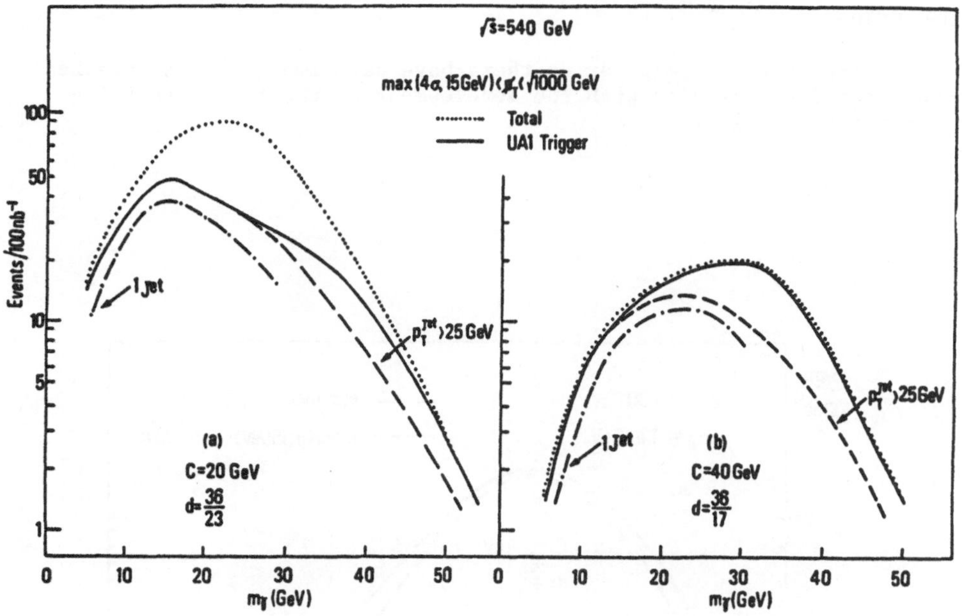

Fig. 3

see instead that the results are much more stable. Gluinos with masses $m_{\tilde{g}} > 40$ GeV are seen to be also allowed.

We now consider the results of the 1984 run which have recently become available [25]. Theoretical predictions for $\sqrt{s} \simeq 630$ GeV, as obtained by our group, are shown in fig. 4. These predictions refer to the 1984 UA1 trigger and the 4σ inclusive selection (i.e. $\not{p}_T > \max(4\sigma, 15\text{ GeV})$). Also shown are the curves that in particular refer to the more restrictive condition $\sqrt{1000}$ GeV $> \not{p}_T > \max(4\sigma, 15\text{ GeV})$. We consider it safest to compare these predictions with the total number of observed events, without distinguishing monojet and multijet events. With an integrated luminosity of 263nb^{-1} a total of about 60 events were observed, of which about 20 with $\not{p}_T < \sqrt{1000}$ GeV.

By taking the square roots of these numbers as a measure of "σ", the standard deviation, and rescaling it to 100nb^{-1}, one obtains that the solid and dashed curves in fig. 4 should be below 6 and 3 events respectively at the 2 "σ" level in the allowed region. One then obtains the result that the 1984 data drastically reduce the window for light gluinos. We now find :

$$m_{\tilde{g}} \gtrsim 55 \text{ GeV} \tag{90}$$

or

$$m_{\tilde{g}} \lesssim 7 \text{ GeV}. \tag{91}$$

Note that at the 1 "σ" level the window would essentially be closed (once the beam dump results are also taken into account).

CONCLUSION

In the last two years or so there have been impressive theoretical advances and new results with the development of the first promising superstring theories.

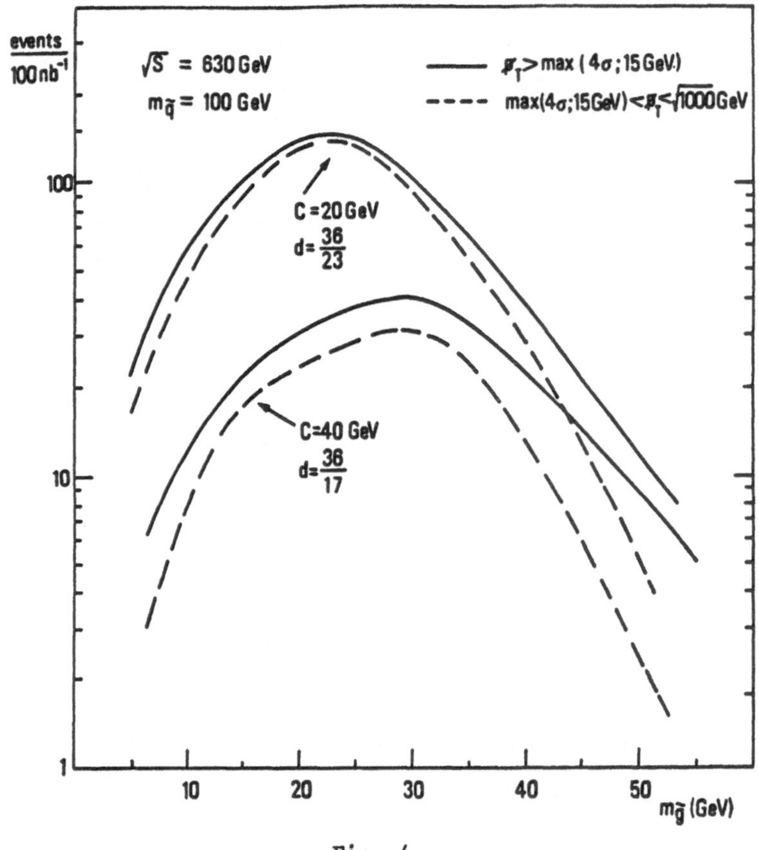

Fig. 4

The already existing indications that important new physics can be revealed by experiments probing the TeV energy region are reinforced by these new developments. I think that, independently of how great the theorists are, the future of particle physics depends on the possibility that present experiments and the ones under development will intend discover departures from the present theory, thus giving the necessary new objective impulse to the field.

REFERENCES

This list, apart from some recent works, is mainly restricted to books and reviews where more complete references can be found.

[1] M. Gell-Mann, Suppl. Nuovo Cimento 9 (1972) 733;
H. Fritzsch, M. Gell-Mann and H. Leutwyler, Phys. Lett. B47 (1973) 365;
S. Weinberg, Phys. Rev. Lett. 31 (1973) 494;
Phys. Rev. D8 (1973) 4482;
D.J. Gross and F. Wilczek, Phys. Rev. Lett. 30 (1973) 1343;
Phys. Rev. Lett. D8 (1973) 3633.

[2] S.L. Glashow, Nucl. Phys. 22 (1961) 579;
S. Weinberg, Phys. Rev. Lett. 19 (1967) 1264;
A. Salam, Proceedings of the VIII Nobel Symposium (Stockholm, 1968) p. 367.

[3] The UA1 and UA2 Collaborations at CERN.

[4] G.G. Ross, "Grand Unification", Benjamin 1985. GUTS were introduced by J.C. Pati, A. Salam, Phys. Rev. Lett. 31 (1973) 661;
Phys. Rev. D10 (1974) 275 and
H. Georgi, S.L. Glashow, Phys. Rev. Lett. 32 (1974) 438.

[5] E. Farhi, L. Susskind, Phys. Rep. 174 (1981) 277.

[6] J. Bjorken, Phys. Rev. D19 (1979) 335;
P. Hung and J. Sakurai, Nucl. Phys. B143 (1981) 81;
H. Terazawa, Prog. Theor. Phys. 64 (1980) 1963;
H. Harari and N. Seiberg, Phys. Lett. B98 (1981) 269;
O.W. Greenberg and J. Sucher, Phys. Lett. B99 (1981) 339;
L. Abbott and E. Fahri, Nucl. Phys. B189 (1981) 547;
H. Fritzsch, R. Kogerler and D. Schildknecht, Phys. Lett. B114 (1982) 157.

[7] P. Fayet, S. Ferrara, Phys. Rep. 32 (1977) 251;
H.P. Nilles, Phys. Rep. 110 (1984) 1;
H.E. Haber, G.L. Kane, Phys. Rep. 117 (1985) 71.

[8] P. Van Nieuwenhuizen, Phys. Rep. 68 (1981) 189;
J. Bagger, J. Wess, Princeton Univ. Press 1983.

[9] E. Cremmer, B. Julia, Nucl. Phys. B159 (1979) 141.

[10] J.H. Schwarz, Phys. Rep. 69 (1982) 223;
M.B. Green, Surveys H. Energy Phys. 3 (1983) 127;
L. Brink, Superstrings, CERN-TH 4006 (1984).

[11] Dual Theory, M. Jacob editor, North-Holland, 1974.

[12] M.B. Green, J.H. Schwarz, Phys. Lett. 149B (1984) 117.

[13] D.J. Gross, J.A. Harvey, E. Martinec, R. Rohm, Phys. Rev. Lett. 54 (1985) 502.

[14] E. Witten, Phys. Lett. 155B (1985) 151;
P. Candelas, G.T. Horowitz, A. Strominger, E. Witten, B258 (1985) 46;
E. Witten, Nucl. Phys. B258 (1985) 46.

[15] J.C. Pati and A. Salam, Phys. Rev. D10 (1974) 275;
R.N. Mohapatra and J.C. Pati, Phys. Rev. D11 (1975) 566;
R.N. Mohapatra and G. Senjanovich, Phys. Rev. Lett. 40 (1980) 912;
Phys. Rev. D23 (1981) 165.
For a recent review see : R.N. Mohapatra, Proceedings of the NATO Summer School, Munich, W. Germany, 1983.

[16] J.A. Bagger, S. Dimopoulos, E. Masso, M.H. Reno, Nucl. Phys. B258 (1985) 565;
G. Senjanovic, F. Wilczek, A. Zee, Phys. Lett. 141B (1984) 389.

[17] A. De Rujula, H. Georgi, S. Glashow in 5th Workshop of Grand Unification ed. by K. Kang et al., World Sci. (1984).

[18] R. Barbieri, D. Nanopoulos, Phys. Lett. 91B (1980) 369.

[19] H. Georgi, H.R. Quinn, S. Weinberg, Phys. Rev. Lett. 33 (1974) 451.

[20] D. Chang, R.N. Mohapatra, J.M. Gipson, R.E. Marshak, M.K. Parida, Phys. Rev. D31 (1985) 1718.

[21] R. Barbieri, S. Ferrara, D. Nanopoulos, Z. Phys. C13 (1982) 267;
Phys. Lett. 116B (1982) 6;
see also R. Barbieri, S. Ferrara, Surveys in H. En. Phys. 4 (1983) 33.
[22] C. Kounnas, A.B. Lahanas, D. Nanopoulos, M. Quiros, Phys. Lett. 132B (1983) 95;
Nucl. Phys. B236 (1984) 438.
[23] G. Altarelli, B. Mele, S. Petrarca, Phys. Lett. 160B (1985) 317 and Proceedings of the EPS Conf. on High En. Phys., Bari 1985.
[24] J. Ellis, H. Kowalski, CERN-TH 4126 (1985). A recent analysis can also be found in R.M. Barnett, H.E. Haber, G.L. Kane, LBL-18990 (SLAC-PUB-3551 (1985)).
[25] C. Rubbia, Proceedings of the Kyoto Symp. on Leptons and Photons, 1985.

THE STANDARD MODEL AND A LITTLE BEYOND

J.-M. Gérard

Max-Planck-Institut für Physik und Astrophysik
-Werner Heisenberg Institut für Physik-
8000 München 40, Fed. Rep. of Germany

ABSTRACT

We briefly review a few theoretical motivations to go beyond the standard model for electroweak interactions. The simplest candidates for such an extension are grand unification, supersymmetry, composite or left-right symmetric models. Could one of them solve the possible crisis arising from the new measurements of the B-meson decays confronted with CP violation in the kaon system ?

A SURVEY OF THE STANDARD MODEL

The standard model for electroweak interactions [1] is based on the $SU(2)_L \times U(1)$ gauge symmetry which is spontaneously broken via the vacuum expectation value (v.e.v.) of a single scalar doublet Φ. It contains the unique scale Λ_w appearing in the scalar potential

$$V(\Phi) = \lambda(\Phi^2 - \Lambda_w^2)^2 \quad (1.1)$$

such that all the dimensional quantities of the theory can be expressed in terms of this Fermi scale ($\Lambda_w = 2^{-1/4} G_F^{-1/2}$) :

- The $W^{\pm}$ and Z masses

$$M_w = M_z \cos\theta_w = \frac{e}{2\sin\theta_w}\Lambda_w. \quad (1.2)$$

- The strength of the weak charged current

$$\mathcal{L}^{current}_{eff.} = \frac{1}{2\Lambda_w^2} J_\mu^{V-A} J_\mu^{V-A}. \quad (1.3)$$

- The fermion mass matrix for each charge sector (Q = -1, 2/3, -1/3)

$$M^Q_{ij} = Y^Q_{ij}\Lambda_w. \quad (1.4)$$

From the paragraph above one notices that the model contains a certain number of free parameters in addition to the Λ_w scale :

- One weak angle θ_w which relates the usual coupling to the electron charge :

$$\sin \theta_w = \frac{e}{g} \tag{1.5}$$

and also defines the photon and Z physical states as linear combinations of W_3 and B :

$$\begin{pmatrix} \gamma \\ Z \end{pmatrix} = \begin{pmatrix} \cos \theta_w & -\sin \theta_w \\ \sin \theta_w & \cos \theta_w \end{pmatrix} \begin{pmatrix} B \\ W_3 \end{pmatrix} \tag{1.6}$$

- Unconstrained complex Yukawa coupling matrices which define the physical mass eigenstates after diagonalization

$$m^Q_{diag.} = U^Q_L \, Y^Q \, U^{Q^+}_R \, \Lambda_w \tag{1.7}$$

and consequently split the fermion families ($m_e \ll m_\mu \ll m_\tau, \ldots$). Moreover a maximal parity violation in the charged current J_μ is ensured by assuming that left-handed (right-handed) quarks and leptons belong to weak isospin doublets (singlets). The resulting quark current

$$J_\mu = \overline{u}^o \, \gamma_\mu (1-\gamma_5) d^o \tag{1.8}$$

has to be reexpressed in terms of the physical states (see Eq. (1.7)) :

$$J_\mu = \overline{u} \, K_L(\theta_i, \delta_j) \gamma_\mu (1-\gamma_5) d \tag{1.9}$$

with $K_L \equiv U^u_L \, U^{d^+}_L$ the Cabibbo-Kobayashi-Maskawa mixing matrix [2] depending on $\frac{n(n-1)}{2}$ angles θ_i and $\frac{(n-1)(n-2)}{2}$ phases δ_j in the case of n quark families.

To summarize the standard model involves a certain amount of arbitrariness which can be classified according to the nature of the associated breaking :

- Λ_w and θ_w are free parameters related to the spontaneous breaking of the local symmetry.

- Parity P, family symmetry and CP are explicitly broken.

Experimentally we know that :

- $\Lambda_w \sim 250$ GeV and $\sin^2 \theta_w \sim 0.22$ are consistent with the measured W and Z masses [3].

- The charged weak current is of V-A type and the Cabibbo-like mixing angles between light and heavy quarks are small. In magnitude we indeed observe the following intriguing pattern :

$$K_L = \begin{pmatrix} K_{ud} & K_{us} & K_{ub} \\ K_{cd} & K_{cs} & K_{cb} \\ K_{td} & K_{ts} & K_{tb} \end{pmatrix} \simeq \begin{pmatrix} 1 & \lambda & < \lambda^3 \\ -\lambda & 1 & \lambda^2 \\ -\lambda^3 & -\lambda^2 & 1 \end{pmatrix} \tag{1.10}$$

where the expansion parameter [4] is $\lambda \equiv \sin\theta_c \sim 0.22$.

A FEW THEORETICAL MOTIVATIONS TO GO BEYOND

Lack of predictive power means disease in theoretical physics, and therefore many attempts to cure it for the standard model have appeared during the last ten years. All these attempts try to give an answer to important questions.

a) Is θ_w calculable ?

Let us assume the existence of a renormalizable extension for the standard model such that θ_w is no longer a free parameter but a fixed angle. In this case any quantum correction to this angle has to be finite since there is no available counterterm to absorb the infinities (in QED we know that the corrections to g-2 of the electron are all finite). In particular the one-loop fermionic correction to relations (1.6) must be finite :

$$\mathrm{Tr}_{\mathrm{fermions}}\ Q(T_3 - s^2\theta_w\ Q) = 2n(1 - \frac{8}{3}s^2\theta_w) = 0 \qquad (2.1)$$

This implies that any attempt to calculate θ_w in a model where the usual quarks and leptons are the only fermions leads to the prediction

$$s^2\ \theta_w = \frac{3}{8} \qquad (2.2)$$

which is a value by far too large compared to the experimental one. We therefore have the following alternative : either to introduce extra (charged) fermions to modify Eq. (2.1) or to assume that relation (2.2) is valid only at some high scale such that it fits with data when renormalization effects are taken into account [5]. The second possibility corresponds to the well-known grand unification approach [6].

b) Is Λ_w under control ?

The classical scalar potential $V(\Phi)$ (Eq. (1.1)) gets additional terms via one-loop quantum corrections :

$$\delta V(\Phi) = \frac{1}{64\pi^2} \sum_J (-1)^{2J}(2J+1)\{[c^t + \log\frac{M_J^2}{\Lambda^2}]\mathrm{Tr}\ M_J^4 + 4\Lambda^2\ \mathrm{Tr}\ M_J^2\} \qquad (2.3)$$

where Λ denotes the regularization cut-off of the tadpole-like diagrams involving scalar (J=0), fermion (J=1/2) and gauge boson (J=1) fields of mass M_J.

If the standard model turns out to be a fundamental theory, this cut-off is unphysical since absorbed in the usual renormalization procedure, Λ_w is just a free scale, and only the finite piece of δV has physical consequences. Indeed, if we assume a light scalar ($M_\phi < M_W$) then we obtain

$$\delta V(\Phi)|_{finite} \sim [-4m_t^2 + 3(2M_W^2 + M_Z^2)]\log \Phi^2. \tag{2.4}$$

The negative sign in front of the top quark contribution (spin-statistics) implies an upper bound on the heaviest fermion [7] :

$$m_t \lesssim 100 \text{ GeV} \tag{2.5}$$

in order to avoid an effective potential unbounded from below. On the other hand, stability of the non symmetric vacuum requires a lower bound on the neutral scalar mass [8]

$$M_\phi \gtrsim 7 \text{ GeV}. \tag{2.6}$$

Finally an effective scale invariant potential would imply an estimate of the scalar mass [9]

$$M_\phi \simeq 10 \text{ GeV}. \tag{2.7}$$

However, if the standard model is a low energy effective theory, Λ corresponds to the physical scale where new physics should appear. Then the quadratic dependence on $\Lambda_{phys.}$ (Eq. (2.3)) constrains the Fermi scale to be around this physical cut-off

$$\Lambda_w \sim \Lambda_{phys.} \tag{2.8}$$

The gravitational interaction whose typical scale $\Lambda_{phys.}$ is defined to be the Planck mass is obviously a rather dangerous candidate which motivated physicists to introduce (solftly broken) supersymmetry [10]. In this framework indeed, the superpartners of the usual particles ensure the identity

$$\sum_J (-1)^{2J}(2J+1)\mathrm{Tr}\, M_J^2(\Phi) = 0 \tag{2.9}$$

killing therefore all possible quadratic divergencies of the theory.

Another way of avoiding the naturalness problem [11] associated to Eq. (2.8) is based on the introduction of a strong fermion-fermion interaction at the Fermi scale. Below this scale, the scalar bound states consist of elementary spin 1/2 fields, $\Phi \sim \overline{\Psi}\Psi$, and may get a v.e.v. of order Λ_w :

$$\Lambda_w \sim \Lambda_{strong}. \tag{2.10}$$

For illustration let us consider one family of massless quark-like fermions (U,D). If we switch-off the electroweak interactions, one can write an effective lagrangian for the $\Phi = \overline{Q}Q$ bound state below the typical scale Λ_{color} :

$$\mathcal{L}_{eff.} \ni y(\overline{P},\overline{N})\Phi\begin{pmatrix}P\\N\end{pmatrix} - \lambda(\Phi^2 \pm \Lambda_c^2)^2 \tag{2.11}$$

where P = UUD and N = UDD are color singlet fermionic bound states. This effective theory has a global $SU(2)_L \times SU(2)_R \times U(1)$ chiral symmetry, a remnant of the fundamental QCD considered.

A relative minus sign in the scalar potential (2.11) triggers the breaking of this chiral symmetry into the diagonal vector subgroup

$SU(2)_V \times U(1)_V$

$$\langle\Phi\rangle = \Lambda_c \sim 0.1 \text{ GeV} \tag{2.12}$$

and implies the existence of three Goldstone bosons. If we now switch on electroweak interactions the gauge symmetry $SU(2)_L \times U(1)$ is spontaneously broken and we obtain the isospin mass relations [12] :

$$M_w = M_z \cos\theta_w = \frac{e}{2\sin\theta_w}\Lambda_c \sim 30 \text{ MeV} \tag{2.13}$$

$$m_P = m_N = y\,\Lambda_c. \tag{2.14}$$

Relation (2.13) requires a rescaling of Λ_c to the value of Λ_w (see Eq. (1.2)). Two interesting scenarii have been proposed. The first one needs a higher color representation for U and D, e.g. sextets in order to preserve asymptotic freedom [13]. A more radical modification is based on the introduction of a new confining interaction called Technicolor [12,14] with typical scale $\Lambda_{tc} \sim \Lambda_w$. However, both approaches suffer from the same disease : Eq. (2.14) obviously implies that the familiar quarks and leptons cannot be identified with P and N-like fermions and are therefore elementary and massless. In the first scenario additional colored fermions [15] give the quarks and leptons a mass but then spoil the needed asymptotic freedom property of the theory. In the second one, the Extended Technicolor interactions [16] introduced to solve this fermion mass generation problem ($\Lambda_{etc} \sim 10\Lambda_{tc}$) violate the experimental constraints on flavor changing neutral currents [17].

Let us therefore assume a positive sign in the effective scalar potential (2.11) such that $\langle\Phi\rangle = 0$, namely $SU(2)_L \times SU(2)_R \times U(1)$ remains unbroken. In this toy model the massless P and N bound states resemble the quarks and leptons which are indeed light with respect to the scale Λ_w. Moreover, the allowed unbroken chiral symmetry at the effective level is strongly constrained by a consistency condition [11] : the associated Adler-Bell-Jackiw anomalies induced by the massless fermionic bound states have to match with the ones present at the fundamental level. This condition provides a genuine information about the dynamics of composite models and cannot therefore be disregarded. In the case of $SU(3)_C$ with two massless flavors (U,D) the unique constraint due to the $[SU(2)_L]^2\,U(1)_V$ triangle anomaly is trivially satisfied if P and N are indeed massless.

We have seen that an $SU(2)_L \times SU(2)_R$ QCD-like theory allows a Nambu-Goldstone as well as a Wigner realization (- or + relative sign respectively in the effective scalar potential), the latter one being promising for composite quark and lepton models. However, this alternative cannot be extended to a larger number of colors and (or) flavors [18], for which only the former realization is then allowed. (For example, the lightness of the strange quark implies a Nambu-Goldstone mode for the usual QCD [11]). There is another, not unrelated, difficulty for composite quark and lepton builders : in vector-like gauge theories (where bare masses are possible for all fermions) neither vectorial global symmetries nor *parity* can be spontaneously broken in the absence of Yukawa couplings [19].

These difficulties can in principle be eluded in the framework of supersymmetric strong interaction models where chiral symmetries are protected [20]. For illustration let us consider a supersymmetric QCD-like theory with N colors and M flavors. For $M \geqslant N$ unbroken chiral symmetries are consistent with extended [21] anomaly matching (see Table 1).

Table 1.

Anomaly matching solutions for supersymmetric QCD with N colors and M massless flavors. (□ denotes the fundamental representation of the group).

	Unbroken flavor symmetries	massless spin 1/2 bound state content
$M < N$	$SU(M)_{L+R} \times U(1)_V$	-
$M = N$	$SU(N)_L \times SU(N)_R \times U(1)_A$	$(\square, \bar{\square}, N)$ $(1, 1, N)$
$M = N+1$	$SU(N+1)_L \times SU(N+1)_R \times U(1)_A \times U(1)_V$	$(\bar{\square}, 1, 1, N)$ $(1, \square, 1, -N)$ $(\square, \bar{\square}, N-1, 0)$
$M > N+1$	$SU(N)_L \times SU(M-N)_L \times SU(M)_R \times U(1)_A \times U(1)_V$	$(\bar{\square}, \square, 1, 0, -M)$ $(\square, 1, \bar{\square}, N, M-N)$ $(1, 1, 1, M, 0)$

Moreover, for $M > N+1$ we find left-right asymmetric solutions [22] which provide an example of dynamical parity violation. These rather interesting properties arise from the fact that the fundamental theory contains massless scalars A_i $(i = 1\ldots M)$, the superpartners of the fermions Ψ_i. On the one hand these scalars actively participate to the formation of massless fermionic bound state $(\Psi A, \ldots)$ and on the other hand they couple to the gauginos $\tilde{G}$ via the important chiral invariant Yukawa-like gauge coupling

$$g\Psi_i \tilde{G} A_i^*. \tag{2.15}$$

Although supersymmetric composite models develop other very interesting features [23] such as the appearance of quasi Goldstone fermions [24], no convincing mechanism concerning the family replication puzzle and the mass generation for quarks and leptons [25] came out, and this was after all the main motivation of this approach.

c) Are CP and P invariance restored at high energy ?

In the standard model CP and P are explicitly broken via complex Yukawa couplings and V-A gauge currents respectively. A spontaneous violation of CP invariance requires additional scalar doublets [26] such that a $U(1)_Y$ gauge transformation cannot simultaneously rotate away all the phases associated to the complex v.e.v's. If we impose natural flavor conservation in neutral currents (NFC) in order to avoid too heavy ($\sim$ 5-10 TeV) neutral scalars, we need at least three scalar doublets such that the CP violating phases are carried by physical charged scalars [27].

However, if we want parity P to be on equal foot with CP, then we have to extend the electroweak gauge symmetry to $SU(2)_L \times SU(2)_R \times U(1)$ [28]. The right-handed fermions transform like doublets under the new $SU(2)_R$ group such that the matter reads now :

$$\begin{pmatrix}\nu\\ e\end{pmatrix}_{L,R} \quad \begin{pmatrix}u\\ d\end{pmatrix}_{L,R} . \tag{2.16}$$

The effective lagrangian (1.3) is therefore modified :

$$\mathcal{L}^{current}_{eff.} = \frac{1}{2\Lambda_w^2}\{J_\mu^{V-A} J_\mu^{V-A} + \left(\frac{M_{W_L}}{M_{W_R}}\right)^2 J_\mu^{V+A} J_\mu^{V+A}\} \tag{2.17}$$

to take into account the exchange of the new charged gauge boson W_R. From the present non observation of this (V+A) impurety one concludes $M_{W_R} \gtrsim 250$ GeV. However, a more restrictive lower bound on the W_R mass is provided by the experimental value of the $K_L - K_S$ mass difference [29] :

$$M_{W_R} \gtrsim 1.7 \text{ TeV}. \tag{2.18}$$

Obviously, in the heavy W_R limit we recover the standard model features (see for example (2.17)). In this limit, the unique (1/2, 1/2) scalar multiplet coupled to matter fields splits into two $SU(2)_L \times U(1)$ doublets, one being superheavy with a mass of the order of M_{W_R}. This means that the minimal left-right symmetric model cannot implement a spontaneous CP violation [30] : maximal P violation in the charged gauge sector implies minimal CP violation. Just like in the standard model we need therefore additional scalars. However, in spite of this, we will see that left-right symmetric models contain a genuine mechanism for spontaneous CP violation.

POSSIBLE INCONSISTENCY IN THE STANDARD MODEL

This chapter is a rough sketch of the present CP violation status for the kaon system in the standard electroweak model, more detailed reviews, including the beon system and the electric dipole moment of the neutron, being now available [31]. As we know the Cabibbo-Kobayashi-Maskawa mixing matrix contains four independent parameters in the case of three quark and lepton families [2]. A phenomenologically useful parametrization based on an expansion with respect to the Cabibbo angle (see Eq. (1.10)) has been recently proposed [4] :

$$K_L \sim \begin{pmatrix} 1 & \lambda & A\lambda^3(\rho-i\eta) \\ -\lambda & 1-i\eta A^2\lambda^4 & A\lambda^2 \\ A\lambda^3(1-\rho-i\eta) & -A\lambda^2 & 1 \end{pmatrix}. \tag{3.1}$$

(Unitarity needs extra terms which are however unimportant for the description of the kaon system).

The quantities measured in the B-meson decays are easily expressed in terms of A, ρ and η. The new world average value [32] of the B lifetime

$$\tau_B = (1.13 \pm 0.16)10^{-12}\text{s} \tag{3.2}$$

together with a spectator model estimate [33] (phase space and QCD corrections included)

$$\tau_B \simeq \frac{1}{3.8}\left(\frac{m_\mu}{m_b}\right)^5 A^{-2} \lambda^{-4} \tau_\mu$$
$$\simeq 1.3\, A^{-2}\, 10^{-12} s \qquad (3.3)$$

fix the absolute value of A to be around 1 and justify therefore "a posteriori" the λ-expansion in Eq. (3.1). On the other hand, the important ratio $\bar{R} \equiv \Gamma(b \to u e \nu)/\Gamma(b \to c e \nu)$ reads

$$\bar{R} \simeq \frac{1}{0.6} \frac{A^2\lambda^6(\rho^2+\eta^2)}{A^2\lambda^4}. \qquad (3.4)$$

The experimental constraint [34] $\bar{R} \lesssim 0.03$ implies the following restriction on the ρ-η space

$$\rho^2 + \eta^2 \lesssim (0.6)^2 \qquad (3.5)$$

to be confronted with the constraint on η due to an observed CP violation in the kaon system [35].

The CP violating measurable quantities in the K-decays are

$$\eta^{+-} \equiv \frac{A(K_L \to \pi^+\pi^-)}{A(K_S \to \pi^+\pi^-)} \sim \varepsilon + \varepsilon'$$
$$\eta^{oo} \equiv \frac{A(K_L \to \pi^o\pi^o)}{A(K_S \to \pi^o\pi^o)} \sim \varepsilon - 2\varepsilon' \qquad (3.6)$$

$K_{L,S}$ being the physical states

$$K_L = \left(\frac{K^o + \bar{K}^o}{\sqrt{2}}\right) + \bar{\varepsilon}\left(\frac{K^o - \bar{K}^o}{\sqrt{2}}\right)$$
$$K_S = \left(\frac{K^o - \bar{K}^o}{\sqrt{2}}\right) + \bar{\varepsilon}\left(\frac{K^o + \bar{K}^o}{\sqrt{2}}\right) \qquad (3.7)$$

namely the eigenvectors of the $K^o - \bar{K}^o$ Hamiltonian. The small $\bar{\varepsilon}$ parameter is the CP violating impurety due to the complex off-diagonal matrix element $M_{12} \equiv \langle K^o|H_{eff}|\bar{K}^o\rangle$:

$$\bar{\varepsilon} \sim \frac{e^{i\pi/4}}{2\sqrt{2}} \frac{\mathrm{Im}\, M_{12}}{\mathrm{Re}\, M_{12}}. \qquad (3.8)$$

This parameter is not measurable in itself, being quark phase convention dependent. A phase redefinition of the strange quark

$$s \to e^{i\xi} s \qquad (3.9)$$

implies indeed (for ξ infinitesimal)

$$\mathrm{Im}\, M_{12} \to \mathrm{Im}\, M_{12} + 2\xi\, \mathrm{Re}\, M_{12}. \qquad (3.10)$$

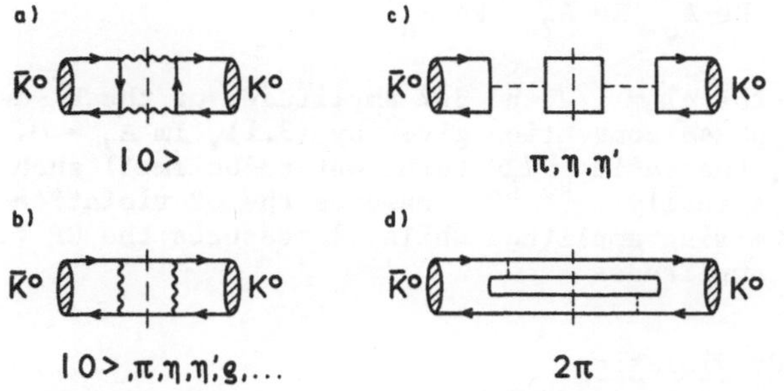

Fig. 1 : Short distance (see Eq. (3.12)) W-exchange (wavy line) contributions (1a-b), and long distance (see Eq. (3.16)) $H^{\Delta S=1}_{eff.}$ (dashed line) contributions (1c-d) to the $\Delta S = 2$ K^o-$\bar{K}^o$ amplitude.

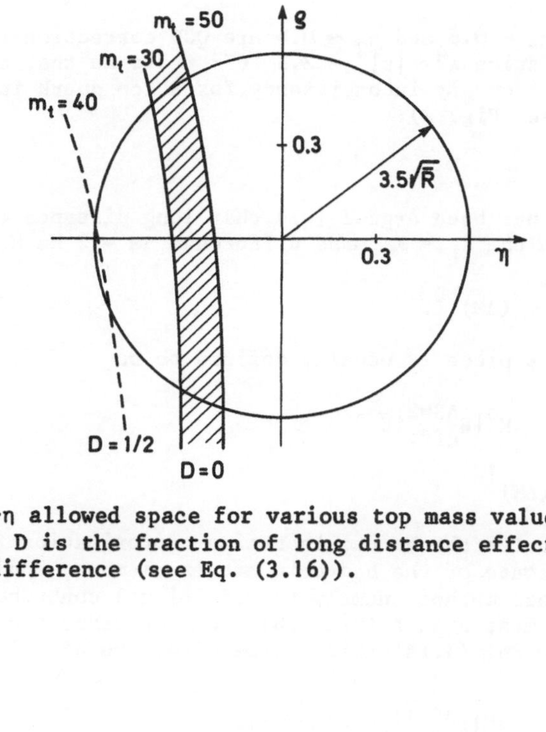

Fig. 2 : The ρ-η allowed space for various top mass values (in units of GeV). D is the fraction of long distance effect in the $K_L - K_S$ mass difference (see Eq. (3.16)).

The two phase convention independent parameters appearing in (3.6) are

$$\varepsilon \sim \frac{e^{i\pi/4}}{2\sqrt{2}} \left[\frac{\text{Im } M_{12}}{\text{Re } M_{12}} + 2\frac{\text{Im } A_o}{\text{Re } A_o}\right]$$
$$\varepsilon' \sim \frac{e^{i\pi/4}}{\sqrt{2}} \frac{\text{Re } A_2}{\text{Re } A_o} \left[\frac{\text{Im } A_2}{\text{Re } A_2} - \frac{\text{Im } A_o}{\text{Re } A_o}\right] \tag{3.11}$$

where $A_{o,2}$ are the $\Delta I = 1/2$ and $3/2$ amplitudes of the K^o-decay into two pions. In the phase convention given by (3.1), Im $A_2 = 0$. Moreover, as we will see, the ratio $\frac{\text{Im } A_o}{\text{Re } A_o}$ turns out to be small such that in the standard model ε really measures the CP violation in the $\Delta S = 2$ $K^o - \overline{K}^o$ mixing amplitude while ε' measures the CP violation in the $\Delta S = 1$ K-decay amplitudes.

a) The $\Delta S = 2$ CP violation

In the standard model the short distance contribution (SD) to M_{12} is mainly due to the box-diagrams (Fig. 1a-b). Assuming $m_t < M_W$ one finds :

$$e^{-i\pi/4}\, \varepsilon^{SD} \sim \frac{1}{2\sqrt{2}} \frac{\text{Im}[M_{12}(c,c) + M_{12}(t,t) + 2\, M_{12}(c,t)]}{\text{Re } M_{12}(c,c)}$$
$$\sim \frac{1}{\sqrt{2}} A^2\lambda^4\eta\left[1 - \frac{\eta_2}{\eta_1}\left(\frac{m_t}{m_c}\right)^2 A^2\lambda^4(1-\rho) - \frac{\eta_3}{\eta_1} \ln\left(\frac{m_t}{m_c}\right)^2\right] \tag{3.12}$$

where $\eta_1 \sim 0.8$, $\eta_2 \sim 0.6$ and $\eta_3 \sim 0.4$ are QCD correction factors [36]. The numerical relation $\lambda^4 \sim |\varepsilon|^3 = 2.3\ 10^{-3}$ tells us that the standard model seems safe from any inconsistency for a top quark in the range (30-50) GeV if (see Fig. 2)

$$\eta \simeq -0.3. \tag{3.13}$$

However, it has been argued [37] that long distance effects (LD) are non negligible in the $K_L - K_S$ mass difference $\Delta M = 2$ Re M_{12} :

$$\Delta M = (\Delta M)^{SD} + (\Delta M)^{LD}. \tag{3.14}$$

The short distance piece is usually defined to be

$$(\Delta M)^{SD} = 2 \langle K^o|H^{\Delta S=2}_{eff.}|\overline{K}^o\rangle$$
$$\equiv B(\Delta M)^{|o\rangle} \tag{3.15}$$

where $(\Delta M)^{|o\rangle}$ denotes the vacuum insertion approximation (V.I.A.) [38] used for the estimate of the hadronic matrix element, and B measures the departure from this method, namely the additional contributions due to intermediate low mass states (Fig. 1b). On the other hand, the long distance piece in Eq. (3.14) arises from effective $\Delta S = 1$ operators and reads

$$(\Delta M)^{LD} = 2 \sum_n \frac{\langle K^o|H^{\Delta S=1}_{eff.}|n\rangle\langle n|H^{\Delta S=1}_{eff.}|\overline{K}^o\rangle}{m_k^2 - m_n^2}$$

$$\equiv D\ \Delta M \tag{3.16}$$

where n represents single intermediate low mass states ($\pi,\eta,\eta',\ldots$) (Fig. 1c) as well as double ones ($2\pi,\ldots$) (Fig. 1d). From Eq. (3.16) we conclude

$$\varepsilon = (1-D)\varepsilon^{SD} \tag{3.17}$$

if we assume no sizeable LD contribution to Im M_{12} [39,40]. We therefore have to rescale the relation (3.13) :

$$\eta \simeq -\frac{0.3}{1-D}. \tag{3.18}$$

From Eqs. (3.5) (3.18) we conclude that a D parameter between 1/2 and 3/2 could be a signal of physics beyond the standard model (see Fig. 2 for m_t = 40 GeV and D = 1/2).

A rather confident estimate of $(\Delta M)^{|0>}$ via the box diagrams implies (see Eqs. (3.14), (3.15), (3.16)) the following relation between D and B :

$$1-D = \frac{(\Delta M)^{|0>}}{\Delta M}\, B \simeq (0.7 - 0.9)B \tag{3.19}$$

the main uncertainties coming from the value of the charm mass and the QCD correction (η_1) appearing in the theoretical calculation. Unfortunately, the theoretical estimates of the B and D parameters are rather taste-dependent. For illustration we give an incomplete list (Table 2) of proposed values derived within various frameworks.

Despite the modesty of the authors we would like to comment briefly of one of them [40], based on the 1/N expansion [51] (N being the number of colors) which provides a complete (diagrammatic) *classification* of the short and long distance effects. In the large N limit, the leading contributions to $(\Delta M)^{SD}$ and $(\Delta M)^{LD}$ are given by Fig. 1a and Fig. 1c respectively (simple quark loop counting). The corresponding hadronic matrix elements being factorized, they are easily estimated. Moreover the large N approximation removes also the ambiguity ($\eta-\eta'$ mixing, glue content) inherent in the estimate of the η' contribution to M_{12}^{LD}, since this pseudoscalar is now just another Goldstone boson on equal foot with the π and η [52]. Last but not least $(\Delta M)^{LD}$ can be expressed in terms of measured quantities

$$(\Delta M)^{LD} = \frac{9}{64}\left(1-\frac{m_\pi^2}{m_k^2}\right)\left(\frac{m_\pi}{m_k}\right)^3 \frac{\Gamma(K_L\to\gamma\gamma)}{\Gamma(\pi\to\gamma\gamma)}\, m_k \tag{3.20}$$

$$= (1.02 \pm 0.08)\ 10^{-15}\ \text{GeV}$$

namely D $\sim$ 0.3. In a world where N = 3 is a large number, happy people do not have to worry about ε and ΔM (B = 3/4) as long as $\bar{R}$ > 0.01 (see Fig. 2). But what about us ?

b) The ΔS = 1 CP violation

The estimate of the ΔS = 1 K-decay amplitudes is deeply connected to our present understanding of the observed ΔI = 1/2 rule :

$$\frac{\text{Re}\ A_2}{\text{Re}\ A_o} \sim \frac{1}{22}. \tag{3.21}$$

Although QCD octet dominance [53], penguins [54] and 1/N expansion [40]

Table 2.

Theoretical estimates of the B and D parameters.

Assumptions	B	D
V.I.A. [38]	1	0
PCAC + SU(3) [41]	$\sim \pm 1/3$	-
Harmonic oscillator quark model [42]	1.4 - 2.8	-
M.I.T. bag model [43]	(-0.4) - 2.5	-
QCD sum rules [44]	$\lesssim 2.0 \pm 0.5$	-
+ 1/N expansion [45]	$\lesssim 1$	-
+ chiral symmetries [46]	$\sim \pm 1/3$	-
Lattice [47]	O(1)	-
Dispersion relations $(\pi,\eta,\eta',\rho,\omega,2\pi)$ [48]	-	$\begin{cases} 1.54 \pm 1.4 \\ 0.56 \pm 1.28 \end{cases}$
(2π) [49]	-	0.46 ± 0.13
$(\pi,\eta,\eta',\rho,\omega,2\pi)$ [50]	0.9 - 1.2	0.1 ± 0.4
Large N limit [40]	3/4	~ 0.3

enhance the $\Delta I = 1/2$ channel with respect to the $\Delta I = 3/2$ one, serious doubts still stand about its final explanation. We therefore use the empiric ratio (3.21) to estimate ε' (3.11) :

$$\varepsilon' \simeq - \frac{e^{i\pi/4}}{\sqrt{2}} \frac{1}{22} \frac{\text{Im } A_o}{\text{Re } A_o} \tag{3.22}$$

where Im A_o arises from the controversial penguin diagram

$$\frac{\text{Im } A_o}{\text{Re } A_o} \simeq f \frac{y_6}{z_6} A^2 \lambda^4 \eta. \tag{3.23}$$

In Eq. (3.23), f is the fraction of the $\Delta I = 1/2$ K-decay amplitude due to the Q_6 penguin operator with μ-dependent Wilson coefficient $R_6 \sim z_6 + i A^2\lambda^4\eta y_6$. Using [55] $1 \lesssim \frac{y_6}{z_6} \lesssim 2$, we obtain for $f = 1/2$:

$$\frac{\varepsilon'}{\varepsilon} \sim \frac{(7.5 \pm 2.5)}{1-D} 10^{-3}. \tag{3.24}$$

Experimentally, measurements [56] of $\left|\frac{\eta^{oo}}{\eta^{+-}}\right|^2 \sim 1 + 6 \text{ Re } \varepsilon'/\varepsilon$ yield rather small central values :

$$\text{Re } \varepsilon'/\varepsilon = (-4.6 \pm 5.3 \pm 2.4)\ 10^{-3} \quad \text{(Chicago-Saclay)}$$

$$= (+1.7 \pm 8.4)\ 10^{-3} \quad \text{(Yale-BNL)}. \tag{3.25}$$

Note that a smaller fraction f brings ε'/ε closer to these experimental bounds (3.25). However, it simultaneously worsens the $\Delta I = 1/2$ rule puzzle. A possible clash between $\Delta I = 1/2$ rule and ε'/ε is emerging. An observed negative value of ε'/ε would be of course the cleanest signal of a new CP violating source though the (most unlikely) possibility of having $D > 3/2$ and $B < 0$ cannot be excluded by Eqs. (3.5), (3.18), (3.19).

NEW SOURCES OF CP VIOLATION BEYOND $SU(2)_L \times U(1)$

Motivated by the number of free parameters in the standard model, in Chapter II we briefly reviewed a few simple extensions. What do they say about CP violation ?

a) Grand Unification Theories

Even in the minimal SU(5) model [6] new physical phases [57] indeed do appear in the matter couplings to leptoquark gauge bosons. However they are unrelated to CP violation in the kaon system and do not cure a possible CP crisis.

b) Composite models for quarks and leptons

In this framework, the full Cabibbo-Kobayashi-Maskawa mixing matrix should emerge uniquely. Unfortunately, a realistic model for quark and lepton families has still to be found.

c) Supersymmetric $SU(3) \times SU(2) \times U(1)$

We all know that the strong coupling of quarks to gluons (G_μ) is flavor-blind :

$$g_s\, \bar{q}^o\, \gamma_\mu\, q^o\, G_\mu = g_s\, \bar{q}\, \gamma_\mu\, q\, G_\mu. \tag{4.1}$$

However, supersymmetry requires another gauge term which describes the strong interaction between superpartners of the quarks and gluons, the squarks ($\tilde{q}$) and gluinos ($\tilde{G}$) respectively :

$$g_s\, \tilde{q}^{o*}\, q^o\, \tilde{G} = g_s\, \tilde{q}^*\, \tilde{U}^q\, U^{q^+}\, q\, \tilde{G}. \tag{4.2}$$

Supersymmetry being obviously broken at low energy, the quark and squark mass matrices are diagonalized by somewhat different unitary matrices (U^q and $\tilde{U}^q$ respectively). Strong interactions induce therefore $\Delta S = 1$ and $\Delta S = 2$ flavor and CP violating processes [58] via superparticle exchanges. Supergravity [10] provides us with an attractive scenario for supersymmetry breaking by soft terms. In this particular framework, quantum corrections induced by charged particles rotate the quark and squark mass matrices in different directions of the flavor space. In the left-handed sector of the theory, $\tilde{U}^q\, U^{q^+}$ is then a function of the Kobayashi-Maskawa mixing matrix and we can estimate the supersymmetric contribution to ε [59] and ε'/ε [60]. For rather light squarks and gluinos (~ 40 GeV) the value of ε allows a lighter top quark [61] and ε'/ε can be very small and even negative [60]. Supersymmetry could solve a CP crisis.

d) Left-right symmetric models

If we assume spontaneous P and CP violation, namely real Yukawa couplings [62], we obtain the following relation between the left-handed and the right-handed Kobayashi-Maskawa mixing matrices [63,64] :

$$K_R = J^u K_L^* J^{d^+} \tag{4.3}$$

where $J^{u,d}$ are unitary diagonal matrices (which can be rotated away via right-handed quark phase redefinitions in $SU(2)_L \times U(1)$). In this left-right symmetric model the number of CP violating phases is therefore $1/2\ (n-1)(n-2)$ for K_L and $1/2\ n(n+1)$ for K_R. For three generations of quarks and leptons ($n = 3$) we recover the usual K_L mixing matrix which is orthogonal :

$$K_L = O_L \tag{4.4}$$

if and only if [65] the quark mass matrices satisfy the condition

$$\det.[m^u m^{u^+}, m^d m^{d^+}] = 0. \tag{4.5}$$

As pointed out in Ref. 64, a very small mixing angle, namely a very small $\overline{R}$ (see Eq. (3.4)) could in fact approach this "ideal" situation (4.4) where CP violation is only due to external phases ($J^{u,d}$). If in addition we neglect the (real) $W_L - W_R$ mixing

$$\theta_{L-R} \ll \frac{M^2_{W_L}}{M^2_{W_R}} \tag{4.6}$$

the phases of A_o and A_2 are trivially tracked and found to be identical. From Eq. (3.11) we conclude that in the "ideal" limits (4.4) (4.6) the model is superweak :

$$\varepsilon' = 0. \tag{4.7}$$

On the other hand an important enhancement [29] ($r = 215 \neq 1$) in the $\Delta S = 2\ K^o - \overline{K}^o$ amplitude due to the $W_L - W_R$ exchange box-diagram with charm quarks prevents ε from vanishing [64] :

$$|\varepsilon| \lesssim \frac{1}{\sqrt{2}} \frac{M^2_{W_L}}{M^2_{W_R}} (r-1). \tag{4.8}$$

Eqs. (2.17) and (4.8) provide us with an interesting link between the strength of P and CP breakings. Moreover from (4.8) we obtain an upper bound on the W_R mass such that the right-handed charged gauge boson should lie in the range [64,66] (see Eq. (2.18))

$$1.7 \text{ TeV} \lesssim M_{W_R} \lesssim 21 \text{ TeV} \tag{4.9}$$

to trigger a genuine mechanism for CP violation in the (light) kaon system without the assistance of a (heavy) third quark family. The real world being not ideal, careful studies [67] without the assumptions (4.4) and (4.6) have been undertaken. They lead to the conclusion that left-right symmetric model remains indeed a serious challenger.

NEW SOURCES OF CP VIOLATION INSIDE $SU(2)_L \times U(1)$

Two obvious ways to provide the standard model with new CP violation sources are the addition of more scalar doublets and the introduction of a fourth quark family [68]. In both cases as we will see, the experimental informations on ε and ε'/ε lead to rather strong constraints on the model.

a) A minimal model with spontaneous CP violation and NFC

As already mentioned in Chapter II this minimal model [27] contains three scalar doublets Φ_a whose (real) Yukawa couplings to quarks read

$$\mathcal{L}_Y = (\bar{u}\ \bar{d})_L \frac{m^d}{v_1} d_R\, \Phi_1 + (\bar{u}\ \bar{d})_L \frac{m^u}{v_2^*} u_R\, \tilde{\Phi}_2 \tag{5.1}$$

with

$$\langle\Phi_a\rangle = v_a\, e^{i\varphi_a} \quad (a = 1,2,3). \tag{5.2}$$

A suitable phase redefinition of the right-handed quark fields renders K_L orthogonal such that CP violation is only due to charged scalar exchanges [69].

If the three v.e.v.'s are assumed (almost) degenerate

$$|v_1| \sim |v_2| \sim |v_3| \tag{5.3}$$

the ε parameter is dominated by the CP violation in the $\Delta S = 1$ K-decays :

$$\frac{\mathrm{Im}\, A_o}{\mathrm{Re}\, A_o} \gg \frac{\mathrm{Im}\, M_{12}}{\mathrm{Re}\, M_{12}} \tag{5.4}$$

and from Eq. (3.11) one concludes [70]

$$\frac{\varepsilon'}{\varepsilon} \simeq -\frac{\mathrm{Re}\, A_2}{\mathrm{Re}\, A_o} \simeq -\ 45\ 10^{-3}. \tag{5.5}$$

Controversial arguments based on long distance contribution due to η' [71] could only reduce ε'/ε to the value [72]

$$\frac{\varepsilon'}{\varepsilon} \sim -\ 15\ 10^{-3}. \tag{5.6}$$

However we could assume some hierarchy among the v.e.v.'s. In fact this is even suggested by Eq. (5.1) and the empiric observation of a large mass splitting in the almost decoupled (see Eq. (1.10)) third family of quarks (t,b). If this splitting is due to a spontaneous isospin breaking effect (same order Yukawa couplings) we obtain

$$\frac{m_b}{m_t} \simeq \left|\frac{v_1}{v_2}\right|. \tag{5.7}$$

A fit of ε and ε'/ε agrees with (5.7) and leads to the following hierarchy [73] :

$$|v_1| \ll |v_2| < |v_3| \tag{5.8}$$

namely the leptons also couple to Φ_1 and the $SU(2)_L \times U(1)$ breaking is mainly due to the scalar doublet Φ_3 uncoupled to the matter fields. The predicted light charged scalar ($M_{H^+} \lesssim 35$ Gev) could be detected in toponium decays if $m_t \gtrsim 40$ Gev.

b) A fourth quark family

The introduction of a new fermion family yields three Kobayashi-Maskawa phases and spoils [74] the analysis presented in chapter III. However a puzzling connection between mixing angles and quark masses appears in the $K^0 - \overline{K}^0$ box diagram. Indeed in the limit of a very heavy up quark mass ($m_{t'} \gg M_W$), the associated amplitude grows like [75]

$$1/4\ K^2_{t'd}\ K^2_{t's}\ m^2_{t'}. \tag{5.9}$$

Therefore contrary to the penguin diagram, no decoupling appears in the $\Delta S = 2$ amplitude *if* the mixing angles are unrelated to quark masses. The experimental value of the $K_L - K_S$ mass difference ΔM is almost saturated by the charm such that

$$\left|\frac{K_{t'd}\ K_{t's}}{K_{cd}\ K_{cs}}\right| \lesssim 2\,\frac{m_c}{m_{t'}}. \tag{5.10}$$

On the other hand a sizeable t' effect in ε needs

$$\left|\frac{K_{t'd}\ K_{t's}}{K_{td}\ K_{ts}}\right| \gtrsim 2\,\frac{m_t}{m_{t'}}. \tag{5.11}$$

If we generalize the λ-expansion (Eq. (1.10)) proposed in Ref. 4, we obtain at least the following fourth family decoupling pattern in K_L

$$|K_L| \sim \begin{pmatrix} 1 & \lambda & \lambda^3 & \lambda^4 \\ \lambda & 1 & \lambda^2 & \lambda^3 \\ \lambda^3 & \lambda^2 & 1 & \lambda \\ \lambda^4 & \lambda^3 & \lambda & 1 \end{pmatrix}. \tag{5.12}$$

This implies the following allowed range for $m_{t'}$:

$$2 \text{ Tev} \lesssim m_{t'} \lesssim 24 \text{ Tev} \tag{5.13}$$

clearly incompatible with perturbation. This naive argument has of course to be taken with some caution. It just reminds us that a fourth fermion family does not necessarily solve the possible CP violation problems.

CONCLUSION

The discovery of CP violation in the kaon system more than twenty years ago [35] still remains a famous challenge for any physicist. Indeed, in the minimal version of the electroweak standard model, the unique CP violating phase generated by the third quark family provides a first open window on the rather unconstrained Yukawa sector. Recent measurements in the B-meson decays could already suggest some inconsistency in which case we should look for new particles to carry CP violation. The known value of ε and the bounds on ε'/ε already impose rather severe theoretical upper limits on the mass of these new particles (e.g. $M_{\tilde{q}} \approx M_{\tilde{G}} \lesssim 40$ Gev, $M_{W_R} \lesssim 20$ Tev, $M_{H^+} \lesssim 35$ Gev). However we should keep in mind that the surprise could arise from the beon system [76] or the electric dipole moment of the neutron [77,78] which are very sensitive to new physics.

REFERENCES

[1] S.L. Glashow, Nucl. Phys. 22 (1961) 579;
S. Weinberg, Phys. Rev. Lett. 19 (1967) 1264;
A. Salam, in Proc. 8th Nobel Symp., ed. N. Svartholm, p. 367 (1968).

[2] M. Kobayashi and T. Maskawa, Prog. Theor. Phys. 49 (1973) 652.

[3] G. Arnison et al., Phys. Lett. 122B (1983) 103 and Phys. Lett. 126B (1983) 398;
M. Banner et al., Phys. Lett. 122B (1983) 476;
P. Bagnaia et al., Phys. Lett. 129B (1983) 130.

[4] L. Wolfenstein, Phys. Rev. Lett. 51 (1983) 1945.

[5] H. Georgi, H.R. Quinn and S. Weinberg, Phys. Rev. Lett. 33 (1974) 451.

[6] J.C. Pati and A. Salam, Phys. Rev. Lett. 31 (1973) 661;
H. Georgi and S.L. Glashow, Phys. Rev. Lett. 32 (1974) 438.

[7] P.Q. Hung, Phys. Rev. Lett. 42 (1979) 873;
H.D. Politzer and S. Wolfram, Phys. Lett. 82B (1979) 242.

[8] S. Weinberg, Phys. Rev. Lett. 36 (1976) 294;
A.D. Linde, JETP Lett. 23 (1976) 64.

[9] S. Coleman and E. Weinberg, Phys. Rev. D7 (1973) 1888.

[10] For a review, see for instance H.P. Nilles, Phys. Rep. 110 (1984) 1.

[11] G. 't Hooft, "Recent Developments in Gauge Theories", Proceedings of Advanced Study Institute, Cargèse 1979, Eds. G. 't Hooft et al., (Plenum Press, N.Y., 1980).

[12] S. Weinberg, Phys. Rev. D19 (1979) 1277;
L. Susskind, Phys. Rev. D20 (1979) 2619.

[13] W.J. Marciano, Phys. Rev. D21 (1980) 2425.

[14] For review see P. Sikivie : Lectures given at the International School of Physics "Enrico Fermi" Varenna 1980, CERN preprint TH-2951.

[15] P.Q. Hung, Fermilab preprint 80/78-THY (1980) unpublished;
G. Zoupanos, Phys. Lett. 129B (1983) 315;
D. Lüst, E. Papantonopoulos and G. Zoupanos, Z. Phys. C25 (1984) 81.

[16] S. Dimopoulos and L. Susskind, Nucl. Phys. B155 (1979) 237;
E. Eichten and K. Lane, Phys. Lett. 90B (1980) 125.

[17] S. Dimopoulos and J. Ellis, Nucl. Phys. B182 (1981) 505.

[18] Y. Frishman, A. Schwimmer, T. Banks and S. Yankielowicz, Nucl. Phys. B177 (1981) 157.

[19] C. Vafa and E. Witten, Nucl. Phys. B234 (1984) 173.

[20] W. Buchmüller and S.T. Love, Nucl. Phys. B204 (1982) 429.

[21] S. Takeshita, Prog. Theor. Phys. 68 (1982) 912;
J.-M. Gérard, J. Govaerts, Y. Meurice and J. Weyers, Phys. Lett. 116B (1982) 29; Nucl. Phys. B234 (1984) 138;
T.R. Taylor, Phys. Lett. 125B (1983) 185.

[22] J.-M. Gérard and H.P. Nilles, Phys. Lett. 129B (1983) 243.

[23] For recent reviews, see :
W. Buchmüller, CERN preprint TH 4004/84 (1984);
R.D. Peccei, Max-Planck-Institute preprint MPI-PAE/PTh 35/84 (1984).

[24] W. Buchmüller, S.T. Love, R.D. Peccei and T. Yanagida, Phys. Lett. 115B (1982) 233;
W. Buchmüller, R.D. Peccei and T. Yanagida, Phys. Lett. 124B (1983) 67; Nucl. Phys. B227 (1983) 503.

[25] For a systematic attempt see A. Masiero, R. Pettorino, M. Roncadelli and G. Veneziano, CERN preprint TH 4166/85 (1985).

[26] T.D. Lee, Phys. Rev. D8 (1973) 1226.

[27] S. Weinberg, Phys. Rev. Lett. 37 (1976) 657.

[28] J.C. Pati and A. Salam, Phys. Rev. D10 (1974) 275;
R.N. Mohapatra and J.C. Pati, Phys. Rev. D11 (1975) 566;
G. Senjanovič and R.N. Mohapatra, Phys. Rev. D12 (1975) 1502.

[29] G. Beall, M. Bander and A. Soni, Phys. Rev. Lett. 48 (1982) 848.

[30] For a technical proof see A. Masiero, R.N. Mohapatra and R.D. Peccei, Nucl. Phys. B192 (1981) 66.

[31] See e.g. : A.J. Buras, Proceedings of the Workshop on the Future of

Medium Energy Physics in Europe, in Freiburg (H. Koch and F. Scheck editors) (1985); MPI-PAE PTh 46/84;
P. Langacker, Invited Talk presented at the Aspen Winter Conference Series, University of Pennsylvania preprint UPR-0276T (1985).
[32] A.J. Buras, MPI preprint PAE/PTh 65/85; to appear in the Proceedings of the Bari Conference (1985).
[33] See for example R. Rückl, "Weak Decays of Heavy Flavors", CERN preprint (1983); to appear in Physics Reports.
[34] A. Chen et al., (CLEO Coll.), Phys. Rev. Lett. 52 (1984) 1084;
C. Klopfenstein et al. (CUSP Coll.), Phys. Lett. 130B (1983) 444.
[35] J.H. Christenson, J.W. Cronin, V.L. Fitch and R. Turlay, Phys. Rev. Lett. 13 (1964) 138.
[36] F.J. Gilman and M.B. Wise, Phys. Rev. D27 (1983) 1128.
[37] L. Wolfenstein, Nucl. Phys. B160 (1979) 501;
C.T. Hill, Phys. Lett. 97B (1980) 275;
G. Ecker, Phys. Lett. 147B (1984) 369.
[38] M.K. Gaillard and B.W. Lee, Phys. Rev. D10 (1974) 897.
[39] J.-M. Frère, J. Hagelin and A.I. Sanda, Phys. Lett. 151B (1985) 161.
[40] A.J. Buras and J.-M. Gérard, MPI preprint MAE/PTh 40/85.
[41] J.F. Donoghue, E. Golowich and B.R. Holstein, Phys. Lett. 119B (1982) 412.
[42] P. Colić, B. Guberina, D. Tadić and J. Trampetić, Nucl. Phys. B221, (1983) 141.
[43] R.E. Shrock and S.B. Treiman, Phys. Rev. D19 (1979) 2148;
P. Colić et al., Ref. 42.
[44] B. Guberina, B. Machet and E. de Rafael, Phys. Lett. 128B (1983) 269.
[45] B. Machet, Z. Phys. C26 (1984) 449.
[46] A. Pich and E. de Rafael, Phys. Lett. 158B (1985) 477.
[47] N. Cabibbo, G. Martinelli and R. Petronzio, Nucl. Phys. B244 (1984) 381;
R.C. Brower, G. Maturana, M.B. Gavela and R. Gupta, Phys. Rev. Lett. 53 (1984) 1318.
[48] I.I. Bigi and A.I. Sanda, Phys. Lett. 148B (1984) 205.
[49] M.R. Pennington. Phys. Lett. 153B (1985) 439.
[50] P. Cea, G. Nardulli and G. Preparata, Phys. Lett. 148B (1984) 477;
P. Cea and G. Nardulli, Phys. Lett. 152B (1985) 251.
[51] G. 't Hooft, Nucl. Phys. B72 (1974) 461; B75 (1974) 461;
G. Rossi and G. Veneziano, Nucl. Phys. B123 (1977) 507;
E. Witten, Nucl. Phys. B160 (1979) 57.
[52] E. Witten, Nucl. Phys. B156 (1979) 269;
G. Veneziano, Nucl. Phys. B159 (1979) 213.
[53] M.K. Gaillard and B.W. Lee, Phys. Rev. Lett. 33 (1974) 108;
G. Altarelli and L. Maiani, Phys. Lett. 52B (1974) 351.
[54] A.I. Vainshtein, V.I. Zakharov and M.A. Shifman, Sov. Phys. JETP 45 (1977) 670.
[55] F.J. Gilman and M.B. Wise, Phys. Rev. D20 (1979) 2392.
[56] R.-H. Bernstein et al., Phys. Rev. Lett. 54 (1985) 1631;
J.K. Black et al., Phys. Rev. Lett. 54 (1985) 1628.
[57] J. Ellis, M.K. Gaillard and D.V. Nanopoulos, Phys. Lett. 88B (1979) 320.
[58] M.J. Duncan, Nucl. Phys. B221 (1983) 285;
J.F. Donoghue, H.P. Nilles and D. Wyler, Phys. Lett. 128B (1983) 55.
[59] J.-M. Gérard, W. Grimus, Amitava Raychaudhuri and G. Zoupanos, Phys. Lett. 140B (1984) 349.
[60] J.-M. Gérard, W. Grimus and Amitava Raychaudhuri, Phys. Lett. 145B (1984) 400.
[61] J.-M. Gérard, W. Grimus, A. Masiero, D.V. Nanopoulos and Amitava Raychaudhuri, Phys. Lett. 141B (1984) 79.
[62] For an alternative, see G. Ecker, W. Grimus and H. Neufeld, Nucl. Phys. B247 (1984) 70.
[63] R.N. Mohapatra, F.E. Paige and D.P. Sidhu, Phys. Rev. D17 (1978) 2462.
[64] G.C. Branco, J.-M. Frère and J.-M. Gérard, Nucl. Phys. B221 (1983) 317.

[65] C. Jarlskog, CERN preprint TH-4242/85 (1985).
[66] H. Harari and M. Leurer, Nucl. Phys. B233 (1984) 221.
[67] D. Chang, Nucl. Phys. B214 (1983) 435;
G. Ecker and W. Grimus, Nucl. Phys. B258 (1985) 328 and Ref. 18 therein.
[68] See also F. del Aguila and J. Cortès, Phys. Lett. 156B (1985) 243.
[69] G.C. Branco, Phys. Rev. Lett. 44 (1980) 504.
[70] A.I. Sanda, Phys. Rev. D23 (1981) 2647;
N.G. Deshpande, Phys. Rev. D23 (1981) 2654;
J.F. Donoghue, J.S. Hagelin and B.R. Holstein, Phys. Rev. D25 (1982) 195;
D. Chang, Phys. Rev. D25 (1982) 1318.
[71] T.N. Pham, Phys. Lett. 145B (1984) 113;
J.F. Donoghue and B.R. Holstein, UMHEP-213 (1984).
[72] J.-M. Frère et al., in Ref. 39;
A.I. Sanda, talk presented at the Fifth Moriond Workshop on Heavy Quarks, Flavour Mixing and CP violation (Jan. 1985).
[73] G.C. Branco, A.J. Buras and J.-M. Gérard, Phys. Lett. 155B (1985) 192; Nucl. Phys. B259 (1985) 306.
[74] M. Gronau and J. Schechter, SLAC preprint PUB-3451 (1984);
U. Turke, E.A. Paschos, H. Usler and R. Decker, Dortmund preprint Do-TH 84/26 (1984);
T. Hayashi, M. Tanimoto and S. Wakaizumi, KTCP-8501 (1985);
I.I. Bigi, Z. Phys. C27 (1985) 303.
[75] T. Inami and C.S. Lim, Progr. Theor. Phys. 65 (1981) 297;
A.J. Buras, Phys. Rev. Lett. 46 (1981) 1354.
[76] J.-M. Gérard, W. Grimus, A. Masiero, D.V. Nanopoulos and Amitava Raychaudhuri, Nucl. Phys. B253 (1985) 93;
M.B. Gavela, A. Le Yaouanc, L. Oliver, O. Pène, J.C. Raynal, M. Jarfi and O. Lazrak, Phys. Lett. 154B (1985) 147;
G. Ecker and W. Grimus, UWThPh-1985-14.
[77] Beall and Deshpande, Phys. Lett. 132B (1983) 427.
G. Ecker, W. Grimus and H. Neufeld, Nucl. Phys. B229 (1983) 421.
[78] W. Buchmüller and D. Wyler, Phys. Lett. 121B (1983) 321;
J. Polchinski and M.B. Wise, Phys. Lett. 125B (1983) 393;
D.V. Nanopoulos and M. Srednicki, Phys. Lett. 128B (1983) 61;
J.-M. Frère and M.B. Gavela, Phys. Lett. 132B (1983) 107;
A.I. Sanda, preprint RU 85/B/121.

ALGEBRA OF ANOMALIES

Michel Talon

Laboratoire de Physique Théorique et Hautes Energies
Université Pierre et Marie Curie
F-75230 Paris, France

ABSTRACT

The algebraic set up for anomalies, à la Stora, is reviewed. Then a brief account is provided of the work of M. Dubois Violette, M. Talon, C. Viallet [1], [2], [3], in which the general algebraic solution to the consistency conditions is described.

INTRODUCTION

The B.R.S. Algebra

Recent progress in the theory of quantum anomalies has shown that chiral anomalies are closely connected to the index theorem of Atiyah-Singer [4]. As a matter of fact, looking at the Dirac operator in external gauge and gravitational fields, L. Alvarez-Gaumé and E. Witten [5,6] have shown that 1-loop chiral anomalies can be extracted from the usual expressions of index theory.

Another approach to the anomaly problem goes back to the work of J. Wess and B. Zumino [7], and of C. Becchi, A. Rouet, R. Stora [8]. They have shown that, perturbatively, a purely algebraic consistency condition appears which severely restricts the possible expressions for anomalies. The coefficients of these expressions can be obtained by a Feynman diagram calculation, or equivalently, application of the index theorem, and depend on the particle content of the theory. Moreover, Novikov and E. Witten have shown that there is a priori quantification of such a coefficient, [9]. This can be derived by a simple global argument about the effective Lagrangian of J. Wess and B. Zumino, in which all perturbative consequences of anomalies are collected [7-10] and supports the non-renormalization theorem for anomalies [11].

Very useful notations for developing this algebra come from differential geometry [12,13]. Let us introduce the gauge potential A, a 1-form (on the principal bundle) with values in the Lie algebra g of the structure group, and its curvature $F = dA + A^2$, a 2-form, g-valued. Then F has covariant derivative $DF = dF + AF - FA = 0$ (Bianchi identity). Conversely, look at the algebra generated by abstract objects A (of odd type) and F (of even type) with values in g, and define an antiderivation d by :

$$dA = F - A^2 \qquad dF = FA - AF$$

One checks easily that $d^2 = 0$. The algebra so obtained is called the Weil algebra of g [14].

In gauge field theory, it is usual to introduce the Faddeev-Popov "ghost field" χ, a g-valued 1-form (on the group of gauge transformations, see [10, 15, 3] and its "derivative" $d\chi$. Then, one defines the B.R.S. symmetry δ [8], which generates infinitesimal gauge variations, and augments the χ-degree by 1, by :

$$\delta A = -(d\chi + \chi A + A\chi) \qquad \delta F = -\chi F + F\chi \qquad \delta\chi = -\chi^2$$

It is easy to check that : $d^2 = \delta^2 = 0 \quad d\delta + \delta d = 0$. This algebra has been extensively used by B.R.S. [8], and J. Dixon, R. Stora [16], and we shall call it the B.R.S. algebra [2]. See also [10].

The Local Cohomology

An anomaly is the fact that a classical symmetry of a theory cannot be maintained at the quantum level. As an example, suppose the classical action L(A) is gauge invariant : $\delta L(A) = 0$. Its quantum analog, $\Gamma(A)$, the generating function of 1.P.I. graphs, is a *non local* object, which may turn out to be non gauge invariant :

$$\delta\Gamma(A) = a(\chi,A)$$

where a is the anomaly, and since $\delta^2 = 0$,

$$\delta a(\chi,A) = 0$$

This is the Wess-Zumino consistency condition [7].

The anomaly a comes (consider for example the Adler triangle anomaly [11]) from incomplete cancellations between divergent part of formally gauge invariant Feynman diagrams (at least at the 1-loop level), leaving as anomaly a *finite, local, polynomial* of the fields, integrated over space-time (The divergent part is a local polynomial).

Moreover, $\Gamma(A)$ after renormalization at the 1-loop level, is defined only up to a finite, integrated, local polynomial b of the fields. If $\Gamma \to \Gamma + b$, obviously $a \to a+\delta b$. Then we have to solve the consistency condition $\delta a = 0$, modulo trivial solutions of the form $a = \delta b$, in the class of integrated local polynomials of the fields. This is called the *cohomology* problem [15].

Note that the same problem occurs, even more obviously, in the framework of perturbative renormalization of gauge theories [17], at all orders, since counter-terms are, by definition, local polynomials. But one has to cope with the technical problem of showing that only such polynomials $a(\chi, A)$ of first degree in χ, not involving other auxiliary fields, give anomalous counter-terms [17].

A nice feature of this cohomological formulation is provided by the recent discovery of L. Faddeev [18], showing that Schwinger terms [11], obey a similar equation $\delta a(\chi,A) = 0$, where a is of second degree in χ, and can be cancelled by redefinition of fields if $a = \delta b$. Then R. Jackiw has shown [19], that a's of third χ-degree can be given physical significance, and one conjectures that all solutions of the local cohomology problem have some physical interpretation.

Reduction to Non-Integrated Form

Supposing the space-time manifold M is compact (and some section of the principal bundle, assumed trivial, has been chosen, so that A,... are defined on M) write :

$$a = \int_M Q^{1}_{2n-2} \quad \begin{matrix}\leftarrow \text{ghost number} \\ \leftarrow \text{M-dimension}\end{matrix}$$

We shall admit that $\delta a = 0$ is equivalent to the existence of a Q_{2n-3} such that :

$$\delta Q^{1}_{2n-2} + dQ^{2}_{2n-3} = 0$$

One says that the polynomial in the fields Q^{1}_{2n-2} is a δ-*cocycle modulo d.* Similarly if $a = \delta b$ and $b = \int_M L^{0}_{2n-2}$ there exists a L^{1}_{2n-3} such that :

$$Q^{1}_{2n-2} = \delta L^{0}_{2n-2} + dL^{1}_{2n-3}$$

and one says that Q^{1}_{2n-2} is a δ-*coboundary modulo d.* So we have to compute the space of δ-cocycles modulo d, quotiented by δ-coboundaries modulo d. We shall call it $H^{1}_{2n-2}(\delta/d)$: the δ-*cohomology modulo d* with ghost degree 1 and space degree 2n-2. Schwinger terms are given by $H^{2}_{2n-3}(\delta/d)$ and more generally we shall compute $H^{k}_{2n-1-k}(\delta/d)$ for all n and all k, with $0 \leqslant k \leqslant 2n-1$.

The Algebraic Hypothesis

All known non-trivial Q's, and all Q's derived from the index theorem [5,20] can be expressed *algebraically* in terms of the forms A, F, χ, $d\chi$, i.e. they live in the B.R.S. algebra. We shall work only *in this class* and this means no higher derivatives of A than the first, automatic antisymmetrizations of algebraic objects. The general problem is still unsolved, in spite of some attempts [21]. This algebraic class seems the natural class for a general space-time (results do not depend on the topology of M).

THE STORA TRICK

The "Secondary" Polynomials

Consider the *invariant polynomials* (under $F \to g^{-1}Fg$, where g lies in the structure group) $\mathrm{Tr}\, F^n$, which are ordinary *scalar* valued forms of degree 2n. They obey : $\delta \mathrm{Tr}\, F^n = n\, \mathrm{Tr}\, \delta F.F^{n-1} = 0$ by the cyclic property of the trace. The same property shows that $d\, \mathrm{Tr}\, F^n = n\, \mathrm{Tr}\, DF.F^{n-1} = 0$ by Bianchi identity. One can show (algebraic Poincaré lemma. See Appendix I) that this implies existence of a polynomial in A and F (Chern-Weil secondary polynomial $Q^{0}_{2n-1}(A,F)$ such that :

$$\mathrm{Tr}\, F^n = d\, Q^{0}_{2n-1}(A,F)$$

Finally, R. Stora has noted that :

$$F = (d + \delta)(A + \chi) + (A + \chi)^2$$

$$(d + \delta)F + (A + \chi)F - F(A + \chi) = 0$$

(easy to check), so the same identities that hold between polynomials in A and F in the Weyl algebra, remain true in the B.R.S. algebra, replacing

A by A+χ, F by F, and d by d+δ. He has called this trick the "Russian formula". Then :

$$\mathrm{Tr}\, F^n = (d + \delta)\, Q^0_{2n-1}(A + \chi, F)$$

The Dixon-Stora Chain of Anomalous Terms

Let us develop $Q^0_{2n-1}(A + \chi, F)$ in powers of χ :

$$Q^0_{2n-1}(A+\chi,F) = Q^0_{2n-1}(A,F) + Q^1_{2n-2}(\chi,A,F) + Q^2_{2n-3}(\chi,A,F)+\ldots + Q^{2n-1}_0(\chi)$$

where Q^k is of degree k in χ. Since δ augments the χ-degree by 1, the above equation gives a whole chain of equations [16, 22, 10]. So the Q^k are δ-cocycles modulo d, but they can be 0 or δ-coboundaries modulo d (see later).

$$\mathrm{Tr}\, F^n = d\, Q^0_{2n-1}$$

$$0 = \delta\, Q^0_{2n-1} + d\, Q^1_{2n-2}$$

$$0 = \delta\, Q^1_{2n-2} + d\, Q^2_{2n-3}$$

$$0 = \delta\, Q^2_{2n-3} + d\, Q^3_{2n-4}$$

$$0 = \delta\, Q^{2n-2}_1 + d\, Q^{2n-1}_0$$

$$0 = \delta\, Q^{2n-1}_0$$

Example : n = 2

By using the technique of Appendix I, it is easy to obtain Q^0_{2n-1} from $\mathrm{Tr}\, F^n$. Then one gets :

$$Q^0_3 = \mathrm{Tr}(AF - 1/3\, A^3) \quad Q^1_2 = \mathrm{Tr}\, \chi(F-A^2) \quad Q^2_1 = -\mathrm{Tr}\, \chi^2 A \quad Q^3_0 = -1/3\, \mathrm{Tr}\chi^3$$

Example : n = 3

Similarly :

$$Q^0_5 = \mathrm{Tr}\, (AF^2 - 1/2\, A^3F + 1/10\, A^5)$$

$$Q^1_4 = \mathrm{Tr}\chi\, (F^2 - 1/2(A^2F + FA^2 + AFA) + 1/2\, A^4)$$

This can be rewritten (First noticed by Stora) :

$Q^1_4 = \mathrm{Tr}\chi\, d(AdA + 1/2\, A^3)$ and this is Bardeen's result [23]

$$Q^2_3 = \mathrm{Tr}\, [-1/2(\chi^2A + \chi A\chi + A\chi^2)F + 1/2(\chi^2A^3 + \chi A\chi\, A^2)]$$

$$Q^3_2 = \mathrm{Tr}\, [-1/2\chi^3F + 1/2(\chi^3A^2 + A\chi A\chi^2)]$$

$$Q^4_1 = \mathrm{Tr}\, 1/2\chi^4\, A \qquad Q^5_0 = \mathrm{Tr}\, 1/10\, \chi^5$$

Of course it is easy to check (but laborious !) that :

$$\delta Q^K + d\, Q^{K+1} = 0$$

The δ Cohomology

The last term of the chain is always a δ-*cocycle*, that is : $\delta Q_0^{2n-1} = 0$. Directly, $\mathrm{Tr}\,\chi^{2n} = 0$, by the cyclic property of the trace, and so $\delta \mathrm{Tr}\,\chi^{2n-1} = 0$. Also, we have seen that $\delta\,\mathrm{Tr}\,F^n = 0$, so, generally :

$$\delta\{ \sum_{n_i m_j} C\{n_i, m_j\} \prod_i \mathrm{Tr}\,F^{n_i} \prod_i \mathrm{Tr}\,\chi^{2m_j-1}\} = 0$$

where the scalar forms $\mathrm{Tr}\,F^{n_i}$ are even, and the scalar forms $\mathrm{Tr}\,\chi^{2m_j-1}$ are odd. The first belong to the algebra $\mathcal{D}_s(g)$ of *invariant polynomials* on g, and the last to the algebra $\mathcal{D}_\wedge(g)$ of *invariant forms* on g (g being, as before, the Lie algebra of the structure group). Finally, $\sum_{n_i m_j}\{\quad\}$ belongs to the tensor product $\mathcal{D}_s(g) \quad \mathcal{D}_\wedge(g)$. As a matter of fact, it can be shown [1], that the δ-cohomology H(δ) (that is δ-cocycles modulo δ-coboundaries) is identical to this tensor product (see a sketch of proof in Appendix I)

$$H(\delta) = \mathcal{D}_s(g) \otimes \mathcal{D}_\wedge(g)$$

THE STRUCTURE OF $\mathcal{D}_s(g)$ AND $\mathcal{D}_\wedge(g)$

The following results have been found by A. Weil, C. Chevalley, H. Cartan, J.L. Koszul [14, 24, 25], in 1949, under the hypothesis that g is a *reductive Lie algebra*, i.e., is a direct product of simple algebras and abelian algebras.

Primitive Forms

There exists a finite number of invariant forms on g, all of odd degree, such that every invariant form is a sum of products of primitive forms, i.e., primitive forms generate $\mathcal{D}_\wedge(g)$. The number of independent primitive forms is equal to the *rank* of g. Examples : For SU(2) → $\mathrm{Tr}\,\chi^3$. For SU(3) → $\mathrm{Tr}\,\chi^3$ and $\mathrm{Tr}\,\chi^5$. For orthogonal groups, the powers jump by 4 because F is an antisymmetric matrix, and for SO(2N), the Pfaffian gives in addition a new primitive form. See the table of primitive forms in Appendix II. For a U(1) factor, just put χ which is a scalar. Then, let P_o be the vector space of primitive forms. They naturally anticommute, so

$$\mathcal{D}_\wedge(g) = \Lambda\, P_o$$

(exterior algebra on P_o). Note this is also the de Rahm cohomology of the structure group [24,31].

Transgressions

For each *primitive* form Q^{2n-1} there exists an invariant polynomial P_n on g (non unique) such that we have a chain of equations, as in II, b) :

$$P_n(F) = d\, Q^0_{2n-1}(A,F)$$

$$0 = \delta\, Q^0_{2n-1} + d\, Q^1_{2n-2}$$

$$0 = \delta\, Q_0^{2n-1}$$

Primitive forms are *characterized* by the existence of such a non-interrupted chain. We *choose* a P_n for each independent Q_0^{2n-1} and denote

$P_n = \tau\, Q_0^{2n-1}$. Then τ extends by linearity to a map

$$\tau : P_o \to \mathcal{D}_s(g)$$

called *transgression*. Examples : For SU(3) choose :

$$\tau(-1/3 \text{ Tr } \chi^3) = \text{Tr } F^2 \qquad \tau(1/10 \text{ Tr } \chi^5) = \text{Tr } F^3$$

Invariant Polynomials

All invariant polynomials on g can be obtained by doing sum of products of elements $\tau\, Q^{2n-1}$. Moreover, they naturally commute and are algebraically independent [25]. So, these sums of products generate the *symmetric* algebra on $\tau\, P_o$ (also called the polynomial algebra on $\tau\, P_o$

$$\mathcal{D}_s(g) = S\, \tau\, (P_o)$$

(Note we have used this connection between forms and polynomials before).

In other words, for SU(3), $\mathcal{D}_s(g)$ is given by all P(Tr F^2, Tr F^3) where P is an ordinary polynomial in two indeterminates. Finally :

$$E_o = H(\delta) = \Lambda\, P_o \otimes S\, \tau\, P_o$$

Note that E_o only contains objects of *even* space-time degree (χ being of degree 0 and F of degree 2 in space-time).

RELATION BETWEEN H(δ/d) and H(δ)

The ∂ Operator

Let Q be a δ-cocycle modulo d. Then : $\delta Q + dQ' = 0$. Taking the δ, $d\delta Q' = 0$, and by triviality of the d (See Appendix I), there exists a Q" such that $\delta Q' + dQ'' = 0$, so, Q' is also a δ-cocycle modulo d. This argument goes back to Stora-Dixon [16]. Moreover, the application $Q \to Q'$ maps δ-coboundaries modulo d to δ-coboundaries modulo d (easy to check) and so, define a map :

$$\partial : H(\delta/d) \to H(\delta/d)$$

by ∂ (class of Q) = class of Q'. Note that δ corresponds to going down by 1 ladder in the chain of equations, in *cohomology*.

The p operator

If Q is a δ-cocycle ($\delta Q = 0$), it is also a δ-cocycle modulo d (take $Q' = 0$), but perhaps a δ-coboundary modulo d. One can define :

$$P : H(\delta) \to H(\delta/d)$$

by p (class of Q in H(δ)) = class of Q in H(δ/d).

The i operator

Let Q be a δ-cocycle modulo d ($\delta Q + dQ' = 0$). Taking the d, one gets :

$$\delta(dQ) = 0$$

so dQ is a δ-cocycle. One can define :

$$i : H(\delta/d) \to H(\delta)$$

by i (class of Q in $H(\delta/d)$) = class of (dQ) in $H(\delta)$.

The Exact Couple

Collecting the above facts, we get :

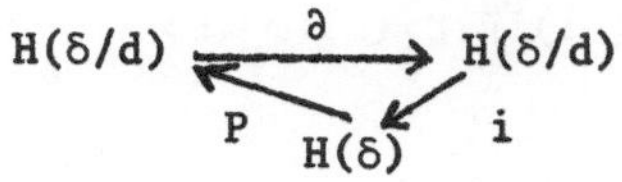

and it is easy to show [2] that this triangle is exact at all 3 corners, i.e., Ker ∂ = Im p, Ker i = Im ∂, Ker p = Im i. Since $H(\delta)$ vanishes in odd space-time degrees, ∂ induces isomorphisms :

$$H_{odd}(\delta/d) \xrightarrow{\sim} H_{even}(\delta/d)$$

So, taking $F_o = H_{even}(\delta/d)$, $E_o = H(\delta)$, $p_o = p$, $i_o = i_o \circ \partial^{-1}$ well defined on $H_{even}(\delta/d)$, one gets the exact triangle (with no more degeneracy) :

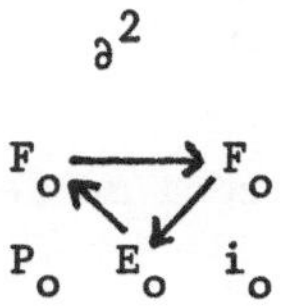

This situation is known in homological algebra as an *exact couple* (Massey [26]), and from it, one can derive a spectral sequence which allows to compute $H(\delta/d)$, knowing $H(\delta)$.

THE GENERALIZED TRANSGRESSION LEMMA

We now prove the key lemma which allows the computation of the appropriate spectral sequence.

The Spaces E_r

We have seen that $E_o = \Lambda P_o \otimes S\tau P_o$, where P_o is the vector space of all primitive forms. Let us introduce the subspace P_r generated by primitive forms *of degree* $\geq 2r+1$. Then define :

$$E_r = \Lambda P_r \otimes S \tau P_r$$

(For technical reasons [2], we shall assume, in what follows, that *constants are excluded* from E_o, E_r, and the $H(\delta/d)$). For example, for SU(3) : $P_o = P_1 = \{\mathrm{tr}\,\chi^3, \mathrm{tr}\,\chi^5\}$, $P_2 = \{\mathrm{tr}\,\chi^5\}$, $P_3 = 0$ so $E_3 = E_4 = \ldots = 0$. An abelian factor has just 1 primitive form χ, belonging to P_o, but not to P_1.

The Lemma

Let us consider a generic element of E_r :

$$X = \prod_i \tau\, \zeta_i(F) \quad \omega_o(\chi) \ldots \omega_n(\chi)$$

where ω_1 and ζ_j are primitive forms of degree $\geq 2r+1$. By transgression, associate to each $\omega_k(\chi)$ some $L_k(A,F)$ such that $\tau\omega_k(F) = dL_k(A,F)$ with $\omega_k(\chi) = L_k(\chi,0)$. By Stora trick, one has :

$$(d+\delta)\, L_k(A + \chi,\ F) = \tau\omega_k(F)$$

Then let us compute :

$$(d+\delta)\{\prod_i \tau\zeta_i(F)L_o(A+\chi,F)\ldots L_n(A+\chi,F)\} = \Sigma_p(-1)^p \prod_i \tau\zeta_i(F)\tau\omega_p(F)\, L_o(A+\chi,F)\ldots\hat{L}_p\ldots L_n(A+\chi,F)$$

and develop in powers of χ, beginning by the highest. One obtains a chain of equations :

$$\delta X = 0$$

$$dX + \delta Q_1 = 0$$

$$dQ_{2r} + \delta Q_{2r+1} = 0$$

$$dQ_{2r+1} + \delta Q_{2r+2} = d_r X$$

where $d_r X$ is the first non vanishing contribution of the second member of the above equation. So

$$d_r X = \Sigma_{\substack{\{p \text{ such that }\}\\ \{\deg\omega_p = 2r+1\}}} (-1)^p \prod_i \tau\zeta_i(F)\tau\omega_p(F)\omega_o(\chi)\ldots\widehat{\omega_p}(\chi)\ldots\omega_n(\chi)$$

(The hat means omission). But then, obviously, $d_r X$ belongs to E_r.

<u>Remarks</u>

Here we see why $\deg\omega_p \geq 2r+1$ is needed. If not, the chain stops before.

The idea of taking products of L_k's originates in Dixon, Stora [16].

Note all the Q_k's are δ-cocycles modulo d for $1 \leq k \leq 2r+1$. We shall show that under appropriate choices of X, they give the *most general solution to the anomaly problem* in the algebraic class. This has been first conjectured in [21]. Of course, there is still no answer in general.

A similar argument shows that, if some ζ_i is of degree $< 2r+1$, these cocycles modulo d are trivial (Apply the above lemma to

$$X = \Pi_{j\neq i}\ \tau\zeta_j(F)\ \zeta_i(\chi)\omega_o(\chi)\ldots\omega_n(\chi)$$

Then

$$d_s X = \Pi_{j\neq i}\ \tau\zeta_j(\chi)\ \tau\zeta_i(\chi)\omega_o(\chi)\ldots\omega_n(\chi)$$

with some $s < r$, is also $dQ_{2S+1} + \delta Q_{2S+2}$, that is a δ-coboundary modulo d. Finally

$$p_o\ (\Pi_j\ \tau\zeta_j(F)\ \omega_o(\chi)\ldots\omega_n(\chi)) = 0\ \ .)$$

This is why we impose deg $\zeta_i \geq 2r+1$.

The Q_K are easily computed if we write :

$$L_k(A+\chi,F) = Q^0 + Q^1 + Q^2 + \dots \leftarrow \chi\text{degree}$$

then develop the product $L_o(A+\chi,F) \dots L_n(A+\chi,F)$ and keep the terms of given χ-degree.

Obviously, if we start from a product $\omega_o(\chi) \dots \omega_n(\chi)$, we shall never be able to climb the chain up to some expression containing F only, i.e., only *primitive* forms can be transgressed.

The formula for $d_r X$ is reproduced by defining the rules for d_r : d_r is an antiderivation defined on E_r by :

- $d_r \ \tau\zeta(F) = 0$
- $d_r \ \omega = \tau\omega$ if $\deg\omega = 2r+1$
- $d_r \ \omega = 0$ if $\deg\omega > 2r+1$

Then it is obvious that $d_r^{\,2} = 0$

Denoting by α the class of Q_{2r} in $H(\delta/d)$, it is clear that

$$d_r X = i \circ \partial^{-1}(\alpha) \qquad \text{in } H(\delta)$$

$$\partial^{2r}\alpha = p(X) \qquad \text{in } H(\delta/d)$$

(because ∂^{2r} is the descent of the chain by 2r ladders, and $\partial^{-1}\alpha$ = class of Q_{2r+1} in $H(\delta/d)$, while i (class of Q_{2r+1}) = class of dQ_{2r+1} in $H(\delta)$ = class of $d_r X$ in $H(\delta)$.).

The Theorem

We can now state the appropriate restrictions on X which give the general solution to the anomaly problem. Let us denote by $E_r^{\,r}$ the subspace of E_r of the elements *containing explicitely at least one form of degree* $2r+1$ *or its transgression.*

So, if P^{2r+1} is the vector space of primitive forms of degree 2r+1, one gets :

$$E_r^{\,r} = \{\oplus_{m+n\geq 1}(S^m \,\tau P^{2r+1}) \otimes (\Lambda^n P^{2r+1})\} \otimes \{\text{anything of higher degree}\}$$

Then, obviously $E_r = E_r^{\,r} \oplus E_{r+1}$ and thus :

$$E_r = E_r^{\,r} \oplus E_{r+1}^{\,r+1} \oplus E_{r+2}^{\,r+2} \oplus \dots$$

Moreover, $d_r : E_r^{\,r} \to E_r^{\,r}$ and $d_r : E_{r+1} \to 0$. It is easy to see that d_r is *trivial* on $E_r^{\,r}$ (that is, if $X \in E_r^{\,r}$ and $d_r X = 0$, there exists $Y \in E_r^{\,r}$ such that $X = d_r Y$. See Appendix I). Then, obviously $H(E_r, d_r) = E_{r+1}$. This shows that the spectral sequence derived from the exact couple (see next section) is just given by the E_r, and, from this, it is easy to show that :

There is a *bijection* between the anomalous terms in $H(\delta/d)$ (even or odd space-time degree) and the Q_K's derived from the X's, *provided we choose X in a supplementary subspace* Σ_r of $d_r(E_r)$ in E_r.

SKETCH OF THE PROOF

Starting from the exact couple of IV,d), homological algebra tells us to form derived exact couples in the following way :

Put $F_1 = \partial^2 F_0 = \mathrm{Im}\partial^2$ and define $d_0 : E_0 \to E_0$ by $d_0 = i_0 \circ p_0$. Then : $d_o = i_o \circ (p_o \circ i_o) \circ p_o = 0$, and so, define $E_1 = H(E_o, d_o)$. It is easy to construct $i_1 : F_1 \to E_1$ and $p_1 : E_1 \to F_1$ which, together with $\partial^2 : F_1 \to F_1$ form another exact couple. By induction, one then defines a rth derived exact couple $(F_r, E_r, i_r, p_r, d_r)$.

The notation suggests, and it is true, that the (E_r, d_r) so defined are identical to the (E_r, d_r) of section V. To show this, note that the $E_r = \Lambda P_r \otimes S\tau P_r$ are included in E_o :

$$E_o \supset E_1 \supset E_2 \supset \ldots$$

and one can define $_{2r}\ p_r = p|E_r$. Then it is easy to check by induction that, on $F_r = \partial^{2r}F_o$, $i_r(\partial^{2r}\alpha)$ is the projection of $i\ (\alpha)$ on E_r, under the decomposition : $E^o = \{E^o_{r-1} \oplus E_1 \oplus \ \ldots \oplus E_{r-1}\} \oplus E_r$
Indeed, let $X \in E^o_r$ and α = class of Q_{2r} Then

$$i_r \circ p_r(X) = i_r(\partial^{2r}\alpha) = \mathrm{proj}_{\ Er}\, i_o(\alpha) = \mathrm{proj}_{\ Er}(d_r\ X)$$

since $$i_o(\alpha) = i \circ \partial^{-1}(\alpha) = d_r\ X.$$

Since dr X ∈ Er one gets $i_r \circ p_r = d_r$, and then $E_{r+1} = H(E_r, d_r)$, as needed.

Having computed the spectral sequence E_r, it is easy to conclude. Obviously, $F_o \approx \mathrm{Ker}\partial^2 \oplus \mathrm{Im}\partial^2 \approx p\ (E_o) \oplus F_1$, and by induction, $F_o \approx \oplus_i p_i(E_i)$. The isomorphism of $p_r(E_r)$ with part of F_o is obtained through the action of ∂^{-2r}, and so, to $X \in E_r$ is associated α = class of $Q_{2r} \in H_{even}(\delta/d)$, to which is associated class of Q_{2r+1} in $H_{odd}(\delta/d)$. Since we have the sum over r, all the Q_k's of the chain appear in $H(\delta/d)$.

Finally, to get the restriction on X, we need to compute Ker P_r. We show that :

$$\mathrm{Ker}\ p_r = \mathrm{Im}\ d_r \oplus \mathrm{Im}\ d_{r+1} \oplus \ldots$$

because if $X \in E_r$ is such that $p_r(X) = 0$, then $d_r X = i_r \circ p_r\ (X) = 0$. Since $E_{r+1} = H(E^r, d_r)$ there exists $Z \in E_{r+1}$ with $X = d_r Y + Z$. Then $p_r(X) = p_r(Z) = 0$ and one can proceed by induction. Then :

$$p_r(E_r) \approx E_r\ /\ \mathrm{Ker}\ p_r \approx \{\oplus_{k \geqslant 0} E_{r+k}{}^{r+k}\}\ /\ \{\oplus_{k \geqslant 0} \mathrm{Im} d_{r+k}\}$$

$$\approx \oplus_{k \geqslant 0} E_{r+k}{}^{r+k}\ /\ \mathrm{Im}\ d_{r+k}$$

from which the stated result is obvious.

CONCLUSION

We have seen that possible anomalous terms are determined only in terms of the cohomology of g (the Lie algebra of the structure group). In addition to previously known terms derived from primitive forms (Stora), there appear new "factorizable anomalies", for sufficiently high ghost number. The proeminent role is played by a purely algebraic structure, the B.R.S. algebra, constructed only in terms of g. For more details on its construction and properties, and its relevance for the anomaly problem, see [2]. Moreover, the elements of B.R.S. algebra have geometrical interpretation given in [3], together with examples of the above construction. For more details on this geometry, see [15,27].

More topological aspects of the anomaly problem are connected with the calculation of the coefficients of possible anomalous terms in a given model. Here, the topology of the gauge group (that is, for trivial bundle, functions $x \in M \to g(x) \in G$) plays a central role [6,27]. Of course, this topology depends on the topology of space-time [28], and perhaps on the topology of the principal bundle and differentiability assumptions on $g(x)$.

These methods should be applicable to gravity and supergravity theories, by considering the bundle of orthonormal frames. Nevertheless, some constructions [29, 30], do not have at present time a clear geometric significance.

Finally, the problem of showing that we have here all the possible anomalies, modulo trivial δ-cocycles like Tr $(F_{\mu\nu}{}^2)$ is still open.

APPENDIX I

Proof that d is Trivial

Take a polynomial $P(A,F,\chi,d\chi)$ and replace $F = dA + A^2$. Everything is expressed as $P(A, dA, \chi, d\chi)$. Define d' by : $d'(A) = 0$, $d'(dA) = A$, $d'(\chi) = 0$, $d'(d\chi) = \chi$. Then $dd' + d'd$ is a derivation, which gives 1 on A, dA, χ, $d\chi$. So, on any monomial of these variables, it gives the degree of this monomial. Then, if $dP = 0$ apply {1/degree . d'd + 1/degree $._0$dd'} to P, and obtain $P = dQ$ with $Q = 1/\text{degree} \,.\, d'P$. In this way, the Q_{2n-1} are easily obtained.

Proof that d_r is Trivial on E_r

Exactly the same proof works.

Sketch of Proof : $H(\delta) \cong \mathcal{D}_\Lambda(g) \boxtimes \mathcal{D}_s(g)$

The algebra generated by A, F, χ, $d\chi$ is also generated by A, δA, F, χ because $\delta A = -d\chi - \chi A - A\chi$. By the same argument as before and Künneth theorem, the part in A, δA can be thrown away for calculating $H(\delta)$, [33].

Then, since $\delta F = -\chi F + F\chi$ and $\delta\chi = -\chi^2$, since χ anticommutes and F commutes, $H(\delta)$ identifies with the cohomology of the Lie algebra g, the part in χ, with values in the module of polynomials of g, the part in F. For a definition which makes this identification obvious, see [14].

Finally, a theorem of Hochschild, Serre [34] shows that, for a reductive Lie algebra, this cohomology is simply

$$\mathcal{D}_\Lambda(g) \boxtimes \mathcal{D}_s(g)$$

APPENDIX II

We shall give here a table of degrees of primitive forms for various simple Lie groups. Note, in particular, that a Tr F^6 term cannot occur for E_8 !

SU(N)	3, 5, 7, ..., 2N-1
SO(2N+1), Sp(2N)	3, 7, 11, ..., 4N-5, 4N-1
SO(2N)	3, 7, 11, ..., 4N-5, 2N-1 (↑ related to the Pfaffian)

(for a discussion of these classical algebras, see [31])

G_2	3, 11
F_4	3, 11, 15, 23
E_6	3, 9, 11, 15, 17, 23
E_7	3, 11, 15, 19, 23, 27, 35
E_8	3, 15, 23, 27, 35, 39, 47, 59

(for a discussion of these exceptional algebras, see [32]).

REFERENCES

[1] M. Dubois-Violette, M. Talon, C.M. Viallet, Phys. Lett. 158B (1985) 231.
C.M. Viallet Symp. on anomalies ... Chicago 1985 W.S.P., Singapour (1985).
[2] M. Dubois-Violette, M. Talon, C.M. Viallet, Com. Math. Phys. 102 (1985) 105.
[3] M. Dubois-Violette, M. Talon, C.M. Viallet, Ann. Inst. H. Poincaré, 14, 1 (1986) 103.
M. Dubois-Violette, Journées relativistes Marseille (1985) to appear.
[4] M. Atiyah, I.M. Singer, Ann. Math. 87 (1968) 484, 87 (1968) 546, 92 (1970) 119.
[5] L. Alvarez-Gaumé, E. Witten, Nucl. Phys. B234 (1984) 269.
see also, on such a connection, earlier papers :
G.'t Hooft, Phys. Rev. Lett. 37 (1976) 8.
C. Bernard, A. Guth, E. Weinberg, Phys. Rev. D17, 4 (1978) 1053.
L. Brown, R. Carlitz, C. Lee, Phys. Rev. D16, 2 (1977) 417.
[6] L. Alvarez-Gaumé, P. Ginsparg, Nucl. Phys. B243 (1984) 449.
[7] J. Wess, B. Zumino, Phys. Lett. 37B (1971) 95.
[8] C. Becchi, A. Rouet, R. Stora, Ann. Phys. N.Y. 98 (1976) 287.
[9] E. Witten, Nucl. Phys. B223 (1983) 422.
[10] B. Zumino in "Les Houches" Summer 1983 (North Holland)
B. Zumino, Nucl. Phys. B253 (1985) 477.
J. Manes, R. Stora, B. Zumino (to appear).
[11] S. Adler in "Brandeis" Summer 1970, M.I.T. Press.
[12] J. Dieudonné, Elements of analysis IV, Chapter XX. Academic Press.
[13] S. Kobayashi, K. Nomizu, Foundations of differential geometry, Chapter II, Interscience.
[14] H. Cartan in Colloque de topologie, Bruxelles (1950), Masson.
[15] L. Bonora, P. Cotta-Ramusino, Comm. Math. Phys. 87 (1983) 589.
[16] R. Stora in "Cargèse" Summer 1976, Plenum.
[17] L. Beaulieu in "Cargèse" Summer 1983, also Phys. Rep. "Perturbative gauge theories" to appear (1985).
[18] L. Faddeev, Phys. Lett. 145B (1984) 81.
[19] R. Jackiw in APS meeting, Santa Fé, (1984) published by Adam Hilger, Bristol (1985).
[20] L. Alvarex-Gaumé, P. Ginsparg, Ann. Phys. N.Y. 161, 2 (1985) 423.

[21] J. Thierry-Mieg, Phys. Lett. 147B (1984) 430.
[22] R. Stora in "Cargèse", Summer 1983, Plenum.
[23] W. Bardeen, Phys. Rev. 184, 5 (1969) 1848.
[24] J.L. Koszul, Bull. Soc. Math. France 78 (1950) 65.
[25] C. Chevalley, Am. J. Math. 77 (1955) 778.
[26] W. Massey, Ann. Math. 56 (1952) 363.
see also S. Mac Lane, Homology Springer (1963).
[27] M. Atiyah, I.M. Singer, Proc. Natl. Ac. Sc. U.S.A. 81 (1984) 2597.
[28] M. Asorey, P.K. Mltter "Cohomology of the Yang-Mills gauge orbit space and Dimensional reduction" Preprint Zaragoza (1985).
[29] L. Beaulieu, M. Bellon, Preprint LPTHE 85/03 (1985).
[30] L. Beaulieu, M. Bellon, Preprint LPTENS 85/7 (1985).
[31] W. Greub, S. Halperin, R. Vanstone "Connections, curvature, cohomology" vol. III, Acad. Press.
[32] C. Chevalley, Int. Math. Congress, Cambridge 1950.
[33] D. Sullivan, I.H.E.S. Pub n° 47 (1977) 269.
[34] G. Hochschild, J.P. Serre, Ann. Math. 57 (1953) 591.

QUANTUM BLACK HOLES

G. 't Hooft

Institute for Theoretical Physics
Princetonplein 5, P.O. Box 80.006
3508 TA Utrecht, The Netherlands

SUMMARY

The conventional way to set up a decent looking theory for interacting elementary particles is to take some set of classical fields, satisfying simple, polynomial field equations. These equations should be compatible with an extremum principle: they are taken to be the Euler-Lagrange equations generated by some Lagrange function L of the fields. One then "quantizes" the theory by replacing Poisson brackets by commutators.

The extremely successful "standard model" was found this way, and its successors, the "grand unified models", with or without supersymmetry, are promising.

Yet, even though gravitational forces among individual particles are extremely weak, no particle theory can be complete without gravity. Now Einstein's theory of gravity, called "general relativity", is also of the Euler-Lagrange form, but the standard quantization procedure fails, for two reasons. The first problem is that not all Euler-Lagrange systems of fields can be quantized: it must be possible to formulate a complete renormalization programme. The interactions give rise to divergent expressions for higher order processes, and only if these higher order interactions can be absorbed into the fundamental ones the theory can be formulated in such a way that all calculations of physically observable effects are finite. In "quantum gravity" the higher order interactions have a dimensionality different from the fundamental ones, because Newton's constant G has dimensions (even if we set h = c = 1) and the renormalization procedure fails.

There are powerful models and ideas designed to overcome the renormalization problems such as "supergravity", Kaluza-Klein mechanisms and the latest, "superstrings". But these models cannot avoid the fact that there are energy and length scales at which G is large, so that a perturbative treatment of gravitation fails. Since all systems mentioned above are perturbative systems the renormalization problem is still a mystery.

Another problem with quantum gravity is, if possible, even more mysterious. Suppose that we had "regularized" the gravitational forces at the small distance end in the way that, for instance, the weak intermediate vector boson "regularized" the fundamental 4-fermion interaction vertex of the weak interactions. Imagine that all excitations of fields and the space time metric $g_{\mu\nu}$ itself with wave length smaller than the Planck length have been suppressed, somehow. Then what we discover is that the gravitational forces are unstable. Given a sufficiently large amount of matter, it can collapse under its own weight.

Classical (i.e. non-quantized) general relativity tells us exactly what will happen: a "black hole" is formed[1)]. But how is this formulated in a quantum theory? A brilliant observation was made by S. Hawking: when a field theory is quantized in the background metric of a black hole he found that the black hole actually emits particles in a completely random ("thermal") way[2)]. His theory permits the calculation of the radiation temperature:

$$kT = 1/8\pi MG \tag{1}$$

where M is the mass of the black hole, in natural units (h = c = 1).

Apparently black holes are just another form of matter, obviously unstable against "Hawking decay". Unfortunately, this picture cannot be complete. Radioactive particles such as uranium nuclei, may emit various other particles ($\alpha,\beta,\gamma,\ldots$), but not exactly thermally. Indeed, the radiation could only be completely thermal if the Hilbert space of all possible black holes would be infinite dimensional, even when total energy and momentum of the black hole are constrained. This is our problem: the quantum version of the black hole has infinite phase space, and other symptoms of a "run-away solution".

Logical consistency of quantum mechanics requires Hilbert space to have a denumrable number of dimensions. It is here that our theory - no matter how it is regularized! - fails. Attempts to cure the theory might even involve a careful reconsideration of the quantum mechanical measuring process. In that case one might find theories[3)] in which the radiation temperature differs from eq. (1), possibly by a factor 2.

Crucial for understanding the depth of the problem is to realize how gravitational interaction takes place between ingoing and outgoing particles[4)]. early ingoing particles and late outcoming particles see each other with an enormous Lorentz contraction factor when they meet, close to the horizon[5)]. This is why mutual gravitational interactions become important there. They are exchanged via shock waves. The horizon, as experienced by one particle is shifted due to the presence of the other, and these shifts can be computed precisely.

If we assume that only particles that can escape to infinity contribute to the physical Hilbert space of a black hole one sees that its dimensionality remains denumerable[6)], but no viable dynamical model could be constructed.

Black holes are the heaviest and most compact forms of matter that we can imagine. A complete paticle theory can have nothing but a spectrum of black hole-like objects at its high-energy end. This is why we believe that a resolution of the black hole problem will in the same time disclose the complete small-distance structure of our world. This is probably also the reason why the problem is so difficult.

We conclude that "unification" of black holes with the other elementary paticles is an essential requirement for a viable theory of elementary interactions.

REFERENCES

1. See for instance: S. Chandrasekhar, Am. J. Phys. 37, 577 (1969).
C.W. Misner, K.S. Thorne and J.A. Wheeler, "Gravitation", Freeman, San Francisco, 1973.
R.M. Wald, "General Relativity", Univ. of Chicago Press, 1984.
2. S.W. Hawking, Comm. Math. Phys. 43, 199 (1975).
J.B. Hartle and S.W. Hawking, Phys. Rev. D13, 2188 (1976).
W.G. Unruh, Phys. Rev. D14, 870 (1976).
R.M. Wald, Comm. Math. Phys. 45, 9 (1975); Phys. Rev. D20, 1271 (1979).
3. G. 't Hooft, J. Geom. Phys. 1, 45 (1984).
4. G. 't Hooft, Nucl. Phys. B256, 727 (1985).
5. T. Dray and G. 't Hooft, Nucl. Phys. B253, 173 (1985); Comm. Math. Phys. 99, 613 (1985).
6. G. 't Hooft, lectures given at the 5th Adriatic Meeting on Particle Physics, Dubrovnik, June 1986.

INDEX